Praise for *Charles C. Mann's*

1493

"Revelatory."
                                    —Lev Grossman, *Time*

"Compelling and eye-opening."
                    —*Publishers Weekly* (Top 100 Books of the Year)

"A book to celebrate. . . . A bracingly persuasive counternarrative to the prevailing mythology about the historical significance of the 'discovery' of America. . . . *1493* is rich in detail, analytically expansive and impossible to summarize. . . . [Mann's book] deserves a prominent place among that very rare class of books that can make a difference in how we see the world, although it is neither a polemic nor a work of advocacy. Thoughtful, learned and respectful of its subject matter, *1493* is a splendid achievement."
                                            —*The Oregonian*

"Despite his scope, Mann remains grounded in fascinating details. . . . Such technical insights enhance a very human story, told in lively and accessible prose."
                                        —*The Plain Dealer*

"Mann's excitement never flags as he tells his breathtaking story. . . . There is grandeur in this view of the past that looks afresh at the different parts of the world and the parts each played in shaping it."   —*Financial Times*

"A muscular, densely documented follow-up [to Mann's 1491]. . . . Like its predecessor, 1493 runs to more than 400 pages, but it moves at a gallop. . . . As a historian Mann should be admired not just for his broad scope and restless intelligence but for his biological sensitivity. At every point of his tale he keeps foremost in his mind the effect of humans' activities on the broader environment they inhabit."   —*The Wall Street Journal*

"Evenhandedness, a sense of wonder, the gift of turning a phrase. . . . Mann loves the world and adopts it as his own."   —*Science*

"Charles C. Mann glories in reality, immersing his reader in complexity. . . . The worn clichés crumble as readers gain introductions to the freshest of the systems of analysis gendered in the first post-Columbian millennium."
          —Alfred W. Crosby, author of *The Columbian Exchange*

"In the wake of his groundbreaking book 1491 Charles Mann has once again produced a brilliant and riveting work that will forever change the way we see the world. Mann shows how the ecological collision of Europe and the Americas transformed virtually every aspect of human history. Beautifully written, and packed with startling research, 1493 is a monumental achievement."
          —David Grann, author of *The Lost City of Z*

"Fascinating. . . . Convincing. . . . A spellbinding account of how an unplanned collision of unfamiliar animals, vegetables, minerals and diseases produced unforeseen wealth, misery, social upheaval and the modern world."
          —*Kirkus Reviews* (starred review)

"A fascinating survey. . . . A lucid historical panorama that's studded with entertaining studies of Chinese pirate fleets, courtly tobacco rituals, and the bloody feud between Jamestown colonists and the Indians who fed and fought them, to name a few. Brilliantly assembling colorful details into big-picture insights, Mann's fresh challenge to Eurocentric histories puts interdependence at the origin of modernity."

—*Publishers Weekly* (starred review)

"[1493] is readable and well-written, based on his usual broad research, travels and interviews. A fascinating and important topic, admirably told." —John Hemming, author of *Tree of Rivers*

"Charles Mann expertly shows how the complex, interconnected ecological and economic consequences of the European discovery of the Americas shaped many unexpected aspects of the modern world. This is an example of the best kind of history book: one that changes the way you look at the world, even as it informs and entertains." —Tom Standage, author of *A History of the World in Six Glasses*

"A landmark book. . . . Entrancingly provocative, *1493* bristles with illuminations, insights and surprises." —*Shelf Awareness*

"Fascinating. . . . Engaging and well-written. . . . Information and insight abound on every page. This dazzling display of erudition, theory and insight will help readers to view history in a fresh way." —*BookPage*

"Spirited. . . . One thing is indisputable: Mann is definitely global in his outlook and tribal in his thinking. . . . Mann's taxonomy of the ecological, political, religious, economic, anthropological and mystical melds together in an intriguing whole cloth." —*The Star-Ledger*

"Mann has managed the difficult trick of telling a complicated story in engaging and clear prose while refusing to reduce its ambiguities to slogans. He is not a professional historian, but most professionals could learn a lot from the deft way he does this. . . . *1493* is thoroughly researched and up-to-date, combining scholarship from fields as varied as world history, immunology, and economics, but Mann wears his learning lightly. He serves up one arresting detail after another, always in vivid language. Most impressive of all, he manages to turn plants, germs, insects and excrement into the lead actors in his drama while still parading before us an unforgettable cast of human characters. He makes even the most unpromising-sounding subjects fascinating. I, for one, will never look at a piece of rubber in quite the same way now. . . . The Columbian Exchange has shaped everything about the modern world. It brought us the plants we tend in our gardens and the pests that eat them. And as it accelerates in the twenty-first century, it may take both away again. If you want to understand why, read *1493*."

—*The New York Times Book Review*

"Although many have written about the impact of Europeans on the New World, few have told the worldwide story in a manner accessible to lay readers as effectively as Mann does here."

—*Library Journal*

"The chief strength of Mann's richly associative books lies in their ability to reveal new patterns among seemingly disparate pieces of accepted knowledge. They're stuffed with forehead-slapping 'aha' moments. . . . If Mann were to work his way methodically through the odd-numbered years of history, he could be expected to publish a book about the global impact of the Great Recession sometime in the middle of the next millennium. If it's as good as *1493*, it would be worth the wait."

—*Richmond Times-Dispatch*

Charles C. Mann

1493

Charles C. Mann, a correspondent for *The Atlantic, Science,* and *Wired,* has written for *Fortune, The New York Times, Smithsonian, Technology Review, Vanity Fair,* and *The Washington Post,* as well as for the TV network HBO and the series *Law & Order.* A three-time National Magazine Award finalist, he is the recipient of writing awards from the American Bar Association, the American Institute of Physics, the Alfred P. Sloan Foundation, and the Lannan Foundation. His *1491* won the National Academies Communication Award for the best book of the year.

www.charlesmann.org

1493

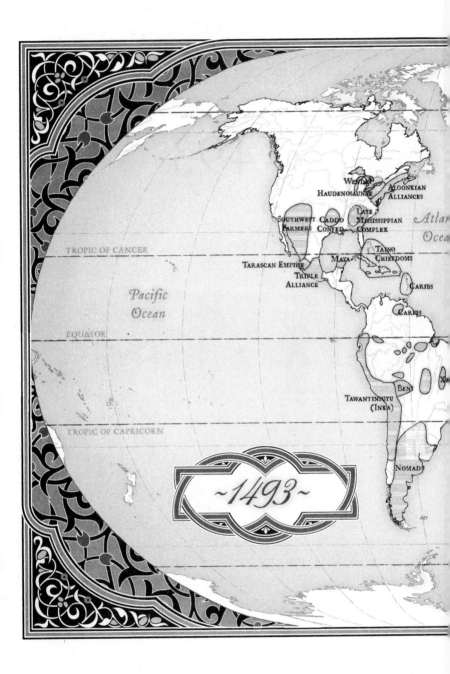

WENDAT
HAUDENOSAUNEE   ALGONKIAN
ALLIANCES

LATE
SOUTHWEST   CADDO   MISSISSIPPIAN
FARMERS   CONFED.   COMPLEX

Atlar
Ocea

TROPIC OF CANCER

TARASCAN EMPIRE   MAYA   TAINO
CHIEFDOMS

TRIPLE
ALLIANCE

CARIBS

Pacific
Ocean

CARIBS

EQUATOR

BENI

TAWANTINSUYU
(INKA)

TROPIC OF CAPRICORN

~1493~

NOMADS

ARCTIC CIRCLE

UNION OF
KALMAR
MUSCOVY

*LAND
*AND
*NGLAND

HOLY
ROMAN
EMPIRE

FRANCE

*UGAL SPAIN

*AGHRIB
AL AQSĀ

SHAYBANIDS

KHANATE
OF THE
OIRATS

OTTOMAN
EMPIRE

CHAGATAI
KHANATE

MAMLUKS

AK KOYUNLU

KOREA
JAPAN

TIMURIDS

TIBET

MING
EMPIRE

SONGHAY

SOUTH ASIAN
STATES

LOANGO
TYO

KONGO
NDONGO

MALAWI

MUTAPA

Indian
Ocean

*lantic

*cean

ANTARCTIC CIRCLE

1493

Uncovering the New World
Columbus Created

CHARLES C. MANN

VINTAGE BOOKS
A DIVISION OF RANDOM HOUSE, INC.
NEW YORK

FIRST VINTAGE BOOKS EDITION, JULY 2012

The author gratefully acknowledges the support of the Lannan Foundation.

Portions of this book have appeared in different form in *The Atlantic*,
National Geographic, *Orion*, and *Science*.

Maps created by Nick Springer and Tracy Pollock, Springer Cartographics
LLC; copyright © 2011 by Charles C. Mann

The Library of Congress has cataloged the Alfred A. Knopf edition as follows:
Mann, Charles C.
1493 : uncovering the new world Columbus created / Charles C. Mann.—
1st ed.
p. cm.
Includes bibliographical references and index.
1. History, Modern.   2. Economic history.   3. Commerce—History.
4. Agriculture—History.   5. Ecology—History.   6. Industrial revolution.
7. Slave trade—History.   8. America—Discovery and exploration—
Economic aspects.   9. America—Discovery and exploration—
Environmental aspects.   10. Columbus, Christopher—Influence.   I. Title.
D228.M36   2011
909'.4—dc22   2011003408

Vintage ISBN: 978-0-307-27824-1

Book design by Virginia Tan
Author photograph © J. D. Sloan

Printed in the United States of America

10   9   8   7   6   5   4   3   2   1

To the woman who built my house,

and is my home

—CCM

✣ CONTENTS ✢

CODA / *Currents of Life*

ࣷ MAPS ࣷ

✂ PROLOGUE ✂

Like other books, this one began in a garden. Almost twenty years ago I came across a newspaper notice about some local college students who had grown a hundred different varieties of tomato. Visitors were welcome to take a look at their work. Because I like tomatoes, I decided to drop by with my eight-year-old son. When we arrived at the school greenhouse I was amazed—I'd never seen tomatoes in so many different sizes, shapes, and colors.

A student offered us samples on a plastic plate. Among them was an alarmingly lumpy specimen, the color of an old brick, with a broad, green-black tonsure about the stem. Occasionally I have dreams in which I experience a sensation so intensely that I wake up. This tomato was like that—it jolted my mouth awake. Its name, the student said, was Black from Tula. It was an "heirloom" tomato, developed in nineteenth-century Ukraine.

"I thought tomatoes came from Mexico," I said, surprised. "What are they doing breeding them in Ukraine?"

The student gave me a catalog of heirloom seeds for tomatoes, chili peppers, and beans (common beans, not green beans). After I went home, I flipped through the pages. All three crops originated in the Americas. But time and again the varieties in the catalog came from overseas: Japanese tomatoes, Italian peppers, Congolese beans. Wanting to have more of those strange but tasty tomatoes, I went on to order seeds, sprout them in plastic containers, and stick the seedlings in a garden, something I'd never done before.

Not long after my trip to the greenhouse I visited the library. I discovered that my question to the student had been off the mark. To begin, tomatoes probably originated not in Mexico, but in the Andes Mountains. Half a dozen wild tomato species exist in Peru and Ecuador, all but one inedible, producing fruit the size of a thumbtack. And to botanists the real mystery is less how tomatoes ended up in Ukraine or Japan than how the progenitors of today's tomato journeyed from South America to Mexico, where native plant breeders radically transformed the fruits, making them bigger, redder, and, most important, more edible. Why transport useless wild tomatoes for thousands of miles? Why had the species not been domesticated in its home range? How had people in Mexico gone about changing the plant to their needs?

These questions touched on a long-standing interest of mine: the original inhabitants of the Americas. As a reporter in the news division of the journal *Science*, I had from time to time spoken with archaeologists, anthropologists, and geographers about their increasing recognition of the size and sophistication of long-ago native societies. The botanists' puzzled respect for Indian plant breeders fit nicely into that picture. Eventually I learned enough from these conversations that I wrote a book about researchers' current views of the history of the Americas before Columbus. The tomatoes in my garden carried a little of that history in their DNA.

They also carried some of the history *after* Columbus. Beginning in the sixteenth century, Europeans carried tomatoes around the world. After convincing themselves that the strange fruits were not poisonous, farmers planted them from Africa to Asia. In a small way, the plant had a cultural impact everywhere it moved. Sometimes not so small—one can scarcely imagine southern Italy without tomato sauce.

Still, I didn't grasp that such biological transplants might have played a role beyond the dinner plate until in a used-book store I came across a paperback: *Ecological Imperialism*, by Alfred W.

Crosby, a geographer and historian then at the University of Texas. Wondering what the title could refer to, I picked up the book. The first sentence seemed to jump off the page: "European emigrants and their descendants are all over the place, which requires explanation."

I understood exactly what Crosby was getting at. Most Africans live in Africa, most Asians in Asia, and most Native Americans in the Americas. People of European descent, by contrast, are thick on the ground in Australia, the Americas, and southern Africa. Successful transplants, they form the majority in many of those places—an obvious fact, but one I had never really thought about before. Now I wondered: Why *is* that the case? Ecologically speaking, it is just as much a puzzle as tomatoes in Ukraine.

Before Crosby (and some of his colleagues) looked into the matter, historians tended to explain Europe's spread across the globe almost entirely in terms of European superiority, social or scientific. Crosby proposed another explanation in *Ecological Imperialism*. Europe frequently had better-trained troops and more-advanced weaponry than its adversaries, he agreed, but in the long run its critical advantage was biological, not technological. The ships that sailed across the Atlantic carried not only human beings, but plants and animals—sometimes intentionally, sometimes accidentally. After Columbus, ecosystems that had been separate for eons suddenly met and mixed in a process Crosby called, as he had titled his previous book, the Columbian Exchange. The exchange took corn (maize) to Africa and sweet potatoes to East Asia, horses and apples to the Americas, and rhubarb and eucalyptus to Europe—and also swapped about a host of less-familiar organisms like insects, grasses, bacteria, and viruses. The Columbian Exchange was neither fully controlled nor understood by its participants, but it allowed Europeans to transform much of the Americas, Asia, and, to a lesser extent, Africa into ecological versions of Europe, landscapes the foreigners could use more comfortably than could their original inhabitants. This ecological imperialism, Crosby argued, provided the British, French, Dutch, Portuguese, and Spanish with the consistent edge needed to win their empires.

Crosby's books were constitutive documents in a new discipline: environmental history. The same period witnessed the rise of another discipline, Atlantic studies, which stressed the importance of interactions among the cultures bordering that ocean. (Recently a number of Atlanticists have added movements across the Pacific to their purview; the field may have to be renamed.) Taken together, researchers in all these fields have been assembling what amounts to a new picture of the origins of our world-spanning, interconnected civilization, the way of life evoked by the term "globalization." One way to summarize their efforts might be to say that to the history of kings and queens most of us learned as students has been added a recognition of the remarkable role of *exchange*, both ecological and economic. Another way might be to say that there is a growing recognition that Columbus's voyage did not mark the discovery of a New World, but its creation. How that world was created is the subject of this book.

The research has been greatly aided by recently developed scientific tools. Satellites map out environmental changes wreaked by the huge, largely hidden trade in latex, the main ingredient in natural rubber. Geneticists use DNA assays to trace the ruinous path of potato blight. Ecologists employ mathematical simulations to simulate the spread of malaria in Europe. And so on—the examples are legion. Political changes, too, have helped. To cite one of special importance to this book, it is much easier to work in China nowadays than it was in the early 1980s, when Crosby was researching *Ecological Imperialism*. Today, bureaucratic suspicion is minimal; the chief obstacle I faced during my visits to Beijing was the abominable traffic. Librarians and researchers there happily gave me early Chinese records—digital scans of the originals, which they let me copy onto a little memory stick that I carried in my shirt pocket.

What happened after Columbus, this new research says, was nothing less than the forming of a single new world from the collision of two old worlds—three, if one counts Africa as separate from Eurasia. Born in the sixteenth century from European desires to join the thriving Asian trade sphere, the economic system for exchange ended up transforming the globe into a single ecological system by the nineteenth century—almost instantly, in biological terms. The creation of this ecological system helped Europe seize, for several vital centuries, the political initiative, which in turn shaped the contours of today's world-spanning economic system, in its interlaced, omnipresent, barely comprehended splendor.

Ever since violent protests at a 1999 World Trade Organization meeting in Seattle brought the idea of globalization to the world's attention, pundits of every ideological stripe have barraged the public with articles, books, white papers, blog posts, and video documentaries attempting to explain, celebrate, or attack it. From the start the debate has focused around two poles. On one side are economists and entrepreneurs who argue passionately that free trade makes societies better off—that both sides of an uncoerced exchange gain from it. The more trade the better! they say. Anything less amounts to depriving people in one place of the fruits of human ingenuity in other places. On the other side is a din of environmental activists, cultural nationalists, labor organizers, and anti-corporate agitators who charge that unregulated trade upends political, social, and environmental arrangements in ways that are rarely anticipated and usually destructive. The less trade, they say, the better. Protect local communities from the forces unleashed by multinational greed!

Whipsawed between these two opposing views, the global network has become the subject of a furious intellectual battle, complete with mutually contradictory charts, graphs, and statistics—and tear gas and flying bricks in the streets where political leaders meet behind walls of riot police to wrangle through international-trade agreements. Sometimes the moil of slogans and counter-slogans, facts and factoids, seems impenetrable, but as I learned more I came to suspect that both sides may be correct. From the outset globalization brought both enormous economic gains *and* ecological and social tumult that threatened to offset those gains.

It is true that our times are different from the past. Our ancestors did not have the Internet, air travel, genetically modified crops, or computerized international stock exchanges. Still, reading the accounts of the creation of the world market one cannot help hearing echoes—some muted, some thunderously loud—of the disputes now on the television news. Events four centuries ago set a template for events we are living through today.

What this book is not: a systematic exposition of the economic and ecological roots of what some historians call, ponderously but accurately, "the world-system." Some parts of the earth I skip entirely; some important events I barely mention. My excuse is that the subject is too big for any single work; indeed, even a pretense at completeness would be unwieldy and unreadable. Nor do I fully treat how researchers came to form this new picture, though I describe some of the main landmarks along the intellectual way. Instead in 1493 I concentrate on areas that seem to me to be especially important, especially well documented, or—here showing my journalist's bias—especially interesting. Readers wishing to learn more can turn to the sources in the Notes and Bibliography.

Following an introductory chapter, the book is divided into four sections. The first two lay out, so to speak, the constituent halves of the Columbian Exchange: the separate but linked exchanges across the Atlantic and Pacific. The Atlantic section begins with the exemplary case of Jamestown, the beginning of permanent English colonization in the Americas. Established as a purely economic venture, its fate was largely decided by ecological forces, notably the introduction of tobacco. Originally from the lower Amazon, this exotic species—exciting, habit-forming, vaguely louche—became the subject of the first truly global commodity craze. (Silk and porcelain, long a passion in Europe and Asia, spread to the Americas and became the next ones.) The chapter sets the groundwork for the next, which discusses the introduced species that shaped, more than any others, societies from Baltimore to Buenos Aires: the microscopic creatures that cause malaria and yellow fever. After examining their impact on matters ranging from slavery in Virginia to poverty in the Guyanas, I close with malaria's role in the creation of the United States.

The second section shifts the focus to the Pacific, where the era of globalization began with vast shipments of silver from Spanish America to China. It opens with a chronicle of cities: Potosí in what is now Bolivia, Manila in the Philippines, Yuegang in southeast China. Once renowned, now little thought of, these cities were the fervid, essential links in an economic exchange that knit the world together. Along the way, the exchange brought sweet potatoes and corn to China, which had accidental, devastating consequences for Chinese ecosystems. As in a classic feedback loop, those ecological consequences shaped subsequent economic and political conditions. Ultimately, sweet potatoes and corn played a major part in the flowering and collapse of the last Chinese dynasty. They played a small, but similarly ambiguous role in the Communist dynasty that eventually succeeded it.

The third section shows the role of the Columbian Exchange in two revolutions: the Agricultural Revolution, which began in the late seventeenth century; and the Industrial Revolution, which took off in the early and mid-nineteenth century. I concentrate on two introduced species: the potato (taken from the Andes to Europe) and the rubber tree (transplanted clandestinely from Brazil to South and Southeast Asia). Both revolutions, agricultural and industrial, supported the rise of the West—its sudden emergence as a controlling power. And both would have had radically different courses without the Columbian Exchange.

In the fourth section I pick up a theme from the first section.

Here I turn to what in human terms was the most consequential exchange of all: the slave trade. Until around 1700 about 90 percent of the people who crossed the Atlantic were African captives. (Native Americans also were part of the slave trade, as I explain.) In consequence of this great shift in human populations, many American landscapes were for three centuries largely dominated, in demographic terms, by Africans, Indians, and Afro-Indians. Their interactions, long hidden from Europeans, are an important part of our human heritage that is just coming to light.

The meeting of red and black, so to speak, took place against a backdrop of other meetings. So many different peoples were involved in the spasms of migration set off by Columbus that the world saw the rise of the first of the now-familiar polyglot, world-encompassing metropolises: Mexico City. Its cultural jumble extended from the top of the social ladder, where the conquistadors married into the nobility of the peoples they had conquered, to the bottom, where Spanish barbers complained bitterly about low-paid immigrant barbers from China. A planetary crossroads, this great, turbulent metropolis represents the unification of the two networks described in the first part of this book. A coda set in the present suggests that these exchanges continue unabated.

In some respects this image of the past—a cosmopolitan place, driven by ecology and economics—is startling to people who, like me, were brought up on accounts of heroic navigators, brilliant inventors, and empires acquired by dint of technological and institutional superiority. It is strange, too, to realize that globalization has been enriching the world for nigh on five centuries. And it is unsettling to think of globalization's equally long record of ecological convulsion, and the suffering and political mayhem caused by that convulsion. But there is grandeur, too, in this view of our past; it reminds us that every place has played a part in the human story, and that all are embedded in the larger, inconceivably complex progress of life on this planet.

. . .

As I write these words, it's a warm August day. Yesterday my family picked the first tomatoes from our garden—the somewhat improved successor of the tomato patch I planted after my visit to the college twenty years ago.

After I planted the seeds from the catalog, it didn't take me long to discover why so many people love puttering in their gardens. Messing around with the tomatoes felt to me like building a fort as a child: I was both creating a refuge from the world and creating a place of my own in that world. Kneeling in the dirt, I was making a small landscape, one that had the comfortable, comforting timelessness evoked by words like *home*.

To biologists this must seem like poppycock. At various times my tomato patch has housed basil, eggplant, bell peppers, kale, chard, several types of lettuce and lettuce-like greens, and a few marigolds, said by my neighbors to repel bugs (scientists are less certain). Not one of these species originated within a thousand miles of my garden. Nor did the corn and tobacco grown in nearby farms; corn is from Mexico, tobacco from the Amazon (this species of tobacco, anyway—there was a local species that is now gone). Equally alien, for that matter, are my neighbors' cows, horses, and barn cats. That people like me experience their gardens as familiar and timeless is a testament to the human capacity to adapt (or, less charitably, to our ability to operate in ignorance). Rather than being a locus of stability and tradition, my garden is a biological record of past human wandering and exchange.

Yet in another way my feelings are correct. Almost seventy years ago the Cuban folklorist Fernando Ortiz Fernández coined the awkward but useful term "transculturation" to describe what happens when one group of people takes something—a song, a food, an ideal—from another. Almost inevitably, Ortiz noted, the new thing is transformed; people make it their own by adapting, stripping, and twisting it to fit their needs and situation. Since Columbus the world has been in the grip of convulsive transculturation. Every place on the earth's surface, save possibly scraps of Antarctica, has been changed by places that until 1492 were too remote to exert any impact on it. For five centuries now the crash and chaos of constant connection has been our home condition; my garden, with its parade of exotic plants, is a small example. How *did* those tomatoes get to Ukraine, anyway? One way to describe this book would be to say that it represents, long after I first asked the question, my best efforts to find out.

INTRODUCTION

In the Homogenocene

Two Monuments

THE SEAMS OF PANGAEA

Although it had just finished raining, the air was hot and close. Nobody else was in sight; the only sound other than those from insects and gulls was the staticky low crashing of Caribbean waves. Around me on the sparsely covered red soil was a scatter of rectangles laid out by lines of stones: the outlines of now-vanished buildings, revealed by archaeologists. Cement pathways, steaming faintly from the rain, ran between them. One of the buildings had more imposing walls than the others. The researchers had covered it with a new roof, the only structure they had chosen to protect from the rain. Standing like a sentry by its entrance was a hand-lettered sign: *Casa Almirante*, Admiral's House. It marked the first American residence of Christopher Columbus, Admiral of the Ocean Sea, the man whom generations of schoolchildren have learned to call the discoverer of the New World.

La Isabela, as this community was called, is situated on the north side of the great Caribbean island of Hispaniola, in what is now the Dominican Republic. It was the initial attempt by Europeans to make a permanent base in the Americas. (To be precise, La Isabela marked the beginning of *consequential* European

Lines of stones mark the outlines of now-vanished buildings at La Isabela, Christopher Columbus's first attempt to establish a permanent base in the Americas.

settlement—Vikings had established a short-lived village in New-foundland five centuries before.) The admiral laid out his new domain at the confluence of two small, fast-rushing rivers: a forti-fied center on the north bank, a satellite community of farms on the south bank. For his home, Columbus—Cristóbal Colón, to give him the name he answered to at the time—chose the best location in town: a rocky promontory in the northern settlement, right at the water's edge. His house was situated perfectly to catch the afternoon light.

Today La Isabela is almost forgotten. Sometimes a similar fate appears to threaten its founder. Colón is by no means absent from history textbooks, of course, but in them he seems ever less admirable and important. He was a cruel, deluded man, today's critics say, who stumbled upon the Caribbean by luck. An agent of imperialism, he was in every way a calamity for the Americas' first inhabitants. Yet a different but equally contemporary perspective suggests that we should continue to take notice of the admiral. Of all the members of humankind who have ever walked the earth, he alone inaugurated a new era in the history of life.

The king and queen of Spain, Fernando (Ferdinand) II and Isabel I, backed Colón's first voyage grudgingly. Transoceanic travel in those days was heart-stoppingly expensive and risky—the equivalent, perhaps, of space-shuttle flights today. Despite relentless pestering, Colón was able to talk the monarchs into supporting his scheme only by threatening to take the project to France. He was riding to the frontier, a friend wrote later, when the queen "sent a court bailiff posthaste" to fetch him back. The story is probably exaggerated. Still, it is clear that the sovereigns' reservations drove the admiral to whittle down his expedition, if not his ambitions, to a minimum: three small ships (the biggest may have been less than sixty feet long), a combined crew of about ninety. Colón himself had to contribute a quarter of the budget, according to a collaborator, probably by borrowing it from Italian merchants.

Everything changed with his triumphant return in March of 1493, bearing golden ornaments, brilliantly colored parrots, and as many as ten captive Indians. The king and queen, now enthusiastic, dispatched Colón just six months later on a second, vastly larger expedition: seventeen ships, a combined crew of perhaps fifteen hundred, among them a dozen or more priests charged with bringing the faith to these new lands. Because the admiral believed he had found a route to Asia, he was sure that China and Japan—and all their opulent goods—were only a short journey beyond. The goal of this second expedition was to create a permanent bastion for Spain in the heart of Asia, a headquarters for further exploration and trade.

The new colony, predicted one of its founders, "will be widely renowned for its many inhabitants, its elaborate buildings, and its magnificent walls." Instead La Isabela was a catastrophe, abandoned barely five years after its creation. Over time its structures vanished, their very stones stripped to build other, more successful towns. When a U.S.–Venezuelan archaeological team began excavating the site in the late 1980s, the inhabitants of La Isabela were so few that the scientists were able to move the entire settlement to a nearby hillside. Today it has a couple of roadside fish restaurants, a single, failing hotel, and a little-visited museum. On the edge of town, a church, built in 1994 but already showing signs of age, commemorates the first Catholic Mass celebrated in the Americas. Watching the waves from the admiral's ruined home, I could easily imagine disappointed tourists thinking that the colony had left nothing meaningful behind—that there was no reason, aside from the pretty beach, for anyone to pay attention to La Isabela. But that would be a mistake.

Babies born on the day the admiral founded La Isabela—January 2, 1494—came into a world in which direct trade and communication between western Europe and East Asia were largely blocked by the Islamic nations between (and their partners in Venice and Genoa), sub-Saharan Africa had little contact with Europe and next to none with South and East Asia, and the Eastern and Western hemispheres were almost entirely ignorant of each other's very existence. By the time those babies had grandchildren, slaves from Africa mined silver in the Americas for sale to China; Spanish merchants waited impatiently for the latest shipments of Asian silk and porcelain from Mexico; and Dutch sailors traded cowry shells from the Maldive Islands, in the Indian Ocean, for human beings in Angola, on the coast of the Atlantic. Tobacco from the Caribbean ensorcelled the wealthy and powerful in Madrid, Madras, Mecca, and Manila. Group smoke-ins by violent young men in Edo (Tokyo) would soon lead to the formation of two rival gangs, the Bramble Club and the Leather-breeches Club. The shogun jailed seventy of their members, then banned smoking.

Long-distance trade had occurred for more than a thousand years, much of it across the Indian Ocean. China had for centuries sent silk to the Mediterranean by the Silk Road, a route that was lengthy, dangerous, and, for those who survived, hugely profitable. But nothing like this worldwide exchange had existed before, still less sprung up so quickly, or functioned so continuously. No previous trade networks included both of the globe's two hemispheres; nor had they operated on a scale large enough to disrupt societies on opposite sides of the planet. By founding La Isabela, Colón initiated permanent European occupation in the Americas. And in so doing he began the era of *globalization*—the single, turbulent exchange of goods and services that today engulfs the entire habitable world.

Newspapers usually describe globalization in purely economic terms, but it is also a biological phenomenon; indeed, from a long-term perspective it may be *primarily* a biological phenomenon. Two hundred and fifty million years ago the world contained a single landmass known to scientists as Pangaea. Geological forces broke up this vast expanse, splitting Eurasia and the Americas. Over time the two divided halves of Pangaea developed wildly different suites of plants and animals. Before Colón a few venturesome land creatures had crossed the oceans and established themselves on the other side. Most were insects and birds, as one would expect, but the list also includes, surprisingly, a few farm species—bottle gourds, coconuts, sweet potatoes—the subject today of scholarly head-scratching. Otherwise, the world was sliced into separate ecological domains. Colón's signal accomplishment was, in the phrase of historian Alfred W. Crosby, to reknit the seams of Pangaea. After 1492 the world's ecosystems collided and mixed as European vessels carried thousands of species to new homes across the oceans. The Columbian Exchange, as Crosby called it, is the reason there are tomatoes in Italy, oranges in the United States, chocolates in Switzerland, and chili peppers in Thailand. To ecologists, the Columbian Exchange is arguably the most important event since the death of the dinosaurs.

Unsurprisingly, this vast biological upheaval had repercussions on humankind. Crosby argued that the Columbian Exchange underlies much of the history we learn in the classroom—it was like an invisible wave, sweeping along kings and queens, peasants and priests, all unknowing. The claim was controversial; indeed, Crosby's manuscript, rejected by every major academic publisher, ended up being published by such a tiny press that he once joked to me that his book had been distributed "by tossing it on the street, and hoping readers happened on it." But over the decades since he coined the term, a growing number of researchers have come to believe that the ecological paroxysm set off by Colón's voyages—as much as the economic convulsion he began—was one of the establishing events of the modern world.

On Christmas Day, 1492, Colón's first voyage came to an abrupt end when his flagship, the *Santa María*, ran aground off the northern coast of Hispaniola. Because his two remaining vessels, the *Niña* and *Pinta*, were too small to hold the entire crew, he was forced to leave thirty-eight men behind. Colón departed for Spain while those men were building an encampment—a scatter of makeshift huts surrounded by a crude palisade, adjacent to a larger native village. The encampment was called La Navidad

(Christmas), after the day of its involuntary creation (its precise location is not known today). Hispaniola's native people have come to be known as the Taino. The conjoined Spanish-Taino settlement of La Navidad was the intended destination of Colón's second voyage. He arrived there in triumph, the head of a flotilla, his crewmen swarming the shrouds in their eagerness to see the new land, on November 28, 1493, eleven months after he had left his men behind.

He found only ruin; both settlements, Spanish and Taino, had been razed. "We saw everything burned and the clothing of Christians lying on the weeds," the ship's doctor wrote. Nearby Taino showed the visitors the bodies of eleven Spaniards, "covered by the vegetation that had grown over them." The Indians said that the sailors had angered their neighbors by raping some women and murdering some men. In the midst of the conflict a second Taino group had swooped down and overwhelmed both sides. After nine days of fruitless search for survivors Colón left to find a more promising spot for his base. Struggling against contrary winds, the fleet took almost a month to crawl a hundred miles east along the coast. On January 2, 1494, Colón arrived at the shallow bay where he would found La Isabela.

Almost immediately the colonists ran short of food and, worse, water. In a sign of his inadequacy as an administrator, the admiral had failed to inspect the water casks he had ordered; they, predictably, leaked. Ignoring all complaints of hunger and thirst, the admiral decreed that his men would clear and plant vegetable patches, erect a two-story fortress, and enclose the main, northern half of the new enclave within high stone walls. Inside the walls the Spaniards built perhaps two hundred houses, "small like the huts we use for bird hunting and roofed with weeds," one man complained.*

---

* Short of water, the expedition drank from rivers. Some researchers believe that Colón and his men thus caught shigellosis, a disease caused by a feces-borne bacterium native to the American tropics. In reaction to the

Most of the new arrivals viewed these labors as a waste of time. Few actually wanted to set up shop in La Isabela, still less till its soil. Instead they regarded the colony as a temporary base camp for the quest for riches, especially gold. Colón himself was ambivalent. On the one hand, he was supposed to be governing a colony that was establishing a commercial entrepôt in the Americas. On the other hand, he was supposed to be at sea, continuing his search for China. The two roles conflicted, and Colón was never able to resolve the conflict.

On April 24 Colón sailed off to find China. Before leaving, he ordered his military commander, Pedro Margarit, to lead four hundred men into the rugged interior to seek Indian gold mines. After finding only trivial quantities of gold—and not much food—in the mountains, Margarit's charges, tattered and starving, came back to La Isabela, only to discover that the colony, too, had little to eat—those left behind, resentful, had refused to tend gardens. The irate Margarit hijacked three ships and fled to Spain, promising to brand the entire enterprise as a waste of time and money. Left behind with no food, the remaining colonists took to raiding Taino storehouses. Infuriated, the Indians struck back, setting off a chaotic war. This was the situation that confronted Colón when he returned to La Isabela five months after his departure, dreadfully sick and having failed to reach China.

A loose alliance of four Taino groups faced off against the Spaniards and one Taino group that had thrown its lot in with the foreigners. The Taino, who had no metal, could not withstand assaults with steel weapons. But they made the fight costly for the Spaniards. In an early form of chemical warfare, the Indians threw gourds stuffed with ashes and ground hot peppers at their attack-

---

bacterium, the body can develop Reiter's syndrome, an autoimmune disease that makes sufferers feel as if large chunks of the body, including the eyes and bowels, are swollen and inflamed—symptoms that afflicted Colón later that summer. Reiter's is always painful and sometimes fatal. If, as these scientists suspect, Reiter's led, years later, to the admiral's death, Columbus himself was an early victim of the Columbian Exchange.

ers, unleashing clouds of choking, blinding smoke. Protective bandannas over their faces, they charged through the tear gas, killing Spaniards. The intent was to push out the foreigners—an unthinkable course to Colón, who had staked everything on the voyage. When the Spaniards counterattacked, the Taino retreated scorched-earth style, destroying their own homes and gardens in the belief, Colón wrote scornfully, "that hunger would drive us from the land." Neither side could win. The Taino alliance could not eject the Spaniards from Hispaniola. But the Spaniards were waging war on the people who provided their food supply; total victory would be a total disaster. They won skirmish after skirmish, killing countless natives. Meanwhile, starvation, sickness, and exhaustion filled the cemetery in La Isabela.

Humiliated by the calamity, the admiral set off for Spain on March 10, 1496, to beg the king and queen for more money and supplies. When he returned two years later—the third of what would become four voyages across the Atlantic—so little was left of La Isabela that he landed on the opposite side of the island, in Santo Domingo, a new settlement founded by his brother Bartolomé, whom he had left behind. Colón never again set foot in his first colony and it was almost forgotten.

Despite the brevity of its existence, La Isabela marked the beginning of an enormous change: the creation of the modern Caribbean landscape. Colón and his crew did not voyage alone. They were accompanied by a menagerie of insects, plants, mammals, and microorganisms. Beginning with La Isabela, European expeditions brought cattle, sheep, and horses, along with crops like sugarcane (originally from New Guinea), wheat (from the Middle East), bananas (from Africa), and coffee (also from Africa). Equally important, creatures the colonists knew nothing about hitchhiked along for the ride. Earthworms, mosquitoes, and cockroaches; honeybees, dandelions, and African grasses; rats of every description—all of them poured from the hulls of Colón's vessels and those that followed, rushing like eager tourists into lands that had never seen their like before.

Cattle and sheep ground American vegetation between their flat teeth, preventing the regrowth of native shrubs and trees. Beneath their hooves would sprout grasses from Africa, possibly introduced from slave-ship bedding; splay-leaved and dense on the ground, they choked out native vegetation. (Alien grasses could withstand grazing better than Caribbean groundcover plants because grasses grow from the base of the leaf, unlike most other species, which grow from the tip. Grazing consumes the growth zones of the latter but has little impact on those in the former.) Over the years forests of Caribbean palm, mahogany, and ceiba became forests of Australian acacia, Ethiopian shrubs, and Central American logwood. Scurrying below, mongooses from India eagerly drove Dominican snakes toward extinction. The change continues to this day. Orange groves, introduced to Hispaniola from Spain, have recently begun to fall to the depredations of lime swallowtail butterflies, citrus pests from Southeast Asia that probably came over in 2004. Today Hispaniola has only small fragments of its original forest.

Natives and newcomers interacted in unexpected ways, creating biological bedlam. When Spanish colonists imported African plantains in 1516, the Harvard entomologist Edward O. Wilson has proposed, they also imported scale insects, small creatures with tough, waxy coats that suck the juices from plant roots and stems. About a dozen banana-infesting scale insects are known in Africa. In Hispaniola, Wilson argued, these insects had no natural enemies. In consequence, their numbers must have exploded—a phenomenon known to science as "ecological release." The spread of scale insects would have dismayed the island's European banana farmers but delighted one of its native species: the tropical fire ant *Solenopsis geminata*.* *S. geminata* is fond of dining

---

* Every species has a scientific name with two parts: the name of its genus—the group of related species it belongs to—and the species name proper. Thus *Solenopsis geminata* belongs to the genus *Solenopsis* and is the species geminata. By convention, the genus is abbreviated after the first time it appears with the species name: *S. geminata*.

on scale insects' sugary excrement; to ensure the flow, the ants will attack anything that disturbs them. A big increase in scale insects would have led to a big increase in fire ants.

So far this is informed speculation. What happened in 1518 and 1519 is not. In those years, according to Bartolomé de Las Casas, a missionary priest who lived through the incident, Spanish orange, pomegranate, and cassia plantations were destroyed "from the root up." Thousands of acres of orchards were "all scorched and dried out, as though flames had fallen from the sky and burned them." The actual culprit, Wilson argued, was the sap-sucking scale insects. But what the Spaniards *saw* was *S. geminata*—"an infinite number of ants," Las Casas reported, their stings causing "greater pains than wasps that bite and hurt men." The hordes of ants swarmed through houses, blackening roofs "as if they had been sprayed with charcoal dust," covering floors in such numbers that colonists could sleep only by placing the legs of their beds in bowls of water. They "could not be stopped in any way nor by any human means."

Overwhelmed and terrified, Spaniards abandoned their homes to the insects. Santo Domingo was "depopulated," one witness recalled. In a solemn ceremony, the remaining colonists chose, by lottery, a saint to intercede with God on their behalf—St. Saturninus, a third-century martyr. They held a procession and feast in his honor. The response was positive. "From that day onward," Las Casas wrote, "one saw by plain sight that the plague began to diminish."

From the human perspective, the most dramatic impact of the Columbian Exchange was on humankind itself. Spanish accounts suggest that Hispaniola had a large native population: Colón, for instance, casually described the Taino as "innumerable, for I believe there to be millions upon millions of them." Las Casas claimed the population to be "more than three million." Modern researchers have not nailed down the number; estimates range from 60,000 to almost 8 million. A careful study in 2003 argued that the true figure was "a few hundred thousand." No matter what the original number, though, the European impact was horrific. In 1514, twenty-two years after Colón's first voyage, the Spanish government counted up the Indians on Hispaniola for the purpose of allocating them among colonists as laborers. Census agents fanned across the island but found only 26,000 Taino. Thirty-four years later, according to one scholarly Spanish resident, fewer than 500 Taino were alive. The destruction of the Taino plunged Santo Domingo into poverty. The colonists had wiped out their own labor force.

Spanish cruelty played its part in the calamity, but its larger cause was the Columbian Exchange. Before Colón none of the epidemic diseases common in Europe and Asia existed in the Americas. The viruses that cause smallpox, influenza, hepatitis, measles, encephalitis, and viral pneumonia; the bacteria that cause tuberculosis, diphtheria, cholera, typhus, scarlet fever, and bacterial meningitis—by a quirk of evolutionary history, all were unknown in the Western Hemisphere. Shipped across the ocean from Europe these maladies consumed Hispaniola's native population with stunning rapacity. The first recorded epidemic, perhaps due to swine flu, was in 1493. Smallpox entered, terribly, in 1518; it spread to Mexico, swept down Central America, and then continued into Peru, Bolivia, and Chile. Following it came the rest, a pathogenic cavalcade.

Throughout the sixteenth and seventeenth centuries novel microorganisms spread across the Americas, ricocheting from victim to victim, killing three-quarters or more of the people in the hemisphere. It was as if the suffering these diseases had caused in Eurasia over the past millennia were concentrated into a span of decades. In the annals of human history there is no comparable demographic catastrophe. The Taino were removed from the face of the earth, though recent research hints that their DNA may survive, invisibly, in Dominicans who have African or European features, genetic strands from different continents entangled, coded legacies of the Columbian Exchange.

TO THE LIGHTHOUSE

A placid, whispering river runs through Santo Domingo, capital of the Dominican Republic. On the west bank of the river stands the stony remains of the colonial town, including the palace of Diego Colón, the admiral's firstborn son. From the east bank rises a vast mesa of stained concrete, a monolith 102 feet high and 689 feet long. It is the Faro a Colón—the Columbus Lighthouse. The structure is called a lighthouse because 146 four-kilowatt lights are mounted on its summit. They point straight up, assaulting the heavens with a fusillade of light intense enough to cause blackouts in surrounding neighborhoods.

Like a medieval church, the lighthouse is laid out as a cross, with a long nave and two short transepts projecting from the sides. At the central intersection, inside a crystal security box, is an ornate golden sarcophagus said to contain the admiral's bones. (The claim is disputed; in Seville, Spain, another ornate sarcophagus also is said to house Colón's remains.) Beyond the sarcophagus are a series of exhibits from many nations. When I visited not long ago, most focused on the hemisphere's original inhabitants, depicting them as the passive or even grateful recipients of European largesse, cultural and technological.

Unsurprisingly, native people rarely endorse this view of their history, and Colón's part in it. An army of activists and scholars has bombarded the public with condemnations of the man and his works. They have called him brutal (he was, by today's standards) and racist (he wasn't, strictly speaking—modern concepts of race had not yet been invented); incompetent as an administrator (he was) and as a seaman (he wasn't); a religious fanatic (he surely was, from a secular point of view); and a greedy monomaniac (a charge, the admiral's supporters would say, that could be leveled against all ambitious souls). Colón, his detractors charge, never understood what he had found.

Completed in 1992, this huge, cross-shaped memorial to Columbus in Santo Domingo was designed by the young Scottish
architect Joseph Lea Gleave, who attempted to capture in stone
what he regarded as Columbus's most important role: the man
who brought Christianity to the Americas. The structure, he
said modestly, would be "one of the great monuments of the
ages."

How different it was in 1852, when Antonio del Monte y
Tejada, a celebrated Dominican *litterateur,* closed the first of
the four volumes of his history of Santo Domingo by extolling
Colón's "great, generous, memorable and eternal" career. The
admiral's every action "breathes greatness and elevation," del
Monte y Tejada wrote. Do not "all nations . . . owe him eternal
gratitude"? The best way to acknowledge this debt, he proposed,
would be to erect a gigantic Columbus statue, "a colossus like
the one in Rhodes," sponsored by "all the cities of Europe and
America," that would spread its arms benevolently across Santo
Domingo, the hemisphere's "most visible and noteworthy place."

A grand monument to the admiral! To del Monte y Tejada,
the merits of the idea seemed obvious; Colón was a messenger
from God, his voyages to the Americas the result of a "divine

decree." Nonetheless, building the monument took almost a century and a half. The delay was partly economic; most nations in the hemisphere were too poor to throw money at a monstrous statue on a faraway island. But it also reflected the growing unease about the admiral himself. Knowing what we know today about the fate of the Indians on Hispaniola, critics asked, should there be any monument to his voyages at all? Given his actions, what kind of person was buried in the golden box at its center?

The answer is hard to arrive at, even though his life is among the best documented of his time—the newest edition of his collected writings runs to 536 pages of small print.

During his lifetime, nobody knew him as Columbus. The admiral was baptized as Cristoforo Colombo by his family in Genoa, Italy, but changed his name to Cristovao Colombo when he moved to Portugal, where he was an agent for Genoese merchant families. He called himself Cristóbal Colón after 1485, when he moved to Spain, having failed to persuade the Portuguese king to sponsor an expedition across the Atlantic. Later, like a petulant artist, he insisted that his signature be an incomprehensible glyph:

$$. S .$$
$$S . A . S$$
$$X \ M \ \overset{.}{Y}$$
$$: \ X\rho o \ FERENS. /$$

(No one is sure what he meant, but the third line could invoke Christ, Mary, and Joseph—*Xristus Maria Yosephus*—and the letters up top may stand for *Servus Sum Altissimi Salvatoris*, "Servant I am of the Highest Savior." Xρo FERENS is probably *Xristo-Ferens*, "Christ-Bearer.")

"A well-built man of greater than average stature," according to a description attributed to his illegitimate son Hernán, the admiral had prematurely white hair, "light-colored eyes," an aquiline nose, and fair cheeks that readily flushed. He was a mercurial man, moody and inconstant one hour to the next.

Although subject to fits of rage, Hernán remembered, Colón was also "so opposed to swearing and blasphemy that I give my word I never heard him say any oath other than 'by San Fernando.'" (St. Ferdinand). His life was dominated by overweening personal ambition and, arguably more important, profound religious faith. Colón's father, a weaver, seems to have scrambled from debt to debt, which his son apparently viewed with shame; he actively concealed his origins and spent his entire adult life striving to found a dynasty that would be ennobled by the monarchy. His faith, always ardent, deepened during the long years in which he was vainly begging rulers in Portugal and Spain to back his voyage west. During part of that time he lived in a politically powerful Franciscan monastery in southern Spain, a place enraptured by the visions of the twelfth-century mystic Joachim di Fiore, who believed that humankind would enter an age of spiritual bliss after Christendom wrested Jerusalem from the Islamic forces who had conquered it centuries before. The profits from his voyage, Colón came to believe, would both advance his own fortunes and fulfill di Fiore's vision of a new crusade. Trade with China would pour so much money into Spain, he predicted, "that in three years the Monarchs will be able to set about preparing for the conquest of the Holy Land."

Integral to this grand scheme were Colón's views on the size and shape of the earth. As a child, I—like countless students before me—was taught that Columbus was ahead of his time, proclaiming the planet to be large and round in an era when everyone else believed it to be small and flat. My fourth-grade teacher showed us an etching of Columbus brandishing a globe before a platoon of hooting medieval authorities. A shaft of sunlight illuminated the globe and the admiral's flowing hair; his critics, by contrast, squatted like felons in the shadows. My teacher, alas, had it exactly backward. Scholars had known for more than fifteen hundred years that the world was large and round. Colón disputed both facts.

The admiral's disagreement with the second fact was minor.

The earth, he argued, was not perfectly round but "in the shape of a pear, which would all be very round, except for where the stem is, where it is higher, or as if someone had a very round ball, and in one part of it a woman's nipple would be put there." At the very tip of the nipple, so to speak, was "the Earthly Paradise, where nobody can go, except by divine will." (During a later voyage he thought he had found the nipple, in what is now Venezuela.)

The king and queen of Spain cared not a whit about the admiral's views of the world's shape or heaven's location. But they were keenly interested in his ideas about its size. Colón believed the planet's circumference to be at least five thousand miles smaller than it actually is. If this idea were true, the gap between western Europe and eastern China—the width, we know today, of both the Atlantic and Pacific oceans and the lands between them—would be much smaller than it actually is.

The notion enticed the monarchs. Like other European elites, they were fascinated by accounts of the richness and sophistication of China. They lusted after Asian textiles, porcelain, spices, and precious stones. But Islamic merchants and governments stood in the way. If Europeans wanted the luxuries of Asia, they had to negotiate with powers that Christendom had been at war with for centuries. Worse, the mercantile city-states of Venice and Genoa had already cut a deal with Islamic forces, and now monopolized the trade. The notion of working with Islamic entities was especially unwelcome to Spain and Portugal, which had been conquered by the armies of Muhammad in the eighth century and had spent hundreds of years in an ultimately successful battle to repel them. But even if they did make arrangements with Islam, Venice and Genoa stood ready to use force to maintain their privileged position. To cut out the unwanted middlemen, Portugal had been trying to send ships all the way around Africa—a long, risky, expensive journey. The admiral told the rulers of Spain that there was a faster, safer, cheaper route: going west, across the Atlantic.

In effect, Colón was challenging the Greek polymath Eratosthenes, who in the third century B.C. had ascertained the earth's circumference by a method, the science historian Robert Crease wrote in 2003, "so simple and instructive that it is reenacted annually, almost 2,500 years later, by schoolchildren all around the globe." Eratosthenes concluded that the world is about twenty-five thousand miles around. The east-west width of Eurasia is approximately ten thousand miles. Arithmetic would require that the gap between China and Spain be about fifteen thousand miles. European shipbuilders and potential explorers both knew that no fifteenth-century vessel could survive a voyage of fifteen thousand miles, let alone make the return trip.

Colón believed that he had, as it were, disproved Eratosthenes. A skilled intuitive seaman, the admiral had plied the eastern Atlantic from Africa to Iceland. During these travels he used a sailor's quadrant in an attempt to measure the length of a degree of longitude. Somehow he convinced himself that his results vindicated the claim, attributed to a ninth-century caliph in Baghdad, that a degree was 56⅔ miles. (It is actually closer to sixty-nine miles.) Colón multiplied this value by 360, the number of degrees in a circle, to calculate the circumference of the earth: 20,400 miles. Coupling this figure with an incorrectly large estimate of the east-west length of Eurasia, Colón argued that the journey across the Atlantic could be as little as three thousand miles, six hundred miles of which could be cut off by setting sail from the newly conquered Canary Islands. This distance could easily be traversed by Spanish vessels.

Crossing their fingers that Colón was right, the monarchs submitted his proposal to a committee of experts in astronomy, navigation, and natural philosophy. The committee of experts rolled its collective eyes. From its perspective, Colón's claim that he—a poorly educated man fumbling with a quadrant on a wave-tossed ship—had refuted Eratosthenes was like someone claiming to have demonstrated in a backwoods shack that gravity didn't pull iron as much as scientists thought, and that one could therefore hoist an anvil with a loop of thread. In the end, though, the king and queen ignored the experts—they told Colón to try the thread.

After landing in the Americas in 1492, the admiral naturally claimed that his ideas had been vindicated.* The delighted monarchs awarded him honors and wealth. He died in 1506, a rich man surrounded by a loving family; nevertheless, he died a bitter man. As evidence had emerged of his failings, personal and geographical, the Spanish court had revoked most of his privileges and shunted him aside. In the anger and humiliation of his later years, he slid into religious messianism. He came to believe that he was God's "messenger," destined to show the world "the new heaven and earth of which Our Lord spoke through Saint John in the Apocalypse." In one of his last reports to the king, the admiral suggested that he, Colón, would be the ideal person to convert the emperor of China to Christianity.

Much the same mix of grandiosity and disappointment characterized the Columbus monument. Del Monte y Tejada's proposal for a memorial to the admiral was finally approved in 1923, at a meeting of the Western Hemisphere's governments. Progress was slow—the design competition wasn't held for another eight years, and the monument itself wasn't built for another six decades. During most of that time the Dominican Republic was ruled by the tyrant Rafael Trujillo. A classic case of narcissistic personality disorder, Trujillo erected scores of statues to himself

---

* It is conceivable that Colón knew before his departure that the Atlantic could be crossed. He wrote in the margin of one of his books that while in Ireland he'd seen "people from Cathay [China]"—"a man and a wife brought in on a couple of logs in an extraordinary manner." Some writers argue that the "logs" were dugout canoes, and the people therefore Inuit or Indians. Most historians do not agree, though, because there is little evidence that Colón visited Ireland, let alone that he saw two Indians there. The couple could have been Sami from Finland, who often have Asian features. In addition it seems implausible that the sole record of this amazing event—Indians paddling a canoe to Europe!—should be a few marginal scribbles in a book.

Every American nation promised to contribute to the Columbus memorial when it was approved in 1923, but the checks were slow in coming—the U.S. Congress, for example, didn't appropriate its share for another six years. In May of 1930 Dominican army head Rafael Trujillo became president in a fraud-ridden election. Three weeks later a hurricane wiped out Santo Domingo, killing thousands. Deciding that the memorial would symbolize the city's revitalization, Trujillo staged a design competition in 1931. On the jury were eminent architects, including Eliel Saarinen and Frank Lloyd Wright. More than 450 entries came in, including these by (clockwise from top left) Konstantin Melnikov, Robaldo Morozzo della Rocca and Gigi Vietti, Erik Bryggman, and Iosif Langbard.

and hung a giant neon sign that read "God and Trujillo" over the harbor of Santo Domingo, which he had renamed Trujillo City. As his reign grew more barbarous, international enthusiasm for the lighthouse waned—supporting the project was seen as endorsing the dictator. Many nations boycotted the inauguration, on October 12, 1992. Pope John Paul II reneged on his promise to celebrate a Mass at the opening, though he did appear nearby a day before. Meanwhile, protesters set police barricades on fire, denouncing the admiral as "the exterminator of a race." Residents of the walled-off slums around the monument told reporters that they thought Colón deserved no commemoration at all.

A thesis of this book is that their belief, no matter how understandable, is mistaken. The Columbian Exchange had such far-reaching effects that some biologists now say that Colón's voyages marked the beginning of a new biological era: the Homogenocene. The term refers to homogenizing: mixing unlike substances to create a uniform blend. With the Columbian Exchange, places that were once ecologically distinct have become more alike. In this sense the world has become one, exactly as the old admiral hoped. The lighthouse in Santo Domingo should be regarded less as a celebration of the man who began it than a recognition of the world he almost accidentally created, the world of the Homogenocene we live in today.

SHIPLOADS OF SILVER

At a busy corner in a park just south of the old city walls in Manila is a grimy marble plinth, perhaps fifteen feet tall, topped by lifesize bronze statues, blackened by pollution, of two men in sixteenth-century attire. The two men stand shoulder to shoulder, faces into the setting sun. One wears a friar's habit and brandishes a cross as if it were a sword; the other, in a military breastplate, carries an actual sword. Compared to the Columbus Lighthouse, the monument is small and rarely visited by tourists. I found no mention of it in recent guidebooks and maps—a historical embarrassment, because it is the closest thing the world has to an official recognition of globalization's origins.

The man with the sword is Miguel López de Legazpi, founder of modern Manila. The man with the cross is Andrés Ochoa de Urdaneta y Cerain, the navigator who guided Legazpi's ships across the Pacific. One way to summarize the two Spaniards' contribution would be to say that together Legazpi and Urdaneta achieved what Colón failed to do: establish continual trade with China by sailing west. Another way to state their accomplishment would be to say that Legazpi and Urdaneta were to economics what Colón was to ecology: the origin, however inadvertent, of a great unification.

Legazpi, slightly the more well known, was born about a decade after the admiral's first voyage. For most of his life he showed no sign of Colón's penchant for maritime adventure. He trained as a notary, inheriting his father's position in the Basque city of Zumárraga, near the border with France. In his late twenties he went to Mexico, where he worked in the colonial administration for thirty-six years. His life was jerked out of its cozy rut when he was approached by Urdaneta, a friend and cousin who was among the few survivors of Spain's failed attempt, in the 1520s, to establish an outpost in the spice-laden Maluku Islands. (Formerly known as the Moluccas, they are south of the Philippines.) Urdaneta had been shipwrecked in the Malukus for a decade, eventually being rescued by the Portuguese. After returning he had sworn off adventures and joined the Augustinian religious order. Thirty years later, the next king of Spain wanted to take another stab at establishing a base in Asia. He ordered Urdaneta to return to sea. Urdaneta's position as a clergyman made him unable by law to serve as head of the expedition. He chose Legazpi for the job, despite his lack of a nautical background. Legazpi's thoughts about the likelihood of success may be indicated by his decision to prepare for the voyage by selling all of his worldly possessions and sending his children and grandchildren to stay with family members in Spain.

Because Portugal had taken advantage of the Spanish failures to occupy the Malukus, the expedition was told to find more spice islands nearby and establish a trade base on them. The king of Spain also wanted them to chart the wind patterns, to introduce the area to Christianity, and to be a thorn in the side of his nephew and rival, the king of Portugal. But the underlying goal was China—"the stimulus that pulled Spain, as the vanguard of Christendom, to search the seaways," as the historian Antonio García-Abásolo put it in 2004. "One cannot overemphasize the continuity of the goals for the actions undertaken by Colón, [conqueror of Mexico Hernán] Cortés and Legazpi." All of them sought China.

Legazpi and Urdaneta left with five ships on November 21, 1564. Reaching the Philippines, Legazpi set up camp on the island of Cebu, midway up the archipelago. Meanwhile, Urdaneta set

about figuring out how to return to Mexico—nobody had ever successfully made the trip. Simply retracing the expedition's westward route was not possible, because the trade winds that had blown the ships from Mexico to the Malukus would impede their return. In a stroke of navigational genius, Urdaneta avoided the contrary currents by sailing far to the north before turning east.

On Cebu, Legazpi was plagued by mutiny and disease and harassed by Portuguese ships. But he slowly expanded Spanish influence north, approaching China. Periodically the Spanish viceroy in Mexico City dispatched reinforcements and supplies. Important among the supplies were silver bars and coins, mined in Mexico and Bolivia, intended to pay the Spanish troops.

A turning point occurred in May 1570, when Legazpi dispatched a reconnaissance mission: two small ships with about a hundred Spanish soldiers and sailors, accompanied by scores of native Filipino Malays in proas (low, narrow outrigger-type boats, rigged with one or two fore-and-aft sails). After two days' northerly sail, they reached the island of Mindoro, about 130 miles south of modern Manila (which is on Luzon, the chain's biggest island). Mindoro's southern coast consists of a number of small bays, one next to another like tooth marks in an apple. The Malays on the expedition learned from local Mangyan people that two Chinese junks were at anchor forty miles away, in another cove—a trading post near the modern village of Maujao (mah-oo-how).

Every spring ships from China traveled to several Philippine islands, Mindoro among them, to exchange porcelain, silk, perfumes, and other goods for gold and beeswax.* Shaded by parasols made of white Chinese silk, the Mangyan descended from

---

* Because China did not make enough beeswax for its needs, many Chinese made candles from a substitute: the lower-quality wax produced by a scale insect. The Philippines house both the Asian honeybee and the giant honeybee; the huge nests of the latter are rich sources of wax.

their upland homes to meet the Chinese, who beat small drums to announce their arrival. Maujao, which has a freshwater spring a few feet back from the beach, had long been a meeting point; local officials told me that archaeology students have found Chinese porcelain there that dates to the eleventh century. Legazpi had ordered the excursion's commander to contact—politely, not aggressively—any Chinese he encountered. Hearing of the junks' presence, the commander sent one of the two Spanish ships and most of the proas to meet the Chinese "and to request peace and friendship with them."

Leading the contact group was Juan de Salcedo, Legazpi's twenty-one-year-old grandson, popular with and respected by the soldiers despite his youth. Unluckily, high winds separated the vessels; Salcedo's ship was pushed badly off course. The vessels spent the night in different harbors, protected from the storm by the high, narrow fingers of rock that define the coves. Temporarily leaderless but eager to gain the riches of China, the Spanish soldiers in the proas moved east at first light. Rounding a narrow, rocky promontory on the southern side of Maujao, they came upon the Mangyan and Chinese. The Chinese put on a show of force, one of Salcedo's men later recalled, "beating on drums, playing on fifes, firing rockets and culverins [a kind of small, portable cannon], and making a great warlike display." Taking this as a challenge, the Spaniards attacked—a rash act, "for the Chinese ships were large and high, while the proas were so small and low that they hardly reached to the lower bollards on the enemy's ships." They raked the junks' decks with musket fire, threw grappling hooks over the sides, clambered onto the decks, and killed lots of Chinese traders. Onboard, the attackers found small quantities of silk, porcelain, gold thread, "and other curious articles."

When Salcedo finally arrived in Maujao, hours after the battle, he was "not at all pleased with the havoc." Far from requesting "peace and friendship," as he had ordered, his men had wantonly slain Chinese sailors and left their ships in ruins. (The chronicle, probably written by Martín de Goiti, Salcedo's right-hand man, makes no mention of the Mangyan, whom the Spaniards didn't care about; one assumes they fled the carnage.) Salcedo apologized, freed the survivors, and returned the meager plunder. The Chinese, the expedition member reported, "being very humble people, knelt down with loud utterances of joy." Still, there was a problem. One of the junks was totally destroyed; the other was salvageable, but the ship rigging was so different from European rigging that nobody in the expedition knew how to mend it. Salcedo ordered some of his troops to help the surviving vessel limp to the Spanish headquarters, where Legazpi's men might be able to help.

The Chinese sailed home in their reconstructed junk and reported that Europeans had appeared in the Philippines. Amazingly, they had come from the east, though Europe lay to the west. And the barbarians had something that was extremely desirable in China: silver. Meanwhile, Legazpi took over Manila and waited for their return.

In the spring of 1572, three junks appeared in the Philippines. They contained a carefully chosen selection of Chinese manufactured goods—a test of what Legazpi would pay for, and pay the most for. It turned out the Spaniards wanted everything, a result, Legazpi's notary reported, that "delighted" the traders. Especially coveted were silk, rare and costly in Europe, and porcelain, made by a technology then unknown in Europe. In return, the Chinese took every ounce they could of Spanish silver.

More junks came the next year, and the year after that. Because China's hunger for silver and Europe's hunger for silk and porcelain were effectively insatiable, the volume of trade grew enormous. The "galleon trade," as it would become known, linked Asia, Europe, the Americas, and, less directly, Africa. (African slaves were integral to Spain's American empire; as I describe later, they and their descendants far outnumbered Europeans there.) Never before had so much of the planet been bound in a single network of exchange—every populous area on earth, every habitable continent except Australia. Dawning with Spain's arrival in the Philippines was a new, distinctly modern era.

That era was regarded with suspicion from the beginning. China was then the earth's wealthiest, most powerful nation. By virtually any measure—per capita income; military strength; average lifespan; agricultural production; culinary, artistic, and technical sophistication—it was equal to or superior to the rest of the world. Much as rich nations like Japan and the United States today buy little from sub-Saharan Africa, China had long viewed Europe as too poor and backward to be of commercial interest. Its principal industry was textiles, mainly wool. China, meanwhile, had *silk*. Reporting to the Spanish king in 1573, the viceroy in Mexico lamented that "neither from this land nor from Spain, so far as can now be learned, can anything be exported thither that they do not already possess." With silver, though, Spain finally had something China wanted. Badly wanted, in fact—Spanish silver literally became China's money supply. But there was an unease about having the nation's currency in the hands of foreigners. The court feared that the galleon trade—the first large-scale, uncontrolled international exchange in Chinese history—would usher in large-scale, uncontrolled change to Chinese life.

The fears were entirely borne out. Although emperor after emperor refused entry to almost all human beings from Europe and the Americas, they could not keep out other species. Key players were American crops, especially sweet potatoes and maize;* their unexpected arrival, the agricultural historian Song Junling wrote in 2007, was "one of the most revolutionary events" in imperial China's history. The nation's agriculture, based on rice,

---

* In the United States the name is "corn." I use "maize" hereafter for two reasons. First, multicolored Indian maize, which was usually eaten after drying and grinding, is strikingly unlike the sweet yellow kernels conjured up in the U.S. by the word "corn." Second, "corn" in Britain refers to a region's most important cereal crop—oats in Scotland, for example.

had long been concentrated in river valleys, especially those of the Yangzi and Huang He (Yellow) rivers. Sweet potatoes and maize could be grown in the dry uplands. Farmers moved in numbers to these areas, which had previously been lightly settled. The result was a wave of deforestation, followed by waves of erosion and floods, which caused many deaths. The regime, already straining under many problems, was further destabilized—to Europe's benefit.

Spain, too, was uneasy about the galleon trade. The annual shipments of silver to Manila were the culmination of a centuries-long quest to trade with China. Nonetheless, Madrid spent almost the entire period trying to limit the exchange. Again and again, royal edicts restricted the number of ships allowed to travel to Manila, cut the amount of allowable exports, set import quotas for Chinese goods, and instructed Spanish merchants to form a cartel to raise prices.

From today's perspective the Spanish discontent is surprising. Both sides gained by the exchange of silk for silver, as economic theory would predict. But it was Europe that emerged in the stronger position. With the galleon trade, declaimed the historian Andre Gunder Frank, "Europeans bought themselves a seat, and then even a whole railway car, on the Asian train." Legazpi's encounter with the Chinese signaled the arrival of the Homogenocene in Asia. And following it, gliding in the slip-stream, came the rise of the West.

The statue of Legazpi and Urdaneta was not intended to commemorate any of these ideas or events. It was proposed in 1892 by Manila's Basque community to celebrate the Basque role in the city's history (Legazpi and Urdaneta were Basques, as were many of their men). By the time Catalan sculptor Agustí Querol i Subirats cast the bronze, the United States had seized the Philippines from Spain. The islands' new rulers had little interest in a monument to dead Spaniards; the statue languished at a customs house until 1930, when it was finally erected.

Walking around the monument, I wished that it were larger,

given that it is the closest equivalent to a formal commemoration of globalization we have today. I also wished it were more complete. To truly mark the galleon trade, Legazpi and Urdaneta would have to be surrounded by Chinese merchants: equal partners in the exchange. Such a monument probably will never be built, not least because the worldwide network is still viewed with unease, even by many of its beneficiaries.

Across the street from the monument is another, more popular park, named after José Rizal, a writer, doctor, and martyred anti-Spanish revolutionary who is a national hero in the Philippines. At the center of Rizal Park is a reflecting pool edged with flower gardens and statuary. All the statues are bronze busts on concrete columns. All depict Filipinos who died fighting Spanish rule.

As close to a monument to globalization as the world is likely to see, this statue to Miguel López de Legazpi and Andrés de Urdaneta, initiators of the silver trade across the Pacific, occupies a little-frequented corner of a park in central Manila.

On the side of the pool facing the Legazpi monument is a bust of Rajah Sulayman, identified by a plaque as "the brave Muslim ruler of the kingdom of Maynila (Manila) who refused the offer of 'friendship' by the Spaniards . . . under Miguel Lopez de Legazpi." (Parentheses in original.) Good editors deride fake quotation marks like those around "friendship" as "scare quotes" and tell reporters not to use them. Here they may be merited.

Legazpi approached Sulayman soon after encountering the Chinese. The Spaniards wanted to use Manila's harbor as a launching point for the China trade. When Sulayman said he didn't want the Spaniards around, Legazpi leveled his principal village, killing him and three hundred of his fellows. Modern Manila was established on the ruins.

Sulayman and the other people around the pool were, in effect, the first antiglobalization martyrs. They have been awarded a place considerably more prominent than the deserted corner given to Legazpi and Urdaneta. In the end, though, they lost, each and every one of them.

Big speakers mounted on iron columns at the corners of the pool issue bulletins from the redoubts of Classic Rock. Walking around the area, I was nearly run over by a train fashioned into a replica of Thomas the Tank Engine, a children's-book and -television character owned by Apax Partners, a British private-equity firm said to be among the world's largest. Over Thomas's smiling, tooting head I could see the towers of the hotels and banks in Manila's tourist district. The birthplace of globalization looked a lot like many other places. In the Homogenocene, Kentucky Fried Chicken, McDonald's, and Pizza Hut are always just minutes away.

REVERSALS OF FORTUNE

The Homogenocene? A new epoch in the history of life, brought into being by the abrupt creation of a world-spanning economic system? The claim seems grandiose. But imagine a thought experiment: flying around the earth in 1642, a century and a half after Colón's first voyage, threescore and ten after the first Chinese silk from Manila arrived in Mexico. Think of it as a round-the-world cruise at 35,000 feet of a planet in the first stages of a great disturbance. The brochure promises that the cruise will hit the highlights of the nascent Homogenocene. What will the passengers see?

One answer would be: a world bound together by hoops of Spanish silver. Silver from the Americas is well on its way to doubling or tripling the world's stock of precious metals. Potosí, in what is now southern Bolivia, is the main source—the biggest, richest strike in history. Begin the cruise here, at this central node in the network. Located more than thirteen thousand feet up the Andes, Potosí sits at the foot of an extinct volcano that is as close to a mountain of pure silver as geology allows. Around it is an almost treeless plateau, strewn with glacial boulders, scoured by gelid winds. Agriculture struggles here, and there is no wood for fire. Nonetheless, by 1642 this mining city had become the biggest, densest community in the Americas.

Potosí is a brawling, bawling boomtown marked by extravagant display and hoodlum crime. It is also a murderously efficient mechanism for the extraction and refining of silver ore in appallingly harsh conditions. Indian workers haul the ore on their backs up crude ladders from hundreds of feet below the surface, then extract the silver by mixing the ore with highly toxic mercury. Smelters on the slopes transform the metal into bars of almost pure silver, typically weighing sixty-five pounds and stamped with sigils guaranteeing their quality and authenticity. Other silver is stamped into coins—the Spanish peso is on its way to becoming a de facto world currency, as the U.S. dollar is today. Battalions of llamas—more sure-footed and altitude tolerant than mules and horses—carry the coins and bars down from the mountains, every dangerous step guarded by men with weapons. They hoist the silver onto ships in Arica, on the Chilean coast, which shuttle it to the great port of Lima, seat of the Spanish colonial government. From Lima the silver is loaded onto the first of a series of military convoys that will transport it across the world.

From the plane, follow the silver fleet as it travels north. To the east of the convoy rise the Andean slopes, gripped in ecological turmoil. Humankind has lived here for many thousands of years, erecting some of the world's first urban complexes in the valleys north of Lima. A hundred and fifteen years before this overflight, smallpox swept in. After it came other European diseases, and then Europeans themselves. Millions died, fearful and suffering, in shattered mountain villages. Now, decades later, slopes terraced and irrigated for centuries remain empty. Shrubs and low trees have overwhelmed abandoned farms. A huge volcanic eruption in 1600 covered central Peru with up to three feet of ash and rubble. Four decades later, little has been cleared away. Andean ecosystems have gone feral. Sailing north, the silver fleet is passing something akin to wilderness, at least in patches.

Some of the vessels anchor in Panama, while others go to Mexico. Watching from the plane, observe that the Panamanian silver crosses the isthmus, bound for Europe, whereas most of the Mexican silver is bound ultimately for Asia. How much goes where is the subject of brisk dispute, both by customs officials in 1642 and by historians today. The Spanish monarchy, perpetually hungry for cash, wants the silver in the home country. Spanish colonists want to send as much as possible to China—coins and bars can be traded there more profitably than anywhere else. The tension leads, inevitably, to smuggling. Official statistics suggest that no more than a quarter of the silver went across the Pacific. In the past historians have largely assumed that government scrutiny kept the smuggling to perhaps 10 percent of the total, meaning that the official statistics were roughly correct. A new wave of researchers, however, argues that smuggling was rampant; China sucked up as much as half of the silver. The debate is not simple pedantry. One side regards European expansion as the primary motivating force in world affairs; the other views the earth as a single economic unit largely driven by Chinese demand.

Follow the Europe-bound silver as it is carried by mule train over the mountains to Portobelo, then Panama's main Caribbean port. Guarded by an armada of galleons, bristling with guns and crewed by as many as two thousand seamen and soldiers, the silver traverses the Atlantic every summer, its departure timed to avoid hurricane season. The convoy bellies up to the mouth of the Guadalquivir, Spain's only major navigable river, and then sixty miles upstream to Seville.

Unloaded onto the quays, the chests of treasure are the emblem of a paradox: silver from the Americas has made the Europe of 1642 affluent and powerful beyond its giddiest fantasy. But Europe itself is plagued from one end to the other by war, inflation, rioting, and weather calamities. Turmoil is nothing new in Europe, which is divided by language, culture, religion, and geography. But this is the first time that the turmoil is intimately linked to human actions on opposite ends of the earth. Trouble volleys from Asia, Africa, and the Americas to Europe, shuttling about the world on highways of Spanish silver.

Cortés's conquest of Mexico—and the plunder that came from it—threw Spain's elite into delirium. Enraptured by sudden wealth and power, the monarchy launched a series of costly foreign wars, one overlapping with another, against France, the Ottoman Empire, and the Protestants in the Holy Roman Empire. Even as Spain defeated the Ottomans in 1571, discontent in the Netherlands, then a Spanish possession, was flaring into outright revolt and secession. The struggle over Dutch independence lasted eight decades and spilled into realms as far away as Brazil, Sri Lanka, and the Philippines. Along the way, England was drawn in; raising the ante, Spain initiated a vast seaborne invasion of that nation: the Spanish Armada. The invasion was a debacle, as was the fight to stop rebellion in the Netherlands.

War spawned war. In 1642, Spain is combating secession in Andalusia, Catalonia, and Portugal, which it has ruled for six decades; France is fighting Spain on its northern, eastern, and southern borders; and Swedish armies are battling the Holy Roman Empire. (Emperor Ferdinand III, the son-in-law of one Spanish king and the future father-in-law of another, is so closely allied with Spain that he has often been called a Spanish puppet.) Almost the only European nation not directly or indirectly at war with Spain is England, which is convulsed by its own civil strife—

the ascetic Puritan rebellion that will soon lead to civil war and the execution of the king.

The costs are staggering. At the height of the Vietnam War, the United States fielded about 500,000 soldiers. If the U.S. had wanted to send out the same proportion of its men that Spain did in its war with the Dutch, according to Dennis Flynn, an economic historian at the University of the Pacific, it would have had to send 2.5 *million*. "Even though all this silver was coming in from Bolivia, Spain didn't have enough money to pay its army in the Netherlands," he told me. "So the men mutinied constantly. I did a count once—there were forty-five mutinies between 1572 and 1607. And that was just *one* of Spain's wars."

To pay for its foreign adventures, the court borrowed from foreign bankers; the king felt free to incur debts because he believed they would be covered by future shipments of American treasure, and bankers felt free to lend for the same reason. Alas, everything cost more than the monarch hoped. Debt piled up hugely—ten or even fifteen times annual revenues. Nonetheless the court continued to view its economic policy in the optative mood; few wanted to believe that the good times would end. The inevitable, repeated result: bankruptcy. Spain defaulted on its debts in 1557, 1576, 1596, 1607, and 1627. After each bankruptcy, the king borrowed more money. Lenders would provide it—after all, they could charge high interest rates (Spain paid up to 40 percent, compounded annually). For obvious reasons the high interest rates made the next bankruptcy more likely. Still the process continued—everyone believed the silver would keep pouring into Seville. Now, in 1642, so much silver has been produced that its value is falling even as the mines slacken. The richest nation in the world is hurtling toward financial Armageddon. Europe is complexly interconnected; Spain's economic collapse is dragging down its neighbors.

The silver trade was not the only cause of this tumult— religious conflict, royal hubris, and struggles among classes all were important—but it was an essential part. The flood of precious metal unleashed by Cortés so vastly increased Spain's money supply that its small financial sector could not contain it. It was as if a billionaire suddenly deposited a fortune into a tiny country bank—the bank would immediately redeposit the cash into other, bigger institutions that could do something with it. American silver overflowed from Spain like water from a bathtub and washed into bank vaults in Italy, the Netherlands, and the Holy Roman Empire. Payments for Spanish military adventures filled coffers across the continent.

Economics 101 predicts what will happen in these circumstances. New money chases after the same old goods and services. Prices rise in a classic inflationary spiral. In what historians call a "price revolution," the cost of living more than doubled across Europe in the last half of the sixteenth century, tripling in some places, and then rose some more. Because wages did not keep pace, the poor were immiserated; they could not afford their daily bread. Uprisings of the starving exploded across the continent, seemingly in every corner and all at once. (Researchers have called it the "general crisis" of the seventeenth century.)

Hope for the peasantry was provided by American crops, which by 1642 have ridden the silver route across the Atlantic. As the plane sweeps over Europe, it descends low enough for passengers to view the marks of the Columbian Exchange: plots of American maize in Italy, carpets of American beans in Spain, fields crowded with the shining, upturned visages of American sunflowers in France. Big tobacco leaves soak up sunlight on Dutch farms; tobacco is so common in Catholic Europe that Pope Urban VIII has this year denounced its use (in Protestant England, it is endorsed even by the nation's most notorious killjoy, Oliver Cromwell). Most important will be the potato, which is beginning to fill bellies in Germany, the Netherlands, and, increasingly, Ireland. In ordinary times, the quickly increasing agricultural productivity would soothe some of the discontent caused by inflation and war. But these are not ordinary times: the plane's instruments reveal that the climate itself has been changing.

For almost a century Europe has experienced frighteningly snowy winters, late springs, and cold summers. Frigid Mays and Junes delay French wine harvests until November; people walk a hundred miles across the frozen sea from Denmark to Sweden; Greenland hunters moor their kayaks on the Scottish shore. After three failed harvests, Catholic mobs in Ireland rise up, robbing and killing the hated English Protestants—attacks those Protestants use as an opportunity to seize Catholic land. Fearing that growing Alpine glaciers will overrun their homes, Swiss villagers induce their bishop to exorcise a threatening ice front—an echo of the Spaniards in Santo Domingo, seeking God's help against the plague of ants. Annual visits from the bishop drive back the glacier by eighty paces. The order of the world seems overturned.

Historians call the freeze the Little Ice Age. Enduring from about 1550 to about 1750 in the Northern Hemisphere, this global thermal anomaly is difficult to pin down; its onset and duration differed from one region to the next. Because few people then kept written records of weather conditions, paleoclimatologists (researchers of ancient climate) must study it with imperfect measures like the thickness of tree rings and the chemical composition of tiny bubbles of gas in polar ice. Based on such indirect evidence, some researchers proposed that the Little Ice Age was attributable to a decline in the number of sunspots known as the Maunder Minimum. Because sunspots are correlated with the sun's energy output, fewer sunspots implies less-intense solar irradiation—enough, these researchers argued, to cool the earth. Other scientists theorized that the temperature drop was due to big volcanic eruptions, which blast sulfur dioxide into the upper atmosphere. High above the clouds, the sulfur dioxide mixes with water vapor to form minute droplets of sulfuric acid—shiny motes in the sky—that reflect some of the sun's light into space. This phenomenon existed in 1642; a massive eruption in the southern Philippines the year before is now thought to have cooled the earth for as long as three years. Both hypotheses have drawn sharp criticism, though. Many scientists believe that the impact of the Maunder Minimum was too small to account for the Little Ice Age. Others argue that a series of individual volcanic eruptions could not have caused a steady temperature drop.

In 2003, William F. Ruddiman, a paleoclimatologist at the University of Virginia, suggested a different cause for the Little Ice Age—an idea that initially seemed outlandish, but that is increasingly treated seriously.

As human communities grow, Ruddiman pointed out, they open more land for farms and cut down more trees for fuel and shelter. In Europe and Asia, forests were cut with the ax. In the Americas before Colón, the primary tool was fire—vast stretches of it. For weeks on end, smoke from Indian bonfires shrouded Florida, California, and the Great Plains. Today, many researchers believe that without regular burning, much of the midwestern prairie would have been engulfed by an invading tide of trees. The same was true for the grasslands of the Argentine pampas, the hills of Mexico, the Florida dunes, and the high plains of the Andes.

American forests, too, were shaped by flame. Indians' "frequent fiering of the woods," remarked English colonist Edward Johnson in 1654, made the forests east of the Mississippi so open and "thin of Timber" that they were "like our Parkes in England." Annual fire seasons removed scratchy undergrowth, burned out noxious insects, and cleared land for farms. Scientists have conducted fewer studies of burning in the tropics, but two California paleoecologists (scientists who study past ecosystems) surveyed the fire history of thirty-one sites in Central and South America in 2008 and found that in every one the amount of charcoal in the soil—an indicator of fire—had increased substantially for more than two thousand years.

Enter now the Columbian Exchange. Eurasian bacteria, viruses, and parasites sweep through the Americas, killing huge numbers of people—and unraveling the millennia-old network of human intervention. Flames subside to embers across the Western Hemisphere as Indian torches are stilled. In the forests,

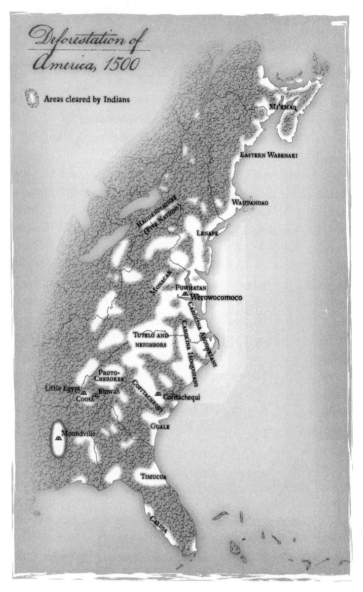

Deforestation of America, 1500

Areas cleared by Indians

MI'KMAQ

EASTERN WABENAKI

WAMPANOAG

HAUDENOSAUNEE (FIVE NATIONS)

LENAPE

MONACAN

POWHATAN

Werowocomoco

CAROLINA ALGONQUIANS

TUTELO AND NEIGHBORS

CAROLINA IROQUOIANS

PROTO-CHEROKEE

Little Egypt

COOSA

Etowah

COFITACHEQUI

Cofitachequi

Moundville

GUALE

TIMUCUA

CALUSA

Using fire, indigenous people in the Americas cleared big areas for agriculture and hunting, as shown in this map of North America's eastern seaboard. European diseases caused a population crash across the hemisphere—and an extraordinary ecological rebound as forests filled in abandoned fields and settlements.

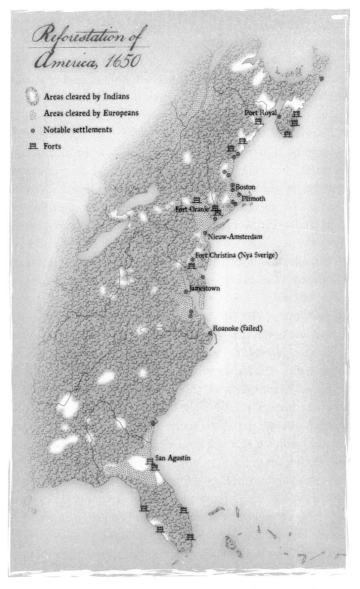

Reforestation of America, 1650

Areas cleared by Indians
Areas cleared by Europeans
Notable settlements
Forts

Port Royal

Boston
Plimoth
Fort Orange
Nieuw-Amsterdam
Fort Christina (Nya Sverige)
Jamestown

Roanoke (failed)

San Agustín

The end of native burning and the massive reforestation drew so much carbon dioxide from the air that an increasing number of researchers believes it was a main driver of the three-century cold snap known as the Little Ice Age.

fire-hating trees like oak and hickory muscle aside fire-loving species like loblolly, longleaf, and slash pine, which are so dependent
on regular burning that their cones will only open and release
seed when exposed to flame. Animals that Indians had hunted,
keeping their numbers down, suddenly flourish in great numbers. And so on.

Indigenous pyromania had long pumped carbon dioxide into
the air. At the beginning of the Homogenocene the pump suddenly grows feeble. Formerly open grasslands fill with forest—a
frenzy of photosynthesis. In 1634, fourteen years after the Pilgrims land in Plymouth, colonist William Wood complains that
the once-open forests are now so choked with underbrush as to
be "unuseful and troublesome to travel through." Forests regenerate across swathes of North America, Mesoamerica, the Andes,
and Amazonia.

Ruddiman's idea was simple: the destruction of Indian societies by European epidemics both decreased native burning and
increased tree growth. Each subtracted carbon dioxide from the
air. In 2010 a research team led by Robert A. Dull of the University of Texas estimated that reforesting former farmland in
American tropical regions alone could have been responsible for
as much as a quarter of the temperature drop—an analysis, the
researchers noted, that did not include the cutback in accidental
fires, the return to forest of unfarmed but cleared areas, and the
entire temperate zone. In the form of lethal bacteria and viruses,
in other words, the Columbian Exchange (to quote Dull's team)
"significantly influenced Earth's carbon budget." It was today's
climate change in reverse, with human action removing greenhouse gases from the atmosphere rather than adding them—a
stunning meteorological overture to the Homogenocene.

Flying the plane back across the Atlantic, the effects of the Little Ice Age are obvious in the Americas, too. Clearly visible from
the air is the filling in of Indian lands by forest—and by snow.
Ice is solid enough that people ride carriages on Boston harbor;
it freezes over most of Chesapeake Bay, and nearly wipes out the

two score French colonists who this year have founded Montreal. Introduced cattle and horses die in snowdrifts in Maine, Connecticut, and Virginia. Other impacts are harder to see. The forests are filling in former Indian lands with cold-loving species like hemlock, spruce, and beech. Vernal pools take longer to evaporate in the canopy they provide in these cool summers. Mosquitoes that breed in those pools thus have an increased chance for survival.

Among these paradoxically cold-loving mosquitoes is *Anopheles quadrimaculatus,* the overall name for a complex of five near-indistinguishable sibling species. Like other *Anopheles* mosquitoes, *A. quadrimaculatus* hosts the parasite that causes malaria—the insect's common name is the North American malaria mosquito. Southeast England at this time is rampant with malaria. Precise documentation will never become available, but there is good reason to suspect that by 1642 malaria has already traveled in immigrant bodies from England to the Americas. A single bite into an infected person is enough to introduce the parasite to its mosquito host, which spreads the parasite far and wide. Virginia and points south have already proven so unhealthy for Europeans that plantation overseers are finding it difficult to persuade laborers to come from overseas to work in the tobacco fields.

Some landowners already have resolved this problem by purchasing workers from Africa. Partly driven by the introduction of malaria, a slave market is beginning to quicken into existence, a profitable exchange that will entwine itself over time with the silver market. As ever, the ships from Africa will form a kind of ecological corridor, through which travel passengers not on any official manifest. Crops like yams, millet, sorghum, watermelon, black-eyed peas, and African rice will follow the slave ships to the Americas. So will yellow fever.

Beyond Chesapeake Bay the airplane flies west, heading toward Mexico. Beneath its wings unfurl the Great Plains. From their southern edge come herds of Spanish horses, scores at a time, brought by silver galleons on the return trip across the Atlantic. Apache and Ute race hundreds of miles south to meet the horses,

followed by Arapaho, Blackfoot, and Cheyenne. As European villagers learned from Mongol horsemen, peasant farmers, tied to
their land, are sitting ducks for cavalry assault. The rush by Indian
nations to acquire horses is thus a kind of arms race. All over the
North American West and Southwest, native farmers are abandoning their fields and leaping onto the backs of animals from
Spain. Long-sedentary societies are becoming wanderers; the
"ancient tradition" of the nomadic Plains Indian is coming into
existence, a rapid adaptation to the Columbian Exchange.

As natives acquire horses, they come into conflict with each
other and the labor force on Spain's expanding ranches. The
ranch workers are Indians, African slaves, and people of mixed
ancestry. In a kind of cultural panic, the colonial government
has created a baroque racial lexicon—mestizo, mulatto, coyote, *morisco, chino, lobo, zambaigo, albarazado*—to label particular
genetic backgrounds. All of these people and more meet in Mexico
City, the capital of New Spain, the richest piece of Spain's American empire. Wealthier and more populous than any city in Spain,
it is an extraordinary jumble of cultures and languages, with no
one group forming the majority. Neighborhoods are divided by
ethnicity—one entire barrio is occupied by Tlaxcalans from the
east. As the back-and-forth continues, engineers struggle to prevent the city from physical collapse. Mexico City has flooded six
times in the last four decades, once remaining inundated for five
years. A troubled, teeming, polyglot metropolis with an opulent
center and seething ethnic neighborhoods at its periphery that is
struggling to fend off ecological disaster—from today's perspective, the Mexico City of 1642 seems strikingly familiar. It is the
world's first twenty-first-century city.

The airplane flies west, to Acapulco, on Mexico's Pacific
coast, the eastern terminus of the galleon trade. Ringed by protective mountains, untroubled by sandbars or shoals, the harbor
is a majestic setting for one of the more listless settlements in the
Americas: several hundred huts scattered like lost clothes at the

edge of the water. Most of Acapulco's few permanent inhabitants are African slaves, Indian laborers, and Asian sailors who jumped ship (the galleons are mainly crewed by Filipinos, Chinese, and other Asians). When the galleons arrive, Spaniards show up, some of them coming from as far as Peru. A market and fair springs into existence; millions of pesos change hands. Then the town empties again as the ships are beached and readied for the next trip across the Pacific.

Follow the silver to its destination in China. The Little Ice Age has taken hold in East Asia, too, though here the impact is typically less a matter of snow and ice than of crashing, copious rain alternating with bouts of cold drought. The five worst years of drought in five centuries occurred between 1637 and 1641. This year, rain is drowning the crops. All the impacts have been exacerbated by a series of volcano eruptions in Indonesia, Japan, New Guinea, and the Philippines. Millions have died. Cold, wet weather and mass deaths ensure that more than two-thirds of China's farmland is no longer being tilled, adding to the famine. Cannibalism is rumored to be frequent. The Ming court—paralyzed by infighting, preoccupied with wars to the north—does little to help the afflicted. It simply doesn't have the funds. Like the Spanish king, the Ming emperor backs his military ventures with Spanish silver, which his subjects must use to pay their taxes. When the value of silver falls, the government runs out of money.

The Ming have long believed their duty is to protect China from malign foreign influence. They have failed. American crops like tobacco, maize, and sweet potato are spreading over hillsides. American silver is dominating the economy. Although the emperors don't know it, American trees are helping to bring the rains. All of these are working against the Ming. Popular discontent is already at such levels that mobs of peasant rebels are tearing violently through half a dozen provinces. Unhappy, unpaid soldiers are mutinying. Flood and famine simply exacerbate the anger. In two years Beijing will fall to a rebellious ex-soldier.

Weeks later, the soldier will be overthrown by the Manchus, who establish a new dynasty: the Qing (pronounced, roughly speaking, "ching").

When Colón founded La Isabela, the world's most populous cities clustered in a band in the tropics, all but one within thirty degrees of the equator. At the top of the list was Beijing, cynosure of humankind's wealthiest society. Next was Vijayanagar, capital of a Hindu empire in southern India. Of all urban places, these two alone held as many as half a million souls. Cairo, next on the list, was apparently just below this figure. After these three, a cluster of cities were around the 200,000 mark: Hangzhou and Nanjing in China; Tabriz and Gaur in, respectively, Iran and India; Tenochtitlan, dazzling center of the Triple Alliance (Aztec empire); Istanbul (officially Kostantiniyye) of the Ottoman empire; perhaps Gao, leading city of the Songhay empire in West Africa; and, conceivably, Qosqo, where the Inka emperors plotted their next conquests. Not a single European city would have made the list, except perhaps Paris, then expanding under the vigorous rule of Louis XII. Colón's world was centered around hot places, as had been the case since *Homo sapiens* first stared in amazement at the African sky.

Now, a century and a half later, that order is in the midst of change. It is as if the globe has been turned upside down and all the wealth and power are flowing from south to north. The once-lordly metropolises of the tropics are falling into ruin and decrepitude. In the coming centuries, the greatest urban centers will all be in the temperate north: London and Manchester in Britain; New York, Chicago, and Philadelphia in the United States. By 1900 every city in the top bracket will be in Europe or the United States, save one: Tokyo, the most Westernized of eastern cities. From the vantage of an extraterrestrial observer, the change would have seemed shocking; an order that had characterized human affairs for millennia had been overturned, at least for a while.

Today the tumult of ecological and economic exchange is like the background radiation of our ever more crowded and unstable planet. It seems distinctly contemporary to find Japanese loggers in Brazil and Chinese engineers in the Sahel and Europeans back-packing in Nepal or occupying the best tables in New York niter-ies. But in different ways all of these occurred hundreds of years ago. If nothing else, the events then remind us that we are not alone in our current jumbled condition. It seems worthwhile to take a look at how we got to where we are today.

PART ONE

Atlantic Journeys

The Tobacco Coast

"LOWLY ORGANIZED CREATURES"

It is just possible that John Rolfe was responsible for the worms. Earthworms, to be precise—the common nightcrawler and the red marsh worm, creatures that did not exist in the Americas before 1492. Rolfe was a colonist in Jamestown, Virginia, the first successful English settlement in the Americas. Most people know him today, if they know him at all, as the man who married Pocahontas, the "Indian princess" in countless romantic stories. A few history buffs understand that Rolfe was a primary force behind Jamestown's eventual success. The worms hint at a third, still more important role: all inadvertently, Rolfe helped to unleash a permanent change in the American landscape.

Like many young English blades, Rolfe smoked—or "drank," as the phrase was then—tobacco, a fad since the Spanish had brought back *Nicotiana tabacum* from the Caribbean. Indians in Virginia also drank tobacco, but it was a different species, *Nicotiana rustica*. *N. rustica* was awful stuff, wrote colonist William Strachey: "poor and weak and of a biting taste." After arriving in Jamestown in 1610, Rolfe talked a shipmaster into bringing him some *N. tabacum* seeds from Trinidad and Venezuela. Six years later Rolfe returned to England with his wife, Pocahontas, and his first big shipment of tobacco. "Pleasant, sweet, and strong," as Rolfe's friend Ralph Hamor described it, Virginia tobacco was a hit.

Exotic, intoxicating, addictive, and disdained by stuffy authorities, smoking had become an aristocratic craze. When Rolfe's shipment arrived, one writer estimated, London already had seven thousand or more tobacco "houses"—café-like places where the city's growing throng of nicotine junkies could buy and drink tobacco. Unfortunately, because the sole source of fine tobacco were the colonies of hated Spain, the weed in England was hard to obtain, costly (the best tobacco sold for its weight in silver), and vaguely unpatriotic. London tobacco houses were thrilled by the sudden appearance of an English alternative: Virginia leaf. They clamored for more. Ships from London tied up to the Jamestown dock and took in barrels of rolled-up tobacco leaves. Typically four feet tall and two and a half feet across at the end, each barrel held half a ton or more. To balance the weight, sailors dumped out ballast, mostly stones, gravel, and soil—that is, for Virginia tobacco they swapped English dirt.

That dirt very possibly contained the common nightcrawler and the red marsh worm. So, almost certainly, did the rootballs of plants the colonists imported. Until the nineteenth century, worms like these were viewed as agricultural pests. Charles Darwin was among the first to realize they were something more; his last book was a three-hundred-page celebration of earthworm power. Huge numbers of these beasts, he noted, live beneath our feet; indeed, the total mass of the earthworms in a cow pasture may be many times the mass of the animals grazing above them. Literally eating their way through the soil, earthworms create networks of tunnels that let in water and air. In temperate places like Virginia, earthworms can turn over the upper foot of soil every ten or twenty years; tiny ecological engineers, they reshape entire expanses. "It may be doubted," Darwin wrote, "whether there are many other animals which have played so important a part in the history of the world, as have these lowly organized creatures."

The exact path of these migrants into North America is impossible to trace. What is clear is that before the arrival of Europeans, New England and the upper Midwest had no earthworms—they were wiped out in the last Ice Age. Earthworms from the south didn't move north after the glaciers melted because the creatures don't travel long distances unless they are transported by human agency. "If they're born in your backyard, they'll stay inside the fence their whole lives," John W. Reynolds, editor of *Megadrilogica*, perhaps the premier U.S. earthworm journal, explained to me. They arrived with Europeans, probably in Virginia, and spread with them. Like the colonists, the worms were conquering a new place. In both cases, the arrival of foreigners was an ecological watershed.

In worm-free woodlands, leaves pile up in drifts on the forest floor. When earthworms are introduced, they can do away with the leaf litter in a few months, packing the nutrients into the soil in the form of castings (worm excrement). As a result, according to Cindy Hale, a worm researcher at the University of Minnesota, "everything changes." Trees and shrubs in wormless places depend on litter for food. If worms tuck nutrients into the soil, the plants can't find them. Many species die off. The forest becomes more open and dry, losing its understory, including tree seedlings. Meanwhile, earthworms compete for food with small insects, driving down their numbers. Birds, lizards, and mammals that feed in the litter decline as well. Nobody knows what happens next. "Four centuries ago, we launched this gigantic, unplanned ecological experiment," Hale told me. "We have no idea what the long-term consequences will be."

In some ways this is unsurprising: Jamestown itself was a case study in unintended consequences. The Virginia colony was an attempt by a group of merchants to snatch up the vast stores of gold and silver they imagined—incorrectly, alas—existed around Jamestown, in the big, shallow estuary of Chesapeake Bay.

Equally important, the merchants wanted to find a route through North America, which they imagined, again incorrectly, to be only a few hundred miles wide, less than a month's journey. And when the colonists came to the Pacific coast, they would be able to sail, possibly with Virginia silver, to the colony's ultimate reason for existence: China. In the anodyne language of economics, Jamestown's founders intended to integrate isolated Virginia into the world market—to globalize it.

Purely as a business venture, Jamestown was a disaster. Despite the profits from tobacco, its backers suffered such heavy losses that their venture collapsed ignominiously. Nonetheless the colony left a big mark: it inaugurated the great struggles over democracy (the colony established English America's first representative body) and slavery (it brought in English America's first captive Africans) that have long marked U.S. history. Rolfe's worms, as one might call them, illustrate another aspect of its course: Jamestown was the opening salvo, for English America, of the Columbian Exchange. In biological terms, it marked the point when *before* turns into *after*. Setting up camp on the marshy Jamestown peninsula, the colonists were, without intending it, bringing the Homogenocene to North America. Jamestown was a brushfire in a planetary ecological conflagration.

STRANGE LAND

On May 14, 1607, three small ships anchored in the James River, at the southern periphery of Chesapeake Bay. In movies and textbooks they are often depicted as arriving in a pristine forest of ancient trees, small bands of Indians gliding, silent as ghosts, beneath the canopy. Implicit in this view is the common description of the colonists as "settlers"—as if the land was unsettled before they came on the scene. In fact, the English ships landed in the middle of a small but rapidly expanding Indian empire called Tsenacomoco.

Three decades before, Tsenacomoco had comprised six small, separate clusters of villages. By the time the foreigners came from overseas, its paramount leader, Powhatan, had tripled its size, to about eight thousand square miles. Tsenacomoco stretched from Chesapeake Bay to the Fall Line, the bluffs at the edge of the Appalachian plateau. In its scores of villages lived as many as fourteen thousand people. Europeans would have been impressed by these numbers; Michael Williams, a historical geographer at Oxford, argued that the eastern U.S. forest may have been more populous in 1600 than even "densely settled parts of western Europe."

The ruler of this land was known by multiple names and titles,

The only known likeness of Powhatan created in his lifetime, this sketch ornamenting a 1612 map by John Smith depicts him in a longhouse, smoking a tobacco pipe while surrounded by wives and advisers.

a hallmark of kings everywhere; Powhatan, the name used most often by the colonists, was also the name of the village in which he was born. Wary, politically shrewd, ruthless when needed, Powhatan was probably in his sixties when the English landed—"well beaten with many cold and stormy winters," according to colonist Strachey, but still "of a tall stature and clean limbs."

His capital of Werowocomoco ("king's house") was on the north bank of the York River, in a little bay where three streams come together. (The York runs more or less parallel to the James and a few miles to its north.) Projecting from the shore was a peninsula dominated by a low rise, twenty-five feet at its highest point, which held most of the village's houses. Behind it, separated by a double moat from the rest of Werowocomoco, was a second, smaller hill, with several structures at its base that combined the function of temples, armories, and treasure houses. Generally closed to commoners, they contained the preserved bodies of important chiefs and priests, mounted on scaffolds and ringed by emblems of wealth and power. Atop the hill was the biggest structure in Tsenacomoco: a great, windowless barrel vault, perhaps 150 feet long, its walls made of overlapping sheets of chestnut bark, with gargoyle-like statues at each corner. At the far end, lighted by torches, was the royal chamber. Inside, the sovereign greeted visitors from a raised, pillow-covered divan, surrounded by wives and advisers, long gray hair tumbling over his shoulders, ropes of fat pearls descending from his neck. Confronted with this regal spectacle, colonist John Smith was awed; the Indian men, who generally had better diets than the English, "seemed like Giants," with deep voices "sounding from them, as a voyce in a vault." Sitting in the center, Powhatan himself, Smith thought, had "such a Majestie as I cannot expresse."

To the English, Powhatan was a recognizable figure: the king of a small domain, with the lofty bearing that they expected from royalty. Any strangeness adhered not to the man in the foreground of the picture but the background against which he appeared: the fields, forests, and rivers of Tsenacomoco. It could hardly have been otherwise. Chesapeake Bay was shaped by ecological and social forces unknown to the colonists. Speaking broadly, the most important ecological force was the region's different tally of plant and animal species; the social force, just as consequential, was the Indians' different land-management practices.

By a quirk of biological history, the pre-Columbian Americas had few domesticated animals; no cattle, horses, sheep, or goats graced its farmlands. Most big animals are *tamable*, in the sense that they can be trained to lose their fear of people, but only a few species are readily *domesticable*—that is, willing to breed easily in captivity, thereby letting humans select for useful characteristics. In all of history, humankind has been able to domesticate only twenty-five mammals, a dozen or so birds, and, possibly, a lizard. Just six of these creatures existed in the Americas, and they played comparatively minor roles: the dog, eaten in Central and South America and used for labor in the far north; the guinea pig, llama, and alpaca, which reside in the Andes; the turkey, raised in Mexico and the U.S. Southwest; the Muscovy duck, native to South America despite its name; and, some say, the iguana, farmed in Mexico and Central America.*

The lack of domestic animals had momentous consequences. In a country without horses, donkeys, and cattle, the only source of transportation and labor was the human body. Compared to England, Tsenacomoco had slower communications (no galloping horses), a dearth of plowed fields (no straining oxen) and pastures (no grazing cattle), and fewer and smaller roads (no carriages to accommodate). Battles were fought without cavalry; winters endured without wool; logs skidded through the for-

---

* In recent years, advanced techniques have let researchers domesticate a few previously undomesticable species in laboratory settings—the silver fox is the most well-known example. In all previous history, though, only about forty large animals were domesticated. (That figure does not include domesticated insects, like the European honeybee and the Mexican cochineal, cultivated as a source of red dye.)

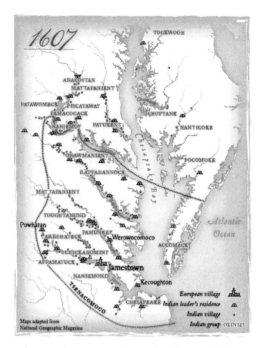

Jamestown was founded inside the small indigenous empire of Tsenacomoco. Most Tsenacomoco villages were located along the rivers that served as the empire's highways. Because the water at the river mouths was brackish, the villages were mostly upstream. The English put Jamestown as far upriver as they could—but not far enough to avoid the bad water.

est without oxen. Distances loomed larger when people had to walk from place to place; indeed, in terms of the time required for Powhatan's orders to reach his minions, Tsenacomoco may have been the size of England itself (it was much less populous, of course).

Just as most Europeans lived in small farm villages, most of Powhatan's people—the "Powhatan Indians," as the newcomers called them—lived in settlements of a few hundred inhabit-

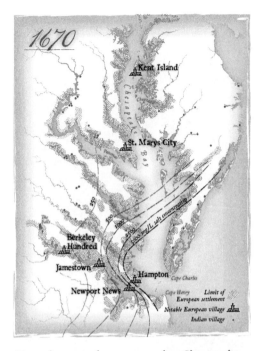

1670

Kent Island

Chesapeake Bay

St. Marys City

3500 mg/L salt concentration

The Falls

Berkeley Hundred

Jamestown

Hampton   *Cape Charles*

Newport News

Cape Henry   Limit of European settlement

Notable European village

Indian village

Even the groundwater was salty. Chesapeake Bay is the remains of a giant meteor crater. The impact shattered rock for miles, letting seawater infiltrate. The U.S. government suggests that salt levels should not exceed 20 milligrams per liter (mg/L); Jamestown water had more than twenty times as much and other settlements had even higher levels.

ants surrounded by large tracts of cleared land: fields of maize and former maize fields. The villages clustered along the three rivers—the Rappahannock, York, and James—that served as the empire's main thoroughfares. Sailing up the James when they arrived, the English saw the banks lined with farms, fields greenly shimmering with newly planted maize, stands of tall trees interspersed among them.

Europe, too, had its own prosperous riverside farms. But

Rather than covering fenced plots with neat rows of wheat, the Powhatan planted many crops at once, as shown in this replica Wendat (Huron) garden in the Crawford Lake Conservation Area in Ontario, Canada. These farms and gardens were so different from anything the English knew that the newcomers often couldn't recognize native fields as cultivated land.

there the similarities ended. To create a farm plot, Europeans cleared forestland, yanked out the stumps with horses and oxen, and plowed the result, again with horses or oxen, until it was a flat expanse of nearly bare soil. In these stripped areas farmers planted single crops: solid rustling expanses of wheat or barley or rye. Fallow plots were used as pasture. Dotting the open areas were patches of forest, clearly demarcated as such, used for hunting and wood.

Lacking draft animals and metal tools, the Powhatan perforce used different methods and obtained different results. They toppled trees by circling their bases with a ring of fire, then laboriously hacked at the burned zone with stone axes until the trunk collapsed. Brush and slash were put to the torch, leaving a heave of blackened stumps. Around the stumps farmers dug shallow holes with long-handled hoes made from bone or clamshells, dropping in each hole a few kernels of maize and several beans. As the maize grew, the young colonist Henry Spellman observed, "the beans run up thereon"—twining themselves around the growing maize. Below the maize grew squash and gourds, pumpkin and melon, common beans and runner beans, ropy vines asprawl in every direction. Here and there patches of thick-leaved tobacco plants stood. The ensemble of charred stumps, hummocked land, and overlapping crops could stretch for considerable distances: "thirty to forty acres of treeless land per capita," in one historian's "conservative" estimate. Smith saw family plots that covered as much as two hundred acres—a third of a square mile.

Except for defensive palisades, Powhatan farmers had no fences around their fields. Why screen off land if no cattle or sheep had to be kept inside? The English, by contrast, regarded well-tended fences as hallmarks of civilization, according to Virginia D. Anderson, a historian at the University of Colorado at Boulder. Fenced fields kept animals in; fenced woodlots kept poachers out. The lack of physical property demarcation signified to the English that Indians didn't truly occupy the land—it was, so to speak, unimproved. Equally unfamiliar was the Powhatan practice of scattering their farm plots within larger cleared areas. To the Indians, fallow lands were a kind of communal larder, a place for naturally occurring useful plants, including grains (little barley, sumpweed, goosefoot), edible greens (wild lettuce, wild plantains), and medicinals (sassafras, dogbane, smartweed). Because none of these species existed in Europe, the English didn't know the groundcover was useful. Instead they saw "unused" land, something that bewildered them. How could Indians go to the trouble of clearing the land but then not use it?

Even Tsenacomoco's streams were different than their English equivalent. English waterways ran swiftly in the spring, scouring away the soil from steep banks, then turned to dribbling trickles in July and August. Beyond the riverbanks the land was drier; one could hike for miles in summer without stepping into mud. Chesapeake Bay was, by contrast, a seemingly endless patchwork of bogs, marshes, grassy ponds, seasonally flooded meadows, and slow-moving streams. It seemed to be wet *everywhere,* no matter what the season. Credit for the watery environment belongs to the American beaver *(Castor canadensis),* which had no real English equivalent. Weighing as much as sixty pounds, these big rodents live in dome-shaped lodges made by blocking streams with mud, stones, leaves, and cut saplings—as many as twenty dams per mile of stream. The dams smear the water across the landscape, so to speak, transforming a rushing rivulet into a series of broad pools and mucky wetlands linked by shallow, multiply branched channels. Indians regarded this as a fine thing—easier to take a canoe through a set of ponds than a narrow, quick-flowing stream. English accounts, by contrast, are filled with descriptions of colonists unhappily stumbling through the sopped countryside.*

The freshwater marshes favored the growth of tuckahoe *(Peltandra virginica,* arrow arum), a semi-aquatic plant found in stands throughout the eastern United States and Canada. Tuckahoe has a bulb-like, underground rhizome (enlarged stem area used for storage) that every spring sends out a thin stalk with a long leaf shaped like a child's sketch of an arrowhead. It was a standing larder for the people of Tsenacomoco, always ready in the spring-

---

* Europeans later hunted the beaver to near extinction—its fur makes especially good felt, then in demand for hats. In this way, they unknowingly replaced one dominant natural engineer with another, the earthworm.

time if they exhausted the maize from the previous fall. Standing shin deep in the marsh, women mucked about with their bare feet and hands, gradually working loose the roots. The work was unpleasant; when I dug up some tuckahoe one warm spring day in Virginia, I ended up perspiring in the heat even as the cold mud numbed my feet. Tuckahoe root contains calcium oxalate, a potentially fatal poison. To break down the toxin, women sliced the peeled root, baked the slices, then ground them into flour with a mortar and pestle. At home I made tuckahoe flour with an oven and a food processor, then added water and boiled up some porridge. One mouthful was enough to tell me why native people preferred maize.

Surrounding the cleared areas and the fruitful marshes was the wood, splendid with chestnut and elm, but hardly untouched. Like the fields, the forest was shaped by native fire. Every fall Indians burned the underbrush, sending ash billowing into the heavens; when ships approached during fire season, the Dutch merchant David Pieterszoon de Vries observed in 1632, "the land is smelt before it is seen." From the embers emerged tender new growth, attracting deer, elk, and moose. These were hunted by fire. Men drove the animals into ambushes with flaming torches, herded them toward waiting archers with strategically placed bonfires, encircled them terrified within mile-long walls of flame. Prowling through the woods one evening, John Smith navigated "by the aboundance of fires all over the woods."

Regular fall burning kept the Maryland forest so open, the Jesuit priest Andrew White wrote in 1634, that "a coach and four horses may travel [through it] without molestation." The statement is hyperbole, but not entirely false—rather than paving roads, Indians used fire to make what the ecological historian Stephen J. Pyne has called "corridors of travel." Well-used paths could be six feet wide, hundreds of miles long, and cleared completely of brush and stones. Occasionally one did find unburned patches of land, Virginia colonist William Byrd warned, and these were dangerous. In those places, "the dead Leaves and Trash of many years are heapt up together, which . . . furnish fewel for a conflagration that carries all before it." Because Indian burning killed underbrush and saplings, the forest encountered by the early English colonists was a soaring space, hushed as a cathedral, formed by widely spaced walnut and oak six feet in diameter—a beautiful sight, but one just as artificial as the burned-off clearings. "Much as cooking helped rework an intractable environment into food and as the forge refashioned rock into metals," Pyne explained, native fire "remade the land into usable forms."

Like the English countryside the colonists left behind them, Chesapeake Bay had been refashioned by its inhabitants into a working landscape. And just as the tidy English checkerboard of fields and woodlots was essential to English culture—indeed, to England's survival—the jumbled patchwork of ecological zones in coastal Virginia was essential to Powhatan culture and survival. But to the newcomers the Virginia coast was not a humanized place. They saw it as a random snarl of marshes, beaver ponds, unkempt fields, and hostile forest. If the English wanted to live and prosper in this new place in their accustomed manner, they would have to transform the land into something more suitable for themselves.

THE RISK POOL

Most accounts of Jamestown focus on John Smith. No surprise: Smith makes great copy. He was a poor boy who made good with luck, nerve, and self-promotion—in just eighteen years he published no less than *five* autobiographical accounts of his deeds. (To be fair, one was printed without his knowledge.) *The True Travels, Adventures and Observations of Captain John Smith* (1630), his major autobiography, is a wild tale of an orphan who left home at thirteen, fought in the Netherlands, lived in a lean-to teaching himself Machiavelli and Marcus Aurelius, battled "a rabble of Pilgrimes of divers Nations going to Rome" aboard a ship in the Mediterranean (they threw him overboard), and became a pirate in the Adriatic—all in the opening chapter. By chapter 4 (its title: "An Excellent Stratagem by Smith") he is using torches to send coded messages between mountaintops—a technique from Machiavelli—as he coordinates a battle in what is now Hungary. Later chapters reveal:

- How Smith served in a Transylvanian army, battling "some Turks, some Tartars, but most Bandittoes, Rennagadoes, and such like."
- How he slew three Turkish aristocrats in single combat before raucous crowds.
- How he was captured and sold into slavery in the Ottoman Empire, where "a great ring of iron" was "rivetted about his necke."
- How he seized the chance to "beat out [his master's] braines" with a farm implement and fled in the man's clothes to Russia, France, and Morocco.
- How in Morocco he joined another band of pirates, preying on Spanish vessels off West Africa.
- How he returned to England and promptly joined the Virginia expedition. He was just twenty-six.

Skeptics have been scoffing at this buckle and swash since 1662, when one noted that the sole record of Smith's adventures is his own writing: "it soundeth much to the dimunition of his deeds, that he alone is the herald to publish and proclaim them." Other writers cheered him as a quintessential American: the original self-made man. During the Civil War, Smith's link with Virginia turned him into a symbol of the Confederate South. Northerners naturally tried to belittle him; after writing an article that highlighted inconsistencies in *True Travels*, historian Henry Adams, a fervent Unionist, crowed that he had executed a "rear attack on the Virginia aristocracy." The cruelest blow came in 1890, when a Hungarian-speaking researcher charged that the people and places

Short, stocky, and homely, John Smith had
a formidable chestnut beard that startled
native people when they encountered it.
He was evidently aware of his unprepos-
sessing appearance: this author's portrait
from his 1624 autobiography was accompanied by a doggerel poem, likely penned by
Smith, claiming that his interior excellence
better than compensated for his less than
handsome exterior.

in Smith's adventures were fictional. Smith, for instance, said he
deployed his "excellent stratagem" at a place called "Olumpagh."
No town named Olumpagh existed in the region. QED: Smith
was a fraud. In the 1950s a second Hungarian-speaking researcher,
Laura Polyani Striker, counterattacked. Smith's places, she said,
were real—the previous researcher had been misled by Smith's
atrocious spelling. Olumpagh, for example, was Lendava, in Slo-

venia, known to Hungarians in those days as "Al Limbach." Such places being unknown in England, Striker argued, Smith indeed must have visited them.

No historians doubt that Smith was at Jamestown. Nor do they dispute that this scrappy, self-confident man befriended Pocahontas, obtained desperately needed food from Powhatan, saved the colony from extinction, and constantly annoyed the colony's leaders, all of whom were his social betters. At the time, English class distinctions were rigid to a degree that is hard now to comprehend; Smith, never one to display deference, so quickly angered Jamestown's gentry that during the voyage from England they threw him in the brig on vague charges. Historians also accept that after landing in Virginia Smith led the search through Chesapeake Bay for a passage to China. But scholarly eyebrows rise in disbelief about what Smith claims happened in December 1607 during one of those expeditions.

Intending to explore the headwaters of the Chickahominy River, Smith went off in a canoe with two Indian guides and two English companions. They ran into a hunting party led by Opechancanough (oh-pee-CHAN-can-oh), Powhatan's younger brother, who was vocally anti-immigrant. He wanted no illegal aliens in Tsenacomoco. During the inevitable skirmish the Indians killed Smith's companions; Smith fell into a swamp and was captured. Opechancanough brought the adventurer to his brother's capital, Werowocomoco. In the most famous version of the story—the one published in *True Travels*—Smith approached Powhatan through a gauntlet: "two rowes of men, and behind them as many women, with all of their heads and shoulders painted red; many of their heads bedecked with the white downe of Birds." The king gave him a public feast. Then, Smith wrote, Powhatan decided to kill him on the spot, in the banquet hall. Executioners "being ready with their clubs to beat out his brains, Pocahontas, the king's dearest daughter," then perhaps eleven years old, suddenly rushed out and cradled Smith's head in her arms "to save him from death." Fondly indulging his daughter's crush, Powhatan commuted Smith's sentence and returned him to Jamestown, where the girl "brought him so much provision, that saved many of their lives, that els[e] for all this had starved with hunger."

Countless romantic novels have been spawned by Smith's tale, but most researchers believe it to be untrue. In his debunking, Henry Adams pointed out that the earliest account of the rescue dates from 1624, in the boastful autobiography Smith published just before the boastful *True Travels*. But Smith also wrote about his abduction in 1608, a few months after it happened, in a report not intended for public view, and said not a word about being saved by a love-smitten Indian maiden. Smith clearly relished the image of infatuated women coming to his rescue—in

John Smith's tale of rescue from execution by the "Indian princess" Pocahontas has proven irresistible to generations of artists, despite historians' disbelief in its veracity. In this 1870 engraving, Pocahontas resembles an opera star, the Powhatan have been given tipi homes like those in the West, and the venue has been transplanted to a hilly and almost treeless expanse unlike anything in coastal Virginia.

True Travels, it happens no less than four times. More damning still, no anthropologist or historian has found any suggestion that the Powhatan ever held feasts for prisoners of war before executing them. Nor were children like Pocahontas admitted to official dinners—they were in the kitchen, washing dishes. "None of the story fits the culture," the anthropologist Helen Rountree told me. "Big meals are for honored guests, not criminals to be executed." In her view, the feast suggests the Indians regarded Smith as a potential treasure trove of data about the foreign invaders. "It's hard to see them killing an intelligence asset," she said.

Historians dislike the Pocahontas-rescue story for another, deeper reason. By pumping up the romance and fanfaronade, it draws attention from what the English were actually trying to accomplish in Virginia—and what happened to Tsenacomoco when they arrived. Brave adventurers like Smith were integral to Jamestown, but the colony was primarily an economic venture. And for all the danger and conflict, its fate was decided less, in the end, by the clash of arms than by impersonal ecological forces—the Columbian Exchange—that nobody in Virginia was then equipped to understand.

Like La Isabela, Jamestown was intended as a trading post, a midway point from which England could seize its share of the China trade. But whereas La Isabela was largely sponsored and controlled by the Spanish monarchy, Jamestown was the creation of private enterprise: a consortium of politically connected venture capitalists known as the Virginia Company. The difference was anything but absolute: Spanish merchants hoped to enrich themselves at La Isabela, and the political ramifications of Jamestown preoccupied the English government. But Jamestown was closer to the capitalist ventures meant in today's discussions of globalization.

The Virginia Company came into existence because English sovereigns—Queen Elizabeth I and her successor, James I—wanted the benefits of trade and conquest but couldn't pay for them. The state had been pushed so deeply into debt by war (in

Elizabeth's case) and profligacy (in James's case) that it could not afford to send ships to the Americas. Nor could it borrow the necessary cash. From moneylenders' point of view, the monarchy was a bad credit risk—it could, and all too often did, assert its prerogative to repudiate its debts. In consequence, they charged it ruinously high interest rates. True, kings and queens had the power to force loans from their subjects, a practice that for obvious reasons was deeply unpopular. But was the certainty of incurring discontent worth the gamble of an American colony?

Elizabeth and James came to the same conclusion: no.

As La Isabela showed, colonization was inherently risky. The English faced the additional danger that most of the Americas already had been claimed by Spain. Hostility between the two nations was intense; indeed, Pope Pius V had practically ordered Catholic monarchs like Spain's Philip II to take up "Weapons of Justice" against Protestant England. ("There is no place at all left for Excuse, Defence, or Evasion," the pope fulminated. Queen Elizabeth, "Slave of Wickedness," had to be overthrown.) Spain sent a fleet to invade England in 1588, England a fleet to invade Spain in the following year. Both attacks failed, in part because of violent weather—a manifestation, perhaps, of the Little Ice Age. Ultimately Elizabeth relied upon a more successful tactic: sponsoring what is remembered in England as "privateering" and in Spain as "terrorism." She authorized English ships to loot any Spanish ships or colonies they came across. After Elizabeth died in 1603, James I ratcheted down tensions. But he knew that installing English colonies in North America would rekindle the conflict. Spain had already planted more than a dozen small colonies and missions on the Atlantic Coast, one of them just miles away from Jamestown's future location (it had failed). The empire would not look favorably on an intrusion into its domain. If that weren't enough, France, too, had claimed North America, setting down five colonies and missions of its own.

Still, the monarchy was unwilling to cede the Americas to the competition. In a kind of white paper to Elizabeth, the influential cleric and writer Richard Hakluyt argued that Christian rulers had a sacred duty to save the souls of "those wretched people"— that is, Indians. "The people of America crye out unto us," he said, to "bringe unto them the gladd tidings of the gospell." Spain, he noted, had already converted "many millions of infidells." And what had been Spain's reward for this deed? God had "open[ed] the bottomles treasures of his riches," letting England's hated adversary acquire vast stores of silver, which in turn had let it open trade with China. Hakluyt pointed out that Spain, formerly a "poore and barren nation," was now so rich that, incredibly, its seamen had almost stopped being thieves. England, by sad contrast, was "moste infamous" for its "outeragious, common, and daily piracies."

And there was opportunity in North America, or so it was thought. Between 1577 and 1580 Sir Francis Drake, England's best-known privateer/terrorist, went on a round-the-world tour, sacking Spain's silver fleet along the way. During this trip he stopped on the west coast of the United States. Exactly what he did there is not known because almost all of the expedition's records have disappeared. But something Drake saw convinced many powerful Londoners that a watery channel cut across North America—it was possible to sail *through* the United States. If so, the Americas could only be a few hundred miles wide. After that short trip one would be on the Pacific shore, ready to sail to China.

Elizabeth and James were wary but persuaded. Unwilling to pay the high interest rates moneylenders charged poor credit risks, though, the sovereigns delegated colonization to an entity that could independently support it: a joint-stock company. An ancestor to the modern corporation, joint-stock companies consisted of groups of wealthy people who pooled their resources to fund a commercial enterprise, being repaid by shares of the proceeds. By working with other investors, members of the company can limit their participation in an uncertain enterprise to a small part of the total sum. If a colony failed, the total loss would be huge but the loss to each individual investor would be tolerable—painful, to be sure, but not disastrous.

As the economic historian Douglass C. North has argued, the joint-stock company was not just a novel means of making money; it was one of many institutional arrangements European societies were developing to mobilize resources efficiently. (North shared the 1993 Nobel Memorial Prize in Economics, largely for working out these ideas.) These institutional arrangements secured property rights (necessary because people will not risk investing if they believe that their gains can be taken away); opened markets (necessary to prevent entrenched interests from stifling innovation); and strengthened democratic governance (necessary to check rulers' excesses). All permitted trade and commerce to be independent, which led to research and investment becoming routine—a constant activity that people could profit from with little state interference. "What counts is work, thrift, honesty, patience, tenacity," wrote the Harvard economist David S. Landes. In his classic *Wealth and Poverty of Nations* (1999), Landes argued that Europe had developed ways of organizing people and resources—private joint-stock companies, for instance—that fostered and rewarded individual initiative, which in turn promoted these virtues. Other places did not develop them. The result of these innovations, North argued, was economic growth so robust that it led to "a new and unique phenomenon": the ascension of European societies to world power.

English joint-stock companies were not immediately successful. The first was created in 1553. Fifty-three years later, when the Virginia Company received its charter, England had just ten. Three of these ventures were created to plant colonies in the Americas. (A fourth American project used a similar risk-sharing arrangement, but was not formalized as a joint-stock company.) Every one of these American enterprises had failed. Soberingly, the attempt, in the 1580s, to take over Roanoke Island, off the North Carolina coast, resulted in great expense—three costly voyages across the Atlantic—and the total obliteration of the colony.*

Despite this dismal record, the Virginia Company believed it worth trying again. At its inception, the company consisted of two investor groups, one in Plymouth and one in London. The Plymouth group focused on what is now New England, and quickly launched a colony on the coast of Maine. It disintegrated within months, and the Plymouth investors threw in the towel. The London group set its sights on Chesapeake Bay and in practice took over the entire venture. Its ships set sail from London on December 20, 1606.

Although Roanoke had been wiped out by its Indian neighbors, the Virginia Company directors reserved their fears for distant Spain. They ordered the colonists—their employees, in today's terms—to reduce the chance of detection by Spanish ships by locating the colony at least "a hundred miles" from the ocean. The instructions didn't mention that this location might already be inhabited. True, the directors viewed conflict with the Indians as unavoidable. But they viewed the conflict as a problem mainly because they feared Indians would "guide and assist any nation that shall come to invade you." That is, they worried about Tsenacomoco not because they feared its citizens would attack the English but because they feared it would help *Spain* attack the English. For this reason, the directors told the colonists to take "Great Care not to Offend the naturals"—*naturals* being a then-common term for native people.

Jamestown was the result. All the good upriver land was already occupied by Indian villages. As a result, the newcomers—

* Roanoke apparently did have one signal impact: introducing England to tobacco. Sir Francis Drake probably brought the plant to the nation in the previous decade—he had acquired it on his round-the-world expedition. But it wasn't widely known until Roanoke colonists returned with strange, fiery clay tubes at their lips. "In a short time," one courtly eyewitness moaned, "many men every-where . . . with insatiable desire and greediness sucked in the stinking smoak."

tassantassas (strangers), the Indians called them—ended up selecting the most upstream uninhabited ground they could find. Their
new home was fifty miles from the mouth of the James. It was
a peninsula near a bend in the river, at a place where the current cut so close to the shore that ships could be moored to the
trees.

Unfortunately for the *tassantassas,* no Indians lived on the
peninsula because it was not a good place to live. The English
were like the last people moving into a subdivision—they ended
up with the least desirable property. The site was boggy and mosquito ridden. Colonists could get water from the James, but it
was not always potable. During the late summer, the river falls as
much as fifteen feet. No longer pushed back by the flow of freshwater, the salty water of the estuary spreads upstream, stopping
right around Jamestown. Because the colonists had arrived in the
midst of a multiyear drought, the summer flow was especially feeble and the concentration of saltwater especially high. The saltwater boundary traps sediments and organic wastes from upstream,
which meant that the English were drinking the foulest water in
the James—"full of slime and filth," complained Percy, the future
colony president. The obvious solution—digging a well—was not
tried for another two years. It was of little help. Chesapeake Bay
is the remains of a huge, 35-million-year-old meteor crater. The
impact-fractured rock at the mouth of the bay lets in the sea, con-
taminating the groundwater with salt. Few Indian groups lived in
the saltwater wedge, presumably for just that reason. Jamestown
was bordered and undergirded by bad water. That bad water, the
geographer Carville V. Earle argued, led to "typhoid, dysentery,
and perhaps salt poisoning." By January 1608, eight months after
landfall, only thirty-eight English were left alive.

Paradoxically, the colony's desperation was its salvation;
Powhatan apparently couldn't bring himself to regard the starving *tassantassas* as a threat. Certain that he could oust the English
at any time, he allowed them to occupy their not-so-valuable real
estate as long as they provided valuable trade goods: guns, axes,

knives, mirrors, glass beads, and copper sheets, the last of which the Indians prized much as Europeans prized gold ingots. After abducting John Smith, this "subtle old fox," as Percy called him, learned enough from his captive to conclude that the profit from trade with the *tassantassas* tomorrow was worth giving them grain today. He sent the foreigner back to Jamestown in January 1608 with enough maize to keep his few remaining companions alive for a while. From Powhatan's point of view, it was a good bet, suggests Rountree, the anthropologist of Tsenacomoco. If the English tried to overstay their welcome, he could simply withhold their food, and the invasion would implode on its own. ("Confidence borne of ignorance," the University of Missouri historian J. Frederick Fausz has noted, characterized the initial attitudes of both English and Indians toward each other.)

After his return from captivity, John Smith took charge of Jamestown. Because he controlled food negotiations with Powhatan, the colony's men of consequence swallowed their displeasure. In any case they could hardly point to a record of success. That spring Smith ordered the survivors to plant crops (they would rather have looked for gold) and rebuild the colony fort (they had accidentally burned it down). He himself continued to explore Chesapeake Bay, persuading himself there was a "good hope" that it stretched to the Pacific.

All the while, Smith negotiated with Powhatan for food. He wanted to dribble out enough knives, hatchets, and iron pots to Tsenacomoco to get the necessary grain shipments but not enough to saturate the Indian demand for English goods. Complicating his task, English demand kept rising; two more convoys in the spring and fall of 1608 increased the number of mouths to about two hundred. Like any good businessman, Powhatan responded to the rising demand by raising maize prices; he asked for guns and swords, rather than hand tools. Smith refused, fearing the consequences of arming the Indians. Powhatan responded by cutting side deals for weapons with Jamestown residents who chafed at Smith's autocratic rule. And he kept the pressure on

Smith by allowing his men to pick off stragglers outside James-town.

Smith left for medical treatment in England in October 1609. Canny but clumsy, he had suffered terrible burns when he accidentally ignited a bag of gunpowder he'd fastened around his waist. For the *tassantassas*, his departure came at a specially bad time. Two months before, yet another convoy had arrived, carrying more than three hundred new colonists, among them another squad of Smith-hating gentlemen. They had persuaded the Virginia Company directors to depose him. Happily for Smith, the ship with the company's written instructions—and his replacement as governor—had been delayed. Still, the scornful newcomers posed an immediate threat to Smith's authority and, to Smith's way of thinking, Jamestown itself. To get them out of his hair, he split up the new arrivals and dispatched them to seek food from several Tsenacomoco groups. This proved to be a mistake.

One party went to the Nansemond, who lived on an island off the opposite, southern bank of the James. When the group's envoys to the Nansemond did not return on time, Percy wrote, the rest of the English "burned [the Indians'] houses, ransacked their temples, took down the corpses of their dead kings from off their tombs, and carried away their [funerary] pearls, copper and bracelets." Smith was appalled. He had berated and bullied and blustered at the Indians, but he also believed that Jamestown should not massacre its food supply. But by then he was too badly injured to force the colonists to apologize.

The incident evidently convinced Powhatan that the *tassantassas'* new leaders had abrogated the pact he had struck with Smith. That winter he struck back, directly and indirectly. On the first, direct track, native fighters cut down seventeen colonists who sought to ransack the village of Kecoughtan for food; killed another party of emaciated *tassantassas* in the forest (as a sign of "contempt and scorn," the Indians left the bodies "with their mouths stopped full of bread [maize]"); wiped out a boatload of soldiers in an upstream outpost established by Smith; and slaughtered a contingent of thirty-three colonists who had been lured to Werowocomoco by promises of grain. The leader of this party, Percy reported, was killed in a fashion that was ghastly, inventive, and slow: "By women his flesh was scraped from his bones with mussel shells and, before his face, thrown into the fire." In the next five years, natives slew as many as one out of every four colonists, Fausz estimated in a history of this "first Indian war."

Powhatan's indirect attack was more deadly still: he stopped sending food. His timing was excellent. Smith left before his official replacement as governor had arrived. His opponents in the colony chose as a temporary leader George Percy, the younger brother of the earl of Northumberland. While under attack, Smith had been unable to force the colonists to maintain Jamestown's

The luckless George Percy, younger son of the earl of Northumberland, in a nineteenth-century copy of a portrait, now lost, made during his lifetime.

gardens or mend the fishing nets. The otiose Percy was even less successful at organizing the colonists—a lack of respect related, one assumes, to his practice of swanning around the muddy encampment in silk garters, gold-banded hats, and embroidered girdles. In consequence, the English had no stockpiled food when Powhatan cut off supplies. As Percy later admitted, they were reduced to eating "dogs, cats, rats and mice," as well as the starch for their Elizabethan ruffs, which could be cooked into a kind of porridge. With famine "ghastly and pale in every face," some colonists stirred themselves to "dig up dead corpse[s] out of graves and to eat them." One man murdered his pregnant wife and "salted her for his food." By spring, only about sixty people had survived what was called the "starving time."

On some level the colony's plight is baffling. Chesapeake Bay was and is one of the hemisphere's great fisheries. Replete with pike, carp, mullet, crab, bass, flounder, turtle, and eel, this long, shallow estuary was so biologically productive that John Smith joked about being able to catch dinner in the frying pan used to cook it. The Atlantic sturgeon that swam in the James grew big enough, one colonist reported, that native boys could loop vines around their tails and be pulled underwater. (I didn't believe this until an archaeologist at Jamestown told me he had uncovered bones from a sturgeon that may have been fourteen feet long.) Oysters grew in such numbers that one mound of discarded shells from native feasts covered nearly thirty acres.

How could the colonists starve in the midst of plenty? One reason was that the English feared leaving Jamestown to fish, because Powhatan's fighters were waiting outside the colony walls. A second reason was that a startlingly large proportion of the colonists were gentlemen, a status defined by not having to perform manual labor. The first three convoys brought a total of 295 people to Jamestown. According to the historian Edmund S. Morgan, fully 92 of them were gentlemen—and many of the rest were "the personal attendants that gentlemen thought necessary to make life bearable even in England." The attendants, too, defined their position by not performing manual labor. But even if they had been able to cast aside their lifelong, ingrained customs, they might not have been able to survive, because the English were unfamiliar with the Virginia environment. They could have tried fishing for bass and catfish, which are common in the lower river at winter. But they didn't know where and when these fish like to feed. As anglers know, fishing in the wrong place at the wrong time is futile. The colonists died of ignorance as much as inanition.

John Rolfe was lucky enough to arrive in Virginia the following spring, after the starving time. Almost a year before, he had left England on the flagship of the expedition that brought the Smith-hating gentry. Rolfe's ship carried Smith's official replacement. Halfway across, a hurricane slammed into the group. The other ships slipped through the storm and made landfall in Virginia, with the results that I described above (attacking the Nansemond, enraging Powhatan, dying in droves). Meanwhile Rolfe's vessel was blown south and nearly sank. For three straight days, one passenger remembered, every person aboard, many "stripped naked as men in galleys," worked bucket chains in chest-deep water. The ship staggered awash to Bermuda, where it wrecked on the northernmost of the country's four main islands. For nine months the survivors remained on the beach, surviving on fish, sea turtles, and the pigs they had brought for Jamestown. They slowly fashioned two smaller vessels from island cedar and the wreckage of their ship. Rolfe's party arrived in Chesapeake Bay on May 23, 1610.

Appalled by the famine and ruin they found, the Bermuda group decided within two weeks to abandon Jamestown. Rolfe and the other newcomers loaded Jamestown's skeletal inhabitants onto their two makeshift vessels and two others at the colony, intending to set off for Newfoundland, where they would beg a ride home from fishing boats that plied the Grand Banks. As they waited for the tide to turn for their departure, a small boat hove into view. It was the longboat preceding yet *another* convoy, this one containing yet another new governor, 250 new colonists, and, most important, a year's worth of food. The previous colonists, despondent, returned to Jamestown and the task of figuring out how to survive.

It wasn't easy. Although they no longer had to depend on Powhatan for food, the Virginia Company later reported, "not less than one hundred and fifty of [the 250 newcomers] died"

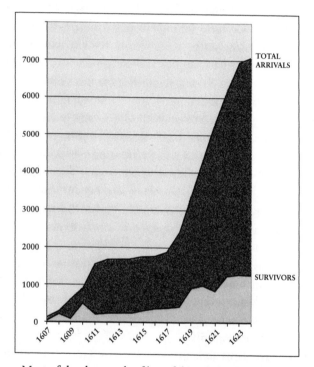

Most of the thousands of hopeful English who came to Virginia quickly died. This chart represents the author's best attempt to calculate the total number of migrants, increasing year by year, and Jamestown's actual population every year. The figures could well be off by several hundred, because the extant records are fragmentary and sometimes contradictory. But the overall picture is clear—and dismaying.

within months, among them Rolfe's young wife. Their fate was anything but atypical. Year after year, the company spent outsize sums to send colonists to Virginia—more than a hundred shiploads all told. Year after year, most of the would-be settlers perished within weeks or months—men and women, rich and poor, child and convict. England shipped about seven thousand people to Virginia between 1607 and 1624. Eight out of ten died.

So unremitting was the parade of death that even today it is painful to pore through the letters, reports, and chronicles Jamestown left behind. From every page dolorous phrases toll. *Few in the Shipp that I came in are left alive. . . .* Many newcomers either *have all perished or have suffered horrible extreamities. . . . In 3 yeares their dyed about* 3000 *p[er]sons.* Reports tally names and fates with the unadorned deadpan of old-fashioned obituary columns. Colony treasurer George Sandys notes that a servant newly shipped from London is *dead before delivered.* Colonist Hugh Pryse is found in the woods *rente in pieces with wolves or other wild Beasts, and his bowels torne out of his body.* In a drunken clash William Epps strikes Edward Stallenge so violently *that he Cleft him to the scull and next day he died.* Surgeon William Rowsley *brought* 10 *men ov[er] w[i]th him to Virginia* but within weeks *all of his servants are dead.* Edward Hill tells his brother in England he remains in Virginia only *to gett what I have lost and then god willing I will leave the Contrey.* (Hill never did leave; unable to recoup his losses, he died in Virginia a year later.) *I am quite out of hart to live in this land,* wails Phoebus Canner, *god send me well out of it.*

On December 4, 1619, John Woodlief landed with thirty-five men at a new plantation, upstream from Jamestown, called Berkeley Hundred. Woodlief had been instructed by his backers to celebrate the day of arrival "as a day of thanksgiving to Almighty god"—the first Thanksgiving in English America. Berkeley Hundred's founders had ordered the date to be observed every year. By the next December 4, thirty-one of the thirty-five *tassantassas* who had landed that day were sleeping in the soil.

Why did the Virginia Company keep trying? "Whatever else

may have entered into the activities of the company," Wesley Frank Craven observed in his history of the company, "it was primarily a business organization with large sums of capital invested by adventurers whose chief interest lay in the returns expected from the investment." Yet the Virginia Company did not *act* like an ordinary business organization. When the initial hope of discovering precious metals and a route to Asia didn't pan out, the company tried wine making, shipbuilding, iron monging, silk weaving, salt panning, and even glassblowing. All failed, at dreadful cost in money and lives. Nonetheless, the firm kept dumping money and people into Virginia. Why didn't the company's backers pull the plug? Why did they keep sending ship after doomed ship?

Equally puzzling, why did Powhatan allow the colony to survive? Jamestown escaped his first assault but remained at the edge of a precipice for years. Why didn't Powhatan push it over, once and for all?

Part of the answer to both questions is the Columbian Exchange.

"ENGLISH FLIES"

Pocahontas probably did not save John Smith when he was captured in 1608, but she did help save Jamestown—by marrying the widower John Rolfe six years later. Evidence suggests she was a curious, mischievous child, one who like all children in Tsenacomoco went without clothing until puberty. After Smith's return from captivity, Pocahontas visited Jamestown, colonist Strachey wrote afterward. The colony's young men turned cartwheels with her, "falling on their hands turning their heels upwards, whom she would follow, and wheel so her self naked as she was all the fort over." Her real name was Mataoka; Pocahontas was a teasing nickname that meant something like "little hellion."

The *tassantassas* liked the girl—but not enough to prevent

them from using her as a hostage. After Smith's departure, when Powhatan had again brought the English to the brink of annihilation, the colony's new leaders decided to counterattack. They put Jamestown under strict martial law—one colonist who stole several pints of oatmeal was chained to a tree until he starved to death—divided the men into military companies, and sent out expeditions to bring Tsenacomoco to heel. Attacking without warning, the colonists razed native villages up and down the James. The Indians repeatedly struck back, picking off colonists one by one, forcing them to retreat behind Jamestown's palisade, where they were claimed by hunger and disease.

It was a classic guerrilla-war stalemate. The *tassantassas* could win every battle, but never obtain a decisive victory; Powhatan's troops could always retreat into the hinterland, then reappear to deadly effect, arrows rushing from the trees in a sudden cloud. Yet Powhatan could not finish off the *tassantassas*, either. He could make the colonists so afraid to venture outside that they couldn't harvest their own crops. But as long as London was willing to keep shipping replacement supplies—and replacement people—the Indians, too, could not win. Both sides were exhausted by March 1613, when Jamestown's military commander, Thomas Dale, ordered a subordinate to trick the teenage Pocahontas into coming aboard an English ship. Then they sailed away with her.

Regarding the young woman as having noble blood, Dale put her under comfortable house arrest at the home of the colony's minister. Meanwhile, he sent a ransom note to Powhatan: to get back his daughter, he would have to return all the swords, guns, and metal tools "he trecherously had stolne," along with all the English prisoners of war. For three months Powhatan refused to negotiate with people he regarded as criminals. Finally he sent back a handful of English captives with an offer: five hundred bushels of maize for the girl. The guns and swords could not be returned, he said, because they had been lost or stolen. Dale scoffed at this claim. Communications ceased for another eight months, during which time some of the freed English captives

Early images of northeastern Indians are rare. This 1616 engraving of Pocahontas (left), executed during her visit to England, is the only known full portrait of a Powhatan. No portraits exist of Opechancanough, though one can imagine him looking something like this shaven-headed man (right), possibly a Virginia Indian visiting London, whose likeness was captured by the Bohemian artist Wenceslaus Hollar in 1645.

ran back to the Indians—they preferred Tsenacomoco with its foreign culture and language to Jamestown with its martial law and famine.

Determined to end the standoff, Dale led Rolfe and 150 musket-wielding *tassantassas* in March 1614 to meet Powhatan. In an angry standoff, several hundred native troops faced Dale's men on the banks of the York River. With both sides fearing a battle that would inflict many casualties, England and Tsenacomoco finally began active parley. Rolfe was on the English negotiating team. Tsenacomoco was represented by Powhatan's brother, Opechancanough, the man who had seized John Smith in the swamp. Over two days they put together an informal pact. Perhaps surprising, a key tenet was that Pocahontas would *not* return home.

After her abduction Pocahontas had been, one colonist reported, "exceeding pensive and discontented." In addition, one assumes, she was bewildered by the *tassantassas*, with their unwieldy clothes, their practice of confining women to the home and garden, their strangely rigid eating habits (at home, people simply dipped into the stewpot when hungry). But over time her attitude changed. Perhaps she was angered by her father's initial refusal to ransom her. Perhaps she liked being treated royally by the English—in her father's house, she was but one of many children from many wives. Perhaps she thought that by staying with the English she could end the war, with its intermittent eruptions of atrocity. Perhaps she simply fell in love with John Rolfe, whom she met while she was in captivity. In any case, she agreed to stay in Jamestown as his bride.

Nobody cared that Pocahontas was already married. Because she was still childless, Rountree says, native custom allowed her to sunder the marital bond at any time. And the English were willing to overlook "savage" marriages—they were un-Christian, and therefore nonexistent. In consequence, both natives and newcomers could treat Pocahontas's wedding to Rolfe as a de facto cease-fire—a "timely and face-saving method of ending the war without capitulation, a written treaty, or a formal winner," as Fausz put it in his history of the strife.

Opechancanough used the suspension of hostilities to take the levers of power from his brother (Powhatan retired in about 1615 and died three years later). Unyielding and methodical, opposed to the *tassantassas* from the day of their arrival, Opechancanough manipulated Jamestown into attacking his native rivals, augmenting his empire even as the English domain expanded. Determined to understand his enemy, the new ruler infiltrated his people into Jamestown. Working in English homes, trading with English ships, and serving in English militias, the Indians studied the ways of the foreigners. Opechancanough's men acquired a stockpile of guns, and trained themselves to use them.

The colonists were blithely unaware of Opechancanough's

schemes. Nonetheless, they initiated, all unintentionally, a devastating countermeasure: the Columbian Exchange. The constant flow of ships to Virginia brought with them an entire suite of new species, opening what would become a multilevel ecological assault. One of the most potent weapons was tobacco.

Even at the height of the war John Rolfe had been experimenting with *N. tabacum*. King James I had initially excoriated smoking as "lo[a]thsome to the eye, hatefull to the Nose, [and] harmefulle to the braine." He thought about banning it but changed his mind—the perpetually cash-short monarch had discovered that tobacco could be taxed. English smokers were relieved, but not happy; the Spaniards kept raising prices. Much as crack cocaine is an inferior, cheaper version of powdered cocaine, Virginia tobacco was of lesser quality than Caribbean tobacco but also not nearly as expensive. Like crack, it was a wild commercial success; within a year of its arrival, Jamestown colonists were paying off debts in London with little bags of the drug. The cease-fire with Powhatan let colonists expand production explosively. By 1620 Jamestown was shipping as much as fifty thousand pounds a year; three years later the figure had almost tripled. Within forty years Chesapeake Bay—the Tobacco Coast, as it later became known—was exporting 25 million pounds a year. Individual farmers were making profits of as much as 1,000 percent on their initial investment.

One thousand percent! And all that was needed was sun, water, and soil! The sums skyrocketed if farmers could afford servants—laborers' annual pay was about £2, but they could grow £100 or even £200 of tobacco in that time. In an object demonstration of the power of economic order to focus the human mind, the *tassantassas* whom John Smith had to order into their fields at gunpoint now became intent on wringing tobacco from the soil. Newcomers poured in, grabbed some land, and planted *N. tabacum*. English-style farms spread like rumors up and down the James and York rivers. So many colonists poured in that the company realized they could not be controlled entirely from across the ocean and created an elected council to resolve disputes—the first representative body in colonial North America. Its opening session lasted from July 30 to August 4, 1619.

Barely three weeks later a Dutch pirate ship landed at Jamestown. In its hold was "20. and odd Negroes"—slaves taken by the pirates from a Portuguese slave ship destined for Mexico. (About thirty more showed up in another ship a few days later.) In their hurry to extract tobacco profits, the *tassantassas* had been clamoring for more workers. The Africans had arrived at harvest time. Without a second thought colonists bought the Africans in exchange for the food the pirates needed for the return trip to Europe. Legally speaking, the "20. and odd" Africans may not have been slaves—their status is unclear. Nevertheless, they were not volunteers; their purchase was a landmark in the road to slavery. Within weeks of each other, Jamestown had inaugurated two of the future United States' most long-lasting institutions: representative democracy and chattel slavery.

Not that the colonists paid attention to these landmarks—they were too busy exporting Virginia leaf. Obsessed by tobacco, some of the leadership complained, the colonists let Jamestown fall once again into ruin: "the Church down, the Palizado's [walls] broken, the Bridge in pieces, the Well of fresh water spoiled; the store-house they used for the Church; the market-place, and streets and all other spare places planted with Tobacco." Massive celebratory drunkenness was common; incoming ships brought liquor and profitably transformed themselves into floating temporary taverns. Dale was forced to issue an order to Virginia's planters: grow food crops, too, or forfeit your tobacco to the colonial government. Few paid attention.*

---

* Tobacco mania spilled over to France. The cash-strapped French court, which ordinarily regarded its American colonies with indifference, even annoyance, spent a fortune to found New Orleans because the crown was appalled by the sums French smokers were spending on English tobacco. In a debt-for-equity swap hatched by the brilliant economist John Law, the court's lenders were allowed to trade government bonds that Paris couldn't pay off for shares in the profits from the new colony, which was envisioned as a huge

Alas, the boom came too late for the Virginia Company. Shipping colonists across the Atlantic only to have them die had exhausted its start-up capital. Company officers persuaded London's powerful clergy that helping Jamestown find more investors was the duty of all English Christians. Sunday after Sunday, ministers urged their parishioners to buy shares in the Virginia Company. "Goe forward," Rev. William Crashaw urged potential "noble and worthy Adventurers," some of whom sat in the pews of his Temple Church, one of the nation's most influential houses of worship. If England did not seize its opportunity in Virginia, Crashaw predicted, future generations would ask, *"Why was there such a pri[z]e put into the hands of fooles who had not hearts to take it?"* (Emphasis in original.)

The tactic worked. Ministers enticed more than seven hundred individuals and companies to put at least £25,000 into the Virginia Company.* (By contrast, historians believe that fewer than a dozen men were the original backers of the company and that they put in just several hundred pounds.) The new sum was enough to send over hundreds of colonists, Rolfe and Dale among them, who eagerly grew tobacco. But even the rush of tobacco profits could not offset the debts from the company's years of losses. The Virginia Company was again running out of money on March 22, 1622, when Opechancanough attacked.

tobacco farm powered by slaves: a replica of Virginia, with Bordeaux instead of London grog. The public rushed to buy shares, driving up their price in a classic speculative bubble. Law hired armed guards to protect himself from importuning would-be investors and was awarded a dukedom in Arkansas by a grateful ruler. The bubble burst in 1720, but the first shipments of would-be nicotine magnates had already left for the Americas, where they learned that New Orleans wasn't, in fact, especially good tobacco land. Exasperated by the continuing losses, France happily handed the city to Spain in 1762 as payment for its aid in the Seven Years' War.

* Equivalents in contemporary money are hard to establish, but this sum surely translates into tens of millions of dollars. Even that vague claim may be misleading, because the pool of investment capital was then much smaller; the capital raised by the Virginia Company was a much bigger percentage of the total available than, say, $50 million would be today.

Although this image is confused in many ways—note the neatly walled fortress in the distance, so utterly unlike Jamestown or any Powhatan settlement—something of the shock caused by the Powhatan attack on Virginia in 1622 was captured in this engraving by the German artist Matthäus Merian.

Early that morning Indians slipped into European settlements, knocking on doors and asking to be let in. Most were familiar visitors. They came unarmed. Many accepted a meal or a drink. Then they seized whatever implement came to hand—kitchen knife, heavy stewpot, the colonists' own guns—and killed everyone in the house. The assault was brutal, widespread, and well planned. So swift were the blows that many colonists died without knowing they were under attack. Entire families fell. Houses burned across what had been Tsenacomoco. At the last minute several Indians told English friends about the attack, providing enough warning to let Jamestown gather its defenses. Nonetheless the attackers killed at least 325 people.

The aftermath claimed as many as seven hundred more. Because the attack disrupted spring planting, the *tassantassas* grew even less maize than usual. Meanwhile, the company tried to rebuild Jamestown by sending over a thousand new colonists. Incredibly, they came to Virginia with no food supplies. Actually, not so incredibly—ship captains were paid by the person transported, so they overloaded their vessels with passengers, carrying as little unprofitable food as possible. The luckless, scurvy-ridden souls aboard were dumped ashore, where they were forced to eat "barks of trees, or moulds [soil] of the Ground." Again colonists scrabbled in rags over handfuls of maize. It was a second "starving time." By spring the survivors were so debilitated, colony treasurer George Sandys wrote, *"the lyveing* [were] *hardlie able to bury the dead."* (Emphasis in original.) Altogether about two out of every three Europeans in Virginia died that year.*

By any measure, Opechancanough was in a commanding position. His forces now more numerous and better supplied than the enemy, they raided English settlements at will. Jamestown's governing council confessed that the colonists couldn't successfully mount a reprisal, "by reasone of theire swyftnes of foote, and advantages of the woodes, to which uppon all our assaultes they retyre." Opechancanough predicted in the summer of 1623 that "before the end of two Moones there should not be an Englishman in all their Countries."

Just as he foresaw, the Virginia Company did not survive. Horrified by the attack, James I created an investigatory commision, which issued a damning report. The company's parliamen-

---

* Not everything went badly for the *tassantassas*. In May 1623, a year and two months after the assault, they staged a counterattack at a peace conference with Tsenacomoco's leadership. At a celebratory toast, one witness recorded, the English passed out poisoned sack (a sherry-like wine), killing "some tooe hundred" Indians. Pursued by a stricken, enraged crowd, the colonists fled to their boats. As they left, they fired into the mob, killing "som 50 more," including, they erroneously believed, Opechancanough. Afterward the English "brought hom parte of ther heades"—that is, they scalped some of their victims.

tary support vanished. Management fought desperately to retain the king's favor. Its investors had sunk into Virginia as much as £200,000, a vast sum at the time. As long as the firm existed the money potentially could be recouped. If James revoked the company charter, it would be beyond recovery. Nevertheless he revoked the charter on May 24, 1624. "Any responsible monarch would have been obliged to stop the reckless shipment of his subjects to their deaths," wrote Morgan, the historian. The wonder was that the king had not done so earlier. Opechancanough had defeated the Virginia Company.

But victory over the company did not mean victory for the Indians. Opechancanough did not launch a final, killing assault, pushing the foreigners into the sea. Indeed, a second coordinated attack didn't take place for *twenty-two years*, when it was far too late. The reason for his hesitation will never be known with certainty, because English accounts provide the great majority of historical records, and the hostilities ensured that the *tassantassas* lost what little view they had into native life. But one possible answer is that Opechancanough had lost Tsenacomoco before his troops fanned out into English homes. By growing tobacco, the English had transformed the landscape into something unrecognizable.

Indians had traditionally raised tobacco, but only in small amounts. The colonists, by contrast, covered big areas with stands of *N. tabacum*. Neither natives nor newcomers understood the environmental impact of planting it on a massive scale. Tobacco is a sponge for nitrogen and potassium. Because the entire plant is removed from the soil, harvesting and exporting tobacco was like taking those nutrients from the earth and putting them on ships. "Tobacco has an almost unique ability to suck the life out of soil," said Leanne DuBois, the agricultural extension agent in James City County, Jamestown's county. "In this area, where the soils can be pretty fragile, it can ruin the land in a couple of years." Constantly wearing out fields, the colonists had to keep moving to new land.

In Tsenacomoco, one recalls, families traditionally farmed their plots for a few years and then let them go fallow when yields declined. The unplanted land became common hunting or foraging grounds until needed again for farms. Because the fallow lands had already been cleared, the foreigners could readily move in and plant tobacco on them. Unlike the Powhatan, the English didn't let their tobacco fields regenerate after they were depleted. Instead, they turned them into maize fields, and then pasture for cattle and horses. Rather than cycling the land between farm and forest, in other words, the foreigners used it continuously—permanently keeping prime farmland and forage land away from the people of Tsenacomoco, pushing the Indians farther and farther away from the shore as they did.

In a decade or two the English had grabbed most of the land cleared by Indians. They moved into the forest, as the environmental historian John R. Wennersten wrote, "using slash-and-burn techniques that had not been seen in Europe for centuries." They felled great numbers of trees, and lavishly used the fallen timber. Farmers marked their property with "worm" fences—zigzag constructions of six to ten interlocking rails—that Wennersten estimates consumed 6,500 long, thick timbers for every mile of fence. Other wood was converted into pitch, tar, turpentine, and wooden planks. The plentiful leftovers were exported, in the form of barrels, casks, kegs, and hogsheads, to timber-starved England. "They have an unconquerable aversion to trees," one eighteenth-century visitor dryly observed. "Not one is spared."

Subject to annual burning, native woodlands had been both open, in that people could freely move around, and closed, in that the canopy of big trees sheltered the land from the impact of rainfall. Taking down the forest exposed the soil. Colonists' ploughs increased its vulnerability. Nutrients dissolved in spring rains and washed into the sea. The exposed soil dried out more quickly and hardened faster, losing its ability to absorb spring rains; the volume and speed of runoff increased, raising river volume. By the late seventeenth century disastrous floods were common. So much soil had washed into the rivers that they became difficult to navigate.

Tobacco from South America was far from the only biological

import. The English brought along all the other species they were accustomed to finding on farms: pigs, goats, cattle, and horses. At first the imported animals didn't fare well, not least because they were eaten by starving colonists. But during the peace after Pocahontas's marriage, they multiplied. Colonists quickly lost control of them. Indians woke up to find free-range cows and horses romping through their fields, trampling the harvest. If they killed the beasts, gun-waving colonists demanded payment. Animal numbers boomed for decades.

The worst may have been the pigs. By 1619, one colonist reported, there were "an infinite number of Swine, broken out into the woods." Smart, strong, and constantly hungry, they ate nuts, fruits, and maize, turning up the marshy soil with their shovel-like noses in search of edible roots. One of these was tuckahoe, the tuber Indians relied upon when their maize harvests failed. Pigs turned out to like tuckahoe—a lot. Traveling through the area in the eighteenth century, the Swedish botanist Peter Kalm found that pigs were "very greedy" for the tubers, "and grow very fat by feeding on them." In places "frequented

After the final defeat of the Virginia Indians in the 1660s, they were required to wear identifying badges—this one belonged to a native leader—if they wanted to enter English settlements.

by hogs," he argued, tuckahoe "must have been extirpated." The people of Tsenacomoco found themselves competing for their food supply with packs of feral pigs.

In the long run, though, the biggest ecological impact may have been wreaked by a much smaller domestic animal: the European honeybee. In early 1622 a ship arrived in Jamestown that was loaded with exotic entities: grapevines, silkworms, and bees. The grapes and silkworms never amounted to much, but the bees thrived. Most bees pollinate only a few plant species and tend to be fussy about where they live. But European honeybees, promiscuous little beasts, pollinate almost anything in sight and reside almost anywhere. Quickly they set up shop throughout the Americas. Indians called them "English flies."

The English imported bees for honey, not to help their crops—pollination wasn't discovered until the mid-eighteenth century—but feral honeybees pollinated farms and orchards anyway. Without them, many of the plants Europeans brought with them wouldn't have proliferated. Georgia probably would not have become the Peach State; Johnny Appleseed's trees might never have borne fruit; Huckleberry Finn might not have had any watermelons to steal. So critical to European success was the honeybee that Indians came to view it as a harbinger of invasion; the first sight of a bee in a new territory, the French-American writer Jean de Crèvecoeur noted in 1782, "spreads sadness and consternation in all minds."

Removing forest cover, blocking regrowth on fallow land, exhausting the soil, shutting down annual burning, unleashing big grazing and rooting animals, introducing earthworms, honeybees, and other alien invertebrates—the colonists so profoundly changed Tsenacomoco that it became harder and harder for its inhabitants to prosper there. Meanwhile, it was easier and easier for Europeans to thrive in an environment that their own actions were making increasingly familiar. Despite starvation, disease, and financial meltdown, immigrants poured into Chesapeake Bay. Axes flashing, oxen straining before the plow, hundreds of new colonists planted spreads of tobacco across every accessible river bluff. When they wore out the soil, they gave the fields over to cattle and then moved on.

Ecologically speaking, Tsenacomoco was becoming ever more like Europe—the hallmark of the nascent Homogenocene. By 1650 the Indian empire was mainly inhabited by Europeans.

"SOE INFINITE A RICHES"

By all accounts, John Ferrar was a modest, pious, hardworking man who spent his life tending the family business. His father, Nicholas, was a cosmopolitan London leather merchant with a

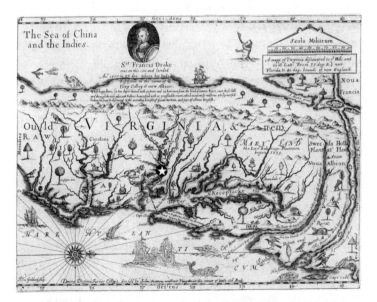

Maps like this one, from 1667, were surprisingly common in seventeenth-century Europe. Depicting North America as a narrow isthmus, it suggested to the Virginia Company's English backers that their colonists at Jamestown (star on map) could easily walk to the Pacific. From there, they could sail to China.

mansion on St. Sythe's Lane, not far from the Bank of England and the Royal Exchange. One of the original stockholders in the Virginia Company, he sank £50 into Jamestown. The investment did not bear fruit, and Nicholas became convinced that the problem lay with the company's well-connected but feckless managers. Rather than pulling out, though, the family invested another £50 in 1618, acquiring a plantation of several thousand acres, administered by another relative whom Nicholas dispatched to Virginia. A few months later, he participated in a sort of shareholders' revolt. New corporate officers were appointed, among them two of Nicholas's other sons: Nicholas Jr., who became the company counsel and secretary, and John, who was given the unsalaried office of deputy treasurer.

Despite his lowly position, John Ferrar found himself effectively in charge of the company's finances—the actual treasurer, an important aristocrat, was too busy harassing the king in Parliament. The firm now was making money from tobacco sales but had piled up so much debt that Ferrar had to scramble to pay creditors. Worse, he claimed, the previous management had embezzled £3,000. Attempts to restore the funds were countered by the thieves' attempts to smear him in court. The intrigue grew so all-consuming that Ferrar held daily crisis meetings at the family manse on St. Sythe's Lane.

In the end, his hard work didn't pay off. Opechancanough's attack in 1622 gave the company's enemies the opening they sought; Nicholas and John, portrayed as reckless swindlers, were briefly thrown in prison. They managed to talk their way free, but cannot have been taken by surprise when the king put an end to the enterprise.

John Ferrar never reconciled himself to the loss. Twenty-five years after the company's demise, he read William Bullock's *Virginia Impartially Examined,* a sixty-six-page tract that blamed him and other managers for Jamestown's troubles. Ferrar filled the margins of his copy with irate rejoinders. Bullock had written that the colony could prosper only by diversifying; rather than focusing exclusively on tobacco, the colonists should have grown wheat and barley. To Ferrar, this was like telling people who were riding off a cliff that they should wear jackets of another color. As far as he was concerned, Virginia's mistake had been to ignore what Sir Francis Drake had learned during the 1570s, when he stopped in California during his round-the-world voyage. Drake had proven—proven!—that the Americas were at most a few hundred miles across. Jamestown's failure to cut through the continent and pioneer a new route to Asia, Ferrar wrote, "is to this day the greatest Error and damadge that hath happened to the Collony all this while." He was certain that only "8 or 10 days March[,] naye it maybe not a 4 days Journy" separated Jamestown and the Pacific. A single expedition west would have discovered "Soe Infinite a Riches to them all as a passadge to a West Sea would prove to them." Instead, they had stupidly filled their days with "Smokey Tobaco."

From today's vantage the story seems more complex. The goal of the Virginia Company had been to integrate Virginia, and thus poor England itself, into the rich new global marketplace. Although Ferrar never recognized it, the company had done exactly that—with "Smokey Tobaco," the first American species to disperse into Europe, Asia, and Africa. Fun, exciting, and wildly addictive, tobacco was an instant hit around the globe—the first time people in every continent simultaneously became enraptured by a novelty. *N. tabacum* was the leading edge of the Columbian Exchange.

By 1607, when Jamestown was founded, tobacco was enthralling the upper classes in Delhi, where the first smoker, to the dismay of his advisers, was none other than the Mughal emperor; thriving in Nagasaki, despite a ban promulgated by the alarmed daimyo; and addicting sailors in Istanbul to such an extent that they were extorting it from passing European vessels. In that same year a traveler in Sierra Leone observed that tobacco, likely brought by slave traders, could be found "about every man's house, which seemeth half their food." Nicotine addiction became so rampant so quickly in Manchuria, according to the Oxford historian Timothy Brook, that in 1635 the khan Hongtaiji discovered that his soldiers "were selling their weapons to buy tobacco." The khan angrily prohibited smoking. On the opposite side of the world, Europeans were equally hooked; by the 1640s the Vatican was receiving complaints that priests were celebrating Mass with lighted cigars. Pope Urban VIII, as enraged as Hongtaiji, promptly banned smoking in church.

From Bristol to Boston to Beijing, people became part of an international culture of tobacco. Virginia played a small but important part in creating this worldwide phenomenon. From today's perspective, though, *N. tabacum* in the end was less important in itself than as a magnet that pulled many other non-human creatures, directly and indirectly, across the Atlantic, of which the most important surely were two minute, multifaceted immigrants, *Plasmodium vivax* and *Plasmodium falciparum*—names little known outside specialist circles, but ones that played a devastating role in American life.

3

Evil Air

In 1985 a bookseller in northeast Spain announced that he had possession of nine letters and reports by Cristóbal Colón, seven of them never seen before, including chronicles of all four of his American voyages. Later that year, Consuelo Varela and Juan Gil, editors of a definitive edition of the admiral's writings, skeptically inspected the papers. Surprising their colleagues, Varela and Gil concluded that the manuscripts were handwritten copies of actual letters and reports by Colón—copies of the type routinely kept by wealthy people in the days before photocopiers. The Spanish government acquired the papers for an undisclosed sum; a facsimile edition was published in 1989. Nine years after that, an English translation appeared.

Because I am interested in Colón, I bought a copy of the translation when I spotted it in a used-book store. Part of a series the Italian state published to honor the five hundredth anniversary of his first voyage to the Americas, the book is a big, lush, cream-colored object that doesn't fit on a standard bookshelf. Disappointing to readers like me, Gil and Varela announced in the introduction that "these previously unknown texts do not present any spectacular revelations" about Colón's life and character. But halfway through the newly revealed chronicle of the admiral's second voyage I came across a curious detail—one that wasn't in the fine biographies by Samuel Eliot Morison and Felipe Fernández-Armesto.

In the translation, Colón explains that after the expedition arrived at La Isabela "all my people went ashore to settle, and everyone realized it rained a lot. They became gravely ill from tertian fever." *Tertian fever*, an old-fashioned term, refers to bouts of fever and chills that occur in a regular forty-eight-hour pattern—a day of sickness followed by a day of quiet, then a day of sickness as the pattern repeats (*tertian*, taken from the Latin for "three days," derives from the Roman custom of counting time from the beginning of one period to the beginning of the next). Tertian fever is the fingerprint of the most important types of malaria, one of humankind's most intractable scourges. Taken literally, Colón seemed to be saying that at La Isabela his men contracted malaria. No wonder the colonists didn't want to work, I thought, and marked the passage with a pencil.

In 2002 Noble David Cook, a historian at Florida International University, in Miami, published an article entitled, alarmingly, "Sickness, Starvation, and Death in Early Hispaniola," which detailed the island's catastrophic history after Colón's landing. Researchers generally agree that human malaria did not exist in the Americas before 1492 (some believe a kind of monkey malaria was present). If Colón's men contracted malaria, Cook explained, they must have brought the disease with them from Spain, which like much of Europe then was rife with the disease. It was a textbook case of the Columbian Exchange, recorded by its progenitor himself.

Remembering the cream-colored book, I hauled it from my bookshelf and turned to the relevant passage. The original Spanish, printed on the facing page, didn't use the Spanish words for malaria or tertian fever. Instead Colón wrote that his men had contracted something called *çiçiones*, a term I had never encountered. Why did Cook and the translator of Colón's letter think this meant malaria?

Çiçiones is hard to find in modern Spanish dictionaries—I consulted the dozen or so in my local library without success. Google, too, was no help. Nor was Colón himself. He provided no description of the symptoms of *çiçiones*, perhaps because he believed they were familiar to his readers. All he said about the disease, in fact, was to guess that it was spread by the native women around La Isabela, "who are abundant there; and since they [that is, the women] were immodest and disheveled, it is no wonder that they [that is, the men] had trouble." To me, this sounded like the admiral thought *çiçiones* was some kind of venereal disease.

But that doesn't jibe with other sources, as I learned when I contacted an expert in sixteenth-century Spanish, Scott Sessions of Amherst College. The first dictionary of the Spanish language appeared in 1611, Sessions told me. In it is an entry for *çiçiones:* "the fever that comes with chills, which is attributed to the *cierzo* [mistral wind], because it is the most acute, cold and penetrating." The next authoritative Spanish dictionary, issued in multiple volumes by the Royal Spanish Academy between 1726 and 1739, similarly defines *çiçiones* as "the fever that starts with chills, which from being acute and penetrating like the mistral wind, as [the first dictionary] says, one derives the word: but it more likely refers to tertian fever"—malaria. Cook and the translator, in other words, were correct: Colón may well have been describing malaria.

The scenario isn't implausible. Malaria can lie dormant in the body for months, only to reemerge at full strength. The disease is transmitted by mosquitoes, which take in microscopic malaria parasites when they drink blood from infected people and pass them on to the next people they bite. Colón left on his second voyage in September 1493. If one of his crew had a malaria relapse after landing in La Isabela, only one bite from the right type of mosquito would be necessary to spread the disease—and those mosquitoes are abundant on Hispaniola.

All of this is highly speculative, to say the least. Today we know that many different diseases cause chills and fevers, including influenza and pneumonia. But for centuries people couldn't distinguish one from another; they didn't understand that malaria was a specific disease. Sessions, the Amherst historian, told me that *paludismo*, the Spanish word for malaria, didn't appear in Royal Spanish Academy dictionaries until 1914. Even then, few realized that it was caused by a mosquito-borne parasite—the 1914 dictionary defined *paludismo* as a "group of deadly phenomena produced by marshy emanations." (The English word "malaria" comes from the Italian *mal aria*, evil or bad air.) Colón was using a word that probably indicates malaria, in other words, but he could well have been describing ordinary chills and fever. A single word is not enough to make a diagnosis.

Yet the impossibility of finding definitive answers does not mean historians should stop seeking them—the question is too important. Despite a global eradication program that began in the 1950s, malaria is still responsible for unimaginable suffering: some three-quarters of a million deaths per annum, the great majority of them children under the age of five. Every year about 225 million people contract the disease, which even with modern medical care can incapacitate for months. In Africa it afflicts so many people so often that economists believe it is a major drag on development; since 1965, according to one widely cited calculation, countries with high rates of malaria have had annual per capita growth rates 1.3 percent less than countries without malaria, enough to ensure that many of the former lost ground to the latter.

As it does today, malaria played a huge role in the past—a role unlike that of other diseases, and arguably larger. When Europeans brought smallpox and influenza to the Americas, they set off *epidemics*: sudden outbursts that shot through Indian towns and villages, then faded. Malaria, by contrast, became *endemic*, an ever-present, debilitating presence in the landscape. Socially speaking, malaria—along with another mosquito-borne disease, yellow fever—turned the Americas upside down. Before these maladies arrived, the most thickly inhabited terrain north of Mexico was what is now the southeastern United States, and the wet forests of Mesoamerica and Amazonia held millions of people. After malaria and yellow fever, these previously salubrious areas became inhospitable. Their former inhabitants fled to safer lands; Europeans who moved into the emptied real estate often did not survive a year.

The high European mortality rates had long-lasting impacts, the Harvard and Massachusetts Institute of Technology economists Daron Acemoglu, Simon Johnson, and James A. Robinson have argued. Even today, the places where European colonists couldn't survive are much poorer than places that Europeans found more healthful. The reason, the researchers said, is that the conquering newcomers established different institutions in disease zones than they did in healthier areas. Unable to create stable, populous colonies in malarial areas, Europeans founded what Acemoglu, Johnson, and Robinson called "extractive states," the emblematic example being the ghastly Belgian Congo in Joseph Conrad's *Heart of Darkness*, where a tiny cohort of high-collared Europeans forces a mass of chained, naked slaves, "shadows of disease and starvation," to build a railroad to ship ivory from the interior.

Tobacco brought malaria to Virginia, indirectly but ineluctably, and from there it went north, south, and west, until much of North America was in its grip. Sugarcane, another overseas import, similarly brought the disease into the Caribbean and Latin America, along with its companion, yellow fever. Because both diseases killed European workers in American tobacco and sugar plantations, colonists imported labor in the form of captive Africans—the human wing of the Columbian Exchange. In sum: ecological introductions shaped an economic exchange, which in turn had political consequences that have endured to the present.

It would be an exaggeration to say that malaria and yellow fever were responsible for the slave trade, just as it would be an exaggeration to say that they explain why much of Latin America is still poor, or why the antebellum cotton plantations in *Gone with the Wind* sat atop great, sweeping lawns, or why Scotland joined England to form the United Kingdom, or why the weak, divided thirteen colonies won independence from mighty Great Britain in the Revolutionary War. But it would not be completely wrong, either.

SEASONING

Malaria is caused by the two hundred or so species in the genus *Plasmodium,* ancient microscopic parasites that plague countless types of reptile, bird, and mammal. Four of those two hundred species target humankind. They are dishearteningly good at their jobs.

Although the parasite consists of but a single cell, its life story is wildly complex; it changes outward appearance with the alacrity of characters in a Shakespearean comedy. From the human point of view, though, the critical fact is that it is injected into our flesh by mosquitoes. Once in the body, the parasite pries open red blood cells and climbs inside. (I am here skipping several intermediate steps.) Floating about the circulatory system like passengers in so many submarines, the parasites reproduce in huge numbers inside the cell. Eventually the burgeoning offspring burst out of the cell and pour into the bloodstream. Most of the new parasites subvert other red blood cells, but a few drift in the blood, waiting to be sucked up by a biting mosquito. When a mosquito takes in *Plasmodium,* it reproduces yet again inside the insect, taking on a different form. The new parasites squirm into the mosquito's salivary glands. From there the insect injects them into its next victim, beginning the cycle anew.

In the body, *Plasmodium* apparently uses biochemical signaling to synchronize its actions: most of the infected red blood cells release their parasites at about the same time. Victims experience

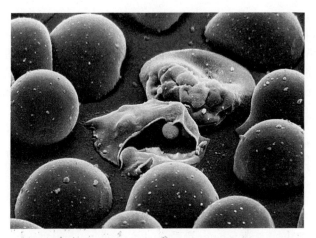

Single-celled *Plasmodium* parasites burst out of dying red blood cells, beginning the assault on the body that leads to full-blown malaria.

these eruptions as huge, coordinated assaults—a single infection can generate *ten billion* new parasites. Overwhelmed by the deluge, the immune system sets off paroxysms of intense chills and fever. Eventually it beats back the attack, but within days a new assault occurs; some of the previous wave of parasites, which have hidden themselves inside red blood cells, have produced a new generation of *Plasmodium,* billions strong. The cycle repeats until the immune system at last fights off the parasite. Or seems to—*Plasmodium* cells can secret themselves in other corners of the body, from which they emerge a few weeks later. Half a dozen episodes of chills and fever, a bit of respite, then another wave of attacks: the badge of full-blown malaria.

If the suffering caused by malaria today is difficult to grasp, it is almost impossible to imagine what it was like when its cause was unknown and no effective treatments existed. One can get a hint by reading the accounts of victims like Samuel Jeake, a seventeenth-century merchant in southeast England, who doggedly recorded every skirmish in his decades-long war with what we now recognize as malaria. To pick an example almost at random, here is Jeake on February 6, 1692, near the end of one six-month bout, stoically recording that he had been "taken ill the Seventh time: with a Tertian Ague [fever]; about 3h p.m. it began, & was of the same nature with my last which I had all January, but this was the worst."

> Feb. 8: A 2d fit which took me earlier & was worse.
> Feb. 10: About noon a 3d fit. which shook me about 3h p.m. a very bad fit & violent feaver. . . .
> Feb. 12: Before noon, a 4th fit. with which I shook about 3h p.m. & then went to bed: where had a very violent Feaver; this being the worst fit of all: my breath very short; & delirious. . . .
> Feb. 14: About noon, a 5th fit. . . . . .
> Feb. 16: About 2h. p.m. a 6th fit, very little, or scarce sensible, but sweat much in the night. And it pleased God that this was the last fit.

The respite lasted just fifteen days.

> Mar. 3: About 4h. p.m. Taken ill the Eighth time: of a Tertian ague, succeeded by a Feaver & sweat in the night. . . .
> Mar. 5: About 3h p.m. A 2d fit; worse than the former.

The attacks stopped nine weeks later. But malaria was not done with Jeake. The parasite, a superbly canny creature, can hide in the liver for as long as five years, periodically emerging to produce full-blown malarial relapses. Six months later, *Plasmodium* again massed in his blood.

Tertian fever of the sort experienced by Jeake is the signature of *Plasmodium vivax* and *Plasmodium falciparum*, which cause the two most widespread types of malaria. Despite the similarity of the symptoms, the two *Plasmodium* species have different effects on the body. After inserting itself inside red blood cells, falciparum, unlike vivax, manages to alter them so that they stick to the walls of the tiny capillaries inside the kidneys, lungs, brain, and other organs. This hides the infected cells from the immune system but slowly cuts off circulation as the cells build up on the capillary walls like layers of paint on an old building. Untreated, the circulation stoppage leads to organ failure, which kills as many as one out of ten falciparum sufferers. Vivax doesn't destroy organs, and thus is less deadly. But during its attacks sufferers are weak, stuporous, and anemic: ready prey for other diseases. With both species, sufferers are infectious while sick—mosquitoes that bite them can acquire the parasite—and can be sick for months.

Plasmodium, a tropical beast, is exquisitely sensitive to temperature. The speed at which the parasite reproduces and develops in the mosquito depends on the temperature of the mosquito, which in turn depends on the temperature outside (unlike mammals, insects cannot control their own internal temperature). As the days get colder, the parasite needs more and more time to develop, until it takes longer than the mosquito's lifespan. Falciparum, the most deadly variety of malaria, is also the most temperature sensitive. Around 72°F it hits a threshold; the parasite needs three weeks at this temperature to reproduce, which approaches the life expectancy of its mosquito host; below about 66°F it effectively cannot survive. Vivax, less fussy, has a threshold of about 59°F.

Unsurprisingly, falciparum thrives in most of Africa but gained a foothold only in the warmest precincts of Europe: Greece, Italy, southern Spain, and Portugal. Vivax, by contrast, became endemic in much of Europe, including cooler places like the Netherlands, lower Scandinavia, and England. From the American point of view, falciparum came from Africa, and was spread by Africans, whereas vivax came from Europe, and was spread by Europeans—a difference with historic consequences.

Human malaria is transmitted solely by the *Anopheles* mosquito genus. In Jeake's part of England the principal "vector," as the transmitting organism is known, is a clutch of tightly related mosquito species known jointly as *Anopheles maculipennis*. The mosquito's habitat centers on the coastal wetlands of the east and southeast: Lincolnshire, Norfolk, Suffolk, Essex, Kent, and Sussex counties. *A. maculipennis*—and the *Plasmodium vivax* it carries—seem to have been uncommon in England until the late sixteenth century, when Queen Elizabeth I began encouraging landlords to drain fens, marshes, and moors to create farmland. Much of this low, foggy terrain had been flooded regularly by the North Sea tides, which washed away mosquito larvae. Draining blocked the sea but left the land dotted with pockets of brackish water—perfect habitat for *A. maculipennis*. Farmers moved into the former marsh, still soggy but now usable. Their homes and barns, heated during cold weather, provided space for the mosquito—and the vivax parasites inside its body—to survive the cold weather, ready to breed and spread in the following spring.

As the British medical historian Mary Dobson has documented, draining the marshes set off an inferno of vivax malaria. Visitors to *maculipennis* habitat recoiled at the wretchedness they encountered. An all-too-typical sight, lamented the Kent writer Edward Hasted in 1798, was "a poor man, his wife, and whole family of five or six children, hovering over the fire in their hovel, shaking with an ague [fever], all at the same time." Curates died in such numbers after being sent to coastal Essex, the writer John Aubrey remarked, that the area was known as "Killpriest." Natives fared no better; babies born in the marshland, Hasted wrote, seldom "lived to the age of twenty-one." Dobson recorded baptisms and burials in twenty-four wetland parishes. In the 1570s, before Queen Elizabeth drained the swamps, baptisms exceeded burials by 20 percent—the population was rising. Two decades later draining was in full swing, and burials outnumbered baptisms by almost a factor of two. Population boomed elsewhere in

As this nineteenth-century copy of a now-lost earlier drawing suggests, malaria was long a constant fear in England's south-eastern marshlands.

England, but these parishes didn't return to their earlier growth rates for two centuries.*

"The marshes would have these bursts of mortality," Dobson told me. "About every ten years they'd have a year in which 10 or 20 percent of the population would die. A few miles away, in higher ground, were some of the healthiest parts of England."

---

* It may seem odd that malaria, a tropical disease, flourished in England of the Little Ice Age. But history is an interplay of social and biological processes. Just as Elizabethan marsh-draining techniques unintentionally helped vivax flourish, the improved drainage methods of the Victorian era dramatically cut malaria, because they didn't leave the brackish pools, thus simultaneously eliminating mosquito habitat and creating better pasturage for cattle, which *A. maculipennis*, if given the choice, prefers to feed upon. Even so, researchers routinely found "thousands" of the insects roosting "in the dark and ill-ventilated pigsties" of poor coastal farmers as late as the 1920s. Today some fear that global warming will foster the spread of malaria. But if people continue destroying mosquito habitat by draining wetlands the hotter weather may have no impact on malaria rates.

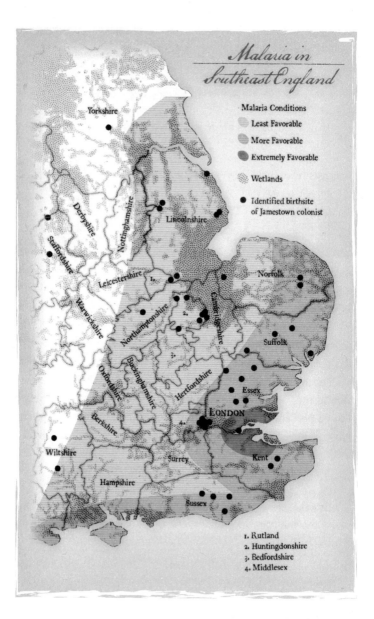

Malaria in
Southeast England

Malaria Conditions

Least Favorable

More Favorable

Extremely Favorable

Wetlands

Identified birthsite
of Jamestown colonist

Yorkshire

Derbyshire

Staffordshire

Nottinghamshire

Lincolnshire

Leicestershire

Warwickshire

1.

Norfolk

Northamptonshire

2.

Cambridgeshire

Suffolk

Oxfordshire

Buckinghamshire

3.

Hertfordshire

Essex

LONDON

4.

Wiltshire

Berkshire

Surrey

Kent

Hampshire

Sussex

1. Rutland
2. Huntingdonshire
3. Bedfordshire
4. Middlesex

Opposite: Tracing the past movement of malaria parasites is difficult—their existence was not discovered until 1880, so all previous data are indirect. By combining health records, estimates of past wetland extent, and early-twentieth-century malaria surveys by the British military, one can see that southeast England must have been seething with malaria. Of the fifty-nine birthplaces of the first Jamestown colonists that have been tracked by the historic-preservation group Preservation Virginia, thirty-five occurred in the regions the military identified as "extremely" or "more" favorable to *Plasmodium*. In addition, all the colonists passed through London and the malarial Thames delta en route. It seems almost certain that some brought the disease with them to Chesapeake Bay.

Inured to the cavalcade of suffering, residents viewed their circumstances with fatalistic cheer. (Readers of Charles Dickens will recall the stoicism of the fen-dwelling Gargerys in *Great Expectations*, raising the child Pip within a short walk of the "five little stone lozenges" that marked the resting places of his "five little brothers.") Traveling in feverish Essex County, writer Daniel Defoe met men who claimed to have "from five or six, to fourteen or fifteen wives." Explaining how this was possible, one "merry fellow" told Defoe that men thereabouts brought in wives from healthier inland precincts.

> [W]hen they took the young lasses out of the wholesome and fresh air, they were healthy, fresh and clear, and well; but when they came out of their native air into the marshes among the fogs and damps, there they presently changed their complexion, got an ague or two, and seldom held it above half a year, or a year at the most; and then, said he, we go to the uplands again, and fetch another.

The marshman laughed as he spoke, Defoe wrote, "but the fact, for all that, is certainly true."

In 1625 the bubonic plague engulfed England. More than

fifty thousand people died in London alone. Many of the urban wealthy fled into the malarial eastern marshes, with results later described by the satirical poet George Wither:

> In Kent, and (all along) on Essex side
> A Troupe of cruell Fevers did reside: . . .
> And, most of them, who had this place [London] forsooke,
> Were either slaine by them, or Pris'ners tooke . . .

In the end, Wither explained, "poorest beggers found more pitty here [London], / And lesser griefe, then richer men had there." The implication is mind-boggling: people who fled to vivax country would have been better off staying home with the bubonic plague.

Data are sketchy and incomplete, but according to the Brandeis University historian David Hackett Fischer about 60 percent of the first wave of English emigrants came from nine eastern and southeastern counties—the nation's *Plasmodium* belt. One example was the hundred-plus colonists who began Jamestown. Fifty-nine of their birthplaces are known, according to Preservation Virginia, the organization that backs Jamestown archaeology; thirty-seven were in malaria-ridden Essex, Huntingdonshire, Kent, Lincolnshire, Suffolk, Sussex, and London. Most of these men, one assumes, set off from higher, inland areas that were less malarial than the coastal wetlands. But many would have come from the marshes. Even those who didn't come from the malaria zone usually passed through it just before departure, their ships waiting for weeks or months at Sheerness, a Kent harbor town near the mouth of the Thames that was a malaria center. Other ships waited at the almost equally pestilential Blackwall, east of London on the same river.

People in malarial paroxysms would have been unlikely candidates for an arduous sea voyage. But *Plasmodium vivax*, one recalls, can hide itself inside the apparently healthy. Colonists could board a ship without symptoms, land in Chesapeake Bay tobacco country, and then be struck by the teeth-chattering chills and sweat-bursting fevers of malaria. At which point, alas, they could unknowingly pass the parasite to every mosquito that bit them.

"In theory, one person could have established the parasite in the entire continent," said Andrew Spielman, a malaria researcher at the Harvard School of Public Health. Almost certainly many of the *tassantassas* at Jamestown were infectious. At some point one of them was bitten by *Anopheles quadrimaculatus*, a cluster of five closely related mosquitoes that is the East Coast's primary malaria vector. "It's a bit like throwing darts," Spielman told me before his death in 2006. "Bring enough sick people in contact with enough mosquitoes in suitable conditions, and sooner or later you'll hit the bull's-eye—you'll establish malaria."

By 1657 the governor of Connecticut colony, John Winthrop, was recording cases of tertian fever in his medical journal. Winthrop, a member of the Royal Society, was one of the most careful scientific observers in New England. "If he said he saw tertian fever, he probably was seeing tertian fever," said Robert C. Anderson, the genealogist who is transcribing Winthrop's medical journal. More than that, Anderson told me, the existence of malaria in the 1650s suggests a date of introduction before 1640—after that year, political convulsions in England shut down emigration to New England for decades. "There were few colonists to bring it over," Anderson said. If *Plasmodium vivax* had come to Connecticut by, say, 1635, I asked Spielman, could one make any inferences about Virginia? "New England is *cold*," he said. "It's hard to believe that malaria would have established itself there before Virginia." Could the parasite have invaded Chesapeake Bay as early as the 1620s? "Given that hundreds or thousands of people from malaria zones came into the area, I wouldn't have trouble believing that," he said. "Once malaria has a chance to get into a place, it usually gets in fast."

Indeed, malaria may have come in *before* 1620. Conditions for the disease were perfect between 1606 and 1612, when tidewater Virginia was struck by drought. (I mentioned the drought in the last chapter.) *A. quadrimaculatus* is happy when wet areas get dry. "In drought years little tributary streams turn into a series of pools," explained David Gaines, public-health entomologist at the Virginia Department of Health. The larvae "thrive in that kind of environment." Quads, as entomologists call them, prefer to breed in open areas rather than shaded forests. After the peace created by Pocahontas's marriage in 1614, colonists cleared land for tobacco—making the environment, Gaines told me, "more quad-friendly, because it would have created those little open pockets of water they love." The *tassantassas* were issuing "an invitation for malaria," he said. "In my experience, malaria takes up invitations right away." If *Plasmodium* arrived with the first colonists, it could help explain, along with salt poisoning, why they were so often described as listless and apathetic; they had malaria.*

Malaria's precise date of arrival will always lie in the realm of speculation. What is clear is that malaria rapidly made itself at home in Virginia. It became as inescapable there as it was in the English marshes—a constant, sapping part of life.

When London investors shipped people to Virginia, Governor George Yeardley warned in 1620, they "must be content to have littell service done by new men the ffirst yeare till they be seasoned"—*seasoning* being the term for the period in which newcomers were expected to battle disease. The prolonged incapacitation of recent migrants was taken for granted. Jamestown minister Hugh Jones wrote a pamphlet in 1724 describing Virginia to Britons. The colony's climate, he incorrectly explained, causes chills and fever, "a severe Fit of which (called a *Seasoning*) most expect, some time after their Arrival in that Climate." Seasoning often was a path to the cemetery; during Jamestown's first half

---

* Early arrival of the parasite could help explain, too, why Opechancanough never expelled the colonists, even after almost wiping them out in 1622. Debilitated by disease, the Powhatan might have had difficulty mounting a sustained war. Unhappily, these intriguing speculations have the disadvantage of having no empirical support.

century, as many as a third of new arrivals died within a year of disembarking. After that, Virginians learned by trial and error to live with vivax, avoiding marshes and staying indoors at dusk; those with acquired immunity carefully tended the sick, most of whom were children as in Africa today. Seasoning deaths fell from 20 or 30 percent around 1650 to 10 percent or lower around 1670—a considerable improvement, but still a level that represented much suffering.

Landon Carter had a prosperous Virginia plantation about sixty miles north of Jamestown. A devoted father, Carter agonized as malaria repeatedly hit his family in the summer and fall of 1757. Worst affected was his infant daughter Sukey, racked by chills and fever in the classic tertian pattern. Like Samuel Jeake, Carter recorded her struggle in a diary:

Dec. 7: Sukey lookt badly all this evening with a quick Pulse.
Dec. 8: 'Tis her usual Period of attack which is now got to every Fortnight. . . . Seems brisk and talkt cheerfully. Her fever not higher.
Dec. 9: Continues better though very pale.
Dec. 10: Sukey a fever early and very sick at her stomach and head ach. This fever went off in the night.
Dec. 11: The Child no fever to day but I thought her pulse a little quick at night.
Dec. 12: Sukey's fever rose at 1 in the night. . . . This Child dangerous ill at 12, dead pale and blue. . . .
Dec. 13: Sukey's fever kept wearing away Yesterday till one in the night when she was quite clear.

To live in Virginia, a heartworn Carter wrote that day, "it is necessary that man should be acquainted with affliction, and 'tis certainly nothing short of it to be confined a whole year in tending one's sick Children. Mine are now never well."

Sukey died the following April, short of her third birthday.

ABOUT-FACE

Malaria had impacts beyond the immediate sufferings of its victims. It was a historical force that deformed cultures, an insistent nudge that pushed societies to answer questions in ways that today seem cruel and reprehensible. Consider the seventeenth-century English entrepreneurs who wanted to make money in North America. Because Chesapeake Bay had no gold and silver, the best way to profit was to produce something else that could be exported to the home country. In New England, the Pilgrims depended on selling beaver fur. In Chesapeake Bay, the English settled on tobacco, for which there was huge demand. To satisfy that demand, the colonists wanted to expand the plantation area. To do that, they would have to take down huge trees with hand tools; break up soil under the hot sun; hoe, water, and top the growing tobacco plants; cut the heavy, sticky leaves; drape them on racks to dry; and pack them in hogsheads for shipping. All of this would require a lot of labor. Where could the colonists acquire it?

Before answering this question make the assumption, abundantly justified, that the colonists have few moral scruples about the answer and are concerned only with maximizing ease and profit. From this point of view, they had two possible sources for the required workforce: indentured servants from England and slaves from outside of England (Indians or Africans). Servants or slaves: which, economically speaking, was the best choice?

Indentured servants were contract laborers recruited from England's throngs of unemployed. Because the poor could not afford the costly journey across the sea, planters paid for the voyage and servants paid off the debt by working for a given period, typically four to seven years. After that, indentured servants were free to claim their own land in the Americas. Slavery is harder to define, because it has existed in many different forms. But its essence is that the owner acquires the right to coerce labor from slaves, and slaves never gain the right to leave; they must work and obey until they die or are freed by their owners. Indentured servants are members of society, though at a low rank. Slaves are usually not considered members of society, either because they were born far away or because they somehow have forfeited their social standing, as in the occasional English practice of turning convicts into slaves.

During the last quarter of the seventeenth century, England chose slaves over servants—indeed, it became the world's biggest slaver. So well known nowadays is the English embrace of slavery that the idea of another path is hard to grasp. But in many respects the nation's turn to bondage is baffling—the institution has so many inherent problems that economists have often puzzled over why it exists. More baffling still is the form that bondage took in the Americas: chattel slavery, a regime much harsher than anything seen before in Europe or Africa.

On the simplest level, slaves were more expensive than servants. In a well-known study, Russell R. Menard of the University of Minnesota tallied up the prices in Virginia and Maryland of slaves and servants whose services had to be sold after their masters' deaths. In the last decades of the seventeenth century, the average price of a prime-age male African slave was £25. Meanwhile, the servants' contracts typically cost about £10. (Technically, I should say that Menard discovered the price was *equivalent to* £25 and £10, because coins were scarce and even illegal in colonial Chesapeake Bay, and people paid their bills with tobacco.) At that time, £25 was a substantial sum: about four years' pay for the typical hired worker in England. The servant was substantially cheaper.

To be sure, servants would eventually be able to leave their master's employ, lowering their value (attempting to take this into account, Menard looked only at servants with more than four years remaining on their contracts). But the longer period of service one could expect from a slave still would not justify slavery economically, the great economist Adam Smith argued.

An inherent flaw with slavery, he maintained, is that slaves made unsatisfactory workers. Because they were usually from distant cultures, they often didn't speak their owners' language and could be so unfamiliar with their owners' societies that they would have to be trained from scratch (Africans, for example, knew only tropical forms of agriculture). Worse, they had every incentive to escape, wreak sabotage, or kill their owners, the people who were depriving them of liberty. Indentured servants, by contrast, spoke the same language, accepted the same social norms, and knew the same farming methods. And their contracts were for a limited time, so they had little reason to run away (unless they thought the planter was going to cheat). Because willing hands are more likely to do their jobs well, Smith reasoned in *The Wealth of Nations*, "the work done by freemen comes cheaper in the end than that performed by slaves." All else being equal, he argued, economics suggests that planters should have chosen the cheaper, easier, less threatening alternative: servants from Europe.

Smith, who hated slavery, was trying to prove that something he detested was not only immoral, but foolish economically. Slavery, in his view, was largely the irrational product of humankind's "love to domineer." But he also believed that people try to find ways around economic problems that stand in the way of their desires. Just as one would expect, slave owners throughout history have created incentives for their slaves to work efficiently: paths to liberty. Work hard and true, masters in effect said, and you will eventually be allowed to walk away. Often, too, slaves were assigned tasks that brought some satisfaction, as in the case of African or Roman armies made up of captive soldiers— the slaves had switched sides, so to speak, but their lives were unchanged in many ways, and there was always the prospect of earning glory.

Slavery in the Americas, though, was something else: a lifelong sentence, in most cases, to awful work in brutal conditions without hope of winning freedom. Every one of Smith's disincentives for effective work was present as rarely before; none of the workarounds developed in the past were employed. The regime was so brutal that it should have generated constant shirking, sabotage, and strife—and, indeed, slaveholder records are endless threnodies of complaint and fear. Why did it arise?

Of all the nations in western Europe, moreover, England would be the last that one would expect to take up this especially brutal form of bondage, because opposition to slavery was more common there than the rest of Europe. If the continent had an antislavery culture, in fact, it was England. This was less a tribute to the nation's moral advancement than an enraged response to the constant targeting of her ships by Barbary pirates, who from the sixteenth to the eighteenth century enslaved tens of thousands of English sailors, soldiers, and merchants. Based in northwest Africa, these Muslim corsairs prowled as far north as the English Channel, ransacking seaside villages and seizing ships at anchor; in just ten days, the mayor of Plymouth complained in 1625, buccaneers lurking outside the harbor took twenty-seven vessels. (Inviting charges of hypocrisy, England lionized Francis Drake, who terrorized Spanish colonies in a similar fashion.) Most English captives were sent to the galleys; many were forcibly converted to Islam; others disappeared into slave caravans bound across the deserts to Ottoman Egypt or sub-Saharan Africa. In those days Algiers alone often held 1,500 English slaves; the Moroccan town of Salé had 1,500 more. Some were sold to Spain and Portugal. Escapees published lurid memoirs of their years under the lash, inflaming the public; churchmen denounced Muslim slavery in the pulpit and took up collections in church to ransom captives. Political leaders, Protestant ministers, and legal experts alike vehemently proclaimed freedom as an English birthright and condemned the pagans and papists (Moroccans and Spaniards) who enslaved them.

Slavery had been widespread in England in medieval times, as it was in the rest of Europe. In Spain and Portugal, beset by conflict with Islam and short of labor for sugar plantations, it continued to be a useful enterprise. (I discuss this further in Chapter 8.) In England, though, it became exceptional—not actually illegal, but rare—for political reasons, for the economic reasons described by Smith, and because slavery as an institution had little appeal in a nation aswarm with mobs of unemployed workers. Publicly outraged by bondage and with no domestic slave industry to protect, the English were Europe's least likely candidates for slavemasters.

In consequence, the English colonies initially turned to indentured servants and largely avoided slaves. Indentured servants comprised between a third and a half of the Europeans who arrived in English North America in the first century of colonization. Slaves were rare—only three hundred lived in all of Virginia in 1650. By comparison, the few Dutch in New Amsterdam, the colonial predecessor to New York, had five hundred slaves. As more English ships came to North America, slaves slowly became more common.

Then, between 1680 and 1700, the number of slaves suddenly exploded. Virginia's slave population rose in those years from three thousand to sixteen thousand—and kept soaring thereafter. In the same period the tally of indentured servants shrank dramatically. It was a pivot in world history, the time when English America became a slave society and England became the dominant player in the slave trade.

What accounts for this about-face? Economists and historians have mulled it over for decades. It was not the lure of profits from the trade itself: the slave business was incredibly important as a historical force and moral stain but not all that important as an economic industry. At its height at the end of the eighteenth century, according to the historians David Eltis and Stanley L. Engerman, slave shipments "accounted for less than 1.5 percent of British ships, and much less than 3 percent of British shipping tonnage." Caribbean sugar, the main slave crop, then accounted for a bit less than 2.5 percent of British GDP, large but not overwhelmingly so; the textile industry, for instance, was more than six times bigger. (Slaves were producing raw materials, not the far more valuable finished industrial goods.)

Some have argued that England changed its collective mind because its American colonies were especially conducive to slavery—they had so much available real estate. Adam Smith predicted in *The Wealth of Nations* that laborers would see the available land around them and leave their jobs, "in order to become landlords themselves." They would hire other workers in turn, who would "soon leave them for the same reason they left their first master." Not for another century did other economists fully draw out the implications of Smith's idea. If employers constantly lost workers to the lure of cheap land, then they would want to restrict their freedom of movement. Bondage was the inevitable end result. Paradoxically enough, America's wide-open frontier was, from this perspective, an incitement to slavery.

On some level, this notion must be true; slavery wouldn't exist if employers didn't want to control workers' movements. But it doesn't explain why slavery was uncommon in the English colonies of New England and New York, which had abundant land, and common in the English colonies of Barbados and St. Kitts, Caribbean islands that had little. In consequence, many researchers turned to a second explanation: England's religious civil war in the mid-seventeenth century, part of the worldwide unrest associated with the Little Ice Age and the uncertainties in the silver trade. The conflict was disastrous; between 1650 and 1680 the country's population fell almost 10 percent. As economics would predict, the decline in the number of English workers drove up English wages, which inevitably increased the price necessary to lure indentured servants across the Atlantic. Meanwhile, the indentured servants who had finished their terms in Massachusetts, Virginia, and Carolina were establishing new plantations and seeking their own indentured servants, increasing demand, which, as one would expect, further lifted prices.

Again, this explanation must be true; any rise in the cost of indentured servants would necessarily make alternatives seem more attractive. But it doesn't explain why colonists chose the alternative they did: captive Africans. Planters could have found labor in Scotland—and, to a lesser extent, Ireland—which in different ways also had been thrown into turmoil by the English civil war. The Little Ice Age had piled on, making the sea too cold for cod, piling up snow in the hills, and walloping the populace with a series of bad harvests. In the worst period, between 1693 and 1700, the Scottish oat harvest failed in every year but one. Desperate Scots fled their homes in huge numbers. Thousands became mercenaries in Russia, Sweden, Norway, and the German principalities; thousands more set up shop in northern Ireland, setting off a cultural collision that endures to this day. Gangs of Scottish refugees roamed London streets, begging for work and food—obvious candidates, or so it would seem, for American colonies. English farmers had employed indigent Scots for centuries. Yet at the very time the supply of desperate Scots was increasing the colonists turned to captive Africans—people who couldn't speak the language, had no wish to cooperate, and cost more to transport. Why?

One way to examine the question would be to evaluate the fortunes of the biggest group of Scots who traveled to the Americas in those years: the Scottish colony in Panama. Organized by an ambitious huckster named William Paterson, the scheme proposed using Panama's strategic location to break Spain's near monopoly on the silk and silver trades. "Seated between the two vast oceans of the universe," Paterson rhapsodized, the colony would control "at least two-thirds of what both Indies [that is, silk-rich Asia and the silver-rich Americas] yield to Christendom." Scottish Panama, he promised, would become "arbitrator of the commercial world," a financial perpetual-motion machine that would endlessly spew riches as it demonstrated that "trade is capable of increasing trade, and money of begetting money to the end of the world."

Dazzled by this vision, more than 1,400 Scots subscribed to a joint-stock company, pledging what has been estimated at between a quarter and a half of the poor nation's available capital. In July 1698 five ships set sail with 1,200 colonists and a year's food supply. They landed on the Panamanian coast and set about clearing the forest to create the port of New Edinburgh. Just eight months later the ragged survivors—fewer than three hundred people—bolted for home, Paterson among them. They arrived just days after the departure of a second Panama expedition: four ships, 1,300 colonists. Nine months later it, too, fled. Not a hundred people made it home. Lost with the dead was every penny invested in the venture.

Calamity usually has many fathers, and Paterson's colony was no exception. Thinking to get started by trading with the local Indians, the Scots had stuffed their ships with the nation's finest woolen hose, tartan blankets, ornamental wigs, and leather shoes—25,000 pairs. Alas, it proved difficult to sell warm socks and itchy blankets in the tropics. Meanwhile, the hard equatorial rain rotted their stores and washed away all efforts to farm. As New Edinburgh grew desperate, William and Mary, rulers of England and Scotland, instructed their other colonies not to help, for fear of offending Spain. Spain for its part knew about the project and periodically attacked.

The main causes of the disaster, though, were malaria, dysentery, and yellow fever. Colonists' accounts record dozens of deaths a week from disease. The first time Spain assaulted New Edinburgh, its soldiers found four hundred fresh graves. The colony had been well supplied, blessed with an adequate water supply, and never troubled by its Indian neighbors. European and African disease had filled that cemetery.

Back in Scotland, the debacle of New Edinburgh set off riots—it had wiped out much of the nation's capital. At the time, England and Scotland remained separate nations despite sharing a monarch. England, the bigger partner, had been pushing a complete merger for decades. Scots had resisted, believing they would become an afterthought in a London-dominated economy. Now England promised to reimburse New Edinburgh's investors as

part of a union agreement. "Even some committed Scottish patri-
ots such as Paterson endorsed the Union Act of 1707," the historian J. R. McNeill wrote in *Mosquito Empires*, a pioneering history
of Caribbean epidemiology, ecology, and war. "Thus Great Britain was born, with assistance from the fevers of Panama."

Beyond that, New Edinburgh showed that Scots—and other
Europeans—died too fast in malarial areas to be useful as forced
labor. Individual Britons and their families continued to make
their own way to the Americas, to be sure, but businesspeople
increasingly resisted sending over large groups of Europeans.
Instead they looked for alternative sources of labor. Alas, they
found them.

"NO DISTEMPERS
EITHER EPIDEMICAL OR MORTAL"

The colony of Carolina was founded in 1670, when about two
hundred colonists from Barbados relocated to the banks of a
river that empties into Charleston Harbor (it was initially called
Charles Town, after the reigning king). Like Virginia, Carolina
was a commercial enterprise, founded by eight powerful English
nobles who hoped to take advantage of the now-established traffic to Virginia by redirecting some of it to the south. The proprietors intended to lease pieces of the colony to would-be planters,
realizing a profit without actually having to expend much effort
or money. Barbados, full of sugar plantations, was crowded.
Some of its English inhabitants, looking to acquire land, decided
to take a flyer on Carolina. Knowing of Virginia's labor problem,
the proprietors promised extra land to anyone who imported
indentured servants, as well as the servants themselves.

Whereas Jamestown had confronted a single Indian empire
under a strong leader, Carolina began amid a chaotic swirl of
native groups. Beginning in about 1000 A.D., hundreds of densely

packed towns—"Mississippian" societies, as archaeologists call them—arose in the Mississippi Valley and the Southeast. Ruled by powerful theocrats who lived atop great earthen mounds, they were the most technologically sophisticated cultures north of Mexico. For reasons that are not well understood, these societies fell apart in the fifteenth century. The disintegration was accelerated by the onset of European diseases. By the time Carolina came into existence, the fragments of Mississippian societies were coalescing into confederacies of allied communities—Creek, Choctaw, Cherokee, Catawba—that were jostling for power across the Southeast.

Slavery occurred in most Indian societies, but the institution differed from place to place. Among Algonkian-language societies like the Powhatan, for instance, slavery was usually a temporary state. Slaves were prisoners of war who were treated as servants until they were either tortured and slain, ransomed back to their original groups, or inducted into Powhatan society as full members. Occasionally, Jamestown's *tassantassas* were able to buy Indian captives for their fields, but they were not generally a source of labor either for the Powhatan or the English. South of Chesapeake Bay was a cultural border where Algonkian societies ran into the nascent confederacies, many of which spoke Muskogean languages. War captives also became slaves in the confederacies, but there slavery was both more common and longer-lasting—traditions dating back to the Mississippians, whose leaders viewed captives as symbols of power and vengeance. Slaves worked in fields, performed menial tasks, and could be given away as gifts; female slaves provided sexual services to honored male visitors (a gesture frequently misunderstood by Europeans, who thought that the Indians were offering their wives). When foreigners appeared in Carolina, the confederacies were endlessly willing to trade surplus captives for axes, knives, metal pots, and, above all, guns.

In the late seventeenth century, the new flintlock rifle was becoming available—the first European firearm that native people regarded as superior to their bows. The matchlocks John Smith brought to Virginia used a lever to lower a burning match onto a small pan of gunpowder; the resultant flash pushed the projectile down the barrel. Heavy and unrifled, matchlocks had to be braced on tripods; because soldiers had to carry around burning fuses to fire them, the weapons were unsuitable for beaver wetlands and almost useless in rain. In optimal conditions, matchlocks could shoot a deadly projectile farther than a bow. But in warfare, conditions are never optimal. Colonial records are replete with descriptions of *tassantassas* unhappily discovering that as a practical matter their weapons were outmatched by native bows—weapons with no moving parts, weapons that could get wet, weapons that could be fired in an instant. Flintlocks, by contrast, ignited the gunpowder by snapping a chunk of flint against a piece of steel, creating a spark. The spark ignited a small charge that in turn set off a bigger charge in the barrel. Smaller, lighter, and more accurate than matchlocks, they could be fired quickly and used in wet weather.

The southeastern confederacies, quickly understanding the new weapons' superiority, determined not to be outgunned, either by the English or their native rivals. An arms race ensued across the Southeast. To build up their stores of flintlocks, native people raided their enemies for slaves to sell—an action that required more firearms. Needing guns to defend themselves, they in turn staged their own slaving raids, selling the captives to Europeans in return for guns. Demand fed demand in a vicious cycle.

Despite the fears of the Virginia Company, Jamestown never was directly threatened by Spain or France. Carolina, closer to Spanish Florida and French Louisiana, had much more reason to worry; indeed, Spain tried to extinguish the colony within months of its founding. Carolina's leaders came up with an elegant scheme; they asked nearby native groups to provide them with slaves by raiding the Indians who were allied with Spain and France, destabilizing their enemies and reducing their labor shortage at the same time.

Economically speaking, indigenous slavery was a good deal for both natives and newcomers. In the Charleston market Indians sometimes could sell a single slave for the same price as 160 deerskins. "One slave brings a Gun, ammunition, horse, hatchet, and a suit of Cloathes, which would not be procured without much tedious toil a hunting," a Carolina slave buyer noted, perhaps with some exaggeration, in 1708. "The good prices The English traders give them for slaves Encourages them to this trade Extreamly."

"Good prices" from the Indian point of view, but cheap to the English. Indian captives cost £5–10, as little as half the price of indentured servants, according to the Ohio State University historian Alan Gallay, author of *The Indian Slave Trade* (2002), a widely lauded account of its rise and fall. More important, the annual cost of ownership was much lower, because slaves did not have to be released after a few years—the purchase price could be amortized over decades. Unsurprisingly, the colonists chose Indian slaves over European servants. A 1708 census, Carolina's first, found four thousand English colonists, almost 1,500 Indian slaves, and just 160 servants, the majority presumably indentured.

In time Carolina grew famous as a slave importer, a place where the slave ships arrived from Africa and the captives, dazed and sick, were hustled to auction. But for its first four decades the colony was mainly a slave *exporter*—the place from where captive Indians were sent to the Caribbean, Virginia, New York, and Massachusetts. Data on Indian shipments are scarce, because colonists, wanting to avoid taxes and regulations, shipped them on small vessels and kept few records. (The big slaving companies in Europe didn't have this choice.) From the fragmentary evidence, Gallay has estimated that Carolina merchants bought between thirty and fifty thousand captive Indians between 1670 and 1720. Most of these must have been exported, given the much lower number found by the Carolina census. In the same period, ships in Charleston unloaded only 2,450 Africans (some came overland from Virginia, though).*

Here notice a striking geographical coincidence. By 1700, English colonies were studded along the Atlantic shore from what would become Maine to what would become South Carolina. Northern colonies coexisted with Algonkian-speaking Indian societies that had few slaves and little interest in buying and selling captives; southern colonies coexisted with former Mississippian societies with many slaves and considerable experience in trading them. Roughly speaking, the boundary between these two types of society was Chesapeake Bay, not far from what would become the boundary between slave and non-slave states in the United States. Did the proximity of Indian societies with slaves to sell help grease the skids for what would become African slavery in the South? Was the terrible conflict of the U.S. Civil War a partial reflection of a centuries-old native cultural divide? The implication is speculative, but not, it seems to me, unreasonable.

In any case, the Indian slave trade was immensely profitable—and very short-lived. By 1715 it had almost vanished, a victim in part of its own success. As Carolina's elite requested more and more slave raids, the Southeast became engulfed in warfare, destabilizing all sides. Victimized Indian groups acquired guns and attacked Carolina in a series of wars that the colony barely survived. Working in groups, Indian slaves proved to be unreliable, even dangerous employees who used their knowledge of the terrain against their owners. Rhode Island denounced the "conspiracies, insurrections, rapes, thefts and other execrable

---

* These figures do not include Indians seized in other colonies. During a vicious Indian war in 1675–76, for instance, Massachusetts sent hundreds of native captives to Spain, Portugal, Hispaniola, Bermuda, and Virginia. And the French in New Orleans seized thousands more. Carolina was a bigger slaver than others, but every English colony in North America was in the same business, with or without the cooperation of local Indians.

crimes" committed by captive Indian laborers, and banned their import. So did Pennsylvania, Connecticut, Massachusetts, and New Hampshire. The Massachusetts law went out of its way to excoriate the "malicious, surly and revengeful" Indian slaves.

The worst problem, though, was something else. As in Virginia, malaria came to Carolina. At first the English had extolled the colony's salubrious climate. Carolina, one visitor wrote, has "no Distempers either Epidemical or Mortal"; colonists' children had "Sound Constitutions, and fresh ruddy Complexions." The colonists decided to use the warm climate to grow rice, then scarce in England. Soon after came reports of "fevar and ague"— rice paddies are notorious mosquito havens. Falciparum had entered the scene, accompanied a few years later by yellow fever. Cemeteries quickly filled. In some parishes, more than three out of four colonists' children perished before the age of twenty. As in Virginia, almost half of the deaths occurred in the fall. (One German visitor's summary: "in the spring a paradise, in the summer a hell, and in the autumn a hospital.")

Unfortunately, Indians were just as prone to malaria as English indentured servants—and more vulnerable to other diseases. Native people died in ghastly numbers across the entire Southeast. Struck doubly by disease and slave raids, the Chickasaw lost almost half their population between 1685 and 1715. The Quapaw (Arkansas) fell from thousands to fewer than two hundred in about the same period. Other groups vanished completely—the last few dozen Chakchiuma were absorbed by the Choctaw. The Creek grew to power by becoming, in the phrase of one writer, "the receptacle for all distressed tribes." It was God's will, Carolina's former governor observed in 1707, "to send unusual Sicknesses" to the Westo Indians, "to lessen their numbers; so that the English, in comparison to the Spaniard, have but little Indian Blood to answer for."

Naturally, the colonists looked for a different solution to their labor needs—one less vulnerable to disease than European servants or Indian slaves.

VILLA PLASMODIA

Like other cells, red blood cells are covered by a surface membrane with proteins, the long, chain-like molecules that are the principal constituents of our bodies. One of these proteins is the Duffy antigen. (The name comes from the patient on whose blood cells the protein was first discovered; an "antigen" is a substance recognized by the immune system.) The Duffy antigen's main function is to serve as a "receptor" for several small chemical compounds that direct the actions of the cell. The compounds plug into the receptor—think of a spaceship docking at a space station, scientists say—and use it as a portal to enter the cell.

The Duffy antigen is not especially important to red blood cells. Nonetheless, researchers have written hundreds of papers about it. The reason is that *Plasmodium vivax* also uses the Duffy antigen as a receptor. Like a burglar with a copy of the front-door key, it inserts itself into the Duffy antigen, fooling the blood cell into thinking it is one of the intended compounds and thereby gaining entrance.

Duffy's role was discovered in the early 1970s by Louis H. Miller and his collaborators at the National Institutes of Health's Laboratory of Parasitic Disease. To nail down the proof, Miller and his collaborators asked seventeen men, all volunteers, to put their arms into boxes full of mosquitoes. The insects were chockablock with *Plasmodium vivax*. Each man was bitten dozens of times—enough to catch malaria many times over. Twelve of the men came down with the disease. (The researchers quickly treated them.) The other five had not a trace of the parasite in their blood. Their red blood cells lacked the Duffy antigen—they were "Duffy negative," in the jargon—and the parasite couldn't find its way inside.

The volunteers were Caucasian and African American. Every Caucasian came down with malaria. Every man who didn't get malaria was a Duffy-negative African American. This was no coincidence. About 97 percent of the people in West and Central Africa are Duffy negative, and hence immune to vivax malaria.

Duffy negativity is an example of *inherited* immunity, available only to people with particular genetic makeups. Another, more famous example is sickle-cell anemia, in which a small genetic change ends up deforming the red blood cell, making it unusable to the parasite but also less functional as a blood cell. Sickle-cell is less effective as a preventive than Duffy negativity—it provides partial immunity from falciparum malaria, the deadlier of the two main malaria types, but its disabling of red blood cells also leads many of its carriers to an early grave.

Both types of inherited immunity differ from *acquired* immunity, which is granted to anyone who survives a bout of malaria, in much the way that children who contract chicken pox or measles are thereafter protected against it. Unlike the acquired immunity to chicken pox, though, acquired malaria immunity is partial; people who survive vivax or falciparum acquire immunity only to a particular strain of vivax or falciparum; another strain can readily lay them low. The only way to gain widespread immunity is to get sick repeatedly with different strains.

Inherited malaria resistance occurs in many parts of the world, but the peoples of West and Central Africa have more than anyone else—they are almost completely immune to vivax, and (speaking crudely) about half-resistant to falciparum. Add in high levels of acquired resistance from repeated childhood exposure, and adult West and Central Africans were and are less susceptible to malaria than anyone else on earth. Biology enters history when one realizes that almost all of the slaves ferried to the Americas came from West and Central Africa. In vivax-ridden Virginia and Carolina, they were more likely to survive and produce children than English colonists. Biologically speaking, they were fitter, which is another way of saying that in these places they were—loaded words!—genetically superior.

Racial theorists of the last century claimed that genetic superiority led to social superiority. What happened to Africans illustrates, if nothing else, the pitfalls of this glib argument. Rather than gaining an edge from their biological assets, West Africans saw them converted through greed and callousness into social deficits. Their immunity became a wellspring for their enslavement.

How did this happen? Recall that vivax, covertly transported in English bodies, crossed the Atlantic early, as I said; certainly by the 1650s, given the many descriptions of tertian fever, quite possibly before. Recall, too, that by the 1670s Virginia colonists had learned how to improve the odds of survival; seasoning deaths had fallen to 10 percent or lower. But in the next decade the death rate went up again—a sign, according to the historians Darrett and Anita Rutman, of the arrival of falciparum. Falciparum, more temperature-sensitive than vivax, never thrived in England and thus almost certainly was ferried over the ocean inside the first African slaves.

Falciparum created a distinctive pattern. Africans in Chesapeake Bay tended to die more often than Europeans in winter and spring—the result, the Rutmans suggested, of bad nutrition and shelter, as well as unfamiliarity with ice and snow. But the African and European mortality curves crossed between August and November, when malaria, contracted during the high mosquito season of early summer, reaches its apex. During those months masters were *much* more likely to perish than slaves—so much more that the overall death rate for Europeans was much higher than that for Africans. Much the same occurred in the Carolinas. Africans there, too, died at high rates, battered by tuberculosis, influenza, dysentery, and human brutality. Many fell to malaria, as their fellows brought *Plasmodium* strains they had not previously encountered. But they did not die as fast as Europeans.

Because no colonies kept accurate records, exact comparative death rates cannot be ascertained. But one can get some idea by looking at another continent with endemic malaria that Europe tried to conquer: Africa. (The idea that one can compare malaria rates in places separated by the Atlantic Ocean is in itself a mark of the era we live in, the Homogenocene.) Philip Curtin, one of slavery's most important historians, burrowed in British records to find out what happened to British soldiers in places like Nigeria and Namibia. The figures were amazing: nineteenth-century parliamentary reports on British soldiers in West Africa concluded that disease killed between 48 percent and 67 percent of them *every year*. The rate for African troops in the same place, by contrast, was about 3 percent, an order-of-magnitude difference. African diseases slew so many Europeans, Curtin discovered, that slave ships often lost proportionately more white crewmen than black slaves—this despite the horrendous conditions belowdecks, where slaves were chained in their own excrement. To forestall losses, European slavers hired African crews.

The disparity between European and African death rates in the colonial Americas was smaller, because many diseases killed Europeans in Africa, not just malaria and yellow fever. But a British survey at about the same time as the parliamentary report indicated that African survival rates in the Lesser Antilles (the southern arc of islands in the Caribbean) were more than three times those of Europeans. The comparison may understate the disparity; some of those islands had little malaria. It seems plausible to say that in the American falciparum and yellow fever zone the English were, compared to Africans, somewhere between three and ten times more likely to die in the first year.

For Europeans, the economic logic was hard to ignore. If they wanted to grow tobacco, rice, or sugar, they were better off using African slaves than European indentured servants or Indian slaves. "Assuming that the cost of maintaining each was about equal," Curtin concluded, "the slave was preferable at anything up to three times the price of the European."

Slavery and falciparum thrived together. Practically speaking, *P. falciparum* could not establish itself for long in Atlantic City, New Jersey; the average daily minimum temperature is above 66 degrees, the threshold for the parasite, for only a few weeks per year. But in Washington, D.C., just 120 miles south, slightly warmer temperatures let it become a menace every fall. (Not for

nothing is Washington called the most northern of southern cities!) Between these two cities runs the Pennsylvania-Maryland
border, famously surveyed by Charles Mason and Jeremiah
Dixon in 1768. The Mason-Dixon Line roughly split the East
Coast into two zones, one in which falciparum malaria was an
endemic threat, and one in which it was not. It also marked the
border between areas in which African slavery was a dominant
institution and areas in which it was not (and, roughly, the division between indigenous slave and non-slave societies). The line
delineates a cultural boundary between Yankee and Dixie that
is one of the most enduring divisions in American culture. An
immediate question is whether all of these are associated with
each other.

For decades an influential group of historians argued that
southern culture was formed in the cradle of its great plantations—
the sweeping estates epitomized, at least for outsiders, by Tara in
the movie *Gone with the Wind*. The plantation, they said, was an
archetype, a standard, a template; it was central to the South's
vision of itself. Later historians criticized this view. Big colonial
plantations existed in numbers only in the southern Chesapeake
Bay and the low country around Charleston. Strikingly, these
were the two most malarial areas in the British colonies. Sweeping drainage projects eliminated Virginia's malaria in the 1920s,
but coastal South Carolina had one of the nation's worst *Plasmodium* problems for another two decades. From this perspective,
the movie's Tara seems an ideal residence for malaria country:
atop a hill, surrounded by wide, smooth, manicured lawns, its
tall windows open to the wind. Every element is as if designed
to avoid *Anopheles quadrimaculatus*, which thrives in low, irregular, partly shaded ground and still air. Is the association between
malaria and this Villa Plasmodia style a coincidence? It seems
foolish to rule out the possibility of a link.

"What would be the attitudes of a population that had a relatively high rate of illness and short life expectancy?" asked the
Rutmans. Some have suggested that the reckless insouciance and

preoccupation with display said to be characteristic of antebellum southern culture are rooted in the constant menace of disease. Others have described a special calm in the face of death. Maybe so—but it is hard to demonstrate that southerners were, in fact, unusually rash or vain or stoic. Indeed, one could imagine arguing the opposite: that the steady, cold breath of mortality on south-erners' necks could make them timid, humble, and excitable.

A different point is more susceptible to empirical demonstra-tion: the constant risk of disease meant that the labor force was unreliable. The lack of assurance penalized small farmers, who were disproportionately affected by the loss of a few hands. Mean-

Tara (shown behind Scarlett O'Hara in this publicity image from *Gone with the Wind*) was created on a studio backlot. None-theless, it was a faithful image of the classic southern plantation. High on a nearly treeless hill, with tall windows to admit the breeze, it was ideally suited to avoid mosquitoes and the dis-eases that accompanied them.

Approximate limit of range
of Plasmodium falciparum

ANOPHELES
QUADRIMACULATUS

Gulf of Mexico

Atlantic
Ocean

ANOPHELES
ALBIMANUS

Caribbean Sea

Pacific
Ocean

ANOPHELES
DARLINGI

American
Anopheles

Range of all
malaria-carrying
Anopheles
species

Approximate limit of range
of Plasmodium falciparum

Opposite: More than four hundred species of mosquito belong to the genus *Anopheles*. Perhaps a quarter can transmit malaria, but only about thirty species are common vectors. Of these thirty, about a dozen exist in the Americas, the most important being *A. quadrimaculatus, A. albimanus,* and *A. darlingi.* Their habitat range and the average temperature go far to explain why the history of certain parts of the Americas—and not others—was dominated by malaria.

while, the Rutmans noted, "a large labor force insured against catastrophe." Bigger planters had higher costs but were better insulated. Over time, they gained an edge; smaller outfits, meanwhile, struggled. Accentuating the gap, wealthy Carolinian plantation owners could afford to move to resorts in the fever-free mountains or shore during the sickness season. Poor farmers and slaves had to stay in the *Plasmodium* zone. In this way disease nudged apart rich and poor. Malarial places, the Rutmans said, drift easily toward "exaggerated economic polarization." *Plasmodium* not only prodded farmers toward slavery, it rewarded big plantations, which further lifted the demand for slaves.

Malaria did not *cause* slavery. Rather, it strengthened the economic case for it, counterbalancing the impediments identified by Adam Smith. Tobacco planters didn't observe that Scots and Indians died from tertian fever and then plot to exploit African resistance to it. Indeed, little evidence exists that the first slave owners clearly understood African immunity, partly because they didn't know what malaria was and partly because people in isolated plantations could not easily make overall comparisons. Regardless of whether they knew it, though, planters with slaves tended to have an economic edge over planters with indentured servants. If two Carolina rice growers brought in ten workers apiece and one ended up after a year with nine workers and the other ended up with five, the first would be more likely to flourish. Successful planters imported more slaves. Newcomers imitated the practices of their most prosperous neighbors. The slave trade took off, its sails filled by the winds of *Plasmodium*.

Slavery would have existed in the Americas without the parasite. In 1641 Massachusetts, which had little malaria, became the first English colony to legalize slavery explicitly. During the mid-eighteenth century, the healthiest spot in English North America may have been western Massachusetts's Connecticut River Valley, according to an analysis by Dobson and Fischer. Malaria there was almost nonexistent; infectious disease, by the standards of the day, extremely rare. Yet slavery was part of the furniture of daily life—at that time almost every minister, usually the most important man in town, had one or two. About 8 percent of the inhabitants of the main street of Deerfield, one of the bigger villages in the valley, were African slaves.

On the other side of the hemisphere's malaria belt, the southern terminus of the habitat for *Anopheles darlingi*, the main South American vector for falciparum, is by the Rio de la Plata (Silver River), the border between Spanish and Portuguese America. South of the river is Argentina. With few mosquitoes to transmit *Plasmodium*, Argentina had little malaria. Yet, like Massachusetts, it had African slaves; between 1536, when Spain founded its first colony on the Rio de la Plata, and 1853, when Argentina abolished slavery, 220,000 to 330,000 Africans landed in Buenos Aires, the main port and capital.

On the other side of the mosquito border were the much bigger Brazilian ports of Rio de Janeiro and São Paulo, where at least 2.2 million slaves arrived. Despite the difference in size, southern Brazil and Argentina were demographically similar: in the 1760s and 1770s, when Spain and Portugal first systematically censused in their colonies, about half of the population in both areas was of African descent. Yet the impact of slavery in them was entirely different. Slavery was never critical to colonial Argentina's most important industries; colonial Brazil could not have functioned without it. Argentina was a society with slaves; Brazil was culturally and economically *defined* by slavery.

All American colonies, in sum, had slaves. But those to which the Columbian Exchange brought endemic falciparum malaria ended up with more. Falciparous Virginia and Brazil became slave societies in ways that non-falciparous Massachusetts and Argentina were not.

YELLOW JACK

In the 1640s a few Dutch refugees from Brazil landed on Barbados, the easternmost Caribbean island. Unlike the rest of the Caribbean, Barbados never had a large Indian population. English colonists moved in, hoping to capitalize on the tobacco boom. When the Dutch refugees arrived the island had about six thousand inhabitants, among them two thousand indentured servants and two hundred slaves. Tobacco had turned out not to grow particularly well on Barbados. The Dutch showed the colonists how to plant sugarcane, which they had learned during an ill-fated venture in Brazil. Europe, then as now, had a sweet tooth; sugar was as popular as it was hard to come by. Barbados proved to be good cane territory. Production rapidly expanded.

Sugar production is awful work that requires many hands. The cane is a tall, tough Asian grass, vaguely reminiscent of its distant cousin bamboo. Plantations burn the crop before harvest to prevent the knifelike leaves from slashing workers. Swinging machetes into the hard, soot-smeared cane under the tropical sun, field hands quickly splattered themselves head to foot with a sticky mixture of dust, ash, and cane juice. The cut stalks were crushed in the mill and the juice boiled down in great copper kettles enveloped in smoke and steam; workers ladled the resultant hot syrup into clay pots, where the pure sugar crystallized out as it cooled. Most of the leftover molasses was fermented and distilled to produce rum, a process that required stoking yet another big fire under yet another infernal cauldron.

The question as ever was where the required labor would

come from. As in Virginia, slaves then typically cost twice as much as indentured workers, if not more. But the Dutch West India Company, a badly run outfit that was desperate for cash, was willing to sell Africans cheap in Barbados. Slaves and indentured servants there were roughly the same price. As one would expect, the island's new sugar barons imported both by the thousands: the sweepings of English streets and luckless captives from Angolan and Congolese wars. Covered in perspiration and gummy cane soot, Europeans and Africans wielded machetes side by side. Then the Columbian Exchange raised the relative cost of indentured servants.

Hidden on the slave ships was a hitchhiker from Africa: the mosquito *Aedes aegypti*. In its gut *A. aegypti* carried its own hitchhiker: the virus that causes yellow fever, itself also of African origin. The virus spends most of its time in the mosquito, using human beings only to pass from one insect to the next. Typically it remains in the body for at most two weeks. During this time it drills into huge numbers of cells, takes over their functioning, and uses the hijacked genetic material to produce billions of copies of itself. These flood the bloodstream and are picked up by biting *aegypti*. For imperfectly understood reasons this cellular invasion usually has little impact on children. Adults are hit by massive internal bleeding. The blood collects and coagulates in the stomach. Sufferers vomit it blackly up—the signature symptom of yellow fever. Another symptom is jaundice, which gave rise to the disease's nickname of "yellow jack." (A yellow jack was the flag flown by quarantined ships.) The virus kills about half of its victims—43 to 59 percent in six well-documented episodes McNeill compiled in *Mosquito Empires*. Survivors acquire lifelong immunity. In Africa yellow fever was a childhood disease that inflicted relatively little suffering. In the Caribbean it was a dire plague that passed over Africans while ravaging Europeans, Indians, and slaves born in the islands.

The first yellow fever onslaught began in 1647 and lasted five years. Terror spread as far away as Massachusetts, which instituted its first-ever quarantine on incoming vessels. Barbados had more Africans and more Europeans per square mile than any other Caribbean island, which is to say that it had more potential yellow fever carriers and potential yellow fever victims. Unsurprisingly, the epidemic hit there first. As it began a man named Richard Ligon landed in Barbados. "We found riding at Anchor, 22 good ships," he wrote later,

> with boats plying to and fro, with Sails and oars, which carried commodities from place to place: so quick stirring, and numerous, as I have seen it below the bridge at *London*. Yet notwithstanding all this appearance of trade, the Inhabitants of the Islands, and shipping too, were so grievously visited with plague (or as killing a disease) that before a month was expired, after our arrival, the living were hardly able to bury the dead.

Six thousand died on Barbados alone in those five years, according to one contemporary estimate. Almost all of the victims were European—a searing lesson for the island's colonists. McNeill estimates that the epidemic "may have killed 20 to 50 percent of local populations" in a swathe from coastal Central America to Florida.

The epidemic didn't kill off the sugar industry—it was too lucrative. Incredibly, Barbados, an island of 166 square miles, was then on its way to making more money than all of the rest of English America. Meanwhile sugar had expanded to nearby Nevis, St. Kitts, Antigua, Montserrat, Martinique, Grenada, and other places. (Cuba had begun growing sugar decades before, but production was small; Spaniards were much too preoccupied by silver to pay attention.) A heterogeneous mass of English, French, Dutch, Spanish, and Portuguese was clearing these islands as fast as possible, sticking cane in the flatlands and cutting trees on the slopes for fuel. Deforestation and erosion were the nigh-unavoidable result; rainfall, no longer absorbed by vegetation, washed soil down the slopes, forming coastal marshes. In the not-too-distant future workers would be ordered to carry the soil in baskets back up the hills—"a true labor of Sisyphus," McNeill remarked in *Mosquito Empires*. McNeill quotes one Caribbean naturalist marveling at "the inconsideration or rather stupidity of west Indian planters in extinguishing many useful woods that spontaneously grow on those islands." Writing in 1791, the naturalist judged that many islands were "almost rendered unfit for cultivation."

Even the worst ecological mismanagement benefits some species. Among the winners in the Caribbean was *Anopheles albimanus,* the region's most important malaria vector. A resident of the bigger Caribbean islands and coastal areas in Yucatán and Central America, *A. albimanus* is a reluctant malaria host, hard for falciparum to infect and slow to pick up vivax (many mosquitoes have bacteria in their gut that inhibit the parasite). It likes to breed in coastal, algae-covered marshes under the open sun. Erosion and deforestation are its friends. Field experiments have shown

Sugar plantations denuded Barbados, as shown in the background of this photograph of workers' huts in the 1890s.

that *albimanus* can reproduce in huge numbers when it has favorable habitat. Given its preferences, the European move into the Caribbean must have marked the beginning of a golden age. As the mosquito population soared, *P. vivax* had more opportunities to overcome the mosquito's reluctance to host it. (Indeed, it may have beaten the mosquito while traveling with Colón; in addition to the reference to *çiçiones* in the admiral's second voyage, his son Hernán later claimed that "intermittent fever" appeared on his fourth voyage.) From the Caribbean, vivax malaria spread into Mexico. Falciparum came much later, the delay partly due to *A. albimanus*'s more complete resistance to the parasite.

Another beneficiary was *Aedes aegypti*, the yellow fever vector. *A. aegypti* likes to breed in small amounts of clear water near human beings; naval water casks are a well-known favorite. Sugar mills abounded with equivalent vessels: the crude clay pots used to crystallize sugar. Plantations had hundreds or thousands of these vessels, which were only used for part of the year and often broken. Today we know that *aegypti* likes to breed in the puddles that collect in the interior of cast-off automobile tires. Sugar pots were a seventeenth- and eighteenth-century equivalent. McNeill noted that the pots would have been full of sugary residue, fodder for the bacteria that *aegypti* larvae feed upon. Sugar plantations were like factories for producing yellow fever.

Incoming Europeans didn't know these details. But they were entirely aware that the Caribbean was, as historian James L. A. Webb wrote in a recent history of malaria, "a lethal environment for non-immunes."

Malaria percolated from the Caribbean into South America, and thence up the Amazon. The river has a plenitude of hosts: a 2008 survey of the Madeira River, an important Amazonian tributary, found no less than nine *Anopheles* mosquito species, all of which carried the parasite. The first Europeans to visit Amazonia described it as a thriving, salubrious place; malaria and, later, yellow fever turned many rivers into death traps. By 1782 the parasite was sabotaging expeditions into the upper reaches of the river basin. For two centuries the disease was a sometime, scattered thing: big stretches of Amazonia, depopulated by smallpox and slavery, had too few inhabitants to sustain the parasite. It may have been more common in the far western tributaries like the Madeira, because they experienced fewer Dutch and Portuguese slave raids, and thus had more people to infect. Malaria nearly killed French naturalist Alcide d'Orbigny in 1832 in the Madeira region, but a decade later another naturalist, the U.S. amateur William Henry Edwards, "encountered but one case" of it on the river, despite camping for days near its mouth.

Much worse was the northeastern bulge of South America, the region the geographer Susanna Hecht has called the Caribbean Amazon. Bounded to the south by the Amazon River in Brazil and to the west by the Orinoco River in Venezuela, it was a watery place that Arawak and Carib people controlled with sprawling networks of dikes, dams, canals, berms, and mounds. Large expanses of forest were managed for tree crops, especially the palms that in tropical places provide fruit, oil, starch, wine, fuel, and building material. Beneath the palms lay scattered patches of manioc (cassava). This landscape of gardens, orchards, and waterways served for centuries as a bridge between the interior and the islands. Such complex arrangements typically are supervised by strong, well-organized governments. Europeans certainly thought the Indians had them—it explained why their repeated efforts to seize this rich agricultural land were repelled. Only in the eighteenth century did the foreigners gain a foothold, aided by the introduction of European diseases: smallpox, tuberculosis, and influenza cleared the way for malaria. Indians retreated into the interior as Europeans seized the coast, creating sugar plantations in what eventually became, after much international squabbling, Guyane (French Guiana), Suriname (formerly Dutch Guiana), and Guyana (formerly British Guiana).

Archetypical may have been Guyane, which was formally acquired by France in a treaty in 1763. Initial colonization efforts proved so disastrous that the nation almost forgot its existence until three decades later, when a military-backed coup overthrew the parliament established by the French Revolution. The new dictatorship piled 328 unwanted deputies, clergymen, and journalists into small vessels and dumped them in the colony. *P. falciparum* greeted them on the shore. Within two years more than half were dead, either killed by malaria or sufficiently weakened by it to be slain by other ailments. Undeterred, the French state kept sending over criminals and undesirables. French prisoners had in the past served as galley slaves on special prison ships in the Mediterranean. After the steam engine made galleys obsolete convicts were dispatched to Guyane. Violent offenders ended up in the infamous prison on Devils Island, seven miles off the coast; the rest joined chain gangs of agricultural labor. Disease claimed so many that Guyane became known as a "dry guillotine"—a blade that killed without needing to wet itself with blood. Perhaps eighty thousand Frenchmen made the passage. Very, very few returned.

Unable to settle in disease zones, Europeans never established communities there. The ideal was offshore ownership. Europeans would remain in the safety of the home country while small numbers of onsite managers directed the enslaved workforce. Because captives would outnumber captors, intimidation and brutality would be necessary to keep the sugar mills grinding. In the realm of falciparum and yellow fever, sugar despotism became the rule: tiny bands of Europeans atop masses of transplanted Africans, angry or demoralized or stoic according to their characters.

Nothing is wrong with offshore ownership per se. If French wine makers buy wineries in California or U.S. wine makers buy wineries in Bordeaux or Burgundy, the acquisitions may sting local pride but are unlikely to have any larger effect on either nation. It is different if foreign wine makers buy *every* winery—or, stronger yet, if people thousands of miles away dominate *every* industry. One all-too-representative example: a single Liverpool firm, Booker Brothers, controlled three-quarters of British Gui-

The French artist Édouard Riou, now best known for his illustrations of Jules Verne, traveled to France's colony of Guyane
in 1862–63. A visit to the prison islands produced this image of
the sea burial of a convict, presumably a victim of malaria or
yellow fever.

ana's economy for almost a century. All the profits ended up
on the other side of the ocean. So did all of the entrepreneurial,
managerial, and technical expertise. Locals provided only labor.
Indeed, they were punished if they tried to do anything else.

As the economists Acemoglu, Johnson, and Robinson noted,
distant and disconnected owners had little interest in building the institutions necessary to maintain complex societies:
schools, highways, sewers, hospitals, parliaments, legal codes,
agricultural-extension agencies, and other governmental systems.
In places with a full array of functioning institutions, locals can
compete economically with foreigners by developing new technologies and new business methods. In extractive states, locals
never got the chance. Most of the English colonists who went to
Virginia or Australia were servants or convicts, the lowest tiers

in the social pyramid. But despite their bottom ranking their status as citizens gave them some ability to use the institutions of the homeland to push back when their leaders tried to oppress them. (Australia's convicts, for example, began winning lawsuits against their would-be abusers almost as soon as they landed.) The slaves in extractive states had no such ability to tap into these institutions. Indeed, elites actively sought to cut off their access. A particular worry was education; echoing many in British Guiana, Booker president Josiah Booker denounced the notion of teaching his company's employees to read because it would encourage them to aspire "far above their station in life." Wrong ideas in the wrong people's hands could put the elites' political power at risk.

History suggests, Acemoglu, Johnson, and Robinson wrote, that industrialization cannot occur without "both investments from a large number of people who were not previously part of the ruling elite and the emergence of new entrepreneurs." Both are next to impossible in extractive states. Over the decades, reformers tried to counteract the system's effects. Missionaries provided education for Guyana's children; the British Anti-Slavery Society thundered unceasingly against mistreatment, launched investigations, and provided aid. "Jock" Campbell, the visionary head of Booker Brothers' corporate successor, spent decades improving sugar workers' conditions. The reformers did everything but change the basic extractive system. When Guyana gained formal independence in 1966, 80 percent of its export earnings were controlled by three foreign companies, one of them Campbell's. The new nation had just one university, a night school established three years before.

WAR AND MOSQUITOES

In malaria zones, the primary victims are children. Adults as a rule have already had the disease and become immune upon survival. The adults who have most to fear are recent arrivals—a lesson that was learned in the Americas again and again, perhaps most dramatically during the U.S. Civil War. Much of the war was fought in the South by troops from the North. Crossing the Mason-Dixon Line, Yankees broke an epidemiological barrier. The effects were enormous.

In July 1861, three months after the conflict began, the Union's Army of the Potomac marched from Washington to the Confederate capital of Richmond, Virginia. It was repulsed at what became known to Yankees as the Battle of Bull Run and to Confederates as the Battle of Manassas. After fleeing to Washington, the generals dragged their feet about further action. President Lincoln railed against their pusillanimity, but they may have had a point. In the year after Bull Run, at least a third of the Army of the Potomac suffered from what army statistics describe as remittent fever, quotidian intermittent fever, tertian intermittent fever, quartan intermittent fever, or congestive intermittent fever—terms generally taken today to mean malaria. Union troops in North Carolina fared still worse. An expeditionary force of fifteen thousand landed at Roanoke Island in early 1862, and spent much of the war enforcing a naval blockade from a fort on the coastline. The air at dusk shimmered with *Anopheles quadrimaculatus*. Between the summer of 1863 and the summer of 1864, the official annual infection rate for intermittent fevers was 233 *percent*—the average soldier was felled two times or more.

From the beginning the Union army was bigger and better supplied than the Confederate army. As at Bull Run, though, the North lost battle after battle. Incompetent generalship, valiant opponents, and long supply lines were partly to blame. But so was malaria—the price of entering the *Plasmodium* zone. During the war the annual case rate never dropped below 40 percent. In one year *Plasmodium* infected 361,968 troops. The parasite killed few directly, but it so badly weakened them that they succumbed readily to dysentery or measles or what military doctors then called "chronic rheumatism" (probably a strep infection). At least 600,000 soldiers died in the Civil War, the most deadly conflict in U.S. history. Most of those lives were not lost in battle. Disease killed twice as many Union troops as Confederate bullets or shells.

Malaria affected the course of the war itself. Sick soldiers had to be carried in litters or shipped out at considerable cost. With so many sick for so long the resource drain was constant. Confederate generals did not control malaria or even know what it was, but it was an extra arrow in their quiver. *Plasmodium* likely delayed the Union victory by months or even years.

In the long run this may be worth celebrating. Initially the North proclaimed that its goal was to preserve the nation, not free slaves; with few dissenting votes, Congress promised rebel states that "this war is not waged" for the "purpose of overthrowing or interfering with [their] rights or established institutions," where "established institutions" was taken to mean slavery. The longer the war ground on, the more willing grew Washington to consider radical measures. Should part of the credit for the Emancipation Proclamation be assigned to malaria? The idea is not impossible.

Plasmodium's contribution to the birth of the United States was stronger still. In May of 1778 Henry Clinton became commander in chief of the British forces during the Revolutionary War. Partly on the basis of inaccurate reports from American exiles in London, the British command believed that the Carolinas and Georgia were full of loyalists who feared to announce their support of the home country. Clinton decided upon a "southern strategy." He would send a force south, which would hold the region long enough to persuade the silent loyalist majority to declare its support for the king. In addition, he promised, slaves who fought for his side would be freed. Although Clinton didn't know it, he was leading an invasion of the malaria zone.

English troops were not seasoned; indeed, two-thirds of the troops who served in 1778 were from malaria-free Scotland. To be sure, many British soldiers had by 1780 spent a year or two in the colonies—but mostly in New York and New England, north

Although almost forgotten today, yellow fever was a terror from the U.S. South to Argentina until the 1930s, when a safe vaccine was developed. This cartoon illustrated a magazine article about an 1873 outbreak in Florida.

of the *Plasmodium* line. By contrast, the southern colonists were seasoned; almost all were immune to vivax and many had survived falciparum.

British troops successfully besieged Charleston in 1780. Clinton left a month later and instructed his troops to chase the Americans into the hinterlands. The man he put in charge of the foray was Major General Charles Cornwallis. Cornwallis marched inland in June, high season for *Anopheles quadrimaculatus*. By autumn, the general complained, disease had "nearly ruined" his army. So many men were sick that the British could barely fight. Loyalist troops from the colonies were the only men able to march. Cornwallis himself lay feverish while his Loyalists lost the Battle of Kings Mountain. "There was a big imbalance. Cornwallis's army simply melted away," McNeill told me.

Beaten back by disease, Cornwallis abandoned the Carolinas and marched to Chesapeake Bay, where he planned to join another British force. He arrived in June 1781. Clinton ordered him to take a position on the coast, where the army could be transported to New York if needed. Cornwallis protested: Chesapeake Bay was famously disease ridden. It didn't matter; he had to be on the coast if he was to be useful. The army went to Yorktown, fifteen miles from Jamestown, a location Cornwallis bitterly described as "some acres of an unhealthy swamp." His camp was between two marshes, near some rice fields.

To Clinton's horror and surprise, a French fleet appeared off Chesapeake Bay, sealing in Cornwallis. Meanwhile, Washington marched south from New York. The revolution was so short of cash and supplies that his army had twice mutinied. Nonetheless an opportunity had arisen. The British army was unable to move; Cornwallis later estimated that only 3,800 of his 7,700 men were fit to fight. McNeill takes pains to credit the bravery and skill of the revolution's leaders. But what he wryly referred to as "revolutionary mosquitoes" played an equally critical role. "*Anopheles quadrimaculatus* stands tall among the Founding Fathers," he said to me. With Cornwallis's troops falling to the Columbian Exchange in ever-greater numbers, the British army surrendered, effectively creating the United States, on October 17, 1781.

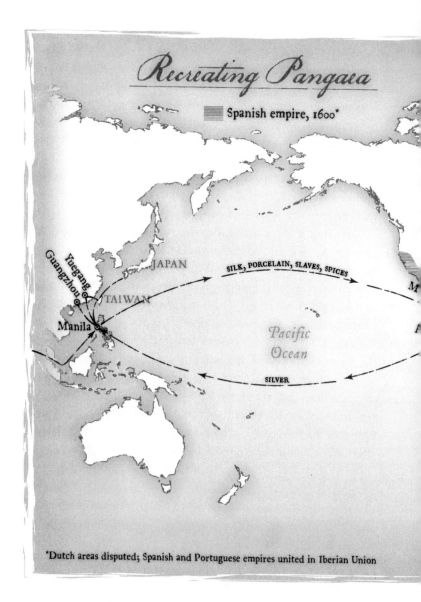

Recreating Pangaea

Spanish empire, 1600*

JAPAN

Yuegang
Guangzhou
TAIWAN

Manila

SILK, PORCELAIN, SLAVES, SPICES

M

A

Pacific
Ocean

SILVER

*Dutch areas disputed; Spanish and Portuguese empires united in Iberian Union

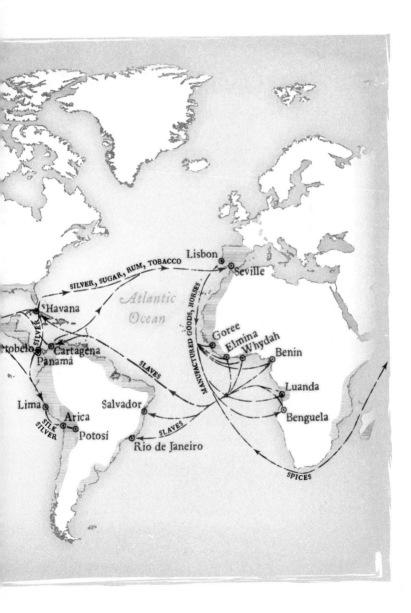

SILVER, SUGAR, RUM, TOBACCO

Atlantic
Ocean

Lisbon

Seville

Havana

SILVER

MANUFACTURED GOODS, HORSES

Goree
Elmina
Whydah
Benin

tobelo
Cartagena
Panamá

SLAVES

Luanda

Lima
Arica
Potosí

SILK
SILVER

Salvador

Benguela

SLAVES

Rio de Janeiro

SPICES

PART TWO

Pacific Journeys

4

Shiploads of Money
(Silk for Silver, Part One)

"THAT EXTRA LITTLE EFFORT"

Vastness was its greatest characteristic, with wonder close behind. The vastness—intimidating, confounding, beyond credence—spoke clearly from a hundred miles away. It is said that kings in their palaces looked over the ocean to see a new mountain range on the horizon: wide-bellied ships by the hundred, rigged fore and aft, soldiers massed at their bulwarks. Strange warlike banners snapped from the topgallants. The armada was larger than any before or since. It must have seemed *geographic*. Wonder attended its sails, followed by capitulation and obeisance. These were the great maritime expeditions sponsored by the Ming emperor Yongle in the early fifteenth century. Such a mark did they leave that some historians believe they were the font of the stories of Sinbad the Sailor.

Built in enormous dry docks, encrusted with precious metals, replete with technical innovations—double hulls, watertight compartments, rust-proofed nails, mechanical bilge pumps—that Europe would not discover for a century, the Chinese ships were

To celebrate the 2010 Olympics, China displayed this exact copy of Zheng He's flagship. Six centuries after the original was built, the ship still was large enough to dazzle crowds.

marvels for their time. The flagship of their commander, Zheng He, was more than 300 feet long and 150 feet wide, the biggest wooden vessel ever constructed. Records claim it had nine masts. Zheng's grandest expedition had 317 ships, an amazing figure even now. The Spanish Armada, then the largest fleet in European history, consisted of about 130; the biggest was half the size of Zheng's flagship.

Zheng himself was among the more unlikely figures to grace Chinese history. Strikingly tall and powerfully built, a Muslim from the backlands, he was captured as a child in 1381 during one of the Yuan dynasty's last battles against the invading Ming. (For a synopsis of Chinese dynastic history, see the chart on p. 162.) Standard treatment by the Ming of enemy boys was castration. The emasculated Zheng was pressed into service at the Ming court and gained a reputation for shrewdness and competence. Displaying his eye for the main chance, he jumped to support a coup in which the monarch's uncle seized power from his nephew. The usurper became the Yongle emperor.* Zheng became one of his most trusted lieutenants. When the ambitious sovereign planned a series of sea expeditions, he put his favorite eunuch in charge.

The voyages began in 1405 and ended in 1433 and took Zheng across the Indian Ocean as far as southern Africa. The Yongle emperor viewed them as a way to throw his weight around, and they were powerfully effective. During these voyages Zheng's fleet subjugated a misbehaving Chinese enclave in Sumatra; intervened in a civil war in Java; invaded Sri Lanka and took its captured ruler to China; and wiped out bandits in Sumatra. Even where no swords were unsheathed Zheng's armadas were a political triumph, scaring the wits out of every foreign leader who saw them. But the voyages were not followed up. They had become a target in political infighting—one bureaucratic faction championing them, another trying to take down the first by decrying their expense. Yongle's son and successor aligned with the faction that opposed his father's policies. He canceled the grand naval adventures on the day he ascended to the throne. Ultimately almost all records of Zheng's travels were suppressed. China didn't again send ships so far outside its borders until the nineteenth century.

Many researchers have seen the failure to continue as emblematic of a fatal insularity in Chinese society. "Why did China not make that extra little effort that would have taken it around the southern end of Africa and up into the Atlantic?" asked Landes, the Harvard historian, in his *Wealth and Poverty of Nations*. Landes's answer: "The Chinese lacked range, focus, and above all, curiosity." Hobbled by Confucian ideology, arrogant and complacent, China was "a reluctant improver and a bad learner."

---

* By convention emperors were referred to not by their personal names, but by the names of their reigns. For example, the usurping uncle, born Zhu Di, chose Yongle ("perpetual happiness") as his title, and thus became the Yongle emperor: the ruler during the Yongle period.

The European Miracle, University of Melbourne historian Eric Jones's celebrated account of the West's climb to political dominance, similarly attributes China's rejection of foreign adventures to "empty cultural superiority" and "self-engrossment." After Zheng He, the empire "retreated from the sea and became inward-looking." China, the McGill University political scientist John A. Hall charged in *Powers and Liberties: The Causes and Consequences of the Rise of the West,* "was stuck in the same stage for over two thousand years, while Europe, in comparison, progressed like a champion hurdler." Bubbling with entrepreneurial vim, Portugal, the Netherlands, Spain, and Britain dragged sclerotic China into the rough-and-tumble of the outside world.

Other scholars disagree with this image of Chinese passivity. Nor do they believe the shutdown of Zheng He's voyages exemplified a cultural lack of curiosity or drive. No matter how far the admiral traveled from home, these writers note, he never encountered a nation richer than his own. Technologically speaking, China was so far ahead of the rest of Eurasia that foreign lands had little to offer except raw materials, which could be obtained without going to the bother of dispatching gigantic flotillas on lengthy journeys. Beijing easily could have sent Zheng past Africa to Europe, observed the George Mason University political scientist Jack Goldstone. But the empire stopped long-range exploration "for the same reason the United States stopped sending men to the moon—there was nothing there to justify the costs of such voyages."

In a broader sense, though, the question remains. Zheng's voyages were an exception to a longer, more consequential trend. During most of the Ming dynasty (1368–1644), Beijing issued edicts that effectively banned private sea trade. The Yongle emperor and a few other rulers opened it up, but they were exceptions; as a rule, the dynasty clamped down on international exploration and exchange. So draconian were the prohibitions that in 1525 the court ordered coastal officials to destroy all private seagoing vessels.

As puzzling in today's context as the shutdown was its rever-
sal. Fifty years after the demolition order another emperor
reversed course. With the court bureaucracy's reluctant blessing,
a new generation of Chinese ships went on the waters. Soon the
Ming had been drawn into a worldwide network of exchange.
In a trice, the Chinese economy became enmeshed with Europe
(a place previously regarded as too poor to be worth bothering
with) and the Americas (a place the emperors hadn't known
existed).

The court had long feared that unrestricted trade would lead
to chaos. Indeed, it did have catastrophic by-effects, though not
those predicted by imperial bureaucrats. I've already described
how the Columbian Exchange across the Atlantic shaped economic and political institutions. Now I turn to the Pacific, where
the economic exchange established itself first and greased the
skids for the Columbian Exchange. Accordingly, this chapter concerns economics and politics. The next chapter describes their
ecological consequences; an environmental convulsion that had
dire economic and political consequences for China—among
them, in part, its later collapse before the West.

"MERCHANTS WERE PIRATES, PIRATES WERE MERCHANTS"

Why did China let in the flood? The decision was driven by two
factors, one largely political, one largely economic. The political
factor was the Ming desire to enhance the power of the state. Beijing's prohibition on private trade had less to do with an abhor-
rence of trade than a desire to control it for the dynasty's benefit.
Unhappily, the attempt backfired—the reaction to the trade ban
ended up weakening government control, rather than strength-
ening it. When Beijing finally admitted this, it abandoned its previous policy. Further driving the emperors to make this decision
was the economic factor: China had severe money woes. Liter-

RECENT CHINESE DYNASTIES

Tang	618–907 A.D.
Chaotic interregnum	907–960
Song	960–1279
Yuan (Mongols)	1279–1368
Ming	1368–1644
Qing (Manchus)	1644–1911

Chinese history is divided into dynasties, beginning before 2000 B.C. The tally here is simplified; the Song dynasty, for instance, is usually split into two eras (it fell apart after an invasion and regrouped with its power center in a different place). And this list doesn't show the messy transitions between dynasties—the Ming dynasty is usually said to have seized power in 1368, but fighting with the Yuan lasted for several decades before and after that date.

ally so—the empire had lost control of its own coinage, and merchants had to buy and sell goods with little lumps of silver. To obtain the necessary silver, China lifted its trade ban, opening itself to the world. Soon the great ships of the galleon trade were carrying silk and silver across the Pacific—the final links in the global economic and ecological network begun by Colón's efforts in the Caribbean islands and Legazpi's sojourn in the Philippine islands.

The Ming trade bans have often been described as emblems of Chinese cultural deficiency (Landes: "the Confucian state abhorred mercantile success"). But they were more complicated than that. The bans did not stop *all* foreign contact. They permitted one exception: "tribute payments," in which foreigners, hosted in designated government hostels, were generously allowed to offer presents to the throne. Then the emperor would, out of politeness, give them Chinese goods in return. He also allowed them to sell anything he didn't want, which was often quite a lot.

Coastal merchants recognized the ban-and-tribute scheme for what it was: a way for the government to control international commerce. It was a busy, lucrative affair—in 1403–04, at the height of the supposed ban on foreign merchants, the Ming court hosted "tribute delegations" from no less than thirty-eight nations. Naturally, the Ming wanted the profits from trade. What the dynasty didn't want was the traders themselves; foreign goods, not foreign people. With a few exceptions, all contact with the world outside was supposed to be supervised by Beijing.*

With bureaucratic logic, court bureaucrats reasoned that because maritime trade was outlawed the nation therefore didn't need a coastal force to police that trade. China reduced its navy to a few vessels, not enough to patrol the nation's long coastline. The entirely unsurprising result was a delirium of smuggling (if business is outlawed, only outlaws will do business).

Wokou filled the southeastern coast. Literally, *wokou* means "Japanese pirates," but most weren't Japanese and many weren't pirates. Although they sometimes had bases in Japan, the majority of the *wokou* groups were led by Chinese traders who turned to smuggling after one Ming edict or another eliminated their livelihoods. Their ships were crewed by a crazy quilt of citizens in trouble: scholars who had failed to obtain an official post; bankrupt businesspeople; draft dodgers; fired government clerks; starving farmers; disgraced monks; escaped convicts; and, of course, actual professional smugglers. Scattered among them were a few skilled sailors lured into piracy by the promise of wealth. When officials tried to stop these people, violence often

* The Ming dynasty's predecessors, the Mongol-led Yuan, had tried to do exactly the same thing, forbidding private overseas trade in 1303, 1311, and 1320. In each case the law was soon repealed. The prospect of monopoly was tempting, but the Yuan always found it more profitable—and *much* less trouble—to tax private trade than to run the trade themselves.

ensued. Every now and then this led to the occupation of a city. "Merchants were pirates, pirates were merchants," Lin Rench-uan, a historian at Xiamen University, told me. They would trade peacefully if they could; not so peacefully if they couldn't.

China's efforts to control piracy were hampered by incompetence at the top. Histories of the late Ming dynasty are like advertisements for the virtues of democracy. One emperor refused to meet with his ministers for twenty years. Another was a drunk. A third ran away from his duties and lived in the palace garden, researching alchemical recipes for immortality and prostituting hundreds of young women. This last was the Jiajing emperor, who reigned from 1521 to 1567. He put the empire into the hands of a cabal of grand secretaries, who concerned themselves with personal advancement, rather than, say, the piracy on the southeast coast.

Worst affected by piracy was the resource-poor province of Fujian, in southeastern China, facing Taiwan across the Taiwan Strait. Most of the province consists of low but craggy mountains

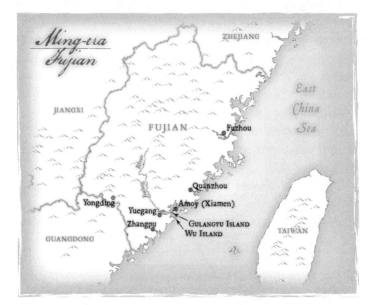

with weathered red soil; flat, arable land is mainly confined to river valleys and a narrow ribbon along the coast. "The mountains peak in rocky summits, and the labor of plowing never ceases," moaned one thirteenth-century Fujianese writer. "The lowlands are salt marshes and cannot be tilled." Famine was a constant risk; despite big terracing and land-reclamation projects, Fujian couldn't grow enough grain to feed itself. Half of the province's rice had to be imported—not an easy task, because the mountains isolate Fujian from the rest of the nation. Among the region's few natural assets are the fine natural harbors that scallop its stony coast. For evident reasons, Fujian depended on the sea. It has long been China's center for maritime trade—which, in the days of sail, meant that it was China's center for international trade. When international trade was officially banned, Fujianese found themselves in an uncomfortable position—there was nothing for them on land.

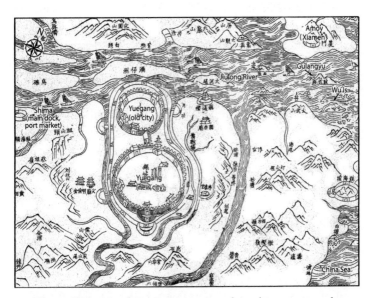

The walled city of Yuegang, portrayed in this seventeenth-century Chinese map, was once one of the world's most important ports. Today its role has been taken by the modern city of Xiamen (then the village of Amoy), on an island in the harbor.

The conflict was particularly intense around the port city of Yuegang. Located at the mouth of the Jiulong River, Yuegang's harbor was full of small islets, sandbars, and other shipping hazards. Because of the area's notorious haze, navigation was difficult—when I puttered about the harbor during my visits, I sometimes couldn't see boats that were only a few hundred yards away. The main docks were several miles up the Jiulong, in water so shallow that ships had to be towed in on the incoming tide. The location was an anti-piracy measure: criminals would not dare raid the docks, because the incoming tide that permitted entrance was too strong to allow escape. At the same time, many Yuegang shipowners *were* pirates—the harbor protected them from people like themselves.

The old city, full of Tang dynasty shrines, was connected by a raised walkway to the newer Ming city, built further inland with larger walls. Inside both were packed huddles of houses—"bandit dens," sneered one official in the 1560s, whose inhabitants "have collaborated with foreigners to spread chaos to the detriment of the local area for a long time now." Indeed, Yuegang was such a pirates' paradise that at one point Beijing divided the populace into groups of ten families that had to account for their members every five days; if one family did something illegal, all ten would be punished.

Imperial China's day-to-day history is largely recorded in the annual gazetteers sent to Beijing from each of the nation's counties. Yuegang's county had so much *wokou* trouble that the gazetteer's compilers eventually devoted a special appendix to it: "Bandit Incursions."

Bandit Incursions began in 1547, when a Dutch merchant/pirate/smuggler group set up a base on Wu Island, a recently shuttered naval base just south of Yuegang's harbor. "Dutch" is a bit of a misnomer; the traders flew a Dutch flag, but they were a hodgepodge of Spanish, Portuguese, and Dutch hustlers with some semi-enslaved Malays. Chinese and Japanese *wokou* happily sent ships to trade with them, as did legitimate businesspeople from Yuegang; a busy, multilingual market sprang up in Wu Island's small but serviceable harbor. Unenthusiastic about the encampment was Zhu Wan, governor for both Fujian and Zhejiang, the province to the north. He dispatched soldiers to drive out the foreigners.

Wu Island consists of two rocky, steep, scrub-covered mounds with a low "saddle" between them. The Dutch had ensconced themselves in an improvised fort atop one of the mounds, forcing the Chinese to attack uphill. In a brief skirmish, the merchant/pirate group beat back the Chinese forces. Zhu changed tactics: he imprisoned ninety local merchants who had traded at Wu Island. In a gesture that even the unsympathetic gazetteer described as altruistic, the Dutch sent emissaries to plead for their allies' lives. Dismissing the entreaties, Governor Zhu beheaded all ninety. The Dutch abandoned Wu Island and gave up their attempt to trade openly; later they roamed the region, preying on the very Fujianese merchants and smugglers with whom they previously had collaborated.

Zhu Wan was anything but satisfied. A rigid, moralistic former magistrate, Zhu irritated his superiors by denouncing corruption at every level in a spray of angry memoranda. He was such a stickler

A former pirate stronghold, Wu Island, in the hazy waters off of Yuegang, is now a center for fishing and aquaculture.

that when his subordinates gave small gifts to his visiting family he punished himself with a hefty fine. Late in 1548 Zhu assaulted a major smuggling base in Zhejiang, scuttling more than 1,200 illicit boats. Led by the infamous "Baldy" Li, *wokou* fled by the hundred to a new base in the extreme southern end of Fujian. Three months later Zhu's men hunted them down there, killing almost 150 and capturing scores of Portuguese, Japanese, and Chinese smugglers.

Many of Li's gang turned out to be from influential Yuegang merchant families.* Angered by this evidence of routine collusion among local elites and foreign smugglers, Zhu ordered all the captives to be summarily executed—the second round of executions in two years. The executions united Zhu's enemies against him. Yuegang's wealthy appealed to Zhu's superiors: the courtiers of the alchemy-besotted Jiajing emperor. Zhu was demoted, then fired, then subjected to politically motivated investigations. Facing indictment, he poisoned himself in January 1550. "Even if the Emperor doesn't kill me," Zhu said, "powerful court officials will kill me. And even if powerful court officials don't kill me, the people of Fujian and Zhejiang will kill me."

Emboldened by Zhu's absence, pirate gangs seized entire towns, pillaging "until the stench of rotten flesh forced them to leave." In one city north of Yuegang more than twenty thousand people died after a pirate assault. Across southeast China, the Ming historian Luo Yuejiong recalled, terrified families "ate without cooking their food, and slept unsoundly on their pillows; farmers left behind their pitchforks and women dozed off on their looms." When the *wokou* attacked, Luo wrote,

> fathers and sons, young and old, were taken prisoner and
> followed the pirates on the road. As for the dead, their

* "Families" is a misnomer. The traders were *gongsi*, which were clan-like groups of related families that often had hundreds of members. I'm reluctant to use the term *gongsi*, though, because it now means "company"—an indication of the familial roots of many Chinese businesses but a source of potential confusion to readers.

heads and bodies were found in different places, bones left out in the grassy swamps, heads stiff. Looking on the horizon, the coastal counties were almost nothing but hilly ruins.

Wokou were "burning homes, seizing women and children, and stealing huge quantities of valuables," wrote the chronicler Zhuge Yuansheng in 1556. "Officials and common people alike were killed with weapons, their bodies, numbering in the tens of millions [an idiom for "huge numbers"], filled ravines. Government troops dared not oppose them." At the mere appearance of *wokou* in an area, he wrote, "people scream in panic and take flight." In a scene straight out of a Stephen Chow martial-arts comedy,

> [a] messenger from Songjiang [near Shanghai] rode at a gallop into town and cried out to his followers, "We're here! we're here!" The locals misunderstood him and thought the [pirates] were coming. Men and women scurried like ants, nothing could stop them. Women and children were separated, families lost countless valuables and possessions. At the time, more than 600 soldiers were garrisoned at the city, stationed on the bastions along the walls; they all threw down their weapons and armor and ran away. Not until the next day did calm return to the town.

In Yuegang the *wokou* didn't strike back at the government until 1557, according to the county gazetteer, when a disgruntled farmer secretly opened the city gates to two pirate gangs. Overwhelming all resistance, the *wokou* "abducted more than a thousand people and burned more than a thousand homes."

Dire as it was, the assault was a sideshow. Even as *wokou* beset Yuegang, twenty-four of the city's merchants pooled resources and built a fleet to work with the pirates in what amounted to an interlocking network of joint ventures. The traders had access to domestic markets; the smugglers, to foreign goods. Known as the Twenty-four Generals, the merchants decided to control access to their home markets by carving up Yuegang, gangland style, into neighborhoods, each dominated by a single "general" in an earthen-walled fortress. Three hundred imperial soldiers were sent to dislodge them. The Twenty-four Generals beat back the attack. Observing this success, other smugglers in other parts of Fujian followed the Generals' lead, forming the Twenty-four Constellations and the Thirty-six Bravos. The region became a bewildering, violent amalgam of overlapping loyalties and betrayals, as business gangs and pirate gangs from different neighborhoods, regions, and nations vied among themselves for control of the smuggling trade.

For Coastal Surveillance Vice Commissioner Shao Pian—the late Zhu Wan's subordinate—the last straw occurred when Fujianese traders invited three thousand Japanese and Portuguese smugglers to reoccupy the former Dutch base at Wu Island. Shao had no good options. Bled by cutbacks, the imperial navy was outgunned and outmanned by the *wokou*—indeed, for its missions it often hired smugglers, who had superior skill and experience. Worse, he could not trust many of his own officers, because they came from the merchant families involved in the smuggling. In a classic move, Shao forged an alliance with—that is, bribed—Hong Dizhen, former leader of the three thousand *wokou* at Wu Island. Hong gathered up a force in 1561 and attacked the smugglers' largest bases in Yuegang. "Countless *wo[kou]* died," the gazetteer states—a face-saving formulation meaning that the pirate gangs, who were allied with the entire local populace, drove Hong back with heavy losses.

Shao effectively capitulated. "Over ten years," the gazetteer reported, "we lost one outpost, two smaller outposts, a prefecture, six counties and no fewer than twenty-some fortified towns. . . . People wailed and ghosts cried out, and the stars and moon gave off no light as the grassy wilderness itself moaned." The world's richest, most technologically advanced nation had utterly lost control of its borders. In 1567 a new Ming emperor threw in the towel and rescinded the ban on private foreign trade.

The government reversed course not only because it recognized its inability to stop smuggling, or because it had begun to appreciate how much Fujian's populace depended on trade. Beijing had come to realize that the nation desperately needed the merchants' most important good: silver.

OUT OF MONEY

Several hundred years before the birth of Christ, the Chinese state began to issue round coins made of bronze, an alloy of copper and tin. Each coin was worth its own weight in bronze and had a square hole in its center. The system had defects. Because bronze was not especially valuable, a single coin wasn't worth much. To create units of larger value, people strung the coins together into groups of one hundred or one thousand.

The strings were heavy, bulky—and still not worth much. Asking large-scale Chinese traders to use them was like asking today's mergers-and-acquisition bankers to buy companies with rolls of quarters. Worse, according to Richard von Glahn, a historian at the University of California at Los Angeles who specializes in the history of Chinese currency, the empire ultimately didn't have enough copper to keep up with the demand for coins. The copper-starved Song dynasty was forced to create a "short-string" standard, in which strings of 770 coins were officially treated as if they contained a thousand.

In 1161 the Song dynasty introduced what would become the first modern paper currency: the *huizi*. Regional governments and powerful merchants had experimented with paper money for two centuries, but the *huizi* was the first nationwide, state-printed banknote. It was denominated in terms of bronze coins; the lowest-value note was worth two hundred coins and the highest was worth three thousand. (The first European banknotes appeared in 1661, five centuries later.)

Theoretically speaking, people could redeem their *huizi* for actual coins. In fact, the Chinese government and Chinese merchants quickly discovered that printing *huizi* would reduce the demand for coins, letting them export the latter to Japan, which used Chinese bronze coins for its currency, too. The more the government printed bills, the greater the number of coins that could be exported. Within a few decades of their creation, *huizi* notes were decoupled, as a practical matter, from coins; no matter what the bills claimed, they couldn't be redeemed for bronze. They had effectively become what economists call *fiat money*.

Fiat money has no intrinsic value, and is worth something only because a government declares it is. The U.S. dollar is an example, as is the euro. As pieces of paper, dollar bills and euro bills are next to worthless. Yet because they are officially printed by government institutions, people can hand these colorful paper rectangles to grocery-store clerks and walk out with bags of food. The silver pesos that circulated in the Spanish empire, by contrast, were *commodity money*: valuable because they were made of a valuable substance. So were Chinese bronze coins, although the bronze wasn't especially precious.

From a government's point of view, commodity money is problematic, because the government does not fully control the money supply—the nation's currency is at the mercy of random shocks. For example, at the time of Colón's voyages cowry shells were used as currency from Burma to Benin.* Then Europeans shipped in vast quantities of shells from the cowry-rich Maldive Islands, in the Indian Ocean. Governments throughout the region

---

* To those accustomed to metal coins, the idea of using shells for money may seem primitive. But they had a signal advantage: unlike the era's coins, which were often debased or faked, shells could not readily be altered or counterfeited.

were overwhelmed. A financial system that had been in place for centuries disintegrated in a flash.

This kind of external pressure has no impact on fiat money. With fiat money, the government has near-complete control over the money supply; it determines how many banknotes are needed and instructs the mints to print them. In theory, politicians can expand or contract the money supply to foster better economic conditions.

Fiat money's greatest defect is the same as its greatest strength: the government decides how many banknotes to print. After introducing paper bills, Song emperors made a stunning discovery: they could buy things simply by stamping patterns of ink onto pieces of paper. For several decades the strategy was successful. As the use of paper money expanded throughout the empire, the nation needed to increase the supply of paper bills, and the emperor's outlays were absorbed in the overall rise. In the early thirteenth century the emperor decided to fight enemies in the north—first the Jin, then the Mongols. To pay for supplies and troops, he turned the printing presses on "high." Inflation was the result. The Song lost to the Mongols before they could set off monetary catastrophe. The Mongols, who became the Yuan dynasty, issued their own paper money—lots of it. To them belongs the honor of inventing hyperinflation. By the 1350s Yuan paper money was practically worthless. In the next decade the dynasty fell to the Ming uprising.

Upon taking the throne, the first Ming ruler, the Hongwu emperor, ordered that new coins be issued in his name—no more worthless paper bills! Alas, the Hongwu emperor discovered that the empire had nearly exhausted its copper mines. Naturally, the price of copper rose; bronze coins ended up costing more to produce than they were supposed to be worth. It was as if every penny cost two pennies to manufacture. Unsurprisingly, not many coins were issued. Ming coins became rarities, so rarely seen that businesspeople hesitated to accept them—merchants had too little experience with the coins to know whether they were genuine or counterfeit.

Quickly the Ming dynasty, like its predecessors, discovered the virtues of an active printing press. Again inflation exploded; the value of the paper bills fell by roughly 75 percent in about a decade. The Hongwu emperor responded by refusing to produce any more coins. Force people to use paper bills—that was the idea. It didn't work. Shutting down the mints increased the scarcity, and hence the rarity, of the new coins, further eroding their value as currency. It also pushed up the value of old coins, which people trusted and understood. And it dramatically increased counterfeiting. The fake coins were for the most part easily distinguishable from real ones. But merchants were so desperate for some way for their customers to pay them that they accepted the counterfeits anyway, although they demanded a premium.

As businesspeople snatched up all the old and counterfeit coins they could find, the value of paper bills continued to fall. In 1394 the government banned the use of its own coins—a policy that "flouted economic realities," wrote von Glahn, the UCLA historian, in *Fountain of Fortune* (1996), a fine history of Chinese money that I have been drawing upon here. As one would expect, the policy failed. The emperors kept trying, prohibiting coins in 1397, 1403, 1404, 1419, and 1425. Every time the ban failed, the emperors would again officially permit coins to circulate—until the next ban. Meanwhile the Ming kept printing paper notes at inflationary rates. All of this may sound completely unhinged, and it was. In the feud- and faction-ridden Ming court, government policies were often accidental by-products of ministerial intrigues, enacted with little regard for their actual effects. The result was that by the time *wokou* were terrorizing the southeast coast, the Chinese empire had no functioning currency.

I am oversimplifying. The currency *did* function—intermittently, unpredictably. Each emperor produced coins with his name stamped on the face. When he died, the succeeding ruler would quickly declare that his predecessor's coins were value-less; only new coins minted by the new emperor would be valid currency. Merchants suddenly saw "their capital evaporate in a single day, often silently mourning their losses before committing suicide," according to the *Ming Shi,* the official history of the dynasty.

Needing something to pay with, merchants and their customers would use old coins from earlier reigns until the new emperor's money arrived; given the lack of copper and dynastic inefficiency, this frequently took years, even decades. Then they would use the new coins until the government suddenly banned them. The result, according to the Taiwanese historian Quan Hansheng, was a constant game of financial hot-potato, with everyone trying to use their coins until just before they lost all value—at which point they would try to unload them onto some hapless sucker.

"Virtually from morning to evening the rules change, and still there is no set policy," moaned one sixteenth-century imperial chancellor. "The people fear that the money they get today will be useless tomorrow and they will no doubt starve. Thus the more the coins change, the more chaos ensues, and the more restrictions there are, the more people panic, so that stores dare not open for business, there is no buying or selling, and cries of anguish ring out."

"Coins received in the morning couldn't be used by evening," explained a central-China gazetteer in 1606. Shopkeepers would suddenly refuse them en masse.

One person would suggest it, and everyone else would agree. Although such actions were strictly forbidden, they had no regard for the law. Before long, merchants from other regions would come to buy old coins, and they would exchange them at a ratio of three to one and cart them away. This is what you call monopolization at its extreme, the power of devious people. Wealthy merchants and powerful brokers sit back and reap heavy profits while the average person suffers. It never ends.

Were these complaints exaggerated? In 1521 the Jiajing emperor began his reign. Still a young man, he was decades away from his prostitute-fueled pursuit of immortality and fiercely determined to regain control of the nation's money supply. He decided to issue new coins that would be of such high quality that the people would reject the old coins and counterfeits. The results were described a century later by the geographer and historian Gu Yanwu in a grandly titled compendium, *The Strategic Advantages and Weaknesses of Each Province in the Empire.* Gu looked at Zhangpu County, about ten miles south of Yuegang. As the Jiajing reign began, Gu reported, the currency preferred by county merchants was, unbelievably, coins from the Song dynasty—the Yuanfeng reign, to be precise, which had ended four centuries before, in 1085 A.D. During the next decade, the Jiajing emperor established mints and punched out coins as fast as possible. The effort made not a jot of difference in Zhangpu County. Year after year, Gu wrote, the preferred money flipped arbitrarily from one Song emperor to another. After each switch, people stuck with the previously favored coins were left high and dry. Not until 1577, five years after the Jiajing emperor's death, did Zhangpu County use legal currency. For the first time in decades, people used coins minted by the current ruler, the Wanli emperor. The reprieve was brief, Gu wrote. "Only one year later, they stopped using Wanli coins."

Silver had long been recognized as a store of value, though rarely used for ordinary, small-scale transactions because it was too scarce and costly. But the uncertainty over bronze coins and paper money grew to the point where desperate merchants took to carrying around little silver ingots, often shaped like shallow bowls one to four inches in diameter. When traders met, they used the ingots to buy and sell, weighing them with jewelers' scales and clipping off needed sums with special shears; to evaluate the ingots' purity, they used *kanyinshi* (silvermasters), who charged a fee for the evaluation and routinely cheated all parties. Awkward as it was, this system was better than using coins that might lose their value at any time. By the end of the *wokou* crisis, one writer complained in 1570, coins were used in fewer than one-tenth of all market transactions. The Chinese government didn't issue the ingots; as if in a libertarian fantasy, the money supply was effectively privatized. Anybody who could lay hands on some silver could get *kanyinshi* to certify it—instant money! Everyone was paying bills with splinters of silver.

Grudgingly and gradually, the Ming emperors adopted this system, too. China's basic tax system—farmers paid a portion of their harvest—hadn't changed for eight hundred years. But over time it had become encrusted with loopholes and extra levies, which created opportunities for corruption. In a series of edicts, Beijing reordered the tax rolls and ordered the citizenry to pay an ever-increasing share of their taxes in raw silver, rather than in kind. By the 1570s, as the Wanli reign began, more than 90 percent of Beijing's tax revenue arrived as lumps of shiny metal.

Called *sycee*, these small silver ingots were used in the Ming and Qing eras instead of coins. The stamps include the mark of the silversmith (difficult to read, but probably Shunxiang Smithy) and the date (the twentieth year of the Guangxu emperor's reign, or 1895).

China was the world's biggest economy. Its "silverization" meant that tens of millions of wealthy Chinese suddenly needed chunks of silver for such basic tasks as paying taxes or running a business. It stoked a voracious demand for the metal. Inconveniently, China's silver mines were just as played out as its copper mines. Businesspeople had trouble laying their hands on enough silver to pay for anything, including their taxes. The sole nearby supply of silver was in Japan. On an official level, China and Japan were not friendly—indeed, the two nations were soon to fight a war in Korea. To get the silver necessary to keep business going, merchants turned to *wokou*. Businesspeople sold silk and porcelain to brutal men with silver, then turned around and used the silver to pay their taxes, which in turn was spent on military campaigns against those brutal men. The Ming government was at war with its own money supply.

Unable to reconcile the contradiction, Beijing finally allowed Fujianese merchants to trade overseas without fear of punishment. Now acting in the open, they sent thousands of people—younger sons from extended families—throughout Asia to establish beachheads for later trading or extortion. Even a backwater like the Malay village of Manila may have had as many as 150 Chinese residents in 1571, when Legazpi showed up. Hundreds more apparently resided elsewhere in the islands. The unexpected discovery of silver-bearing foreigners in the Philippines was, from the Chinese point of view, a godsend. The galleons that brought over Spanish silver were ships full of *money*.

"THE TREASURE OF THE WORLD"

How did silver get on those galleons? According to the stories, it began with a man named Diego Gualpa or Hualpa in April 1545. He was out walking at thirteen thousand feet, possibly looking for a lost llama, on a plateau in the Andes Mountains, at the southern tip of Bolivia. (The altitude, amazingly, was not extreme for the Andes, where most of the population lives on high plains that are almost at that level.) No trees, no animals, no crops, no homes—just a bare, dome-shaped hill, clawed by wind and snow, surrounded by still taller mountains splashed with ice. Stumbling on a high ridge, he steadied himself by seizing a shrub. It came out of the shallow soil. Beneath it, in the hole made by its roots, was a metallic sparkle. Gualpa or Hualpa was on a ledge of silver ore three hundred feet long and thirteen feet wide and three hundred feet deep—the biggest silver strike in history.

Typical ores are at most a few percent silver. The ledge was as much as 50 percent. It was so rich that the Spaniards didn't know how to purify it—they kept boiling away the silver. Andean Indians had some of the world's most advanced metallurgy. Locals were able to do what the foreigners couldn't, in low-temperature smelters fueled by dry grass and llama dung. Soon thousands of native smelters sent their smoke into the chill Andean air. By the early 1560s, two decades after the first strike, the Imperial Villa of Potosí, to give the new boomtown its formal name, had a population of as much as fifty thousand. It would have had even more if Spain hadn't done everything it could to keep people out. Despite these efforts the count grew to 160,000 by 1611. Potosí was as big as London or Amsterdam. It was the highest, richest city in the world.

Lawless, louche, and luxurious, Potosí set the template for countless boomtowns afterward. Courtesans in Chinese silk walked on Persian carpets in rooms sprinkled with scented water. Miners gave fortunes to beggars and spent fortunes on swords and clothes and elaborate celebrations. In a market-stall bidding war, two men drove the price for a single fish to five thousand silver pesos, many years' income for most Europeans. Another man showed up for a duel in "a brocaded tunic the color of mother-of-pearl, studded with diamonds, emeralds, and strands of pearl." At one celebration a city street was actually paved with silver bars. "I am rich Potosí," crowed the city coat of arms, "the treasure of the world, the king of the mountains, the envy of kings."

Shown in this drawing from 1768, Potosí spread across the plains below the silver mountains. Cold, crowded, and violent, it was the highest city in the world and probably the richest.

Enviable, perhaps, but also uncomfortable. Wind and altitude conspire to make the town amazingly cold and the terrain almost lifeless. The air is so thin that the first time I visited I got woozy carrying my suitcase up a flight of stairs. Humiliatingly, my host's ten-year-old sister scooted to my side, grabbed the bag, and ran with it to my room. During the silver era every cup of flour, every piece of clothing, and every scrap of wood had to be carried into the city by llama. Now Bolivia has cars and trucks, but many houses in Potosí still lack heat, as they did in centuries past. In the morning my blanket crackled with frost. Seeing my blue lips, my host's mother kindly melted a cup of coca tea.

Almost as important as the mountain of Potosí was a second Andean peak, Huancavelica, eight hundred miles northwest, which gleamed with mercury deposits. In the 1550s Europeans in Mexico discovered a way to use mercury, rather than heat, to purify silver ore. (Rediscovered, actually—the technique had been known in China for centuries.) Miners pulverized silver ore, spread the powdery result over a flat surface, typically a stone patio, then used rakes and hoes to mix in saltwater, copper sulfate, and mercury, forming a stiff cake. Men, mules, and horses walked over the cake, their footfalls providing the energy for a complex reaction that slowly forced the mercury to combine with the silver in the ore, forming a sticky amalgam. Workers poured water over the cake, washing away everything but the amalgam, which was then scraped into cloth sacks. When the amalgam was heated, the mercury—which boils at just 670 degrees—steamed away, leaving pure silver. After watching a demonstration of the technique, Viceroy Francisco de Toledo seized the Huancavelica mines for the crown, thus arranging what he called "the most important marriage in the world, between the mountain of Huancavelica and the mountain of Potosí."

As long as the mercury lasted, the viceroy realized, the mines would no longer be dependent on Indian technology, which in turn meant that Spaniards could treat natives wholly as a source of labor. Andean peoples had a tradition of communal work that had been co-opted by the Inka to build a great highway system. Taking a page from the Inka playbook, Viceroy Toledo forced natives to deliver, as a tribute, weekly quotas of men to the silver and mercury mines—at the start, roughly four thousand a week each for Potosí and Huancavelica. As lagniappe, mineowners also imported several hundred African slaves each year. Sometimes it is said that the mines killed three to eight million people. This is an exaggeration. Still, conditions were appalling, especially at Huancavelica.

The entrance to the mercury mine was a great archway with pilasters and the royal coat of arms cut into the living rock of the mountain. Inside, the tunnels rapidly narrowed and spread out like jellyfish tentacles. Candles strapped to their foreheads, Indians hauled ore through cramped tunnels with next to no ventilation. Heat from the earth vaporized the mercury—a slow-acting poison—so workers stumbled through the day in a lethal steam. Even in cooler parts of the mine they were hacking away at the ore with picks, creating a fug of mercury, sulfur, arsenic, and silica. The consequences were predictable. Workers served in two-month shifts, often several times a year; after a single stint, many shook from the initial effects of mercury toxicity. Foremen and supervisors died, too—they also spent too much time in the mine. So determined were natives to avoid the mercury pits that parents maimed their children to prevent them from having to serve.

Huancavelica ore was refined in a ceramic oven; the mercury boiled off and condensed on the inside surfaces. If the oven were opened before it was cool—something mineowners, eager to start the next refining cycle, often insisted upon—the result was a faceful of mercury vapor. Numerous official inspectors urged the crown to shut down Huancavelica. But reasons of state always won out; the need for silver was too great. As the mineshafts went deeper into the mountain the inspectors urged that the state dig ventilation shafts. The first was not created for eight decades. Officials who dug up graves in 1604 reported that when miners' corpses decomposed they left behind puddles of mercury.*

Conditions at Potosí were less lethal, but no less inhumane. In near-complete darkness gangs of conscripted Indians carried hundred-pound loads of ore up and down dangling rope-and-leather ladders. Like ants on a string, one chain of men descended one side of the ladders while another chain climbed up the other. Initially Indians were given two weeks' rest above ground for every week of work beneath the surface. Later the rest periods vanished. When miners hit a patch of low-quality ore, they were forced to work harder to make their quota of silver. Failure to

---

* Mercury poisoning was not the sole cause of death. Equally lethal were pneumonia, tuberculosis, silicosis (lesions in the lungs caused by inhalation of silica dust), and asphyxiation (breathing carbon dioxide in badly ventilated tunnels). In 1640 a royal inspector saw three Indians fall into a pit so filled with carbon dioxide that candles couldn't burn (carbon dioxide, which is heavier than air, pools in low areas). Although the pit wasn't deep, the workers did not get up. Their bodies were not retrieved; descending into the pit was too dangerous.

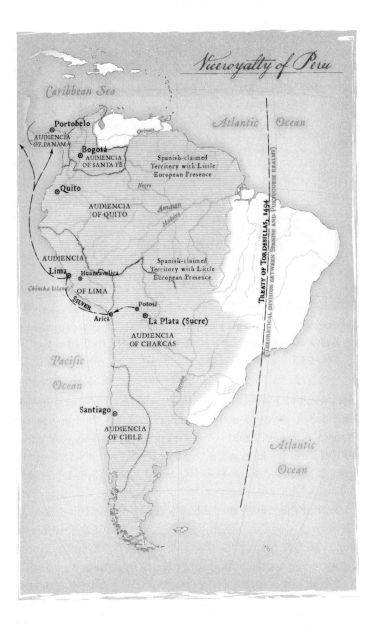

Viceroyalty of Peru

Caribbean Sea

Atlantic Ocean

Portobelo
AUDIENCIA OF PANAMA

Bogotá
AUDIENCIA OF SANTA FÉ

Orinoco

Spanish-claimed Territory with Little European Presence

Quito

Negro

AUDIENCIA OF QUITO

Amazon

Madeira

AUDIENCIA

Lima

Huancavelica

Spanish-claimed Territory with Little European Presence

Chincha Islands

OF LIMA

SILVER

Potosí

Arica

La Plata (Sucre)

AUDIENCIA OF CHARCAS

Pacific Ocean

TREATY OF TORDESILLAS, 1494
(THEORETICAL DIVISION BETWEEN SPANISH AND PORTUGUESE REALMS)

Santiago

AUDIENCIA OF CHILE

Atlantic Ocean

The itinerant artist and editor Theodorus de Bry never saw the mines of Potosí, but he captured something of their cruelty in this engraving from the 1590s.

meet the quota would be punished by whips, clubs, and stones. Horrified antislavery activists denounced the "hellish pits" of Potosí. "If twenty healthy Indians enter on Monday, half come out on Saturday as cripples," one outraged priest wrote to the Spanish royal secretary. How, he asked, could Christian leaders allow this?

Part of the reason that the rule of law broke down beneath the surface was that it had broken down above the surface as well. Violence of every conceivable variety flourished in Potosí. Construction workers found murder victims stuffed into walls or shoved under rocks. Tailors rioted after a guild election, forcing one faction leader to seek shelter in an Augustinian monastery. When the government sent agents to arrest him, the friars jumped on them with drawn swords. City council members wore chain mail at meetings and carried swords and pistols. Political disputes were sometimes resolved by duels fought right in the room. As one may imagine, the ambience was hostile to family life. Despite Potosí's huge population, no child was born there to a European for more than half a century. So unexpected was the first birth that the baby's arrival—on Christmas Eve, 1598— was widely attributed to the miraculous intervention of Nicola da Tolentino, the patron saint of infants.

Potosí was as conflict prone as pirate-ridden Yuegang, but the battles were regarded differently, at least by their chroniclers. The main Chinese accounts of the *wokou*—county gazetteers and official reports—are terse and matter-of-fact, whereas Potosí's most important chronicler, Bartolomé Arzáns de Orsúa y Vela (1676– 1736), spent three decades writing a massive, 1,300-page history of the city that is, among other things, a breathless paean to romantic honor of the sort mocked in *Don Quixote*. Arzáns never published his book, partly because he was afraid to go public—local families might not have liked his descriptions of their forefathers' mayhem, no matter how glorified.

Despite the golden haze Arzáns cast over events, one can see from his account how the city's violence evolved from cinematic face-offs between alpha males into full-fledged battles between ethnic groups. In 1552, seven years after Gualpa or Hualpa discovered silver, the bellicose adventurer Pedro de Montejo arrived in Potosí. In Arzáns's telling, Montejo put up placards challenging one and all to combat, "spear against spear." Such fights "were an admirable thing in Potosí," Arzáns explained. In a city with a permanent European population that almost exclusively consisted of young men on the make, "killing and hurting each other was the sole entertainment."

By general consensus, Montejo had one obvious opponent: the equally bellicose Vasco Gudínez, who had already established a reputation as the city's go-to man for threats and mayhem. Early on Easter morning both men, accompanied by their seconds, rode to the battleground, followed by a crowd of layabouts. After an exchange of insults, Arzáns recounted, the two men "charged at each other and collided so hard that it was like bringing together two rocks." Gudínez, badly wounded,

> withdrew some distance and threw his spear at Montejo with such violence that he did not have time to dodge, and it struck his buckler, passing entirely through it and wounding him in the arm, breaking through the chain mail and the steel plate, and much of the tip went into his body. . . . Montejo, fatally injured and without the defense of his buckler, attacked his opponent with diabolical force with the tip of his sword; he responded to Gudínez's parry with his shield, and raising his arm brave Montejo unleashed a fierce blow to the head that dazed Gudínez and, worse, wounded and knocked his horse to the ground, spilling much blood. Now Montejo had him down and was about to cut off his head, but at the first step he [Montejo] fell dead, pierced through the chest. Gudínez got up with alacrity, and stumbled to the corpse and put his sword to its neck, thinking that he wasn't yet dead.

Arzáns evidently embellished the details of this encounter—he claimed that the two men's seconds thereupon engaged in a three-hour battle to the death, which the wounded Gudínez tarried to watch en route to his sickbed. Arzáns may even have gotten some basic facts wrong (no record exists of any Potosino named Pedro de Montejo, for instance). But the underlying scenario seems indisputable: the city was chock-full of brutal thugs. To reestablish control, the provincial government in Lima sent in troops. After a round of skirmishes, Arzáns wrote, Gudínez's second, a specially vicious hoodlum, was drawn and quartered; Vasco Gudínez himself was jailed.

Vasco means "Basque," and the name was no accident—a disproportionate number of Potosinos were from the Basque country in Spain's Atlantic coast. Culturally, linguistically, and geographically isolated from the rest of Spain, mountainous and agriculturally unpromising, the Basque region was, so to speak, the Fujian of Spain—a center for nautical trade and emigration. Two-thirds of Potosí's mines and the municipal council were controlled by Basques by 1602. Basque leaders bribed royal bureaucrats to look the other way when taxes were due; if non-Basque miners posed a competitive threat, Basque gangs provided muscle. When royal officials tried to sell off a Basque mineowner's lease for unpaid taxes, a Basque gang stabbed to death the would-be purchaser in Potosí's central square. Resentment grew among colonists from other parts of Spain, many of whom lived in wildcat mining camps outside the city. In furtive meetings anti-Basque miners took to identifying themselves with caps made from vicuña wool (the vicuña is a relative of the llama) and called themselves Vicuñas. Basques had no need to identify themselves by dress; they spoke their homeland's native tongue, Euskara, which is unrelated to Spanish.

The struggle gained intensity in August 1618, when a new lawman came to town. In the loosely governed city, he was that most terrifying figure, a tax inspector. "Punctual and tidy, intelligent and modest, he enjoyed nothing more than fulfilling his duties," wrote the Bolivian historian Alberto Crespo of the inspector. "His name was Alonso Martínez Pastrana and he was not Basque." This humorless bean-counter soon learned that Potosinos had been cheating massively on their taxes. The king was supposed to be paid one-fifth of the silver from the mines, as well as part of the revenue from mercury sales and coin minting. Martínez Pastrana calculated that Potosinos had collectively shorted the king 4.5 million pesos, a sum above the mines' official annual output. Because Basques owned the biggest mines and dominated the city government, they were responsible for most of the fraud. Eighteen of the twenty-four members of the municipal council owed back taxes, the inspector said; eleven of the offenders were Basque. Three years later, after a battle with corrupt treasury officials, Martínez Pastrana finally was able to ban overt tax cheats from membership on the municipal council.

In June 1622 a Basque gang leader was found dead on the street, his hands and tongue cut off and minced. Vicuñas correctly were blamed. Furious Basque goon squads roamed the squares, threatening to lynch the "Moors, treasonous Jews and cuckolds" responsible for the murder. If they met a stranger in the street, one account claims, they challenged him in Basque; anyone who responded in Spanish would perish. After a round of murders, a stone-throwing Vicuña mob converged on the home of Domingo de Verasátegui, head of a powerful Basque family—he was one of four wealthy brothers, two of whom were on the municipal council. He was saved only by the sudden appearance of the head of the royal court, who personally escorted him to the safety of the city jail. Verasátegui died a few months later of natural causes, unusual in Potosí.

The crown appointed a new governor for Potosí the following May. (The governor, or *corregidor,* was the highest district-level authority.) Felipe Manrique was a violent man with a short fuse—years before, in a moment of rage, he had slain his wife. On his journey to Potosí the widower met and was smitten by Verasátegui's widow, inflaming Vicuña suspicions. They razed the governor's house, shooting Manrique four times in the process. A full-fledged riot exploded two months later when a Basque tipped his hat to two Vicuñas "in a very arrogant manner." Manrique dispatched military patrols but couldn't stop several thousand Vicuñas from pillaging the homes of prominent Basques.

Seventy years before, Fujian's Zhu Wan had learned the hard way that incorruptible pursuit of duty is not always a means for successful career advancement in government service. Zhu was driven to suicide. The implacable tax collector Martínez Pastrana was luckier: he escaped with his life, though not his career. His superiors bowed to pressure and ended his mission in August 1623, a few weeks before the Vicuñas burned down Governor Manrique's house. He ended up in bitter retirement in Lima, where a street bears his name.

By contrast, the all-too-corruptible Governor Manrique left office on February 19, 1624. The next day he married Verasáte-gui's widow and moved into her splendid manor—"serving to eliminate every doubt," observed Crespo, the Bolivian historian, about "the badly concealed connections that linked the governor and the Basques." Manrique moved to Cuzco (the Spanish name for the former Inka capital of Qosqo), a wealthy man who would become wealthier. With the departure of both men, passions ebbed. Vicuñas disappeared into the countryside, where they robbed travelers with impunity for years.

Incredibly, the Basque-Vicuña war had almost no effect on the flow of silver. Even as Basques and Vicuñas fought in the streets, they cooperated on mining and refining the silver, then shipping it from Potosí. The last was a huge task. One account describes how a single shipment of 7,771 bars left the city in 1549, four years after the lode's discovery. Each bar was about 99 percent silver and weighed about sixty pounds. All were stamped with serial numbers by the foundry and marked with the owner's stamp, the foundry stamp, and the tax man's stamp. By the time the assayer individually certified its purity with his stamp, the bar looked as if it had been graffiti-tagged by a demented numerologist. Each llama could carry only three or four bars. (Mules are bigger than llamas, but need more water and are less surefooted.) The shipment required more than two thousand of the beasts. They were watched by more than a thousand Indian guards who in turn were watched by scores of Spanish guards.

Despite these obstacles the Americas produced a river of silver—150,000 tons or more between the sixteenth and eighteenth centuries, according to the silver trade's most prominent historians, Dennis O. Flynn and Arturo Giráldez of the University of the Pacific. For those centuries Spanish silver washed around the planet—it was 80 percent or more of the world's output—overwhelming governments and financial institutions everywhere it landed. "Right at the beginning there was this *shot* of

silver into Europe," Flynn told me over the course of a long conversation. "We can't be sure about the numbers, but the amount
of silver in Europe may have doubled."

The Spanish silver peso became a universal currency, link-
ing European nations much as the euro does today. (It was
called, famously, a "piece of eight," because it was worth eight
reales—reales were then the basic Spanish coin.) Pesos were the
main currency in the Portuguese, Dutch, and British empires
and widely used in France and the German states. "Because silver was the money supply," Flynn said, "there was an uncontrolled jump—an explosion, really—in the money supply across
Europe." Flynn was trained as an economist. "Rapid, unplanned
shocks to the money supply are generally a bad idea," he said.
Inflation and financial instability were the result.

After sixty years of frenzied production, Flynn and Giráldez
wrote, the world had accumulated so much silver that its value
began to fall. A million pesos in 1640 was worth about a third of
what a million pesos had been worth in 1540. The impact was
multifarious and planet-wide. As the price slid, so did the profits
from silver mining—the mining that was the financial backbone

Much of the silver from Potosí and Mexico
was transformed into "cob" coins, hammered
between crudely engraved dies. This four-real
coin was made in Potosí in the 1570s, before the
coins received dates. The "L" is the initial of the
mint assayer.

of the Spanish empire. Spain did not adjust its tax rates for currency fluctuations (in modern terms, they weren't indexed to inflation). The king collected the same amount of taxes in silver as he had before, but its value plunged, throwing the government into crisis. Spain's economy turned to ash, followed by the economies of a dozen other states equally dependent on Spanish silver, one after another like a chain of firecrackers. The well-off felt beggared; the beggars felt desperate. With nothing to lose, they picked up stones from the streets and looked for targets. Ruin was followed by riot and revolution.

American silver was not the sole cause of the upheaval; still, threads of silver link the revolts against Spain in the Netherlands and Portugal, the ruinous Fronde civil war in France, and even the Thirty Years' War. Flynn and Giráldez said that one of their contributions was to point out that the turbulence in Europe, though devastating, was "a kind of sideshow—most of the silver actually went to Asia." And not just to some generalized part of Asia. A disproportionate share of the silver ended up in a single port in a single Chinese province: Yuegang, in Fujian.

"A FINE BOATLOAD OF WOODEN NOSES"

Once one of the world's most important ports, Yuegang has become a nondescript industrial suburb. The sole remaining sign of its former prominence is a three-story, hexagonal tower that was once part of the city walls. When I visited not long ago, the gate was locked; I had to wait for a neighbor to show up with a key. Inside the tower were signs of occupation by a homeless man: grubby blankets, empty ramen packets, girlie magazines. All I could see from the top of the tower was a printing plant and a smoky trash dump and the long rectangles of spinach and tobacco fields. The docks where chroniclers had described junks "packed together like fish scales" were almost empty. Only the geography was unchanged: beyond the harbor was the Taiwan Strait, Taiwan itself, and the South China Sea.

By the mid-1580s, barely a decade into the galleon trade, Yuegang was sending twenty or more big junks to the Philippines every March, at the start of the rainy season. (Before the silver boom, just one or two small ships went there, even during the intervals when trade was legal.) As many as five hundred merchants crammed aboard each ship with every imaginable commodity: silk and porcelain, of course, but also cotton, iron, sugar, flour, chestnuts, oranges, live poultry, jam, ivory, gems, gunpowder, lacquerware, tables and chairs, cattle and horses, and whatever else the Chinese thought Europeans might want. "Some just brought little bits of stuff," said Li Jinming, a historian at Xiamen University who is the author of a history of Yuegang. "Whatever they could get their hands on. They could sell it at a big premium." Merchants with little capital could only obtain goods to sell by borrowing at high interest rates. "They had to leave their wives and children with the lender as security," Li told me. "If a trader died, the family was out of luck." Lenders took everything they had to repay the contract. If that was not enough, he said, "the wives and children would become servants. The lender could sell their labor to someone else—it was like slavery."

Typically each ship was chartered by a wealthy trader, who rented space in the hold to the others, usually for 20 percent of the merchant's gross sales. Belowdecks was a warren of sealed, watertight compartments, windowless and barely the size of a closet, in which traders stored their goods. Porcelain was packed tightly in cases, Li said, with rice separating the plates and bowls. "They injected water on all sides, then set down the case in a humid place. It glued the ceramics into a solid, unbreakable mass." Theft was rare aboard ship—thieves couldn't escape with their loot. Nonetheless, merchants brought their own food, slept atop their goods, and stayed inside their dark, noisome compartment during the entire ten-day passage to Manila.

"If they could, they went only once," Li said. Small-scale traders tried to avoid repeat voyages—"the trip was too dangerous." The small islands and shallow water in the harbor restricted shipping to several narrow channels, along which oceangoing vessels had to be slowly pulled by smaller boats. *Wokou* lurked in the fog. To draw pirates from their hiding places, merchants sent out scouts in fast, maneuverable galleys. If they spotted *wokou,* they could skip away with a warning. Because the scouts could not travel as far as the Philippines, the last stage of the outward journey was particularly dangerous. Dutch pirates routinely ambushed Chinese ships on the approach to Manila, seizing everything aboard.

The merchants usually docked at Cavite, a long, skinny peninsula five miles from Manila, on the south side of the great bay.* A crowd of Chinese men—sales agents—awaited them. Cautiously the traders would disembark from their cubicles, blinking in the sun, looking for an agent from their extended family. Agents knew how much silver was in the most recent galleon and could raise or lower prices quoted to the Spaniards accordingly; they also had contacts necessary to bribe colonial inspectors. For their services they charged 20 to 30 percent of the sales price. Only after all the Yuegang traders had chosen agents would the ship be inspected by customs agents, who collected a tax—"three percent on everything to his Majesty," as one Manila governor put it. Then the dickering would begin. Everyone had at most two months to make a deal, because the galleons began leaving in mid-June to avoid typhoon season.

Spanish buyers usually met the agents in the Parián, a Chinese ghetto that was a kind of metastasis of Yuegang, full of Fujianese washed by the silver trade to the Philippines. Located in a swamp outside Manila's walls, the Parián was created in 1583

---

* The tip of the peninsula is Sangley Point, *sangley* (a Fujianese word for "traveling merchant") being a pejorative reference to Filipinos of Chinese descent. A typical use of the term is a Manila church official's complaint in 1628 about the "great danger" posed by the "swarms of abandoned heathen sangleys."

by Spanish officials in an attempt to control the growing number of Chinese, whom they regarded as conniving, job-stealing illegal immigrants. Initially it consisted of nothing but four big shed-like buildings Yuegang traders had built to store their goods. To encourage Manila's Chinese residents to leave their homes and move into their warehouses, the Spaniards announced that any non-Spaniards found outside the Parián after sunset would be executed. In some sense, the quarantine was tit for tat: Europeans were not allowed to set foot in China, so Chinese were restricted from the little piece of Europe in Manila.

Denied permanent access to the European town, the Chinese built their own. Around the warehouses grew a maze of arcade-like shopping areas crammed with intensely competitive stores, teahouses, and restaurants. The narrow streets between were jammed at all hours with men in long, floppy-sleeved robes, embroidered silk shoes, and high round caps. Doctors and apothecaries hawked jars of unguents, tisanes, and medicinal roots. People were buying, selling, and making, arguing over tiny cups of Fujianese tea, racing about with piles of carefully packed bundles, eating foods that appalled the Europeans (a Yuegang favorite: chicken embryos baked inside the egg by burying the eggs in piles of salt and exposing the piles to the sun). It was the first Chinatown in the orbit of the West.

For Spain, the Parián was an oddity and a humiliation. From the beginning, when Spain ejected Muslims and Jews from its kingdom, the empire had what it thought of as a civilizing mission: universal conversion to Christianity. Manila was thronged by missionaries, heads afire with the zeal to bring the Roman Catholic church to Asia. They forced Filipino and Malay natives to adopt the cross, but this was a side project. The true goal, at least at the beginning, was to conquer and convert China. Believing that Cortés (conqueror of Mexico) and Pizarro (conqueror of Peru) had needed only small bands of committed men to seize entire empires for Christ, these clerics and soldiers initially imagined that a few thousand Spaniards could repeat these feats in

Frightened by the crowded Chinese ghetto called the Parián, Manila's few hundred resident Spaniards literally walled themselves off from it. To enter Manila proper, Parián residents had to walk across this moat and through a heavily guarded gate.

China. In Manila, the Ming realm seemed so near—vast riches, spiritual and material, almost close enough to touch. Wiser counsel eventually prevailed, as Manila's governors and the Spanish court concluded that China was too big to conquer. Indeed, the Spaniards in the colony began to worry that China might conquer *them*. Fearing annihilation, they allowed the Chinese an otherwise unthinkable concession: to live in their own infidel quarter, worshipping their own un-Christian idols. They even allowed it to have its own *gobernadorcillo*—a mini-governor.

Parián artisans and shopkeepers sold the Spaniards everything from roof tiles to marble statues of Baby Jesus—"much prettier articles than are made in Spain, and sometimes so cheap that I am ashamed to mention it," wrote Domingo de Salazar, bishop of the Philippines. Colonists flocked to the Chinese ghetto, where stores purveyed the latest European styles. European merchants griped about the competition. The monarchy ordered the shops moved further away, but Spaniards kept coming to them, attracted by the low prices.

The trades "pursued by Spaniards have all died out," Salazar lamented, "because people buy their clothes and shoes from the [Parián]." As a warning, he told the story of a Spanish bookbinder and his Chinese apprentice. After carefully observing the master at his work, the apprentice set up his own shop in the Parián, driving his former master out of business. "His work is so good there is no need of the Spanish tradesman." The Chinese were not universally successful, of course. One shopkeeper sold a wooden nose to a Spaniard who had lost his in a duel. He tried to capitalize on his success by importing "a fine boatload of wooden noses." Sales were poor.

By 1591, twenty years after Legazpi entered Manila, the Parián had several thousand inhabitants, dwarfing the official city, which had only a few hundred European colonists. For the Chinese, the arrangement was convenient. They had created a Chinese city outside of China, where the nominal presence of the Spanish authorities insulated them from the scrutiny of the Ming. To the Spaniards, the ghetto was alarming, alien, an unwelcome necessity. And it was *big*, especially when compared to Manila. Despite constant exhortation, Spaniards refused to settle there in any numbers. The city was too remote, too hot, and, above all, too full of disease, especially what we now know as malaria. European residents often sought cooler air by building homes in the hills around town. By bad luck, the hills are the habitat of the mosquito that is the islands' main malaria vector. The more Europeans escaped the heat, the more they got sick.

The only reason Manila attracted any Europeans at all was because it represented an extraordinary opportunity: China would pay twice as much for Spanish silver as the rest of the world. And its merchants were willing to sell silk and porcelain amazingly cheaply. "The prices of everything are so moderate, it's almost for free," one Spaniard had crowed when the Chinese first arrived in Manila. Yet somehow the deals rarely were as lucrative as the newcomers wanted. To their dismay, the Chinese were always able to play them off against each other, bargaining them down time after time. Sitting in the nexus of exchange made Manila's colonists wealthy, but not as wealthy as they wanted. "Among all those one hundred and fifty families who are settled at Manila, there are not two who are *very* rich," groused the Spanish admiral Hieronimo de Bañuelos y Carrillo in 1638.

Trying to regain the advantage, the Manila government imposed taxes, freight charges, and registration fees on Chinese merchants; they were effectively forced to pay soldiers to stand guard over their property. Angered, the Chinese staged an Ayn Rand–style producers' strike, starving Manila of supplies, and the Spaniards backed down. Frustrated, the king ordered the colony to create a kind of cartel: it would buy all incoming Chinese goods at a single price and "distribute [them] fairly among the citizens." In theory, this would wipe out all the Chinese retailers, which in turn would greatly reduce the Parián, which in turn would greatly reduce Spanish anxieties.

Economics 101 says that cartels rarely work, because individual cartel members will cheat and cut side deals. In this case, Economics 101 was correct. Spaniards made secret arrangements with Chinese traders, paying higher prices for better-quality silk or the first chance to select pieces of porcelain. When the galleons left Manila for Mexico, they met Spanish dories full of contraband silk and silver a few miles outside the harbor.

Madrid was dismayed by the magnitude of the galleon trade—too much silver was going out, and too much silk and porcelain was coming in. Exact figures are not possible to calculate, but somewhere between a third and a half of the silver mined in the Americas went to China, either directly via the galleon trade or indirectly, via Europeans' purchases of Chinese goods shipped overland by Central Asian traders or around Africa by the Dutch and the Portuguese. The monarchy was furious, because the king wanted the silver to buy supplies and pay troops in Spain's innumerable wars. ("The Manila galleon's most fearsome adversary was beyond doubt the Spanish administration itself," the French historian Pierre Chaunu observed.) To prune back the galleon trade, officials cut the number of ships allowed to cross the Pacific to two per year. In response the galleons became enormous, ballooning to two thousand tons. Built by conscripted Malays out of tropical hardwoods, they were castles of the sea. On the Manila-bound lap they carried more than fifty tons of silver—equal, Flynn and Giráldez have calculated, to the combined annual exports of the Dutch East India Company, the English East India Company, and the Portuguese Estado da India.

Much or most of that silver was illegal. Worried Mexican officials informed the monarchy in 1602 that the galleons that year had exported almost four hundred tons of silver—eight times the declared amount. Furious imprecations from Madrid changed nothing; smuggling was too lucrative. "The king of China could build a palace with the silver bars which have been carried to his country . . . without their having been registered," Admiral Bañuelos y Carrillo complained thirty-six years later. In 1654 the *San Francisco Javier* sank near Manila Bay. Its official manifest claimed that it carried 418,323 pesos. Centuries later, divers found 1,180,865 aboard. Even if one assumes, absurdly, that the divers found every last coin, the cargo was almost two-thirds contraband.

To restrict trade on the other side, the government issued import quotas. If the junks brought too much silk or porcelain to Manila, customs officials were supposed to send it back. To get around the quotas, Chinese traders arranged to have their agents meet the junks as they approached the Philippines. Much of the onboard merchandise had been ordered the year before, by Spaniards looking at samples. In a mirror image of the Spanish practice of loading illicit silk and porcelain onto galleons after they left Manila, the Chinese offloaded illicit silk and porcelain from their junks before they arrived. Only after these transactions did the ship officially enter the harbor and let the Spanish harbor patrol guide it to its berth.

Spain had its own silk weavers and dressmakers, as did its colony in Mexico. But the scale of Chinese textile production was so much bigger that Europeans couldn't compete. Indeed, the silver-hungry Ming dynasty actually *forced* farmers to plant mulberry trees, the food for silkworms. Landowners with between five and ten *mu* (one *mu* is about one-sixth of an acre) had to plant, the official history of the dynasty says, "half a *mu* each of mulberry and cotton." Those with ten *mu* or more had to plant "twice as much." Farmers who didn't plant mulberry had "to pay one bolt of silk." Spurred by these decrees, farmers in eastern China covered the hills with mulberry trees. By the 1590s, the Fujianese writer Xie Zhaozhe was reporting areas with "mulberries planted on every foot or inch." Rich farmers, he claimed, devoted "more than a million *mu*" (roughly 130,000 acres) to mulberry trees—entire landscapes of a single species. Working in a frenzy, farmers upriver from Yuegang harvested silk five times a year.

To their north, villages in the lower Yangtze became congested hives of small silk factories, attracting workers from other parts of China and spewing out goods at frightening volume. Yuegang merchants sold this silk in Manila, making profits of 30 to 40 percent. Spanish merchants doubled, tripled, or even quadrupled the price and still sold their goods in the Americas for a third the cost of Spanish textiles. Incredibly, they sold silk from China—silk that had crossed two oceans!—in Spain for less than silk produced in Spain. So much raw silk poured into Mexico that a secondary industry sprang up, with thousands of weavers and dressmakers making clothes from Chinese silk and exporting them throughout the Americas and across the Atlantic.

Yuegang merchants initially exported silk as bolts of fabric. But as they got to know their customers, according to Quan Hansheng, the Taiwanese historian, they acquired samples of Spanish clothing and upholstery and in China made perfect knockoffs of the latest European styles. Into the galleons went stockings, skirts, and sheets; vestments for cardinals and bodices for coquettes; carpets, tapestries, and kimonos; veils, head-dresses, and passementeries; silk gauze, silk taffeta, silk crepe, and silk damask. Packed alongside them were women's combs and fans; spices and incense; gems cut and uncut and mounted into rings and pendants; and, alas, Malay slaves.

Alarmed Europeans saw their textile mills threatened—and fought a covert regulatory war against Chinese competition. They importuned the king to restrict silk imports to bolts of fabric, rather than finished clothing. They insisted that he block direct travel between Manila and any place in the world except Acapulco, so that Chinese imports could be monitored. They demanded that he set import quotas by restricting incoming silk to a given number of chests of a specified size. Chinese merchants evaded every trade barrier, often aided by Spaniards in Manila. They built special chests with false bottoms and sides to conceal pre-made clothing. They sent agents to Acapulco to facilitate smuggling on the Mexican end of the trade. They designed special presses to mash huge quantities of silk into the chests, packing them so tightly, according to Li, the Fujianese scholar, "that a single sea-chest had to be carried by six men."

MAGIC MOUNTAIN

Commercial and political imperatives constantly collided. In 1593 the Manila governor, Gómez Pérez Dasmariñas, decided to fulfill Madrid's long-held dream of conquering the Maluku Islands, the spice centers that Legazpi had failed to seize. The supply of European sailors in Manila being inadequate for the task, Pérez Dasmariñas abducted 250 Fujianese merchants from incoming junks to serve as galley slaves. Protests from Manila's Chinese tradespeople led the governor to promise to release the sailors—and seize the necessary men instead from the Parián. "The next day, all their windows were closed, and the merchants closed their shops," the historian Bartolomé Leonardo de Argensola reported a few years later. "The community was deprived of the provisions which they supplied to it."

After more threats from Pérez Dasmariñas, the Chinese caved in, allowing him to conscript more than four hundred men. In return, he promised to treat the men well. The expedition left in October 1593. Contrary winds and currents made the passage to the Malukus difficult. Fearing the expedition would not reach its destination as fast as he wanted, the governor ordered the conscripts chained to their galley benches, where sailors whipped them to greater effort. As further motivation, he cut their elaborately braided hair. "Such an insult among the Chinese is worthy of death, for they place all their honor in their hair," Argensola wrote. "They keep it carefully tended and gaily decked, and esteem it as highly as ladies in Europa." In a well-planned mutiny, the enslaved Chinese in the flagship killed Pérez Dasmariñas and his crew while they slept, then rowed for Fujian.

To Spaniards, the lesson from Pérez Dasmariñas's death was clear: the Chinese were untrustworthy and dangerous. The Manila government evicted twelve thousand of them in 1596. In a few years they were as numerous as before and the government was planning more deportations. Anti-immigration firebrands like Bishop Salazar's successor, Miguel de Benavides, wanted to eject every Chinese person from the islands. There can be no exceptions, he told the king. Spanish businesses would take advantage of any loopholes and hire illegal immigrants. If a hundred Chinese were legally allowed in the Philippines, Benavides predicted, "ten thousand will remain."

Into the festering situation blithely sailed three Chinese high officials. Arriving without notice in May 1603, they emerged from a Chinese warship in sedan chairs ringed by bodyguards and led by drummers and musicians; at the head of the parade marched two men who cried, "Make way for the mandarins!" If Parián residents failed to prostrate themselves, one eyewitness reported, the bodyguards flogged them. The three visitors were the chief military official in Fujian, the county magistrate for Yuegang, and a high-ranking eunuch from Beijing. They had been sent by the emperor to present a letter to Manila's governor, Pedro Bravo de Acuña. It is hard to imagine what Acuña thought as he heard the contents of the letter. Rumors were circulating in China, it explained, of a magic mountain in Cavite, loaded with gold and silver, all free for the taking. The three visitors had been sent to ascertain whether the mountain existed.

To judge by the appalled reports in Chinese records, the expedition seems to have originated in some daffy con job that bubbled through the court bureaucracy—not the only incident of its type in the Ming dynasty. But to the Spaniards, who watched the mandarins comb the colony for gold and silver, the visitors looked like a scouting party for an invasion—a Ming-style Trojan horse. Surely these people could not be the pack of bumblers they appeared to be—they must be part of a sinister plot. While Gov-

Potosí's mines still operate, though at a low level. Scouring the mountain for its last grains of metal, poor miners hack away in lightless tunnels. Conditions are dismayingly like those in centuries past, except that the men are mining zinc and tin, rather than silver.

ernor Acuña debated whether to kill the officials, they apologized for the mix-up and suddenly left.

Fearing the departure signaled an imminent invasion, Acuña ordered his forces to demolish Chinese houses that were too close to Manila's defensive wall, register every Chinese person in the Parián—and buy or confiscate every Chinese weapon.

What happened next is difficult to sort out, because Spanish and Chinese accounts of events differ radically. In the Spanish version, angry Chinese gathered outside the city walls to protest. Acuña sent seventy soldiers, led by his nephew, to quell the protest. Unprovoked, the Chinese mob attacked the soldiers, killing all but four, and fled to the hills outside Manila. After restoring order to the Parián, the government sent a peace emissary to the rioters. The Chinese treacherously slew him and went on a rampage. Naturally, the official reports explain, the government had to protect the citizenry. It sent out more troops. The Chinese in the hills resisted. But they had few weapons and inevitably suffered heavy losses.

Eleven years after the killings, the Ming geographer Zhang Xie wrote *Studies on the East and West Oceans,* a summary of Chinese foreign relations. In it was an account of the incident from the point of view of Parián residents—an account that included a few details that Spanish officials had neglected to mention. Zhang agreed that the Spaniards had entered the Parián and "bought every bit of iron in Chinese hands at a hefty price," supposedly to make cannons. But from that point his account was quite different. There was no angry mob outside the walls, no unprovoked slaughter of soldiers. Instead the government, having effectively disarmed the Chinese, announced a formal residency check, during which they divided the ghetto into groups of three hundred, placed each group in a separate courtyard—and slaughtered it. As word of the massacre leaked out, Zhang wrote, thousands of Chinese fled to the hills outside Manila. After an inconclusive clash, the Spaniards sent a peace emissary. "The Chinese worried that it was a trick to lure them out, and killed the emissary," Zhang admitted. "The barbarians grew angry, and set up an ambush outside the city." When the Chinese ran out of food, they decided to take it from Manila—and walked into the Spanish ambush. Three hundred Spaniards died in the ensuing battle. So did as many as twenty-five thousand Chinese, most of them Fujianese. According to Zhang, only three hundred Chinese survived. A second wave of deaths followed, as some of the new widows in Fujian, many of whom now faced bondage to pay their husbands' debts, killed themselves.

Incredibly, the massacre had no real consequences. Just months after wiping out the Chinese in Manila, the city fathers put down the welcome mat for new immigrants. And Spanish merchants were begging the junks to return—they wanted to buy cheap Chinese silk. "The Spaniards who had seen the Chinese as such a grave threat that their survival depended on their complete disappearance didn't hesitate to do everything possible to ensure the massive repopulation of the Parián," Manel Ollé Rodríguez, a Barcelona historian of Spanish-Chinese relations, observed in 2007.

In Beijing, the Wanli emperor decided that the three treasure mountain officials were responsible for the explosion in the Parián, and ordered them beheaded. Although he accused the Spaniards of "murder[ing] people without license," he also conceded that the dead Fujianese "were base people, ungrateful to China, their native country, to their parents, and to their relatives, since so many years had passed during which they had not returned to China." Translation: they weren't worth the cost of a punitive expedition. Besides, the government still needed the silver.

Within two years the galleon trade and the Parián were almost back to normal. "The Chinese gradually flocked back to Manila," reported the *Ming Shi,* the official history of the dynasty. "And the savages [Spaniards], who saw profit in the commerce with China, did not oppose them. The Chinese population began to grow once more."

Because the situation had reverted to its pre-massacre state, the Spaniards in Manila were as few, dependent, and scared as ever they had been. Eventually they again tightened restrictions on the Chinese. Rebellions flowered in the Parián, followed by expulsions and massacres. The cycle repeated itself in 1639, 1662, 1686, 1709, 1755, 1763, and 1820, each time with an awful death toll.

To modern eyes, the scenario is hard to credit: why would the Chinese keep returning? It is one thing to take a onetime risk, as small-time Fujianese traders did when they made their single visits to Manila; it is another to set up an establishment that will be sacked every few years, with great loss of life. During these incidents the Parián Chinese frequently managed to kill a third or more of the Europeans in the Philippines, as they did in 1603. Yet Manila's merchants invariably invited them back, even smuggling their potential executioners past customs. Why would they repeatedly set up a situation in which they had an excellent chance of being killed?

In *Power and Plenty* (2007), a history of trade in the last millennium, Ronald Findlay and Kevin O'Rourke suggest one way to think about this situation. When economics textbooks describe trade, Findlay and O'Rourke write, they describe two countries "endowed with a certain amount of the various factors of production—land, labor, capital, and so on." The two nations have technologies that transform those factors into goods, "together with a set of preferences over those goods." Private entities within each nation "trade with each other, or not as the case may be, and the consequences of trade are derived for consumers and producers alike."

Typically one nation (the United States, for example) can produce Good A (grain, say) more cheaply than another nation (Japan), while the second more efficiently makes Good B (consumer electronics). By exchanging Good A for Good B (that is, wheat for TV sets) both nations will be better off—a true win-win situation. This is the theory of "comparative advantage," a building block of economics. Vast amounts of evidence support the theory's veracity, which is why almost all economists firmly believe it, and firmly support free trade, which maximizes the potential for all sides to benefit.

In the textbooks, government appears mainly as an outside factor that imposes tariffs, quotas, levies, and so on, influencing the outcome of private trade, often reducing the net economic benefit. But the state does this because trade has two roles: one highlighted in economics textbooks, where private markets allow both sides to gain economically, and one that rarely appears in those textbooks, in which trade is a tool of statecraft, the goal is political power, and both sides usually do not win. In this second role, the net economic benefit of trade is much less important than the political benefit to each side, and the government interventions that exasperate economists can be useful, even vital tools to achieve national preeminence.*

The greatest expansions of world trade have tended to occur when both roles are aligned and commercial ambitions can be enforced, as Findlay and O'Rourke put it, "with the barrel of a Maxim gun, the edge of a scimitar, or the ferocity of nomadic horsemen." Today violence is less common, if only because powerful weapons are so cheap that all sides have them, and states tend to make do with tools like subsidies to industry, exchange-rate manipulation, and export and import regulations. But still today trade expands when government sees it as a way to project and increase power—witness the recent history of Japan and China.

At the same time, the two roles often conflict, and the conflict leads, as in Manila, to a considerable amount of profoundly frac-

---

* In practice, the picture is complicated by business's attempts to manipulate government for their own ends, often to the detriment of state policies, and by groups within the state that use power for private gain. Nevertheless, the distinction between trade as a private exchange between willing parties and trade as a tool of state aggrandizement is useful. Indeed, one reason for the conflict between today's free traders and anti-globalization activists is that the former regard the first role as paramount whereas the latter think in terms of the second.

tured cerebration. For Spain, Manila was both a trading post and a projection of Spanish power in the Pacific. Its traders wanted to generate as much profit as possible by importing as much silk as possible; its political rulers, by contrast, wanted to seize Asian lands, convert Asians to Christianity, thwart Dutch and Portuguese ambitions—and have as much of the silver as possible come to Spain, because the state needed it to pay for wars in Europe. Considered purely as a trading entrepôt, Manila should house as few Spaniards as possible—they were expensive to send over and kept dying of disease—and let Chinese people do all the work. To serve best as an imperial outpost, though, the Spaniards needed to ensure that all vital civic functions were in loyal Spanish hands, and minimize the number and influence of the Chinese. Every step to satisfy one imperative worked against the other.

Like the Spanish court, the Ming court struggled to reconcile the divergent roles of trade. On the one hand, silver from the silk trade became a source of imperial wealth and power. American silver helped pay for huge military projects, including much of the Great Wall of China, which the Ming were revamping and extending. And it fueled an explosion of commerce within China, which led to an economic boom. On the other hand, the money that enabled business to grow also set off inflation, which had its worst impact on the poor. And silver was ever a political threat to the dynasty, because it controlled neither the trade nor the source. Alarmingly, the emperors could not restrict the flow of silver into Fujian, even if they wanted to, because of rampant smuggling. In the eyes of the court, the Fujianese merchants were people of dubious loyalty who had created in the Parián an important Chinese city that was outside imperial supervision. They were becoming wealthy and powerful in a way that was hard for the court to control. Little wonder Beijing kept a wary eye on Yuegang!

There is little evidence, though, that Beijing anticipated the worst consequences. As in Europe, so much silver flooded into China that the price eventually dropped. By about 1640 silver was worth no more in China than it was in the rest of the world. At this point the Ming government was tripped up by an error it had committed decades in the past.

When the court had ordered citizens to pay their taxes in silver, it had set up the tax rolls in terms of the *weight* of silver people had to pay, not its *value*. As with Spain, the taxes were not indexed for inflation. As with Spain, the same amount of tax was worth less money when the price of silver dropped. The Ming dynasty had a revenue shortfall. Not having paper currency, the government couldn't print more money—deficit spending was impossible. Suddenly it couldn't pay for national defense. It was a bad time to run out of money for the military: China was then under assault by the belligerent northern groups now called Manchus. According to William Atwell, a historian at Hobart and William Smith Colleges, the Chinese government's dependence on the silver trade helped push it over the edge. The takeover by the Manchus—they became the Qing dynasty—took decades and was bloody even by the tough standards of Chinese history. Nobody knows how many millions died.

Atwell's contentions have been vigorously debated, yet there is little doubt that China's entry into the galleon trade had consequences of a sort rarely discussed in freshman economics textbooks. Flynn and Giráldez point out that China devoted a big fraction of its productive base to acquiring the silver needed for commerce and government. For hundreds of years, China produced silk, porcelain, and tea to acquire a commodity, silver, which was needed to replace the paper notes that the government had made valueless. It was as if to buy a newspaper for a dollar one first had to make and sell something else to get the dollar banknote. Actually, it was worse: the silver stocks had to be constantly replenished, incurring further costs, because the metal was constantly worn away as it passed from hand to hand. (Paper money wears out, too, but costs next to nothing to replace.)

Given the circumstances, acquiring the silver was entirely rational—it provided monetary stability. But it was also extremely costly. "Rather than pull silver out of their own ground (had China contained rich silver deposits, which it did not), the Chinese produced exports to buy silver that was pulled out of the ground somewhere else," Flynn wrote in an e-mail to me. "Even scholars tend to impute mystical qualities to commodity monies like silver and gold, but we must recognize them as physical products that involve massive production costs. A significant hunk of the GDP of China—then the world's biggest economy—was surrendered in order to secure a white metal that was produced mostly in Spanish America and Japan. Some people made enormous profits from doing this, but think about what else those resources could have been used for."

There was a related, equally large consequence: the Columbian Exchange.

5

Lovesick Grass, Foreign Tubers, and Jade Rice
(Silk for Silver, Part Two)

HIDDEN PASSENGERS

Trade brought more than silver across the Pacific. Tobacco may have led the parade. Somehow Portuguese ships brought the species across oceans and borders to Guangxi, in southern China, where archaeologists have unearthed locally made tobacco pipes dating back to 1549.* Little more than two decades later, the plant arrived in the southeast, aboard a silver ship from Manila. Not long after that, it filtered into the northeast, probably from Korea.

---

* The reader will have noted that I barely mention Dutch and Portuguese trade in Asia, which centered on spices, and focus on Spain and the galleon trade. This is partly to simplify a complex narrative, but mainly because the Spanish empire, the first truly global enterprise, is more germane to this book. In addition, the Netherlands and Portugal were entangled with that empire: the former not wresting full independence from it until 1648; the latter, long independent, forced by dynastic mishaps to accept a Spanish king from 1580 to 1640.

Nicotiana tabacum was as much an object of fascination in Yue-gang as in London and Madrid. "You take fire and light one end [of the pipe] and put the other end in your mouth," explained the seventeenth-century Fujianese poet Yao Lü. "The smoke goes down your throat through the pipe. It can make one tipsy." Writing not long after the smoking weed arrived in Fujian, Yao was amazed by its rapid spread across the province. "Now there is more here than in the Philippines," he marveled, "and it is exported and sold to that country."

Then as now, smoking was made to order for the boredom and inertia of army life. Tobacco was embraced by Ming soldiers, who disseminated it as they marched around the empire. In the southwestern province of Yunnan, one physician reported, Chinese soldiers "entered miasma-ridden [malarial] lands, and none of them were spared disease except for a single unit, whose members were in perfect health. When asked the reason, the answer was that they all smoked." (Mosquitoes dislike smoke, so smoking actually may have provided some protective effect against malaria-carrying insects.) From that point, the account continued, "smoking spread . . . and now in the southwest, whether old or young, they cannot stop smoking from morning until night." As a child in the 1630s, the writer Wang Pu had never heard of tobacco. When he grew to adulthood, he later recalled, "customs suddenly changed, and all the people, even boys not four feet tall, were smoking."

"Tobacco is everywhere," announced what was apparently China's first smoking how-to book. Calling the plant "golden-thread smoke" and "lovesick grass"—the latter a nod to its penchant for hooking the user—the Qing dynasty's legions of smokers may have been the planet's most enthusiastic nicotine slaves. An ostentatious addiction to tobacco became the hallmark of the fashionable rich. Men boasted of their inability to eat, converse, and even think without a lighted pipe. Women carried special silk tobacco purses with elaborate jeweled fastenings; to protect their delicate feminine essences from the harsh spirit of tobacco, they smoked extra-long pipes, some so big that they had to be lugged around by servants. A new poetic sub-genre emerged among China's wealthy aesthetes: the hymn to tobacco.

> *Puffing fragrance, exhaling the Sage's vapor;*
> *Bluish tendrils born from the subtle Smoke.*
> *The Gentleman's Companion, it warms my heart*
> *And leaves my mouth feeling like a divine furnace.*

Late-waking aristocratic women slept with their heads elevated on special blocks so that attendants could do their hair and makeup while they were unconscious—it shortened the time between waking and the first tobacco of the day. "The scene is a little hard to imagine," remarked Timothy Brook, the Canadian historian whose studies of Chinese tobacco I am drawing upon here.

Brook found the tale of the sleeping smokers in Chen Cong's *Yancao pu* (Tobacco Manual), a learned collection of tobacco-related poetry and prose from 1805. An even more recondite compendium, Lu Yao's *Yan pu* (Smoking Manual), appeared around 1774. Lu, a former provincial governor, laid down the rules for nicotine consumption in aristocratic circles. Like a modern etiquette handbook, the manual provided a set of smoking do's and don'ts:

> *Do* smoke: after waking up; after a meal; with guests; while writing; when growing tired from reading; while waiting for a good friend who hasn't shown up yet.
> *Don't* smoke: while listening to a zither; feeding cranes; appreciating orchids; observing plum blossoms; making ancestral offerings; attending the morning court assembly; sleeping with a beautiful woman.

From today's perspective, the Chinese courtier's ornate surrender to tobacco seems absurd, but it had many equally odd counterparts abroad. At the same time that Lu Yao was laying out smoking etiquette, wealthy English were taking snuff (finely ground tobacco stems) in public sessions heavy with ritual. Opening their silver or ivory snuffboxes—"a fetish of the eighteenth century," as the anthropologist Berthold Laufer put it—fashionable young blades scooped out measures of fresh-ground snuff with finger-length ladles made of bone. Parties fell quiet as groups of men in embroidered waistcoats simultaneously inserted tiny pucks of ground tobacco into their noses, then whipped out lace handkerchiefs to muffle the ensuing volley of sneezes. Mastering the arcana of snuff was, for the addict, worth the bother: snorted tobacco delivers nicotine to the bloodstream faster than cigarette smoke. Few were more enraptured by the ritual than the celebrated London dandy Beau Brummell, who claimed to have a different snuffbox for every day of the year. Brummell instructed his fellow gallants in the subtle art of using only one hand to open the box, extract a pinch of snuff, and stick it in a nostril. The injection had to be accomplished with a rakish tilt of the head to avoid unsightly brown drips.

Snuff mania had few consequences in England other than interrupted party chatter, high laundry bills, and nasopharyngeal cancer. China's tobacco addiction occurred in an entirely different context, and thus had an entirely different impact. *N. tabacum* was part of an unplanned ecological invasion that shaped, for better and worse, modern China.

At the time, China had roughly a quarter of the world's population, which had to provide for itself on roughly a twelfth of the world's arable land. Both figures are imprecise at best, but there is little dispute that the nation has long had a lot of people and that it always has had relatively little land to grow crops to feed them. In practical terms, China had to harvest huge amounts of food—half or more of the national diet—from areas with enough water to grow rice and wheat. Unluckily, those areas are relatively small. The nation has many deserts, few big lakes, irregular rainfall, and just two major rivers, the Yangzi and the Huang He

(Yellow). Both rivers run long, looping courses from the western mountains to the Pacific coast, emptying into the sea scarcely 150 miles from each other. The Yangzi carries mountain runoff into the rice-growing flats near the end of its course. The Huang He takes it into the North China Plain, then as now the center of Chinese wheat production. Both areas are vital to feeding the nation; there are no other places in China like them. And both are prone to catastrophic floods.

Song and Yuan, Ming and Qing—every dynasty understood both this vulnerability and the concomitant necessity of maintaining China's agricultural base by controlling the Yangzi and Huang He. So important was water management that European savants like Karl Marx and Max Weber identified it as China's most important institution. Creating and operating huge, complex irrigation systems, Weber claimed, required organizing masses of laborers, which inevitably created a powerful state bureaucracy and subjugated the individual. In an influential book from 1957, the historian Karl Wittfogel built on Marx to describe China and places with similar water-control needs as "hydraulic societies." Wittfogel's view of these societies can be gathered from the title of his book: *Oriental Despotism*. Europe, to his mind, avoided despotism because farmers didn't need irrigation. They fended for themselves, which created traditions of individualism, entrepreneurship, and technological progress that China never had. In recent years this thesis has fallen out of favor. Most Sinologists today believe that hydraulic Asia was just as diverse, individualistic, and market oriented as anywhere else, including non-hydraulic Europe. But this image still remains influential, at least in the West, where China is all too often viewed as an undifferentiated mass of workers, moving ant-wise to the directives of the state.

None of the challenges to past thinkers dispute that China had a relative dearth of land suitable for rice and wheat. From this perspective, the Columbian Exchange was a boon, and China raced to embrace it. "No large group of the human race in the

Old World was quicker to adopt American food plants than the Chinese," Alfred W. Crosby wrote in *The Columbian Exchange.* Sweet potatoes, maize, peanuts, tobacco, chili peppers, pineapple, cashew, manioc (cassava)—all poured into Fujian (via the galleon trade), Guangdong (the province southwest of Fujian, via Portuguese ships in Macao), and Korea (via Japan, which took them from the Dutch). All became part of the furniture of Chinese life—who can imagine Sichuan (Szechuan) food today without heaps of hot peppers? "While men who stormed Tenochtitlan with Cortés still lived," Crosby said, "peanuts were swelling in the sandy loams near Shanghai; maize was turning fields green in south China; and the sweet potato was on its way to becoming the poor man's staple in Fujian." Today China is the world's biggest sweet potato grower, producing more than three-quarters of the global harvest; it is also the world's second-biggest maize producer.

Epitomizing China's readiness to experiment was the Yuegang merchant Chen Zhenlong, who came across sweet potatoes (*Ipomoea batatas*) during a visit to Manila in the early 1590s. Prob-

Sweet potatoes in China are often eaten raw, the skin whittled off in a fashion that makes them somewhat resemble ice cream cones.

ably native to Central America, *I. batatas* had been encountered by Colón on his first voyage; Spaniards had brought the species to the Philippines, where it was quickly adopted by Malays, who already grew the tuber crop taro. Liking the taste, Chen decided to take sweet potatoes home with him. "He bribed the barbarians to get segments of their vines several feet in length," reported his great-great-great-grandson in *True Account of the Story of Planting Sweet Potatoes in Qinghai, Henan, and Other Provinces* (1768), a book-length essay devoted to bragging about the sweet potato feats of the author's ancestors. Chen hid the vines by twisting them around ropes and tossing the ropes into a basket. Spanish customs agents noticed nothing. (They weren't trying to stop the export of sweet potatoes per se, so much as trying to prevent the Chinese from getting their hands on *anything* from which they might derive commercial advantage.) In this way Chen smuggled sweet potatoes into China. "Even though the vines were withered," his great-great-great-grandson wrote later, "they flourished after he stuck cuttings in infertile ground."

The 1580s and 1590s, an intense point in the Little Ice Age, were two decades of hard cold rains that flooded Fujianese valleys, washing away rice paddies and drowning the crop. Famine shadowed the rains. Poor families were reduced to eating bark, grass, insects, and even the seeds found in wild-goose excrement. Chen Zhenlong and his friends seem initially to have thought of the *fanshu*—foreign tubers—as an amusing novelty; they gave them away as presents, a slice or two at a time, neatly wrapped in a box. (Botanically speaking, *fanshu* is a misnomer; *I. batatas* actually has a modified root, rather than a proper tuber.) As hunger tightened its grip, Chen's son, Chen Jinglun, showed the *fanshu* to the provincial governor, to whom he was an adviser. The younger Chen was asked to conduct a trial planting near his home. Successful results persuaded the governor to distribute cuttings to farmers and instruct farmers how to grow and store them. "It was a great fall harvest; both near and far food was abundant and disaster was no longer a threat," exulted the great-great-great-grandson. Near

Yuegang, as much as 80 percent of the locals were living on sweet potatoes.*

Governmental promotion of foreign crops was nothing new in Fujian. Sometime before 1000 A.D., Fujianese merchants brought in a novel type of rice—early-ripening Champa rice—from Southeast Asia. Because the new rice matured quickly, it could be planted in areas with shorter growing seasons. After intensive breeding, farmers created varieties that grew quickly enough to let them plant two crops a year in the same field—one of rice, then a second of wheat or millet. Harvesting twice as much from the same amount of land, Chinese farms became more productive than farms elsewhere in the world. The then-ruling Song dynasty actively promoted the new rice, distributing free seeds, publishing illustrated how-to brochures, sending out agents to explain cultivation techniques, and even providing some low-interest loans to help smallholders adapt. This aggressive adaptation and promotion of a new technology was a key reason for the nation's subsequent prosperity, and its preeminence.

Still, Fujian was lucky that sweet potatoes arrived when they did. The crop spread through the province just in time for the fall of the Ming dynasty, which ushered in decades of violent chaos. Incoming Manchu forces seized Beijing in 1644, beginning a new dynasty: the Qing. The last Ming emperor hanged himself, and pretenders emerged to lead a rump state. Initially it was based in Fujian. In a disordered interlude, pieces of the Ming mil-

---

* Chen was not the only sweet-potato smuggler. According to a nineteenth-century gazetteer, the Chinese doctor Lin Huailan successfully treated a sick Vietnamese princess in 1581. At a banquet in his honor, he was served sweet potatoes. Vietnam had banned exporting the tuber to China "on penalty of death," the gazetteer recounted, but Lin decided to take some anyway. "While crossing the border, he was questioned by a [Vietnamese] border official. Lin answered truthfully, and requested that the officer secretly let him through. The officer said: 'As for what happens today, being a servant of the country, it would be disloyal of me to let you pass; however, being grateful for your virtues, to deny you would be unrighteous.' He then drowned himself. Lin returned, and the tuber spread across Guangdong."

itary splintered away and became, in effect, *wokou*. Meanwhile
actual *wokou* took advantage of the confusion. To deny supplies
to the Ming/*wokou*, the Qing army forced the coastal population
from Guangdong to Shandong—the entire eastern "bulge" of
China, a 2,500-mile stretch of coastline—to move en masse into
the interior.

Beginning in 1652, soldiers marched into seaside villages and
burned houses, knocked down walls, and smashed ancestral
shrines; families, often given only a few days' warning, evacuated
with nothing but their clothes. All privately owned ships were set
afire or sunk. Anyone who stayed behind was slain. "We became
vagrants, fleeing and scattering," one Fujianese family history
recalled. People "simply went in one direction until they halted,"
another said. "Those who did not die scattered over distant and
nearby localities." For three decades the shoreline was emptied
to a distance inland of as much as fifty miles. It was a scorched-
earth policy, except that the Qing scorched the enemy's earth,
not their own.

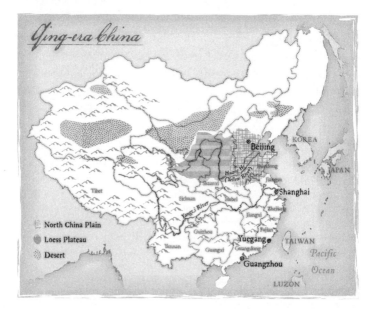

For Fujian, the coastal evacuation amounted to a spectacularly harsh version of the Ming dynasty's ban on overseas trade. In the 1630s, before the political convulsions and the trade bans, twenty or more big junks went to Manila every year, each carrying hundreds of traders. During the evacuation, the number fell to as low as two or three, all illicit. Like the Ming trade bans, the Qing coastal clearance effectively turned over the silver trade to *wokou*.

As it happened, the trade was turned over to one pirate in particular: Zheng Chenggong, known in the West as Koxinga (the name is a corruption of a Chinese honorific). Born in Japan to a Japanese mother and a Fujianese Christian father who was a prominent pirate, Zheng had spent his life flouting Ming law. When the Qing came in, he realized that *wokou* were better off with the lackadaisical, corrupt Ming. He became an admiral in the rump Ming state and led an enormous sea-based assault on the Qing that came close to toppling the new regime. Afterward, he returned to piracy, amassing a fleet that one eyewitness, a Dominican missionary in Fujian, estimated at fifteen to twenty thousand vessels and an army of "a hundred thousand men at arms, all the necessary sailors, and eight thousand horses." Based in a palace in Amoy (now called Xiamen), a city across the river from Yuegang, Zheng controlled the entire southeast coast—a true pirate king.

Manila's traders, having no alternative, begged Zheng in 1657 to buy their silver. When his ships appeared in the harbor, the galleon trade resumed. Perhaps distracted by his running battle with the Qing, Zheng took longer than one would expect to realize that a) the Spaniards in the Philippines had no source but him for silk and porcelain; and b) he, Zheng, was a pirate with a large army. Not until 1662 did he dispatch the Dominican missionary—dressed in the rich robes of an imperial emissary—to Manila with a proposal to change the terms of trade. The Spaniards would give him all their silver, as before. In return, Zheng would not kill them. Panicked, the Spanish governor decided to expel the Chinese in the Philippines. Refusing to be ejected, they barricaded themselves in the Parián. As was by that point customary, Spanish troops forcibly rounded up the Chinese, slaughtering many and forcing the rest to leave Manila on jam-packed ships. The precaution turned out to be needless; just two months later, Zheng died unexpectedly, probably of malaria. His sons fought over their inheritance, and the Manila trade was left alone.

The Qing had ordered the coastal evacuation, but it had disastrous consequences for them, too. As the treasury official Mu Tianyan complained, closing down the silver trade effectively froze the money supply. Because silver was always being wasted, lost, and buried, the pool of Chinese money was actually shrinking. "Every day there is less and less to meet the demand, with no way to restore it," Mu wrote to the emperor. When the money supply falls, each unit becomes more valuable; prices fall in a classic deflationary spiral. To stop the importation of silver "yet desire the wealth of fortune and convenience of use," Mu asked, "how is this different from blocking a source of water while expecting to benefit from its flow?" Reluctantly agreeing, the Qing lifted the ban in 1681.

Meanwhile, though, coastal people had flooded into the mountains of Fujian, Guangdong, and Zhejiang. Inconveniently, these areas were already inhabited. Most of the inhabitants belonged to a different ethnic group, the Hakka, famed for their *tulou*—fortress-like complexes, usually but not always circular, whose earthen outer walls contain scores of apartments, all facing onto a central courtyard. (Today these amazing structures are a tourist attraction.) Decades before the expulsion, the Fujianese scholar Xie Zhaozhe had observed that the Hakka in the hills were packed into every scrap of available real estate:

> There is not an inch of open ground. . . . Truly as someone once said, "Not a drop of water goes unused, and as much as possible even the most rugged parts of mountains

Thousands of *toulou*, clan dwellings of the Hakka, still dot the mountains of Fujian. Made from rammed earth mixed with rice stalks, they had no windows on lower floors as a defensive measure.

are cultivated." One could say that there isn't a bit of land left.

Unable to support themselves, poor Hakka and other mountain peoples had been emigrating north and west for a century, renting uninhabited highland areas—terrain too steep and dry for rice—in neighboring provinces. They cut and burned the tree cover and planted cash crops, mainly indigo, in the exposed earth. After a few years of this slash-and-burn the thin mountain soil was exhausted and the Hakka moved on. ("When they finish with one mountain, they simply move on to the next," the geographer Gu Yanwu complained.) As coastal refugees poured into the mountains, the highland exodus accelerated.

Landless and poor, the Hakka refugees were mocked as *pengmin*—shack people. Strictly speaking, shack people were not vagabonds; they rented land in the heights that was owned but not used by farmers in the more fertile valleys. Shifting from one temporary home to the next, *pengmin* eventually occupied a crooked, 1,500-mile stretch of montane China from the sawtooth hills of Fujian in the southeast to the silt cliffs around the Huang He in the northwest.

Neither rice nor wheat, China's two most important staples, would grow in the shack people's marginal land. The soil was too thin for wheat; on steep slopes, the irrigation for rice paddies requires building terraces, the sort of costly, hugely laborious capital improvement project unlikely to be undertaken by renters.

Almost inevitably, they turned to American crops: maize, sweet potato, and tobacco. Maize (*Zea mays*) can thrive in amazingly bad land and grows quickly, maturing in less time than barley, wheat, and millet. Brought in from the Portuguese at Macao, it was known as "tribute wheat," "wrapped grain," and "jade rice." Sweet potatoes will grow where even maize cannot, tolerating strongly acid soils with little organic matter and few nutrients. *I. batatas* doesn't even need much light, as one agricultural reformer noted in 1628. "Even in low, narrow, damp alleys, where there is only a few feet of ground, if you can look up and see the sky, you can plant them there."

In the south, many farmers' diets revolved around the sweet potato: sweet potatoes baked and boiled, sweet potatoes ground into flour for noodles, sweet potatoes mashed with pickles or deep-fried with honey or chopped into stew with turnips and soybean milk, even sweet potatoes fermented into a kind of wine. In the west, China was a land of maize and another American import: potatoes, originally bred in the Andes Mountains. When the wandering French missionary Armand David lived in a hut in remote, scraggy Shaanxi, his meal plan would not have been out of place, except for a few garnishes, in the Inka empire. "The only plant cultivated near our cabin is the potato," he noted in 1872. "Maize flour, along with potatoes, is the mountain peoples' daily diet; it's usually eaten boiled and mixed with the tubers."

Nobody knew how many shack people were in the hills. Hoping, perhaps, that hiding the problem would avoid the need to solve it, Qing bureaucrats left them out of census reports. But all evidence suggests that the number was not small. In Jiangxi, Fujian's western neighbor, the rigid, nit-picking provincial treasurer insisted in 1773 that the shack people, many of whom had lived in Jiangxi for decades, counted as actual inhabitants of the province and therefore should be included in the reports sent to Beijing. He dispatched field workers to enumerate every Hakka head and every Hakka shack. In rugged Ganxian County, they tallied 58,340 settled inhabitants, most in the main town of Ganzhou—and 274,280 shack people in the surrounding slopes. In county after county the story was repeated, sometimes with a few thousand wanderers, other times a hundred thousand or more. Hidden from the government, more than a million shack people had been slashing and burning their way across Jiangxi. And that, as the Qing court must have realized, was only one medium-sized province.

Coupled with the outflow of shack people was a second, parallel, even bigger wave of migration into the parched, moun-

tainous, thinly settled west. In their quest for social stability, the
Ming had prohibited people from leaving their home regions.
Reversing course, the Qing actively promoted a westward movement. Much as the United States encouraged its citizens to move
west in the nineteenth century and Brazil provided incentives to
occupy the Amazon in the twentieth, China's new Qing masters
believed that filling up empty spaces was essential to the national
destiny. ("Empty," that is, from the Qing point of view; dozens
of non-Chinese peoples—Tibetans, Yao, Uighurs, Miao—lived in
them. By sending in people from the center, the Qing were forcibly incorporating these previously autonomous cultures into the
nation.)* Lured by tax subsidies and cheap land, migrants from
the east swarmed into the western hills. Most of the newcomers
were, like the shack people, poor, politically luckless, and scorned
by urban elites. They looked at the weathered, craggy landscape,
so unwelcoming to rice—and they, too, planted American crops.

China's fifth-largest province is Sichuan, adjacent to Tibet
and nearly as alpine. Back in 1795, according to Lan Yong, a historian at Sichuan's Southwest University, it was a big, roomy
place: more land than California, a population as low as 9 million.
Just 2,300 square miles of its surface, an area half the size of Los
Angeles County, were considered arable. During the next twenty
years, Lan has written, American crops moved into the ridges and
highlands, increasing the pool of farmland to almost 3,700 square
miles. As Sichuan's agricultural capacity increased, its population
increased in tandem, to 25 million. Something similar occurred in
Shaanxi Province, Sichuan's even emptier neighbor to the northeast. Migrants poured into the steep, arid hills along the border
between them, knocking down the trees that clung to the slopes
to make room for sweet potatoes, maize, and, later, potatoes.
The amount of cropland soared, followed by the amount of food

* The ethnic group generally indicated by the word "Chinese" is the Han.
The Manchu were pushing Han from the Chinese core into peripheral areas
settled by other peoples.

grown on that cropland, and then the population. In some places the number of inhabitants increased a hundredfold in little more than a century.

For almost two thousand years, China's numbers had grown very slowly. That changed in the decades after the violent Qing takeover. From the arrival of American crops at the beginning of the new dynasty to the end of the eighteenth century, population soared. Historians debate the exact size of the increase; many believe the population roughly doubled, to as much as 300 million people. Whatever the precise figure, the jump in numbers had big consequences. It was the demographic surge that transformed the nation into a watchword for crowding.

China was not the only Asian nation transformed by the Columbian Exchange. Sweet potatoes became a staple in a broad swath extending from Tahiti to Papua New Guinea, and from New Zealand to Hawai'i. Surprisingly, *I. batatas* was known in much of this area before Columbus—archaeologists have found burned remains of the plant dating back as early as 1000 A.D. in Hawai'i, Easter Island, the Cook Islands, and New Zealand.

Maize at the edge of the Gobi Desert, in Inner Mongolia

(Some researchers view the species's movement across the Pacific as evidence of contact between ancient Polynesians and American Indians; others argue that the seeds, which are contained in small, buoyant, spherical capsules, must have floated across the sea.) Initially it had little impact. But about the time that the Spaniards arrived in Manila *I. batatas* was displacing native crops like yam, sago, and banana. As had the Chinese, islanders were using sweet potato's high yields and tolerance of bad soil to move into highland areas that had been lightly settled before. New Guinea was so transformed that some archaeologists speak of an "Ipomoean revolution." Still, the impact in China was bigger, if only because China is so big, and because the country had a centralized government that could enforce policies that spread sweet potato.

Were maize, potatoes, and sweet potatoes entirely responsible for China's population boom? No. American plants arrived as the Qing were transforming China. Ambitious on many fronts, the dynasty fought disease and hunger, the nation's two major killers, by enacting a program, the world's first, of smallpox inoculation; expanding a nationwide network of granaries that bought surplus grain and sold it at low, state-controlled prices during shortages; and implementing what were, for the era, sophisticated disaster-relief programs (some were as simple as a halt in the collection of grain levies in famine-struck areas). At the same time, the Qing campaigned against the nation's traditional population-control method: female infanticide. Many Chinese men had spent their days as bachelors, because infanticide removed women from the population. Now more could marry and have babies; now their babies were less likely to die from smallpox and starvation. Now, too, farm families were less likely to be driven into penury by the state: the Kangxi emperor promised in 1713 that the dynasty would never raise the basic tax on cropland, even though it was making massive investments in transportation networks so that farmers could sell their harvests, raising their incomes. Happily, those harvests were likely to grow; the Little Ice Age was waning. Some of these policies had been first emplaced by the Ming; the Qing merely executed them efficiently. But all helped raise the number of children, and the proportion of that number who survived to adulthood.

Still, as noted by Lan, the Sichuan historian, most of the increase took place in the areas with American crops. The families that Qing policies encouraged to move west needed to eat, and what they ate, day in and day out, was maize, potatoes, and sweet potatoes. Part of the reason China is the world's most populous nation is the Columbian Exchange.

MALTHUSIAN INTERLUDE

Hong Liangji was born in 1746 near the mouth of the Yangzi, into a family that slowly went on the skids after the unexpected death of his father. Brilliant but volatile, tall and red-faced, Hong "was in his element when he was singing and drinking," one friend recalled. He was often upbraided at school for drunken antics even as he won prizes for his intellect and prose style. An intense, impatient, and easily infuriated man, he would grab his interlocutor's wrist, lean in close, and hammer his point home with febrile intensity. "His eyes would narrow and you could see his neck redden with anger," another friend said, remembering political discussions. "Then he was extremely unsociable." His friends put up with him because he was a fine poet, a lively essayist, and a noted scholar who studied waterways, reconstructed administrative boundaries, and assembled a comprehensive geography of the Qing empire. His greatest intellectual feat, though, passed almost unnoticed. Sometime in 1793 Hong Liangji thought of an idea that may never have occurred to anyone else before.

After finally winning a place in the Qing bureaucracy at the age of forty-four—Hong had failed the civil service exam four times—he was sent as an education inspector to Guizhou Province, in the southwestern hinterland. Essentially a sloping, heavily eroded limestone shelf, the province is a humid jumble of steep gorges, protuberant hills, and long caverns. It was another target for Qing occupation, thronged with migrants from central China who were pushing out its original inhabitants, the Miao. The newcomers were climbing up the hills, planting maize, and beginning families. Hong wondered how long the boom could last.

"Today's population is five times as large as that of thirty years ago," he wrote, with perhaps pardonable exaggeration, "ten times as large as that of sixty years ago, and not less than twenty times as large as that of a hundred years ago." He imagined a man with "a ten-room house and 100 *mu* [about seventeen acres] of farmland." If the man married and had three adult sons, then eight people—the four men and their wives—would live on the parents' farm.

> Eight people would require the help of hired servants; there would be, say, ten people in the household. With the ten-room house and the 100 *mu* of farmland, I believe they would have just enough space to live in and food to eat, although barely enough. In time, however, there will be grandsons, who, in turn, will marry. The aged members of the household will pass away, but there could still be more than twenty people in the family. With more than twenty people sharing a house and working 100 *mu* of farmland, I am sure that even if they eat very frugally and live in crowded quarters, their needs will not be met.

Hong conceded that the Qing had opened up new land to support China's population. But the amount of farmland had

> only doubled or, at the most, increased three to five times, while the population has grown ten to twenty times. Thus farmland and houses are always in short supply, while there is always a surplus of households and population. . . .

Question: Do Heaven-and-earth have a way of dealing with this situation? Answer: Heaven-and-earth's way of making adjustments lies in flood, drought, and plagues.

Five years later, in England, a similar notion came to another man: Rev. Thomas Robert Malthus. A shy, kindly fellow with a slight harelip, Malthus was the first person to hold a university position in economics—that is, the first professional economist—in Britain, and probably the world. He was impelled to think about population growth after a disagreement with his father, a well-heeled eccentric in the English style. The argument was over whether the human race could transform the world into paradise. Malthus thought not, and said so at length—55,000 words, published as an unsigned broadside in 1798. Several longer versions followed. These were signed; Malthus had become more confident.

"The power of population," Malthus proclaimed, "is indefinitely greater than the power in the earth to produce subsistence for man." In textbooks today this notion is often depicted by recourse to a graph. One line on the graph represents the total food supply; it slowly rises in a line from left to right as people clear more land and farm more efficiently. Another line starts out low, quickly curves to meet the first, then soars above it; that line represents human population, growing exponentially. Eventually the gap between the two lines cannot be bridged, and the Four Horsemen of the Apocalypse pay a call. Every effort to increase the food supply, Malthus argued, will only lead to an increase in population that will more than cancel out the increase in the food supply—a state of affairs today known as a *Malthusian trap*. Forget Utopia, Malthus said. Humanity is doomed to exist, now and forever, at the edge of starvation. Forget charity, too: helping the poor only leads to more babies, which in turn produces increased hardship down the road. No matter how big the banquet grows, there will always be too many hungry people wanting a seat at the table. The Malthusian trap cannot be escaped.

The reaction was explosive. "Right from the publication of the *Essay on Population* to this day," the great economic historian Joseph Schumpeter declared, "Malthus had the good fortune—for this *is* good fortune—to be the subject of equally unreasonable, contradictory appraisals." John Maynard Keynes regarded Malthus as the "beginning of systematic economic thinking." Percy Bysshe Shelley, on the other hand, derided him as "a eunuch and a tyrant." John Stuart Mill viewed Malthus as a great thinker. To Karl Marx he was a "plagiarist" and a "shameless sycophant of the ruling classes." "He was a benefactor of humanity," Schumpeter wrote. "He was a fiend. He was a profound thinker. He was a dunce."

Hong, by contrast, was ignored. Unlike Malthus, he never developed his thoughts systematically, in part because he devoted his energy to criticizing the corrupt officials who he believed were looting the Qing state. Appalled at the government's brutal, incompetent reaction to a rebellion by starving peasants in Sichuan and Shaanxi, Hong quit his job in 1799. On his way out, he shot off a rambling but remarkably blunt letter to the crown prince, who passed it to the Jiaqing emperor (not to be confused with the alchemy-crazed Jiajing emperor, who ruled two centuries before). The angered emperor sentenced Hong to life in exile, silencing him.

The lack of recognition was unmerited; Hong apparently captured the workings of the Malthusian trap better than Malthus. (I use the hedge word "apparently" because he never worked out the details.) The Englishman's theory made a simple prediction: more food would lead to more mouths would lead to more misery. In fact, though, the world's farmers have more than kept pace. Between 1961 and 2007 humankind's numbers doubled, roughly speaking, while global harvests of wheat, rice, and maize tripled. As population has soared, in fact, the percentage of chronically malnourished has *fallen*—contrary to Malthus's prediction. Hunger still exists, to be sure, but the chance that any given child will be malnourished has steadily, hearteningly declined. Hong, by contrast, pointed to a related but more complex prospect. The continual need to increase yields, Hong presciently suggested, would lead to an ecological catastrophe, which would cause social dysfunction—and with it massive human suffering.

Exactly this process is what researchers today mean when they talk about the Malthusian trap. Indeed, one way to summarize today's environmental disputes is to say that almost all boil down to the question of whether humankind will continue to accumulate wealth and knowledge, as has been the case since the Industrial Revolution, or whether the environmental impacts of that accumulation—soil degradation, loss of biodiversity, consumption of groundwater supplies, climate change—will snap shut the jaws of the Malthusian trap, returning the earth to pre-industrial wretchedness. Alarming in this context, China provides an example of the latter, at least in part. In the decades after American crops swept into the highlands, the richest society in the world was convulsed by a struggle with its own environment—a struggle it decisively lost.

"THE MOUNTAINS REVEAL THEIR STONES"

Between the 1680s, when the Qing resumed the silver trade, and the 1780s, the price of rice in Suzhou, a rice-trading center near modern Shanghai, more than quadrupled. Incomes did not keep up—a recipe for social unrest. As if on cue, rebellions exploded across China; the convulsion that dismayed Hong is alone said to have led to several million deaths. Part of the reason for the price hike was the influx of silver to Fujian, according to Quan Hansheng, the economic historian, which drove up Chinese food prices in exactly the same way that the influx of silver to Spain had earlier driven up European prices. The population boom presumably increased demand, putting further pressure on the price. State purchases for granaries sometimes had the same effect.

But a big reason for the price rise was that many farmers simply stopped growing rice.

Qing emperors had made a priority of improving transportation networks so that farmers could sell crops profitably. The intent was to facilitate the movement of staple foods; the new roads would help merchants ship rice and wheat from places with abundant harvests to places that needed supplies. Instead smallholders discovered they could make more money by switching from rice and wheat to sugarcane, peanuts, mulberry trees, and, most of all, tobacco.

Initially the Qing court cracked down on this shift, insisting that peasant farmers practice "correct agriculture"—that is, grow rice and wheat. "Tobacco is not healthy for the people," the Yongzheng emperor proclaimed in 1727. "Because cultivating tobacco requires using rich land, its cultivation is harmful for growing grain." But as the court grew more insular and debased—seemingly the fate of all Chinese dynasties—it lost interest in enforcing agricultural correctness.

Farmers seized their opportunity. Tobacco required four to six times more fertilizer and twice as much labor as rice, but was more profitable; China's growing battalions of nicotine addicts were willing to pay more for their pipes than their food. (Some were doubly addicted: they cut their tobacco with opium.) Tobacco appeared in almost every corner in China, according to Tao Weining, an agricultural historian in Guangdong. And it was a big presence in those places: in two typical hilly areas examined by Tao, "nearly half" of the total farmland was devoted to *N. tabacum*. In consequence, the local price of rice doubled, as did the price of most common vegetables and fruits. Farmers ended up spending their tobacco profits on food expensively imported from other parts of China. As in Virginia, tobacco drained the land. When farmers exhausted the soil from one former rice paddy, they went to the next. And when they ran out of rice paddies, they went into the hills.

The same phenomenon is still occurring today. When two

Even four centuries after its introduction tobacco remains so profitable in China that villagers still turn rice paddies into tobacco plots. These Fujianese farmers are drying tobacco in 2009.

friends and I visited the *tulou* houses in Fujian, we walked around the mountain hamlet of Yongding. Generations past, the villagers' ancestors had hacked small, semicircular rice terraces out of the slopes, fertilizing the thin red earth with manure and night soil, then filling the paddies by diverting mountain streams. At the edge of the village a sign proclaimed that China Tobacco, a state monopoly, had contracted with Yongding's farmers to convert their paddies to tobacco. The company had built a new road to facilitate harvest. From atop the terraces we looked down on horizontal arcs of splayed, fleshy green arrows: *N. tabacum*.

In Yongding, the villagers had replaced some of the lost rice with maize, shoving plants into the ground everywhere they could find a scrap of plausible land: roadside ditches, backyard plots, the walls of the gullies below the houses. Somebody had stuck maize seedlings into a pickup-sized heap of dirt and gravel left by a recent landslide. During the eighteenth century, the same kind of thing took place all over China. Jamming maize and sweet potatoes into every nook and crevice, shack people and migrants almost tripled the nation's cultivated area between 1700 and 1850. To create the necessary farmland, they knocked down centuries-old forests. Bereft of tree cover, the slopes no longer retained rainwater. Soil nutrients washed down the hills. Eventually the depleted land would not support even maize and sweet potatoes. Farmers would clear more forest, and the cycle would begin anew.*

Some of the worst devastation was in the steep, crabbed hills of eastern central China, home of the shack people. Heavy, hammering rains, common in this area, constantly flush out minerals and organic matter. The weathered soil can't hold water—"if it doesn't rain for ten days," one local writer said in 1607, "the soil becomes dry and scorched and cracks like the lines on a tortoise's back." The land was arable, in the sense that maize and sweet potatoes would grow in it. But harvesting them for more than a season or two was next to impossible without shoveling in generous amounts of lime or ashes to reduce acidity, manure to boost organic matter, and fertilizer to increase nitrogen and phosphorus. This had to be done every year, because rain kept leaching nutrients.

Shack people, one recalls, rented their farms from landowners in the valleys below. Renting for short, fixed periods, they had no incentive to fertilize, and little means to do it even should they have wanted to. Because the crop was new to their experience, they made beginners' mistakes. Maize is planted in widely spaced rows, unlike wheat and millet, which is grown across solid blocks. Many farmers did not realize for a long time that maize therefore

* Agriculture was not the only cause of deforestation. China consumed huge quantities of timber as fuel and building material. To get the wood, platoons of workers went to distant places, where they wiped out entire forests. Alas, so much lumber was lost, damaged, and stolen during shipping, reported Yang Chang, a historian in Hubei Province's Huazhong Normal University, that less than 2 percent of it was actually used by its intended recipients.

left more of the soil uncovered and hence exposed to rain. And some didn't understand that planting the maize in rows straight up and down the hills, rather than across the slope, would channel that rain down the slope, increasing erosion.

Even if one fertilized the upland soil and minimized the impact of rain, upland deforestation could still cause disaster below, according to Anne R. Osborne, a historian at Rider University, in New Jersey, whose studies of the shack people I am relying on for this account. "The narrowness of the valley plains and basins meant that human settlement and most food production were concentrated along the edges of the rivers," Osborne explained. When the uplands were covered with vegetation, they released rainwater slowly; floods were rare. Replacing stands of trees on steep slopes with temporary plots of maize and sweet potatoes reduced the mountains' water-storage capacity. Rainfall went down the hills in sheets, setting off floods. "Flood waters pouring out of the highlands met almost flat land on the neighboring basins and plains," Osborne wrote. "Slowing suddenly, they dropped their loads of silt, in the river channels or over the farmers' fields, destroying fertile fields and obstructing the channels for future drainage."

Floods were especially problematic for rice farmers, even though their livelihood depended on flooding. Paddies require a continuous trickle of incoming water. If the flow is too weak, the water evaporates; if the flow is too fast, the paddy spills over its banks, carrying away nutrients and possibly the rice itself. Farmers used upstream dikes to hold back water until needed, controlling irrigation levels by adjusting gates. In a flood, the sudden gush of water could wipe out both the dikes and the paddies they fed, bringing down the whole system. Paradoxically, the deluges drowned the rice crop—and then, later, dried out the paddies because the dikes no longer held water for them. By cutting down the forests, the shack people were not only laying waste to the land around them, they were helping to devastate the agricultural infrastructure miles downstream. Because this was occurring in the lower Yangzi, the shack people were wrecking a chunk of the nation's agricultural heartland.

Some locals wholly understood the problem. When the scholar Mei Zengliang paid a nostalgic visit in 1823 to the mountain town in which he had spent his childhood, he asked his former neighbors about the shack people. No ecologist today would have much to add to their response.

> On uncleared mountains [the villagers told him], the soil is firm and the rocks hold fast; grass and trees are thick, years of rotting leaves cover the ground to depths of as much as 2 to 3 *cun* [three to four inches]. Whenever it rains, the rainwater runs off the trees and onto the rotten leaves, then into the soil and rocks, before seeping through cracks in the rocks to form streams. This water flows slowly, and as it flows downward the soil does not go with it. . . . Today [shack people] strip the mountains with blades and axes, and loosen their soil with shovels and hoes, so that before even one rainfall has finished, the sand and rocks wash down with the water, quickly flowing into ravines.

Erosion from the heights drowned the rice paddies in the lower Yangzi valleys, further driving up the price of rice, which encouraged more maize production in the heights, which drowned more rice in the valleys.

As shack people moved into the mountains, floods became ever more frequent. In the Song dynasty (960–1279 A.D.), major floods occurred somewhere in the empire at an average clip of about three every two years. Some farmers, many of them Hakka, illegally moved into the hills during the Ming dynasty (1368–1644), removing trees as they did. Predictably, the pace of deluges increased to almost two per year. The Qing (1644–1911) actively promoted moving peoples into mountain forests. As night follows day, the surge in migration led to a surge in deforestation; the flood rate more than tripled, to a little more than six major floods a year. Worse, the floods mostly targeted China's agricultural centers. Poring through personal diaries, county gazetteers, provincial archives, and imperial disaster-relief records, the historian Li Xiangjun found that 16,384 floods had occurred during the Qing dynasty. The great majority were small. But 13,537 of them occurred in the rich farmlands in the lower Yangzi and Huang He. And the floods kept growing. Between 1841 and 1911, the Qing faced more than thirteen major floods a year—a Katrina every month, as one historian put it to me. "The government had constant disasters in the most populous parts of the realm," he said. "The areas that were most important to feeding everyone. It was not good."

In the 1970s a team of researchers at China's central meteorological bureau pored through huge numbers of local records, looking for descriptions of rainfall and temperature in past centuries. As one might expect, the researchers found few scientific measurements, but many verbal accounts. When they encountered phrases like—to use their examples—"10 consecutive days of heavy summer rain caused rivers to overflow," "spring and summer floods drowned countless people and animals," "summer and fall floods washed away the seedlings of cereal crops," "several days of heavy rains such that boats could travel over land," and "massive winds and heavy rains inundated fields and houses," the researchers concluded that the area had experienced a flood, and marked the map with a 1 in the corresponding area. Descriptions of severe drought were marked with a 5. They gave conditions in between 2, 3, or 4. Although the resultant maps were subjective, the overall course of events was clear. Flipping through the maps in the meteorological bureau book was like watching an animated movie of environmental collapse.

Overwhelmed by the detail on the maps, I decided to look at four rice centers on the lower Yangzi: the cities of Nanjing, Anqing, and Wuhan, and the upper Han River, an important northern tributary of the Yangzi. Between 1500 and 1550, these areas had sixteen number 1s: sixteen major floods. Between 1600 and 1650,

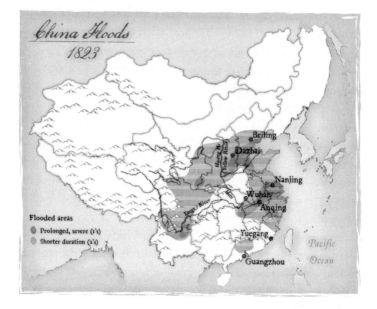

they had eighteen—roughly the same number. Between 1700 and 1750, at the height of the colder, wetter Little Ice Age, there were twenty-seven. Then the Little Ice Age ended, the weather became drier, and there was less rain and snow. But the number of 1s in these parts of China's agricultural core kept increasing. Between 1800 and 1850 these four places alone had thirty-two major floods. Some of the floods extended for hundreds of miles along the river, the 1s inundating city after city, each digit standing for thousands of wrecked lives.

Officials in Zhejiang Province, dismayed by the mounting problems, announced in 1802 that the government would begin sending the despised shack people "back to their native places." They also banned planting maize in the mountains. Almost nothing happened. The officials tried again in 1824, banning the species outright—Zhejiang was supposed to be a maize-free zone. Again nothing happened. The imperial government had a network of "censors" entrusted with rooting out incompetence and corruption. Zhejiang's censors repeatedly asked Beijing to send troops to rip out maize. There was no response. In the kind of phenomenon that makes one despair of the human race's ability to govern itself, the pace of land clearing actually accelerated in the first part of the nineteenth century.

Zhejiang censor Wang Yuanfang couldn't understand it. In the past, he knew, landlords hadn't understood that renting their unused upland property would have disastrous consequences. *"Now* [in 1850] *the waterways are filled with mud, the fields are buried under sand, the mountains reveal their stones and the officials and people know of the great disaster, but they do nothing to stop it. Why?"* (Emphasis in original.)

In part, the failure was due to an inherent problem with mass illegal immigration. It is not easy to deport huge numbers of people—tearing them from homes and families built up over years—without mass suffering. Governments that seek popular support shrink from inflicting this kind of agony (unless the loss of support from one group is made up for by increased support from another). Logistically, there is also the problem of finding a destination for people who have left their original homes decades before. In the case of the shack people, Osborne argued, neither governmental queasiness nor confusion was the chief obstacle. The main problem was that the erosion represented a classic collective-action problem. A legal loophole ensured that rental income, unlike farm income, was tax free. Landowners with rentable property in the highlands thus had an easy source of untaxable income. The ensuing deforestation might ravage their own fields in the valleys, but the risks would be spread across an entire region, whereas the landowners' profits were theirs alone. Absorbing all of the gain and only a fraction of the pain, local business interests beat back every effort to rein in shack people.

In an environmentalists' nightmare, the shortsighted pursuit of small-scale profit steered a course for long-range, large-scale disaster. Constant floods led to constant famine and constant unrest; repairing the damage sapped the resources of the state.

American silver may have pushed the Ming over the edge; American crops certainly helped kick out the underpinnings of the tottering Qing dynasty.

Other factors played their part, to be sure. A rebellion led by a Hakka mystic tore apart the nation, briefly setting up a state of shack people in the Hakka hills of the southeast. A series of weak emperors allowed the bureaucracy to wallow in inanition and corruption. The empire lost two wars with Great Britain, forcing it to cede control of its borders. British forces freely disseminated the opium that the government had gone to war to exclude. And so on—catastrophe, like success, has many progenitors. Unknown to the rampaging European armies, though, their path had been smoothed by the Columbian Exchange.

UNLEARNING FROM DAZHAI

For two generations, one of the most celebrated places in China was Dazhai. A hamlet of a few hundred souls in the dry, knotted hills of north-central China, Dazhai was ravaged by floods in 1963. Standing in the wreckage with his signature sweat-absorbing towel around his head, the local Communist Party secretary refused aid from the state and instead promised that Dazhai would rebuild itself with its own resources—and create a newer, more productive village at the same time. Harvests soared, despite the flood and the area's infertile soil.

Delighted by the increase, Mao Zedong bused thousands of local officials to the village and instructed them to emulate what they saw. Mainly, they saw spade-wielding peasants working in a fury to terrace the hills from top to bottom; rest breaks occurred while reading Mao's *Little Red Book* of revolutionary proverbs. The atmosphere was cult-like: one group walked for two weeks to see the calluses on a Dazhai laborer's hands. China needed to produce grain from every scrap of land, the officials learned. Slogans, ever present in Maoist China, explained how to do it:

Move Hills, Fill Gullies and Create Plains!
Destroy Forests, Open Wastelands!
In Agriculture, Learn from Dazhai!

Filled with excitement, lashed on by local authorities, villagers fanned out across the hills, cutting the scrubby trees on the pitches, slicing the slopes into earthen terraces, and planting what they could on every newly created flat surface. Despite heat and hunger, people worked all day and then lighted lanterns and worked at night. The terraces converted unplantable steep slopes into new farmland. In one village that I visited farmers increased the area of cultivable land by about 20 percent, which seems typical.

Dazhai is in a geological anomaly called the Loess Plateau. For eon upon eon winds have swept across the deserts to the west, blowing grit and sand into central China. Millennia of dust fall have covered the region with vast heaps of packed silt—"loess," geologists call it—some of them hundreds of feet deep. The Loess Plateau is about the size of France, Belgium, and the Netherlands combined.

Loess doesn't form soil so much as pack together like wet snow. For centuries, people in the plateau have dug caves in the loess and lived in them. *Yaodong,* as these cave dwellings are called, are quite cozy—the one I stayed in had a heated platform bed cut from a block of loess. An adjacent woodstove vented through the platform, warming the bed in winter. Looking at the *yaodong* walls that night, I realized that the room, like a scientific probe, revealed the earth's workings. As a rule, soil has three layers: a thin scrim of dead leaves, bits of wood, and other organic matter on top; a band of dark topsoil, usually no more than a foot deep, shot through with humus (partly decomposed organic matter); and a stratum of subsoil below, lighter colored but rich with iron, clay, and minerals. Loess is different; my bedroom walls, carved from a giant heap of mashed-together grit, were uniform from top to bottom.

As every child who plays with mud knows, dust piles are easy to wash away. Silt grains "act like single particles," said Zheng Fenli, a soil scientist at the Institute of Soil and Water Conservation, in the Loess Plateau city of Yangling. They don't clump together firmly. If knocked free by flowing water, Zheng told me, silt grains "are very easy to transport." Washed down steep hills, they can be carried great distances. The Huang He makes a big loop right through the Loess Plateau. It carries an enormous burden of silt—more than any other river in the world—into the North China Plain, China's agricultural heartland.

Because the plain is flat, the river slows down. As the current falters, the silt in the water deposits on the river bottom and along the banks. The silt replenishes the soil—a main reason for the area's farming primacy—but it also builds up the riverbed. In consequence, the Huang He rises one to three inches a year. Over time, it has lifted itself as much as forty feet over the surrounding land. When farmers harvesting wheat fields want to see the river, they look *up*. Moving high in the air, the river wants (so to speak) to overflow its banks, spilling into the North China Plain and creating a ruinous flood.

Such disasters have been a threat for millennia—"two breaks every three years and a channel change every century," the Chinese used to say of the Huang He. But in the eighteenth and nineteenth centuries, erosion drove the breaks and channel changes to be more lethal. In an attempt to subdue the floods, the Qing established a corps of engineers who maintained a five-hundred-mile line of dikes, a network of spillways, locks, and dams, and an array of as many as sixteen secondary channels into which the river could be divided—a hydraulic infrastructure easily as impressive as the Great Wall, and one that was more important to the life of the nation. Not only did the system control a staggeringly complex irrigation network, it connected the river to the Grand Canal, a 1,103-mile passage between Beijing and Hangzhou (a port south of modern Shanghai) that is the longest artificial waterway in the world. Qing emperors may have spent 10 percent or more of the imperial budget on the Huang He.

Nonetheless the system was constantly overwhelmed. As the Chinese weather bureau maps show, excess silt made the Huang He spill over its banks a dozen times between 1780 and 1850—about once every six years. All of the floods were huge. One deluge in 1887 was among the deadliest ever recorded; estimates of the dead range up to a million.

The cause of the flooding—deforestation in the Loess Plateau—was well understood. But Beijing did little about it, even though much of the land clearance had its roots in Qing policies, and the floods were blows to imperial legitimacy. The court's failure to act was not foreordained. Neither was the myopia of the landlords who rented to the shack people. Nobody will ever know whether decisive action could have resolved the nation's ecological problems, because it wasn't tried. Instead the floods continued until the dynasty fell, an event the floods had helped to bring about.

Which made it all the more incredible when Mao Zedong ordered *more* land clearing in the Loess Plateau. Most of the region was already deforested, but the steepest slopes—land too steep to farm—were still covered by low, scrubby growth that held back erosion. Exactly this land was targeted in the 1960s and 1970s for conversion, Dazhai style, into terraces. The terrace walls, made of nothing but packed earth, constantly fell apart; in one Loess Plateau village that I visited after a rainfall, half the population seemed to be shoring up crumbling terraces by pounding the walls flat with shovels. Even when the terraces didn't crumble, rains sluiced away the nutrients and organic matter in the soil. Zuitou, the village, is nestled into steep hills along the Huang He. Walking along the steep paths between *yaodong,* I could almost watch the terraces sliding into the water.

Because erosion removed nutrients, harvests in the newly planted land dropped quickly. To maintain yields, farmers cleared

and terraced new land, which washed away in turn—a perfect example of a "vicious circle," according to Vaclav Smil, a University of Manitoba geographer who has long studied China's environment (his first book on the subject, *The Bad Earth,* appeared in 1984). Erosion into the Huang He went up by about a third during the Dazhai era, Chinese researchers reported in 2006.

The consequences were dire and everywhere apparent. Declining harvests on worsening soil forced huge numbers of farmers to become migrants. Zuitou lost half of its population. "It must be one of the greatest wastes of human labor in history," Smil told me. "Tens of millions of people being forced to work night and day, most of it on projects that a child could have seen were a terrible stupidity. Cutting down trees and planting grain on steep slopes—how could that be a good idea?"

In the most marginal areas farmers planted maize. North of Zuitou, at the edge of the Gobi Desert, I walked around plots of maize growing in almost pure sand. Until the 1960s the region had been covered with thorny forest scrub. Then Mao ordered aggressive planting. It was like forcing people to farm a beach. Astounding to me, the locals had actually coaxed some maize

Beginning in the 1960s, farmers throughout China's Loess Plateau stripped the forest and carved earthen terraces out of the hills (opposite). Because the loess erodes easily, every rain caused terraces to erode (above); maintenance was a constant issue. Eventually the terraces on the steepest slopes collapsed completely (left), and farmers found themselves trying to eke out a living on hills almost too steep to stand on.

from the sand—drying cobs made little yellow heaps on rooftops and barren yards. On carts hauled by tiny Chinese motorbikes men were driving around piles of maize stalks tall as two-story buildings. In the light wind the air was intensely gritty. The Loess Plateau, which once caught dust from the desert, was now producing it.

The People's Republic had initiated plans to halt deforestation. In 1981 Beijing ordered every able-bodied citizen older than eleven to "plant 3–5 trees per year" wherever possible. Three years before, Beijing had initiated what may be the planet's biggest ecological program, the "Three Norths" project: a 2,800-mile band of trees running like a vast screen across China's north, northeast, and northwest, including the frontier of the Loess Plateau. Scheduled to be completed in 2050, this Green Wall of China will, in theory, slow down the winds that drive desertification and dust storms.

Despite their ambitious scope, these efforts did not directly address the soil degradation that was the legacy of Dazhai. Confronting the destruction was politically difficult, though: it had to be done without admitting that Mao had made mistakes. (When I asked local officials if the Great Helmsman had erred, they politely changed the subject.) Only in the last decade did Beijing chart a new course.

Today many of the terraces Zuitou's farmers hacked out of the loess are reverting to nature. In what locals call the "3-3-3" system, farmers replant one-third of their land—the steepest, most erosion-prone slopes—with grass and trees, natural barriers to erosion. They cover another third of the land with harvestable orchards. The final third, mainly plots on the gully floor that have been enriched by earlier erosion, is cropped intensively. By concentrating their limited supplies of fertilizer on that land, farmers can raise yields enough to make up for the land they have sacrificed—that's the theory, anyway. To help the transition along, farmers are compensated with an annual delivery of grain and a small cash payment for up to eight years. By 2010, the program covered more than 56,000 square miles of gully villages, an area the size of Iowa.

At first glance, it seems that a dictatorship would be perfectly suited to accomplish this task. The government can simply order loess dwellers to stop growing millet and plant almonds without worrying about property rights or political protest. It can direct whole villages to go into the hills en masse and plant saplings, millions upon millions of them, in small pits shaped like fish scales. And when the farmers and fields are shifted around, the planners can point to their accomplishments with pride.

Things look different on the ground. Provincial, county, and village officials are rewarded if they plant the number of trees envisioned in the plan, not whether they have chosen tree species suited to local conditions (or listened to scientists who say that trees are not appropriate for grasslands to begin with). Farmers who reap no direct benefit from their work—they are installing trees that do not produce fruit, cannot be cut for firewood, and supposedly stop erosion miles from their homes—have little incentive to take care of the trees they are forced to plant. The entirely predictable result is visible on the back roads of Shaanxi: fields of dead trees, each in its fish-scale pit, lining the roads for miles. "Every year we plant trees," the farmers say, "but no trees survive."

During my visit the lines of dead trees dotted the slopes like contour marks, stretching for miles. The harvest was over, and farmers were about to be marched back in for another try. Tree by tree, the government was trying to undo the accidental legacy of the global silver trade.

PART THREE

Europe in the World

6

The Agro-Industrial Complex

POTATO WARS

When potato plants bloom, they send up five-lobed flowers that bob in fields like fat purple stars. According to tradition, Marie Antoinette liked the blossoms so much that she put them in her hair. Her husband, Louis XVI, supposedly put one in his buttonhole, inspiring a brief vogue in which the French aristocracy swanned around with potato plants on their clothes. Potatoes belong to the nightshade family, which means they are cousins to tomatoes, eggplant, tobacco, sweet peppers, and deadly nightshade. The tubers are not roots but modified stems that store nutrients underground; the eyes, from which new potatoes sprout, are descended from the leaves that grew on the stem. Potato fruits look like green cherry tomatoes but are full of solanine, a poison that is part of the plant's defense system—it prevents pests from eating the seeds. As a rule modern farmers ignore the seed, instead cutting up tubers and planting the pieces. In a testament to linguistic confusion, tubers used for this purpose are called "seed potatoes."

Today the potato is the fifth most important crop worldwide, surpassed in harvest volume only by sugarcane, wheat, maize, and rice. Originally it came from the Andes—not only *Solanum* *tuberosum,* the potato found in supermarkets, but many other types of potato that are eaten only in Ecuador, Peru, and Bolivia. There are also scores of wild potato species that can be found everywhere from Argentina to the southwestern United States. Despite similarities of name and appearance, not one of these potatoes is related to the sweet potato, which belongs to a different botanical family. The two have long been confused; the word "potato" is derived erroneously from *batata,* the Taino name for sweet potato (and the source of its scientific name, *Ipomoea batatas*). The mix-up rankled the early English botanist John Gerard, who complained in 1597 that "those [who] vulgarly impose names upon plants have little either judgement or knowledge of them." Intending to clear up the matter definitively in his "generall historie of plantes," Gerard used the term "Virginia potato" for the ordinary potato, which is not from Virginia. He called sweet potatoes "common potatoes."*

Potatoes are about three-quarters water and one-quarter starch but have vitamins enough to prevent scurvy if consumed in quantity. For 167 days in 1925 two Polish researchers ate almost nothing but potatoes (mashed with butter, steamed with salt, cut with oil into potato salad). At the end they reported no weight gain, no health problems, and, improbably, "no craving for change" in their diet. Historically speaking, the scientists' regimen was not extreme; two British inquiries in 1839 intimated that the average Irish laborer's per capita daily consumption of potatoes was twelve and a half pounds. Ireland was notorious for its potato habit, but the tubers had become so essential to all of northern Europe that Prussia and Austria fought a "potato war" in 1778–79 in which the two armies spent most of their time scrambling to get food for themselves and deny it to the enemy. Only when every potato in Bohemia had been consumed did hostilities end.

Compared to grains, tubers are inherently more productive.

---

* Gerard did not contribute to a third source of confusion: the common practice of referring to sweet potatoes as yams. Yams originated in Asia and Africa and belong to yet another biological family.

If the head of a wheat or rice plant grows too big, the plant will fall over, with fatal results. Modern plant breeders have developed wheat and rice varieties with shorter, stronger stalks that can bear heavier loads of grain. But even they could not support something as heavy as an Idaho potato. Growing underground, a tuber is not limited by the rest of the plant—there are no worries about its architecture. In 2008 a Lebanese farmer dug up a potato that weighed nearly twenty-five pounds. Photographs showed a man holding a tuber bigger than his head.

Many scholars believe that the introduction of *S. tuberosum* to Europe was a key moment in history. This is because their widespread consumption largely coincided with the end of famine in northern Europe. (Maize, another American crop, played a similar but smaller role in southern Europe.) More than that, the celebrated historian William H. McNeill has argued, *S. tuberosum* led to empire: "[P]otatoes, by feeding rapidly growing populations, permitted a handful of European nations to assert dominion over most of the world between 1750 and 1950." Hunger's end helped create the political stability that allowed European nations to take advantage of American silver. The potato fueled the rise of the West.

As important in the long run, the European and North American adoption of the potato set the template for modern agriculture—the agro-industrial complex, as it is sometimes called. Celebrated by agronomists for its bounteous harvests and denounced by environmentalists for its toxicity, the agro-industrial complex rests on three pillars: improved crops, high-intensity fertilizers, and factory-made pesticides. All three are entwined with the Columbian Exchange, and with the potato.

Not only did the Columbian Exchange carry the ultra-productive potato to Europe and North America, it also brought ultra-productive Andean potato-cultivation techniques, including the world's first intensive fertilizer: Peruvian guano. Andean peoples had mined it for centuries from great excremental deposits seabirds left on coastal islands. Fertilizer ships crossed the Atlantic by the hundreds, brimming with guano—and, many researchers believe, a fungus-like organism that blighted potatoes, causing a famine in Ireland that by some measures was the worst in the historical record.

Not long after, potatoes fell to the attack of another imported species, the Colorado potato beetle. Panicked farmers turned to the first inorganic pesticide: a widely available form of arsenic, sprayed with enthusiasm over the field. Competition to produce ever-more-effective arsenic compounds launched the modern pesticide industry—the third component of modern agribusiness. Brought together systematically in the 1950s and 1960s, improved crops, high-intensity fertilizers, and artificial pesticides created the Green Revolution, the explosion of agricultural productivity that transformed farms from Illinois to Indonesia—and set off a political argument about the food supply that grows more intense by the day.

SEA OF GENES

In 1853 an Alsatian sculptor named Andreas Friedrich erected a statue of Sir Francis Drake on a marble plinth in the center of Offenburg, a small city in southwest Germany. Friedrich portrayed Drake staring into the horizon in orthodox visionary fashion. His left hand rested on the hilt of his sword. His right gripped a potato. "Sir Francis Drake," the base proclaimed,

> *disseminator of the potato in Europe*
> *in the Year of Our Lord 1586.*
> *Millions of people*
> *who cultivate the earth*
> *bless his immortal memory.*

The statue was pulled down by the Nazis in early 1939, a small piece of the violent frenzy set off by the anti-Semitic riots of

Kristallnacht. Destroying the statue was a crime against art, not history: Drake almost certainly did not introduce the potato to Europe. Even if he *had* introduced it, though, the statue would be misguided. Credit for *Solanum tuberosum* surely belongs most to the Andean peoples who domesticated it.

Geographically, the Andes were an unlikely place for the creation of a major staple food. The second-biggest mountain range on the planet, the chain of peaks forms an icy barrier on the Pacific Coast of South America that is 5,500 miles long and in many places more than 22,000 feet high. Active volcanoes are scattered along its length like molten jewels on a belt. Ecuador alone had seven eruptions in the last century; San José, on Chile's western border, has gone off seven times since 1822. The volcanoes are linked by geologic faults, which push against each other isometrically, trig-

The Offenburg memorial to Sir Francis Drake's introduction of the potato was destroyed by the Nazis.

gering earthquakes, floods, and landslides. Even when the land is seismically quiet the climate is active. Temperatures in the highlands can fluctuate from 75°F to below freezing in the space of a few hours—the air is too thin to hold the heat. Sudden hailstorms splinter windows and drive vehicles off the road. Famously, El Niño—the name itself is an Andean coinage—brings floods to the coast and drought to the high plains. El Niño episodes can last for years.

The main part of the range consists of three roughly parallel mountain chains separated by high tablelands known as the altiplano. The altiplano (average altitude: about twelve thousand feet) holds most of the region's arable land; it's as if Europe had to support itself by farming the Alps. The sheer eastern face of the Andes catches the warm, humid winds from the Amazon, and consequently is beset by rain; the western, ocean-facing side, shrouded by the "rain shadow" of the peaks, contains some of the earth's driest lands. The altiplano between has a dry season and a wet season, with most of the rain coming between November and March. Left to its own devices, it would be covered by grasses in the classic plains pattern.

From this unpromising terrain sprang, remarkably, one of the world's great cultural traditions—one that by 1492 had reached, according to the University of Vermont geographer Daniel W. Gade, "a higher level of sophistication" than any of the world's other mountain cultures. Even as Egyptian kingdoms built the pyramids, Andean societies were erecting their own monumental temples and ceremonial plazas. Contentious imperia jostled for power from Ecuador to northern Chile. Nasca, with its famous stone lines and depictions of animals; Chavín, with its grand temples at Chavín de Huántar; Wari, landscape engineers par excellence; Moche, renowned for ceramics depicting every aspect of life from war and work to sleeping and sex; Tiwanaku, the highest urban complex ever built (it was centered on Lake Titicaca, the highest navigable lake on the planet); Chimor, successor to Moche, with its sprawling capital of Chan Chan—the tally is enormous. Most famous today are the Inka, who seized much of the Andes in a violent flash, built great highways and cities splendid with gold, then fell to Spanish disease and Spanish soldiers.

The history of the civilizations of the Middle East and Egypt is entwined with the development of wheat and barley; similarly, indigenous societies in Mexico and Central America were founded on maize. In Asia, China's story is written on paper made from rice. The Andes were different. Cultures there were nourished not by cereal crops like these but by tuber and root crops, the potato most important.

Archaeologists have turned up evidence of people eating potatoes thirteen thousand years ago in southern Chile—not the modern *Solanum tuberosum,* but a wild species, *S. maglia,* which still grows on the coast. Geneticists remain uncertain, though, of the exact pathway by which Andean cultures created the domestic potato. Because early Andean natives mainly grew their tubers from seed and apparently planted multiple species of *Solanum* in the same garden, they would have produced countless natural hybrids, some of which presumably gave rise to the modern potato. One often-cited analysis tried to nail down the process; after much study, its author declared that today's potato was bred from four other species, two of which bore the label "unknown." Timing, too, is unclear: archaeologists have established only that Andean peoples were eating wholly domesticated potatoes by 2000 B.C.

Potatoes would not seem obvious candidates for domestication. Wild tubers are laced with solanine and tomatine, toxic compounds thought to defend the plants against attacks from dangerous organisms like fungi, bacteria, and human beings. Cooking often breaks down a plant's chemical defenses—many beans, for example, are safe to eat only after being soaked and heated—but solanine and tomatine are unaffected by the pot and oven. Andean peoples apparently neutralized them by eating dirt: clay, to be precise. In the altiplano, guanacos and vicuñas (wild relatives of the llama) lick clay before eating poisonous plants.

The toxins in the foliage stick—more technically, "adsorb"—to the fine clay particles. Bound to dirt, the harmful substances pass through the animals' digestive system without affecting it. Mimicking this process, Indians apparently dunked wild potatoes in a "gravy" made of clay and water. Eventually they bred less lethal varieties, though some of the old, poisonous tubers still remain, favored for their resistance to frost. Bags of clay dust are still sold in mountain markets to accompany them on the table.

Andean Indians ate potatoes boiled, baked, and mashed as people in Europe and North America do. But they also consumed them in forms still little known outside the highlands. Potatoes were boiled, peeled, chopped, and dried to make *papas secas;* fermented for months in stagnant water to create sticky, odoriferous *toqosh;* ground to pulp, soaked in a jug, and filtered to produce *almidón de papa* (potato starch). The most ubiquitous concoction was *chuño,* made by spreading potatoes outside to freeze on cold nights. As it expands, the ice inside potato cells ruptures cell walls. The potatoes are thawed by morning sun, then frozen again the next night. Repeated freeze-thaw cycles transform the spuds into soft, juicy blobs. Farmers squeeze out the water to produce *chuño:* stiff, Styrofoam-like nodules about two-thirds smaller and lighter than the original tubers. Long exposure to the sun turns them gray-black; cooked into a spicy Andean stew, they resemble gnocchi, the potato-flour dumplings favored in central Italy. *Chuño* can be kept for years without refrigeration, meaning that it can be stored as insurance against bad harvests. It was the food that sustained the conquering Inka armies.

Then as now, farming the Andes was a struggle against geography. Because the terrain is steeply pitched, erosion is a constant threat. Almost half the population cultivates some land with a slope of more than twenty degrees. Every cut of the plow sends dirt clods tumbling downhill. Many of the best fields—those with the thickest soil—sit atop ancient landslides and hence are even more erosion prone than the norm. Problems are exacerbated by the tropical weather patterns: a dry season with too little water, a rainy season with too much. During the dry season, winds scour away the thin soil. Heavy rainfall in the wet season sheets down hills, washing away nutrients, and floods the valleys, drowning crops.

To manage water and control erosion, Andean peoples built more than a million acres of agricultural terraces. Carved like stairsteps into the hills, the Spanish voyager Pedro Sarmiento de Gamboa marveled in 1572, were "terraces of 200 paces more or less, and 20 to 30 wide, faced with masonry, and filled with earth, much of it brought from a distance. We call them *andenes"*

Using a foot plow, Andean Indians break up the ground in this drawing from about 1615 by Felipe Guaman Poma de Ayala, an indigenous noble. Women follow behind to sow seed potatoes.

(platforms)—a term that may have given its name to the Andes. (Fifteenth-century Indians used more appropriate methods than those ordered by Mao in the twentieth century, and had much better results.)

On the flatter, wetter land around Lake Titicaca indigenous societies built almost five hundred square miles of raised fields: rectangular hummocks of earth, each several yards wide and scores or even hundreds of yards long. Separating each platform from its neighbor was a trench as much as two feet deep that collected water. During the night the trench water retained heat. Meanwhile, the complex up-and-down topography and temperature variation of the surface created slight air turbulence that mixed the warmer air in the furrows and the colder air around the platforms, raising the temperature around the crops by as much as 4°F, a tremendous boon in a place where summer nights approach freezing.

In many places raised fields were not possible and so Indians constructed smaller *wacho* or *wachu* (ridges), parallel crests of turned-up earth perhaps two feet wide, separated by shallow furrows of equal size. Because the Americas had no large domesticable animals—llamas are too small to pull a plow or carry human beings—farmers did all the work with hoes and foot plows, long wooden poles with short hafts and sharp stone, bronze, or copper tips and footrests above the tip. Making a line across the field, village men faced backward, lifting up their foot plows and jabbing them into the soil, then stamping on the footrest to gouge deeper. Step by backward step, they created ridges and furrows. Each man's wife or sister faced him with a hoe or mallet, breaking up the clods into smaller pieces. Placed in holes atop the *wacho* were potato seeds or whole small tubers (each had to have at least one eye, from which the new potato would sprout). Sacred songs and chants paced the labor as the line of workers moved methodically down the field. Breaks were accompanied by mugs of *chicha* (maize beer) and handfuls of coca leaves to chew. When one field was done, villagers moved to the next, until everyone's fields were ready—a tradition of collective work that is a hallmark of Andean societies.

Four or five months later, farmers swarmed into the fields, digging up the tubers and leveling the *wacho* for the next crop—often quinoa, the native Andean grain. Every scrap of the potato plant was consumed except the toxic fruits. The foliage fed llamas and alpacas; the stalks became cooking fuel. Some of the fuel was used on the spot. Immediately after harvest, families piled hard clods of soil into igloo-shaped ovens eighteen inches tall. Inside the oven went the stalks, as well as straw, brush, and scraps of wood (after the Spaniards came, people used cattle manure). Fire heated the earthen ovens until they turned white. Cooks pushed aside the ashes and placed freshly harvested potatoes inside for baking. Villagers in the heights still do this today—the stoves glow in the twilight, dotting the hills. Steam curls up from hot food into the clear, cold air. People dip their potatoes in coarse salt and edible clay. Night winds carry the bakery smell of roasting potatoes for what seems like miles.

The potato roasted by precontact peoples was not the modern spud. Andean peoples cultivated different varieties at different altitude ranges. Most people in a village planted a few basic types, but everyone also planted others to have a variety of tastes, each in its little irregular patch of *wacho,* wild potatoes at the margins. The result was chaotic diversity. Potatoes in one village at one altitude could look wildly unlike those a few miles away in another village at another altitude.

When farmers plant pieces of tuber, rather than seeds, the resultant sprouts are clones; in developed countries, entire landscapes are covered with potatoes that are almost genetically identical. By contrast, a Peruvian-American research team found that families in a mountain valley in central Peru grew an average of 10.6 traditional varieties—landraces, as they are called, each with its own name. Karl Zimmerer, now at Pennsylvania State University, visited fields in some villages with as many as twenty landraces. The International Potato Center in Peru has sampled and preserved more than 4,900. The range of potatoes in a single Andean field, Zimmerer observed, "exceeds the diversity of nine-tenths of the potato crop of the entire United States." (Not all varieties grown are traditional. The farmers produce modern, Idaho-style breeds for the market, though they describe them as bland—they're for yahoos in cities.)

In consequence, the Andean potato is less a single identifiable species than a bubbling stew of many related genetic entities. Sorting it out has given decades of headaches to taxonomists (researchers who classify living creatures according to their presumed evolutionary relationships). Learned studies of cultivated potatoes in Andean fields have divided them, variously and contradictorily, into twenty-one, nine, seven, three, and one species, each further sliced into multiple subspecies, groups, varieties, and forms. Four is probably the most commonly used species number today, though the dispute is anything but resolved. As for *S. tuberosum* itself, the most widely accepted recent study parcels it into eight broad types, each with its own name.

Andean natives bred hundreds of different potato varieties, most of them still never seen outside South America.

The potato's wild relatives are no less confounding. In *The Potato*, a magnum opus from 1990, the potato geneticist J. G. Hawkes proclaimed the existence of some 229 named species of wild potato. That did not lay the matter to rest. After analyzing almost five thousand plants from across the Americas, Dutch researchers in 2008 winnowed Hawkes's 229 species down to just ten fuzzily defined entities—"species groups," as they put it—that bob like low, marshy islands in a morass of unclassifiable hybrids that extends from Central America down the Andes to the tip of South America and "cannot be structured or subdivided" into the classic species in biology textbooks. The description of the wild potato as a trackless genetic swamp was, the Dutch admitted, a view that their colleagues might find "difficult to accept."

None of this was apparent, of course, to the first Spaniards who ventured into the Andes—the band led by Francisco Pizarro, who landed in Ecuador in 1532 and attacked the Inka. The conquistadors noticed Indians eating these round objects and despite their suspicion sometimes emulated them. News of the new food spread rapidly. Within three decades, Spanish farmers as far away as the Canary Islands were growing potatoes in quantities enough to export to France and the Netherlands (then part of the Spanish empire). The first scientific description of the potato appeared in 1596, courtesy of the Swiss naturalist Gaspard Bauhin, who awarded it the name of *Solanum tuberosum esculentum,* which later became the modern *Solanum tuberosum.*

Folklore credits Francis Drake with stealing potatoes from the Spanish empire during a bout of piracy/privateering. Supposedly he gave them to Walter Ralegh, founder of the luckless Roanoke colonies.* (Drake rescued the survivors.) Ralegh asked a gardener on his Irish estate to plant them. His cook is said to have served the toxic berries at dinner. Ralegh ordered the plants yanked from

---

* Ralegh and his coevals spelled his name in many ways, including Rawley, Ralagh, and Raleigh. Although the last is most common today, he himself generally used "Ralegh."

his garden. Hungry Irish picked them up from the refuse—hence, apparently, the statue of Drake in Germany. On its face, the tale is unlikely; even if Drake snatched a few potatoes while marauding in the Caribbean, they would not have survived months at sea.

The first food Europeans grew from tubers, rather than seed, the potato was regarded with fascinated suspicion; some believed it to be an aphrodisiac, others a cause of fever, leprosy, and scrofula. Ultraconservative Russian Orthodox priests denounced it as an incarnation of evil, using as proof the undeniable fact that potatoes are not mentioned in the Bible. Countering this, the pro-potato English alchemist William Salmon claimed in 1710 that the tubers "nourish the whole Body, restore in Consumptions [cure tuberculosis], and provoke Lust." The philosopher-critic Denis Diderot took a middle stance in his groundbreaking *Encyclopedia* (1751–65), Europe's first general compendium of Enlightenment thought. "No matter how you prepare it, the root is tasteless and starchy," he wrote. "It cannot be regarded as an enjoyable food, but it provides abundant, reasonably healthy food for men who want nothing but sustenance." Diderot viewed the potato as "windy" (it caused gas). Still, he gave it the thumbs-up. "What," he asked, "is windiness to the strong bodies of peasants and laborers?"

With such halfhearted endorsements, it is little wonder that the potato spread slowly outside of the Spanish colonies. When Prussia was hit by famine in 1744, King Frederick the Great, a potato proponent, had to order the peasantry to eat potatoes. In England, farmers denounced *S. tuberosum* as an advance scout for hated Roman Catholicism. "No Potatoes, No Popery!" was an election slogan in 1765. As late as 1862, the British cookbook and household advice writer Isabella Beeton was warning her readers not to drink "the water in which potatoes are boiled." France was especially slow to adopt the new crop. Into the fray stepped nutritionist, vaccination advocate, and potato proselytizer Antoine-Augustin Parmentier, the Johnny Appleseed of *S. tuberosum*.

Trained as a pharmacist, Parmentier served in the army and was captured five times by the Prussians during the Seven Years' War. As a prisoner he ate little but potatoes for three years, a diet which to his surprise kept him in good health. His effort to understand how this could have happened led Parmentier to become a pioneering nutritional chemist, one of the first to try to figure out what is in food and why it sustains the body. When unseasonable rain and snow in 1769 and 1770 led to crop failures in parts of eastern France, a local academy announced a competition for "Plants that Could in Times of Scarcity be Substituted for Regular Food to Nourish Man." Five of the seven entries touted the potato. Parmentier's essay, the most impassioned and well documented, won the competition. It was the beginning of his career as a potato activist.

His timing was good. Four years after the famine, one of the first acts of the newly anointed king, Louis XVI, was to lift price controls on grain. Bread prices shot up, sparking what became known as the Flour War: more than three hundred civil disturbances in eighty-two towns. Throughout the disturbances Parmentier tirelessly advocated the potato as the solution. Proclaiming that France would stop fighting over bread if the French would eat potatoes, he set up one pro-spud publicity stunt after another: persuading the king to wear potato blossoms; presenting an all-potato dinner to high-society guests;* planting forty acres of potatoes at the edge of Paris, knowing that famished sansculottes would steal them. His efforts were successful. "The potato," announced a later supplement to Diderot's *Encyclopedia,* "is the fruit that feeds more than half of Germany, Switzerland, Great Britain, Ireland and many other countries."

In extolling the potato, Parmentier unwittingly changed it.

---

* Supposedly one guest was Thomas Jefferson, then U.S. ambassador to France. He is said to have liked one potato dish so much that he served it in the White House. In this way Jefferson introduced the United States to French fries.

All of Europe's potatoes descended from a few tubers sent across the ocean by curious Spaniards. From a genetic point of view, the European stock had been created by dipping a teaspoon into the sea of genes in Peru and Bolivia. Parmentier was urging his countrymen to cultivate this limited sample on a massive scale. Because potatoes are grown from pieces of tuber, he was unknowingly promoting the notion of planting huge areas with clones—a true monoculture. The potato fields he was envisioning were thus radically different from their Andean forebears. One was a crazy gumbo, its ingredients unclear; the other was an orderly array of identical parts.

The effects of this transformation were so striking that any general history of Europe without an entry in its index for *S. tuberosum* should be ignored. Hunger was a familiar presence in the Europe of the Little Ice Age, where cold weather killed crops even as Spanish silver drove up prices. Cities were provisioned reasonably well in most years, their granaries monitored by armed guards, but country people teetered on a precipice. When harvests failed, food riots ensued; thousands occurred across Europe between 1400 and 1700, according to the great French historian Fernand Braudel. Over and over, rioters, often led by women, broke into bakeries, granaries, and flour mills and either stole food outright or forced merchants to accept a "just" price. Ravenous bandits swarmed the highways, seizing grain convoys to cities. Order was restored by violent action.

Braudel cited an eighteenth-century tally of famine in France: forty nationwide calamities between 1500 and 1778, more than one every decade. This appalling figure actually understates the level of scarcity, he wrote, "because it omits the hundreds and hundreds of *local* famines." France was not exceptional; England had seventeen national and big regional famines between 1523 and 1623. Florence, hardly a poor city, "experienced iii years when people were hungry, and only sixteen 'very good' harvests between 1371 and 1791"—seven bad years for every bumper year.

The continent could not feed itself reliably. It was caught in the Malthusian trap.

As the sweet potato and maize did in China, the potato (and maize, to a lesser extent) helped Europe escape Malthus. When the agricultural economist Arthur Young toured eastern England in the 1760s he saw a farming world that was on the verge of a new era. A careful investigator, Young interviewed farmers, recording their methods and the size of their harvests. According to his figures, the average yearly harvest in eastern England from an acre of wheat, barley, and oats was between 1,300 and 1,500 pounds. By contrast, an acre of potatoes yielded more than 25,000 pounds—about eighteen times as much.* Growing potatoes especially helped England's poor, Young believed. "It is to be wished, that all persons who have it in their power to render this root more common among them, would exert themselves in it." Potatoes, he proclaimed, "cannot be too much promoted."

Potatoes didn't replace grain but complemented it. Every year, farmers left fallow as much as half of their grain land, to replenish the land and fight weeds (they were plowed under in summer). Now smallholders could grow potatoes on the fallow land, controlling weeds by hoeing. Because potatoes were so productive, the effective result was, in terms of calories, to double Europe's food supply. "For the first time in the history of western Europe, a definitive solution had been found to the food problem," the Belgian historian Chris Vandenbroeke concluded. (The German historian Joachim Radkau was blunter: the key environmental innovations of the eighteenth century, he

---

* This comparison overstates the case. Compared to grains, potatoes have more water, which is nutritionally useless. In the past potatoes were about 22 percent dry matter; wheat, by contrast, was about 88 percent. Thus the 25,620 pounds/acre yield of potatoes found by Young was equivalent to 5,636 pounds/acre of dry matter. Similarly, wheat's 1,440 pounds/acre yield would be 1,267 pounds/acre of dry matter. For this reason, it is fairer to say that potatoes were about four times more productive than wheat.

wrote, were "the potato and coitus interruptus.") Potatoes (and, again, maize) became to much of Europe what they were in the Andes—an ever-dependable staple, something eaten at every meal. Roughly 40 percent of the Irish ate no solid food other than potatoes; the figure was between 10 and 30 percent in the Netherlands, Belgium, Prussia, and perhaps Poland. Routine famine almost disappeared in potato country, a two-thousand-mile band that stretched from Ireland in the west to Russia's Ural Mountains in the east. At long last, the continent could, with the arrival of the potato, produce its own dinner.

Although the potato raised farm production overall, its greater benefit was to make that production more reliable. Before *S. tuberosum,* summer was usually a hungry time, with stored grain supplies running low before the fall harvest. Potatoes, which mature in as little as three months, could be planted in April and dug up during the thin months of July and August. And because they were gathered early, they were unlikely to be affected by an unseasonable fall—the kind of weather that ruined wheat harvests. In war-torn areas, potatoes could be left in the ground for months, making them harder to steal by foraging soldiers. (Armies in those days did not march with rations but took their food, usually by force, from local farmers.) Young's interview subjects used most of their potatoes for animal feed. In bad years, they had been forced to choose whether to feed their animals or themselves. Now they didn't have to make the choice.

The economist Adam Smith, writing a few years after Young, was equally taken with the potato. He was impressed to see that the Irish remained exceptionally healthy despite eating little else: "The chairmen, porters, and coal-heavers in London, and those unfortunate women who live by prostitution—the strongest men and the most beautiful women perhaps in the British dominions—are said to be, the greater part of them, from the lowest rank of people in Ireland, who are generally fed with this root." Today we know why: the potato can better sustain life than any other food when eaten as the sole item of diet. It has all essential nutrients except vitamins A and D, which can be supplied by milk; the diet of the Irish poor in Smith's day consisted largely of potatoes and milk. And Ireland was full of poor folk; England had conquered it in the seventeenth century and seized much of the best land for its own citizens. Many of the Irish were forced to become sharecroppers, paid for their work by being allowed to farm little scraps of wet land for themselves. Because little but potatoes could thrive in this stingy soil, Ireland's sharecroppers were among Europe's most impoverished people. Yet they were also among its most well nourished, because they ate potatoes. Smith drew out the logical consequences: if potatoes ever became, "like rice in some rice countries, the common and favourite vegetable food of the people," he wrote, "the same quantity of cultivated land would maintain a much greater number of people." Ineluctably, Smith believed, "Population would increase."

Smith was correct. At the same time that the sweet potato and maize were midwifing a population boom in China, the potato was helping to lift populations in Europe—the more potatoes, the more people. (The worldwide population boom was a sign and effect of the onset of the Homogenocene.) In the century after the potato's introduction Europe's numbers roughly doubled. The Irish, who ate more potatoes than anyone else, had the biggest boom; the nation grew from perhaps 1.5 million in the early 1600s to about 8.5 million two centuries later. (Some believed it reached 9 or even 10 million.) The increase occurred not because potato eaters had more children but because more of their children survived. Part of the impact was direct: potatoes prevented deaths from famine. The greater impact, though, was indirect: better-nourished people were less likely to die of infectious disease, the era's main killer. Norway was an example. Cold climate had long made it vulnerable to famine, which struck nationwide in 1742, 1762, 1773, 1785, and 1809. Then came the potato. The average death rate changed relatively little, but the big spikes vanished. When they were smoothed out, Norwegian numbers soared.

Such stories were recorded all over the continent. Hard hit by the shorter growing seasons of the Little Ice Age, mountain hamlets in Switzerland were saved by the potato—indeed, they thrived. When Saxony lost most of its agricultural land to Prussia in 1815, refugees filled its towns. To keep up with the rising numbers, farmers ripped out wheat and rye and planted potatoes. The potato harvest was enough to feed Saxony's growing population but not enough for good nutrition—there wasn't enough milk. Farmers in central Spain cut down olive and almond trees and planted potatoes. Village prosperity rose, followed by village numbers. And so on.

Just as American crops were not the only cause of China's population boom, they were not the only reason for Europe's population boom. The potato arrived in the midst of changes in food production so sweeping that some historians have described them as an "agricultural revolution." Improved transportation networks made it easier to ship food from prosperous areas to places with poor harvests. Marshlands and upland pastures were reclaimed. Shared village land was awarded to individual families, dispossessing many smallholders but encouraging the growth of mechanized agriculture (the new owners could be guaranteed of keeping the returns if they invested in their farms). Reformers like Young popularized better cultivation methods, especially the use of manure from stables as fertilizer. Farmers learned to plant fallow fields with clover, which recharges the soil with nutrients. First domesticated by the Moors in Spain, clover helped prevent Europeans from destroying their pastureland soil by overgrazing. The advances were not confined to agriculture. American silver let Europeans build ships to increase trade, raising living standards. Some improvements occurred in the continent's governance and even in its abysmal hygiene standards. As in China, the Little Ice Age began to wane.

In 2010 two economists at Harvard and Yale attempted to account for such factors by comparing events in parts of Europe that were similar except for their suitability for potatoes; any systematic differences, they argued, would be due to the new crop. According to the two researchers' "most conservative" estimate, *S. tuberosum* was responsible for about an eighth of Europe's population increase. Put baldly, the figure may not seem high. But the continent's long boom had many causes. One way to think of this calculation is to say that it suggests the introduction of the potato was as important to the modern era as, say, the invention of the steam engine.

THE GUANO AGE

It was said that the islands gave off a stench so intense that they were difficult to approach. They were a clutch of dry, granitic mounds thirteen miles off the Peruvian shore, about five hundred miles south of Lima on the west coast of South America. Almost nothing grew on them. Called the Chincha Islands, they were never inhabited by Indians—not for long, anyway. Their sole distinction is their population of seabirds, especially the Peruvian booby, the Peruvian cormorant, and the Peruvian pelican. The birds are attracted by the strong coastal current, which pulls cold water from the depths. Phytoplankton feast on the nutrients that rise with the water. Zooplankton eat the phytoplankton and in turn are the primary food of the anchoveta fish, a cousin to the familiar anchovy. Anchoveta live in vast schools that are preyed upon by other fish. Predators and prey both are preyed upon by the Peruvian booby, cormorant, and pelican. All three have nested on the Chincha Islands for millennia. Over time they have covered the islands with a layer of guano as much as 150 feet thick.

Guano makes excellent fertilizer. Fertilizer is, at base, a mechanism for providing nitrogen to plants. Plants need nitrogen to make chlorophyll, the green substance in their leaves that absorbs the sun's energy for photosynthesis. Nitrogen is also a key building block for both DNA and the proteins for which DNA is the template. Although more than three-quarters of the atmosphere is made up of nitrogen gas, from a plant's point of view nitrogen is scarce—the gas is made from two nitrogen atoms that cling to each other so tightly that plants cannot split them apart for use. In consequence, plants seek nitrogen from the soil, where it can be found in forms that they can break down: ammonia (NH_3, or one nitrogen atom and three hydrogen atoms), nitrites (compounds that include NO_2, a group of one nitrogen atom and two oxygen atoms), and nitrates (compounds that include NO_3, a group of one nitrogen atom and three oxygen atoms). All are in less supply than farmers would like, not least because bacteria in the soil constantly digest nitrates and nitrites, turning the nitrogen back into unusable nitrogen gas. Land that has been farmed repeatedly always risks nitrogen depletion.

Unlike mammalian urine, bird urine is a semisolid substance. Because of this difference, birds can build up reefs of urine in a way that mammals cannot (except, occasionally, for big colonies of bats in caves). Even among birds, though, Chincha-style guano deposits—heaps as big as a twelve-story building—are uncommon. To make them, the birds must be relatively large, form big flocks, and defecate where they live (gulls, for instance, release their droppings away from their breeding grounds). In addition, the area must be dry enough not to wash away the guano. The waters off the Peruvian coast receive less than an inch of rain a year. The Chinchas, the most important of Peru's 147 guano islands, house hundreds of thousands of Peruvian cormorants, the most prolific guano producers. According to *The Biogeochemistry of Vertebrate Excretion,* a classic treatise by G. Evelyn Hutchinson, a cormorant's annual output is about thirty-five pounds. Arithmetic suggests that the Chincha cormorants alone produce thousands of tons per year.

Centuries ago Andean Indians discovered that depleted soils could be replenished with guano. Llama trains carried baskets of Chincha guano along the coast and perhaps into the mountains. The Inka parceled out guano claims to individual villages, levying penalties for disturbing the birds during nesting or taking guano allocated to other villages. Blinded by the shine from Potosí silver, the Spaniards paid little attention to conquered peoples' excremental practices. The first European to observe guano carefully was the German polymath Friedrich Wilhelm Heinrich Alexander von Humboldt, who traveled through the Americas between 1799 and 1804. A pioneer in botany, geography, astronomy, geology, and anthropology, Humboldt had an insatiable curiosity about everything that crossed his path, including the fleet of native guano boats that he saw skittering along the Peruvian coast. "One can smell them a quarter of a mile away," he wrote. "The sailors, accustomed to the ammonia smell, aren't bothered by it; but we couldn't stop sneezing as they approached." Among the thousands of samples Humboldt took back to Europe was a bit of Peruvian guano, which he sent to two French chemists. Their analysis showed that Chincha guano was 11 to 17 percent nitrogen—enough to burn plant roots if not properly applied. The French scientists touted its potential as fertilizer.

Few took their advice. Supplying European farmers with guano would involve transporting large quantities of excrement across the Atlantic, a project that understandably failed to enthuse shipping companies. Within several decades, though, the picture changed. Agricultural reformers throughout Europe had begun to worry that the ever-more-intense agriculture necessary to feed growing populations was exhausting the soil. As harvests leveled off and even decreased, they looked for something to restore the land: fertilizer.

At the time, the best-known soil additive was bone meal, made by pulverizing bones from slaughterhouses. Bushels of bones went to grinding factories in Britain, France, and Germany. Demand ratcheted up, driven by fears of soil depletion. Bone dealers supplied the factories from increasingly untoward sources, including the recent battlefields of Waterloo and Austerlitz. "It is now ascertained beyond a doubt, by actual experiment upon an extensive scale, that a dead soldier is a most valuable article of commerce," remarked the London *Observer* in 1822. The newspaper noted that there was no reason to believe that grave robbers were limiting themselves to battlefields. "For aught known to the contrary, the good farmers of Yorkshire are, in a great measure, indebted to the bones of their children for their daily bread."

From this perspective, avian feces began to seem like a reasonable item of commerce. A few bags of guano appeared in European ports in the mid-1830s. Then Justus von Liebig weighed in. A pioneering organic chemist, Liebig was the first to explain plants' dependence on nutrients, especially nitrogen. In his treatise *Organic Chemistry in Its Application to Agriculture and Physiology* (1840), Liebig criticized the use of bone fertilizer, which has little nitrogen. Guano was another story: "It is sufficient to add a small quantity of guano to a soil consisting only of sand and clay, in order to procure the richest crop of maize." Liebig was enormously respected; he was an avatar of the Science that had brought new, productive crops like the potato and maize, and new ways of thinking about agriculture and industry. *Organic Chemistry* was quickly translated into multiple languages; at least four English editions appeared. Sophisticated farmers, many of them big landowners, read Liebig's encomium to guano, flung down the book, and raced to buy it. Yields doubled, even tripled. Fertility in a bag! Prosperity that could be bought in a store!

Guano mania took hold. In 1841, Britain imported 1,880 tons of Peruvian guano, almost all of it from the Chincha Islands; in 1843, 4,056 tons; in 1845, 219,764 tons. In forty years, Peru exported about 13 million tons of guano, receiving for it approximately £150 million, roughly $13 billion in today's dollars. It was the beginning of today's input-intensive agriculture—the practice of transferring huge amounts of crop nutrients from one place to another, distant place according to plans dictated by scientific research.

Hoping to take maximum advantage of the guano rush, Peru nationalized the Chinchas. Soon it discovered that nobody wanted to work on the islands. Except for birds, their only inhabitants were bats, scorpions, spiders, ticks, and biting flies. Not a single plant grew on their barren slopes. Worse, the islands had no water; every drop had to be shipped in. Because the land was blanketed in guano, miners worked, ate, and slept on shelves of ancient excrement. So little rain fell that the soluble materials in the guano never washed away—it remained studded with crystals of ammonia nitrate, which broke in corrosive clouds around miners' shovels. Powdery and acrid, the guano went into miners' carts, which were pushed up rails to a depot atop one of the seaside cliffs. From the cliff, men dumped tons of excrement through a long canvas tube directly into the bellies of vessels below. Slamming into the hold, guano dust exploded from the hatchways, shrouding the ship in a toxic fog. Workers wore masks made from hemp smeared with tar, one visitor noted,

> but the guano mocks at such weak defenses. . . . [T]hey are unable to remain below longer than twenty minutes at one time. They are then relieved by another party, and return on deck perfectly naked, streaming with perspiration, and with their brown skins thickly coated with guano.

The government could have paid high wages to get workers to endure these terrible conditions, but that would have cut into profits. Instead it stocked the islands with a mix of convicts, army deserters, and African slaves. This arrangement proved unsatisfactory: the convicts and deserters killed each other, and the slaves were so valuable that their mainland owners did not wish to part with them. In 1849 Peru gave up trying to run the mines itself and awarded an exclusive concession to Domingo Elías, Peru's biggest cotton grower and one of its principal slave owners. Politically savvy and manically ambitious, Elías had been prefect of Lima; during a time of civil unrest, he briefly declared himself ruler of the nation. In return for the monopoly, Elías was supposed to mine guano with his own slaves, but he, too, was reluctant to take them away from his cotton fields. He induced the government to subsidize merchants who imported immigrants. Promi-

Thousands of Chinese slaves mined the guano of Peru's Chincha Islands, shown here in 1865, for export to Europe as fertilizer. The islands, home for millennia to seabirds, were covered with a layer of guano as much as 150 feet deep.

nent among these subsidized importers was Domingo Elías. By the time the law passed his agents were already in Fujian, waving labor contracts in the faces of illiterate villagers.

In standard indenture practice, the contracts promised the Chinese would pay for their passage by working, typically for eight years, in the newly discovered California gold fields. (The actual destination, the guano archipelago, was not mentioned.) The ruse was plausible: agents for U.S. firms were in Fujian at the same time, telling a similar lie as they sought indentured servants to build railroads. People who signed the bogus Peruvian contract were conducted to bleak human warehouses in Amoy (now called Xiamen, on an island across the river from Yuegang) and, later, Macao. People who refused to sign often were kidnapped and shipped to the same warehouses. In these dark confines slavers burned the letter *C*—for California, their ostensible destination—into the backs of their ears. No longer were the men described as workers. Their new name was *zhuzai*, little pigs. "None were let outside," wrote the Shanghai historian Wu Ruozeng. "Those who resisted were whipped; any who tried to escape were killed."

Peru was not the only destination in the mid-century Chinese diaspora. A quarter of a million or more *zhuzai*, almost all of them men, ended up—more or less willingly, more or less knowingly—in Brazil, the Caribbean, and the United States. But Peru represented the worst passage, the direst conditions, the most dreaded destination. Ultimately at least 100,000 Chinese were taken there. Conditions en route can be compared to those in the transatlantic slave trade. Perhaps one out of eight *zhuzai* died. As on the Atlantic slave ships, revolts were common. Eleven mutinies are known to have occurred on Peru-bound vessels; at least five bloodily succeeded.

Most of the Chinese ended up working in the sugar and cotton plantations on the coast. Some built the railroads that the Peruvian government was constructing with guano money. At any given time between one and two thousand were on the Chincha Islands. In classic divide-and-conquer fashion, Elías forestalled rebellion by setting his African slaves as overseers over his Chinese slaves and holding both to strict deadlines. Spasms of cruelty, slave upon slave, were the inevitable result. Guano miners swung their picks up to twenty hours a day, seven days a week, to fulfill their assigned daily quotas (as much as five tons of guano); two-thirds of their pay was deducted for room (reed huts) and board (a cup of maize and some bananas). Failure to meet the daily quota was rewarded with a five-foot rawhide whip. Minor infractions were punished by torture. Escape from the islands was impossible. Suicide was frequent. One overseer told a *New York Times* correspondent that

> more than sixty had killed themselves during the year, . . . chiefly by throwing themselves from the cliffs. They are buried, as they live, like so many dogs. I saw one who had been drowned—it was not known whether accidentally or not—lying on the guano, when I first went ashore. All the morning, his dead body lay in the sun; in the afternoon, they had covered it a few inches, and there it lies, along with many similar heaps, within a few yards of where they were digging.

So many Chinese died that the overseers marked off an acre of guano as a cemetery.

Journalistic exposés of guano slavery created an international scandal that gave the Lima government an excuse to eject Elías and renegotiate the guano contract with someone else, thus procuring a second round of bribes. Fulminating against the evils of official corruption, Elías sought to regain his lucrative concession by twice staging a coup d'état. Both attempts failed. In 1857 he tried the legal route, running for president without success.

All the while guano flowed to Europe and North America. In addition to signing an exclusive mining concession with Elías,

Peru had awarded a monopoly on shipping guano internationally to a company in Liverpool. With demand outstripping supply, Peru and its British consignees were able to charge high prices. Their customers reacted with fury to what they viewed as extortion. Decrying the "powerful monopoly" on guano, the British *Farmer's Magazine* laid out its readers' demands in 1854. "We do not get anything like the quantity we require; we want a great deal more; but at the same time, we want it at a lower price." If Peru insisted on getting a lot of money for a valuable product, the only fair solution was invasion. Seize the guano islands!

From today's perspective, the outrage—threats of legal action, whispers of war, editorials about the Guano Question—is hard to understand. But agriculture was then "the central economic activity of every nation," as the environmental historian Shawn William Miller has pointed out. "A nation's fertility, which was set by the soil's natural bounds, inevitably shaped national economic success." In just a few years, agriculture in Europe and the United States had become dependent on high-intensity fertilizer—a dependency that has not been shaken since. Britain, first to adopt guano and by far the largest user, was both the most dependent and the most resentful. Much as oil buyers today begrudge the member nations of OPEC, Peru's British customers ranted about the guano cartel. They were apoplectic as Peru's guano barons sauntered around Lima in the latest Parisian fashions, bejeweled trollops on their arms.

Britons were almost entirely silent about Peru's British agents in Liverpool, who used their share of the Peruvian monopoly profits to construct one of the biggest houses in England. Americans were not silent. They fumed as the British gave priority to their British customers, leaving Americans at the end of the guano line. Spurred by their fury, Congress passed the Guano Islands Act in 1856, authorizing its citizens to seize any guano islands they saw. The biggest loads came from Navassa, an island fifty miles west of Haiti, which the United States took in 1857.

After the Civil War its workforce consisted largely of freed slaves. Conditions gradually deteriorated; the former slaves rebelled twice, killing some of their jailers, and the enterprise fell apart in a cloud of scandal. Under the aegis of the Guano Islands Act, merchants claimed title to ninety-four islands, cays, coral heads, and atolls between 1856 and 1903. The Department of State officially recognized sixty-six as U.S. possessions. Most proved to have little guano and were quickly abandoned. Nine remain under U.S. control today.

Guano set the template for modern agriculture. Ever since Liebig, farmers have treated the land as a medium into which they dump bags of chemical nutrients. The nutrients are shipped from far-off places or synthesized in distant factories. Farming is the act of transferring those external nutrients to crops in the field: high volumes of nitrogen go in, high volumes of maize and potatoes go out. Because the harvests in this system are enormous, the crops are no longer vehicles for local subsistence, but products destined for an international market. To maximize output, they are grown in ever-larger, single-crop fields—industrial monoculture, as it is called.

Today scholars often describe the "Green Revolution" after the Second World War—the combination of high-yield crops, agricultural chemicals, and intensive irrigation—as the moment when humankind triumphantly escaped, at least for a while, the limits set by small-scale farms and local resources. But as the Amherst College historian Edward D. Melillo has argued, the arrival of guano ships in Europe and the United States marked an earlier, equally profound Green Revolution, the first in a series of technological innovations that transformed life across the planet.

Before the potato and maize, before intensive fertilization, European living standards were roughly equivalent with those today in Cameroon and Bangladesh; they were below Bolivia or Zimbabwe. On average, European peasants ate less per day than hunting-and-gathering societies in Africa or the Amazon.

Industrial monoculture with improved crops and high-intensity fertilizer allowed billions of people—Europe first, and then much of the rest of the world—to escape the Malthusian trap.* Incredibly, living standards doubled or tripled worldwide even as the planet's population climbed from fewer than 1 billion in 1700 to about 7 billion today.

Along the way guano was almost entirely replaced by nitrates mined from vast deposits in the Chilean desert. The nitrates in turn were replaced by artificial fertilizers, made in factories by a process invented and commercialized in the early twentieth century by two Nobel-winning German chemists, Fritz Haber and Carl Bosch. No matter what their composition, though, fertilizers remain just as critical to agriculture, and through agriculture to contemporary life. In a fascinating 2001 study of the impact of factory-made nitrogen, Vaclav Smil, the University of Manitoba geographer, estimated that two out of every five people on earth would not be alive without it.

By any measure these were amazing accomplishments. Yet like all human endeavors the rise of intensive agriculture had its downside. The guano trade that launched modern agriculture was also the beginning, via the Columbian Exchange, of one of its worst pitfalls: the intercontinental transport of exotic pests. Proof will never be found, but it is widely believed that the guano ships carried a microscopic hitchhiker: *Phytophthora infestans*. *P. infestans* causes late blight, a plant disease that exploded through Europe's potato fields in the 1840s, killing as many as two million people, half of them in Ireland, in what came to be known as the Great Hunger.

---

* This may understate the impact. The historian Kenneth Pomeranz has argued that "some of the most intensely farmed soils of Europe (including in England) faced serious depletion by the early nineteenth century." If guano had not arrived, Pomeranz believes, the consequences may not have been simply remaining at the same level but a full-scale disaster across much of the continent.

THOROUGHLY MODERN FAMINE

The name *Phytophthora infestans* means, more or less, "vexing plant destroyer," a censure that is wholly deserved. *P. infestans* is an oomycete, one of seven hundred or so species sometimes known as water molds. From a biologist's point of view, oomycetes can be thought of as cousins to algae. From a gardener's point of view, *P. infestans* looks and acts like a fungus. It sends out tiny bags of six to twelve spores that are blown on the wind, usually for no more than twenty feet, occasionally for as much as half a mile or even further. When the bag lands on a susceptible plant, it hatches, so to speak, releasing what are technically known as zoospores: mobile, two-tailed cells that slowly swim through moisture on the leaf or stem, looking for the tiny respiratory holes called stomata. If the day is warm and wet enough, the zoospores germinate, sending long, threadlike filaments through the stomata into the leaf. Extensions from the filaments infiltrate leaf cells, hijacking the mechanisms inside; the plant ends up nourishing the invader, rather than itself. The first obvious symptoms—purple-black or -brown spots on the leaves—are visible in about five days. By that time it is often too late. Filaments lace through much of the plant. The oomycete is already generating new bags of spores.

Water is the blight's friend—zoospores cannot germinate on dry leaves. Rain washes zoospores from the leaves onto the soil, letting them attack roots and tubers as much as six inches below the surface. Especially vulnerable are the tuber's eyes. *P. infestans* strikes from the outside in, turning the potato's outer flesh into dry, grainy, red-brown rot. Extensions of blight reach like dark claws toward the center of the tuber. Because the boundary between diseased and healthy tissue is indistinct, the entire potato must usually be thrown away. Care must be taken with disposal: a single infected tuber can generate a million spores.

P. infestans preys on members of the nightshade family: potatoes, tomatoes, eggplants, sweet peppers, and weeds like hairy nightshade and bittersweet nightshade. When shocked European researchers first observed the carnage in potato fields, they naturally assumed the agent responsible came from Peru, the land of potatoes. Seventy years ago most changed their minds. Typically biologists view a species's "center of diversity"—the place where it has the widest array of forms—as its ancestral home. Mexico has hundreds of varieties of maize seen nowhere else, suggesting that the species originated there. Africans are more genetically diverse than Caucasians or Asians; Africa is the cradle of humankind. And so on. In central Mexico, *P. infestans* seemed more varied than anywhere else. Notably, the species occurs in two types—one could think of them as male and female, except that oomycetes have no sexual characteristics—that can combine their DNA, creating an egg-like entity known as an oospore. In other words, *P. infestans* can reproduce both asexually and "sexually," with the quotation marks as a reminder that these creatures are not male and female.* But only in Mexico did the oomycete reproduce sexually, because the rest of the world lacked one of the two forms. Scientists argued that this and other types of diversity indicated that *P. infestans* originated in Mexico—even though there is no evidence of *S. tuberosum* there until the eighteenth century. Alexander von Humboldt, visiting Mexico in 1803 with his samples of guano, made the first certain observation of a potato in Mexico. Humboldt assumed that Spaniards had imported the tuber from the Andes. The potato blight had existed for millennia, in this view, before it encountered a potato. A final detail:

---

* Reproducing both sexually and asexually sounds odd to big, clumsy mammals like us, but it is a canny survival strategy in much of the microworld (malaria-causing *Plasmodium* parasites reproduce both ways, for example). Asexual reproduction is useful in good times, because it produces offspring that are exactly as well adapted genetically to their environment as their parents. Sexual reproduction is valuable when the environment changes, because the sexual shuffling of genes creates variability, which helps the offspring survive in altered circumstances.

because blight was spotted in the United States before Europe, some researchers suggested that it spread there first, then hopped a boat across the Atlantic.

In a series of experiments culminating in 2007, a team led by University of North Carolina plant geneticist Jean Ristaino overturned these ideas. Ristaino's team used the tools of DNA analysis to examine blight from 186 infected potatoes in herbariums, the botanical storehouses in museums. The youngest sample was from 1967; three were collected in Europe in 1845–47, the time of the Great Hunger. Ristaino's scheme was complex in detail, but simple in principle. Because *P. infestans* usually reproduces asexually, the progenitor oomycete and its offspring usually have identical genetical endowments, except for the infrequent occasions when a mutation scrambles DNA. Organisms with similar DNA patterns belong, as geneticists say, to the same "haplogroup." If two individuals belong to the same haplogroup, it is molecular evidence that they share a recent ancestor. Similarly, different haplogroups are a sign of the lack of a recent common ancestor. Ristaino's team found that potato blight from the Andes had a greater number of haplogroups than Mexican blight—it was fundamentally more diverse. Moreover, DNA from the old blight in herbariums—samples preserved for as long as a century and a half—was nearly identical to DNA from Andean blight. "The U.S. and Irish populations were not genetically differentiated from the Peruvian populations," the scientists wrote. Blight from the Andes "initiated epidemics first in the U.S. and then Ireland that led to the famine."

Most likely the blight traveled from Peru to Europe aboard a guano ship to Belgium, probably to Antwerp, the area's most important port. Farmers in the adjacent province of West Flanders were having trouble with their potatoes. In what would now be described as a demonstration of the power of evolution, European plant pathogens—viruses and fungi—were adapting to the new crop. In July 1843 the provincial council of West Flanders voted to import new varieties of potato from North and South

America, hoping some would prove to be less susceptible to the diseases. No record exists of their origins or the means by which they were shipped. It would be odd, though, if the South American potatoes had not come from the Andes.

Almost certainly the potatoes made the journey on a guano ship. Between 1532 and 1840 few ships passed directly from Peru to Europe, because Spain, protective of its silver in Potosí, tightly controlled traffic. As Potosí ran out of ore, the silver ships sailed less frequently. In the 1820s Bolivia and Peru gained their independence and Spanish shipping there closed down altogether. European ships were then free to sail to Lima, but few did: the new nations had little to offer and were politically chaotic to boot. In its first two decades, Peru had more than one change of government per year; it also fought five foreign wars. A direct shipping line between Peru and Britain did not open until 1840. It carried guano. As guano frenzy set in, ships by the score sailed from Europe to the Chincha Islands. One traveler there in 1853 saw 120 vessels clustered about the guano docks. Another, later voyager saw 160. Chances are high that one of these ships unknowingly carried blighted potatoes to Belgium—and infected a continent.

Field trials of West Flanders's new potatoes began in 1844. That summer a nearby French botanist observed a few potato plants with strange, bruise-dark spots. The following winter was extremely cold, which should have killed any blight spores or eggs in the soil. But the experimenters may have stored a few contaminated potatoes and unknowingly planted them the next spring. In July 1845 the West Flanders town of Kortrijk, six miles from the French border, became the launchpad for Europe's first widespread epidemic of potato blight. Carried by windblown spores, the oomycete hopscotched to farms around Paris by August. Weeks later, it was in the Netherlands, Germany, Denmark, and England. Governments panicked and ordered more potatoes from abroad.

Blight was first reported in Ireland on September 13, 1845. By mid-October the British prime minister was privately describ-

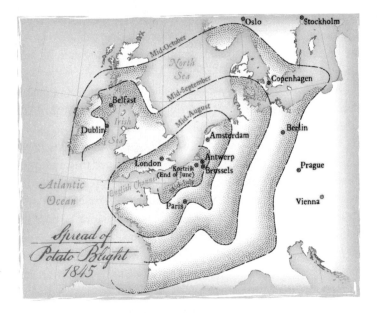

Mid-October

North Sea

Oslo

Stockholm

Copenhagen

Mid-September

Belfast

Mid-August

Dublin

Irish Sea

Berlin

Amsterdam

London

Antwerp

Kortrijk
(End of June)

Brussels

Prague

Atlantic Ocean

English Channel

Mid-July

Paris

Vienna

*Spread of
Potato Blight
1845*

ing the epidemic as a national disaster. Within another month between a quarter and a third of the crop had been lost. Cormac Ó Gráda, an economist and blight historian at University College, Dublin, has estimated that Irish farmers planted about 2.1 million acres of potatoes that year. In two months *P. infestans* wiped out the equivalent of between half and three-quarters of a million acres in every corner of the nation. The next year was worse, as was the year after that.

Recall that almost four out of ten Irish ate no solid food except potatoes, and that the rest were heavily dependent on them. Recall, too, that Ireland was one of the poorest nations in Europe. At a stroke, the blight removed the food supply from half the country—and there was no money to buy grain from outside. The consequences were horrific; Ireland was transformed into a post-apocalyptic landscape. Destitute men lined the roads in their rags, sleeping in crude shelters dug into roadside ditches. People ate dogs, rats, and tree bark. Reports of cannibalism were frequent

In early 1847 the *Illustrated London News* asked the artist James Mahoney to tour the famine-wracked Irish countryside. His articles and illustrations depicted a landscape of ruins and starving beggars—and did much to bring the crisis to the attention of the English public.

and perhaps accurate. Entire families died in their homes and were eaten by feral pets. Disease picked at the survivors: dysentery, smallpox, typhus, measles, a host of ailments listed in death records as "fever." Mobs of beggars—"homeless, half-naked, famishing creatures," one observer called them—besieged the homes of the wealthy, calling for alms. So many died that in many western towns the bodies were interred in mass graves.

As resources vanished, life became a struggle of all against all. Starving men stole into fields to steal turnips from the ground. Farmers dug mantraps in their fields to stop them. Landlords evicted tenants in huge numbers, tore down their homes, then went bankrupt themselves. Neighbor fought neighbor for food and shelter. Crime levels exploded, the murder rate almost doubling in two years. Some hungry people stole to put food on the table, others to be fed while incarcerated. In one case two men released from prison were sent back the next day for trying "to break into jail." The only violent crime to decline was rape, because potential perpetrators lacked the energy.

Hundreds of thousands of desperate people fled the country in what became known as "coffin ships." One passenger remembered bodies "huddled together without light, without air, wallowing in filth and breathing a fetid atmosphere, sick in body, dispirited in heart." The ships marked their passage with a trail of dead slid into the sea. Most migrants went to the United States and Canada. Multitudes of sick and starving filled the quarantine area at Grosse Île, in the St. Lawrence River by Quebec. A mass grave there contains thousands of bodies. They died an ocean away from Ireland but were as much victims of *P. infestans* as if they had never left.

Britain mounted the biggest aid program in its history, but it was catastrophically insufficient—largely, Irish nationalists charge, because London treated the crisis as a chance to expand its efforts to transform Ireland's "primitive" subsistence farming to export-oriented agriculture. Instead of simply providing food, the British pulled people off the farm, massed them in workhouses, and fed them from soup kitchens; meanwhile, the farms were consolidated into bigger, more export-friendly units. Other critics point to the export of food from Ireland during the famine: 430,000 tons of grain in 1846 and 1847, the two worst years. "The Almighty indeed sent the potato blight," nationalist leader John Mitchel thundered, "but the English created the famine."

Examples of British callousness were indeed thick on the ground. Some politicians welcomed the depopulation, which would, one cabinet minister's agent promised, "give us room to become civilized." Others said that giving food to soup kitchens actually did harm; if "large numbers would have perished of starvation," one banking official reasoned, "the material relations of the survivors would have been re-established."

England's defenders retort that although anti-Irish politicians said awful things they were ignored. In practice, the starving had to be massed if they were to be fed; delivering huge amounts of food to scattered families is not easy, even today. The exporters, moreover, were mainly Irish farmers who sold pricy meat and grain to buy cheap food for their families. Failure to accomplish the unprecedented, however dire the consequences, is not a moral lapse—or so the argument goes.

No matter what degree of culpability should be assigned to Britain, the consequences of the famine to Ireland are indisputable: it broke the nation in half. At a million or more fatalities, it was one of the deadliest famines in history, in terms of the percentage of population lost. A similar famine in the United States today would kill almost forty million people. Only the famine of 1918–22 in the Soviet Union may have been worse. Within a decade of the blight another two million fled Ireland. Many more followed in subsequent decades, inexorably driving its population down. The nation never regained its footing. As late as the 1960s its population was half what it had been in 1840. Today Ireland has the melancholy distinction of being the only nation in Europe, and perhaps the world, to have fewer people within the same boundaries than it did more than 150 years ago.

LAZY BEDS

The Great Hunger left such a scar in Ireland that historians could barely bring themselves to look at it for more than a century. Since the 1970s, though, the famine has been the subject of hundreds of books and articles. In all the outpour, though, surprisingly little attention has been devoted to its cause, *P. infestans*—unfortunately, because the oomycete was the protagonist in the first calamity of modern commodity agriculture.

P. infestans came to Ireland with surprising alacrity and overtook the country with puzzling speed. Ireland, an island nation, is eight hundred miles away from West Flanders. Between them are the North Sea and the Irish Sea. Blight spores are fragile: a single hour of exposure to the ultraviolet radiation in sunlight is enough to cut their likelihood of germinating by 95 percent. Even a light rain knocks them from the air. A widely cited ecological model suggests that as a practical matter they cannot surf the winds further than twenty to thirty miles. After tests in Washington State, three scientists concluded that in perfect conditions—strong winds, cool temperatures, no direct sunlight or rain—blight spores might be able to move seventy miles, though fewer than 5 percent would survive. Except around Northern Ireland, the Irish Sea is wider than seventy miles. If the researchers are correct, blight spores could only have been blown to Ireland by traveling from southeast to northwest England, then floating over the North Channel to Belfast—a remarkable journey. (Technically, they are not spores but "sporangia," the bags of spores released by blight, but I am ignoring this distinction for the moment.)

Rain fell, sometimes heavily, on twenty-four of the thirty days after September 13, 1845, when the blight was first reported in Ireland. Yet despite the rain *P. infestans* swept across the country, striking with a remorselessness not seen anywhere else. Something about Ireland was uniquely vulnerable—but what? One part of the answer was the sheer number of potatoes, a fat target for the blight. Another part was the uniformity of the crop. According to Ó Gráda, the blight historian, about half of Ireland was dominated by a single, outstandingly productive variety: the Lumper. Many Irish lived in clusters of farmhouses called *clachans* surrounded by tightly packed, communally owned farmland. Encircled by clones of a single variety of a single tuber species, the *clachans* of western Ireland were among the most uniform ecosystems on the planet.

Irish farmers for centuries had grown crops by cutting out blocks of sod, flipping them upside down, and piling them into long, broad ridges separated by deep furrows—"lazy-bed" farming, as the system was known. (The name may come from an occasional English epithet for the potato: the "lazy root.") Typically four feet wide, the ridges loomed a foot or more above the furrows. They looked strikingly like *wacho,* the ridged fields in Andean societies. Like *wacho,* they were built in boggy soils; the ridges warmed up more quickly in the morning and retained heat longer in the evening than the surrounding flatlands, an advantage in cold places like the Andes and Ireland. Constructed from several layers of sod, the ridges represented concentrations of good soil; farmers could plant them densely, which naturally stifled weeds. Because the ridges were not plowed, they had intact root systems that resisted erosion; the roots also ensured that grass returned quickly after harvest, restoring nutrients.

Unaware of these advantages, eighteenth-century agricultural reformers denounced the lazy-bed/*wacho* method as inefficient, an unproductive obstacle to modernization. Activists like Andrew Wight and Jethro Tull wanted farmers to release soil nutrients by deep, thorough plowing; to plant every possible bit of terrain; to charge the land with fertilizer (manure and then, when it became available, guano); to protect growing crops with ruthless weeding; and to maximize yields by efficient harvesting. Believers in technology, they viewed the newest factory-made harrows, drillers, and harvesters as God-given tools to accomplish these goals. Because these machines needed level land—they couldn't climb up and down ridges—the lazy-beds had to go. On top of everything else, reformers said, the furrows between the ridges were a waste of space.

Wacho occupied a swath of northern Europe that reached from France to Poland and included Britain, Ireland, Scandinavian countries, and Baltic states. As the new methods took hold after about 1750, they disappeared. *Wacho* had almost vanished from Ireland by 1834, when reform enthusiast Edmund Murphy took a cross-country "professional tour" between Dublin on the east coast and Galway on the west, taking "particular notice of the potato crop." Seeing few lazy-beds in areas where they had once been ubiquitous, he crowed that they were "completely superseded. . . . Nothing shows the rapid improvement in agriculture which is at present extending in this country more clearly."

To examine the consequences of the shift to modern cultivation methods, Michael D. Myers, then at the University of Texas in Austin, experimentally created six fields in Northern Ireland: three lazy-beds and three of the level fields that replaced them. He discovered that the simple ridges and furrows created a complex geography, with surprisingly sharp temperature and humidity differences between the top of the ridge and the bottom of the furrow. Plant-disease specialists describe the temperature and humidity conditions that favor *P. infestans* in terms of "blight units"—the higher the number of blight units, the better the chance that blight zoospores on potato leaves will be able to germinate. Myers's lazy-beds had roughly half as many blight units as the level fields. Blight spores were less likely to germinate in the comparatively warm, dry conditions atop the ridges. Because water drained into the furrows, it flowed away and beneath the growing tubers, carrying zoospores away from them. In addition, they had fewer noxious weeds and needed less fertilizer.*

---

* The campaign against lazy-bed farming may not have been reformers' only contribution to destruction. *P. infestans* exploded across Europe so fast that one wonders whether the blight was accidentally distributed by human

Murphy, denouncer of lazy-beds, took his professional tour because disease had been striking Ireland's potatoes. This was in 1834, a decade before the blight; the diseases he was concerned with were viruses, bacteria, nematodes, and so on—ordinary pests that were adapting to the new crop. As the pests evolved, they caused crop failures; fourteen occurred between 1814 and 1845. (None of these incidents was anywhere as severe as the Great Hunger.) Myers, the University of Texas researcher, came to believe these failures were due in part to the abandonment of lazy-bed cultivation, which inadvertently fostered plant disease. (It is worth noting that the Andes did not have such widespread potato epidemics.) The blight was simply the latest and worst pathogen to take advantage of the new scientific agriculture: one kind of potato, on a terrain shaped for technology, rather than biology.

The Great Hunger was the first truly contemporary agricultural disaster. Without the improvements wrought by modern science and technology, the blight would have had far less impact. Alarmed by the blight, governments in France, Belgium, Britain, and the Netherlands quickly asked biologists for help. But in its surge and sweep it was like nothing they had ever seen.

---

action. Ecological models suggest that blight is "more likely to be spread by people than by passive dispersal through the atmosphere." At least one new product suddenly appeared in farms across much of Europe in the early 1840s: guano. On the passage from Lima to Liverpool, one can easily imagine blighted potatoes spilling from a broken barrel, spreading spores into the loose mass of guano in the hold. Blight spores can survive in soil for as much as forty days. If the soil were infected toward the end of the trip, that would allow more than enough time to distribute it. Ireland had been the site of much guano experimentation. By 1843, trials had occurred in at least eleven of its thirty-two counties. Farmers were swapping and borrowing samples with equal vim the next year. It is tempting to wonder whether *P. infestans* was less imported with the guano than imported *in* the guano. (Another pest, the potato cyst nematode, invaded Japan in exactly this way.) After the blight hit, some of Ireland's most progressive farmers advocated a means for returning potato yields to normal: higher doses of guano. All through the Great Hunger the fertilizer ships came.

Many Andean peoples have long grown potatoes in parallel ridges called *wacho* (bottom, in Bolivia near Titicaca), a practice that has been shown to discourage fungal diseases by drying wet soil. Lazy-bed cultivation, as it was called in English, was common in Ireland (top, in northern Ireland in the 1920s) until the early nineteenth century. Recent research suggests the abandonment of lazy-beds helped potato blight race through the countryside, exacerbating the great Irish famine.

During the next forty years, researchers attributed the blight to ozone, air pollution, static electricity, volcanic action, smoke from steam locomotives, excessive humidity or heat, gases from the recently introduced sulfur match, an emanation from outer space, various insects (aphids, ladybugs, tarnished plant bugs), and the potato's own internal debilitation. Edward Hitchcock, a renowned natural historian at Amherst College, assigned blame to an "atmospheric agency, too subtle for the cognizance of our senses." A few thought the cause was a fungus, but they were shouted down. No useful countermeasures were proposed. The plea for help went out to Science, but Science couldn't answer.

"WAR UPON THE BEETLES"

In August 1861 beetles invaded a ten-acre garden in northeastern Kansas that belonged to a potato farmer named Thomas Murphy. His name was appropriate: Murphy, a common Irish surname, was also a slang term for potatoes. Murphy's potatoes—Murphy's Murphys—were overrun by so many beetles that he could barely see the leaves through the swarm of tiny glittering bodies. He knocked the insects from the plants into a basket, he wrote later, and "in a very short time gathered as many as two bushels of them"—remarkable, given that each insect was barely a third of an inch long. In a different context, perhaps, Murphy might have thought the beetle was beautiful, with its yellow-orange body and its forewings marked tigerishly with thin black stripes. But they were devouring his potato plants as fast as they came up.

Murphy had never seen the beetle before its hordes suddenly attacked his potatoes. Nor had his neighbors who also were visited by it, or the farmers in Iowa and Nebraska whom it invaded that summer. The insect marched steadily north and east, expanding its range by fifty to a hundred miles a year, shocking potato growers at every step. It reached Illinois and Wisconsin in 1864; Michigan, by 1870. Seven years later it was attacking potatoes from Maine to North Carolina. The little insects swarmed potato fields in such profusion, according to one widely repeated story, that they stopped nearby trains. Their bodies covered the tracks in a layer deep enough to make the wheels slip "as if oiled, so that the locomotive was powerless to draw the train of cars." Strong winds blew the beetles into the sea, from which they washed ashore in a glittering, yellow-orange carpet that fouled beaches from New Jersey to New Hampshire. Farmers had no idea where the creature had come from or how to stop it from eating their potato fields to the ground.

The Great Hunger still a vivid memory, Europeans cringed to hear the reports of potato devastation. Companies produced thousands of small insect models to help farmers identify Murphy's beetle. Germany imposed what may have been the world's first-ever agricultural quarantine, against U.S. potatoes, in 1870; France, Russia, Spain, and the Netherlands followed suit. Great Britain, which had the most to fear, did not ban U.S. potatoes—it didn't want to set off a trade war. Traveling in ship holds, the beetle kept appearing in European fields, only to be expunged. The First World War distracted governments from the task of monitoring insect movements. Seizing the moment, the beetle established a beachhead in France, then moved east. Today it occupies a swath of Europe that reaches from Athens to Stockholm. In the Americas its realm extends from south-central Mexico to north-central Canada. Many biologists fear that it will spread into East and South Asia, completing a round-the-world journey.

Murphy's beetle is known to entomologists as *Leptinotarsa decemlineata* and to gardeners as the Colorado potato beetle. It is not from Colorado. Nor at first did it have any interest in potatoes. It originated in south-central Mexico, where its diet centered on buffalo bur (*Solanum rostratum*), a weedy, knee-high potato relative with leaves that somewhat resemble oak leaves. From the human point of view, the plant is annoyingly spiny, with barbed seedpods that stick in hair and clothes and are hard to remove without gloves. Biologists believe that buffalo bur was confined to Mexico until Spaniards, agents of the Columbian Exchange, carried horses and cows to the Americas. Quickly realizing the usefulness of these foreign mammals, Indians stole as many as they could, sending them north for their families to ride and eat. Buffalo bur apparently came along, tangled in horse manes, cow tails, and native saddlebags. The beetle followed in its path, hopping along a chain of corrals and stock pens. After arriving in Texas, the bur also could have been carried by bison, which migrate from south to north in the spring. By 1819 the beetle had arrived in the Middle West, where a naturalist observed it feeding on buffalo bur along the Missouri River. In this area it first encountered the cultivated potato.

Chance intervened. In Mexico the beetle, specialized on buffalo bur, finds it easy to ignore the delights of *S. tuberosum;* placed on a potato leaf, it will seek sustenance elsewhere. But one midwestern beetle in the mid-nineteenth century was born with a tiny mutation—perhaps, according to one suggestive study, a slight shift at a particular spot in its second pair of chromosomes, a snippet of DNA that flipped end to end. The mutation was not enough to make the beetle look different or affect its ability to reproduce. But it may have been enough to widen its focus from buffalo bur to a relative, the potato.

"The progeny of one pair, if unmolested for a year, would amount in the aggregate to over 60,000,000 of individuals," the *New York Times* calculated in 1875. The actual figure is more like sixteen million, but the point is valid—a single genetic accident in a single individual was enough to generate a worldwide problem. The beetle is the potato's most devastating pest to this day. "One of the worst features of the present visitation," the newspaper continued, "is that the Colorado beetle is noted for its permanency, and rarely abandons localities until it has ravaged them for several seasons in succession. . . . Under such circumstances, the only resource is to commence an aggressive war upon the beetles."

War with what weapon? Farmers tried everything they

could think of: picking off and crushing beetles with special pin-
cers; trying to find less-attractive potato varieties; encouraging
the insect's natural predators (ladybirds, soldier beetles, certain
species of tiger beetle); moving potato fields every season, thus
avoiding beetles overwintering (an insect version of hibernation)
in the soil; surrounding their plots with buffalo bur, "so as to concentrate the insects, and thus more readily destroy them"—here I
am quoting Charles Valentine Riley, founder and longtime head
of the U.S. Entomological Commission. An Iowa man touted
his horse-drawn beetle remover, which raked the insects into a
box dragged behind. Potato growers doused plants with lime,
sprinkled sulfur, spread ashes, sprayed with tobacco juice. They
mixed coal tar with water and splashed that on the beetles. Some
farmers reportedly tried wine. Others tried kerosene. Nothing
worked.

Insects have bothered farmers since the first planting of crops
in the Neolithic era. But large-scale industrial agriculture changed
the incentives, so to speak. For millennia the potato beetle had
made do with the buffalo bur scattered through the Mexican hills.
By comparison, an Iowa potato farm—hundreds of orderly rows
of a single type of a single species—was an ocean of breakfast. By
adapting to the potato, the beetle could command many more
resources for reproduction than it had ever possessed before; its
numbers naturally exploded. So did those of other pests—the
potato blight is a notable example—that were able to take advantage of the same opportunities. Each of the massive new farms
was a fabulous storehouse of riches for the species that could
exploit it.

Those farms were ever more similar, a hallmark of the
Homogenocene. Because growers planted just a few varieties of
a single species, pests had a narrower range of natural defenses
to overcome. If a species was able to adapt to the potatoes in one
place, it would not have to adapt to those in others. It could simply jump from one identical food pool to the next—a task that
was easier than ever, thanks to modern inventions like railroads,

As this cover illustration on an 1877 number of the London newspaper supplement *Funny Folks* suggests, British farmers feared the arrival of the Colorado potato beetle.

steamships, and refrigeration. Not only did industrial agriculture present insects with a series of rich, identical targets; these faster, denser transportation networks made it ever easier for faraway species to exploit them. In 1898, L. O. Howard, Riley's successor, calculated that at least thirty-seven of the seventy worst insect pests in the United States were recent imports (he wasn't sure of the origins of six others).

The late nineteenth century was, in consequence, a time of insect plagues. The boll weevil, slipping over the border from Mexico, wiped out so much cotton in the South that the governor of South Carolina proclaimed a day of public prayer and fasting to fight the bug. The cottony cushion scale, an Australian insect, swept through California's citrus industry. A European import, the elm leaf beetle, ravaged elm trees in U.S. cities; Dutch elm disease, introduced from Asia despite the name, would arrive later and more or less wipe out all elms east of the Mississippi. Returning the favor, the United States exported phylloxera, an aphid that wrecked vineyards in most of France and Italy.

For the wine industry, the solution was discovered by Riley, the Entomological Commission head: grafting European grape vines onto U.S. grape roots, which resist the aphid. For decades afterward, most French and many Italian grapevines had American roots. For the potato, the solution was more consequential: Paris Green.

Paris Green's insecticidal properties were supposedly discovered by a farmer who finished painting his shutters and in a fit of annoyance threw the remaining paint on his beetle-infested potato plants. The emerald pigment in the paint was Paris Green, made largely from arsenic and copper. Developed in the late eighteenth century, it was common in paints, fabrics, and wallpaper. Farmers diluted it heavily with flour and dusted it on their potatoes or mixed it with lots of water and sprayed.

Paris Green was a simple, reliable solution: buy the pigment, mix in flour or water according to the manufacturer's instructions, apply it with a sprinkler or dust box, and watch potato beetles die. To potato farmers, Paris Green was a godsend. To the nascent chemical industry, it was something that could be tinkered with and extended and improved. If arsenic killed potato beetles, why not try it on other pests? Why not spray Paris Green to combat cotton worm, apple cankerworm, apple codling moth, elm leaf beetle, juniper webworm, and that plague of blueberries, the northern walkingstick? Arsenic killed them all. It was a godsend to cotton farmers reeling from the boll weevil. Eager scientists and engineers invented foggers and pumpers, sprayers and dusters, pressure valves and adjustable brass nozzles. The dust changed to liquid; the copper-arsenic mix changed to a lead-arsenic mix and then a calcium-arsenic mix.

If Paris Green worked, why not market another arsenic-containing pigment, London Purple? Why not other chemicals for other agricultural problems? In the mid-1880s a French researcher discovered that the "Bordeaux mixture"—copper sulfate, used to keep children from eating fruit—would kill downy mildew on grapevines. Given a new chemical weapon, researchers pointed it at other pests and hoped it would prove as deadly as Paris Green. Quickly they found that copper sulfate was—oh, happy day!—the long-sought remedy for potato blight. Spraying potatoes with Paris Green, then copper sulfate, would eliminate both the beetle and the blight.

From the beginning, farmers knew that Paris Green and copper sulfate were toxic. Even before the discovery of its insecticidal properties, many people had got sick from living in homes with wallpaper printed with Paris Green. The thought of spraying food with this poison made farmers anxious. They dreaded the prospect of letting pesticides and fungicides build up in the soil. They worried about exposing themselves and their workers to dangerous chemicals. They were alarmed by the cost of all the technology. All of these fears came true, but all could be adjusted for, at least in part. For a long time, farmers didn't know about the most worrisome issue of all: inevitably, the chemicals would stop working.

Colorado potato beetles are, genetically speaking, unusually diverse, which means that they have an unusually wide range of resources in their DNA. When their populations confront new threats—pesticides, for instance—some individuals are likely to be unaffected by them. To farmers' misfortune, this means that the species can quickly adapt. As early as 1912 a few beetles showed signs of immunity to Paris Green. Farmers didn't notice, though, because the pesticide industry kept coming up with new arsenic compounds that kept killing potato beetles. By the 1940s growers on Long Island found themselves having to use ever-greater quantities of the newest arsenic variant, calcium arsenate, to maintain their fields. Luckily for them, Swiss farmers spent the Second World War testing an entirely new type of pesticide on the potato beetle: DDT, a chemical bug killer with unprecedented range and sweep. Farmers bought DDT and exulted as insects vanished from their fields. The celebration lasted about seven years. The beetle adapted. Potato growers demanded new chemicals. The industry provided dieldrin. It lasted about three years. By the mid-1980s, each new pesticide in the eastern United States was good for about a season.

In what critics call the "toxic treadmill," potato farmers now treat their crops a dozen or more times a season with an ever-changing cavalcade of deadly substances. Many writers have decried this, perhaps none more elegantly than Michael Pollan in *The Botany of Desire*. As Pollan observed, large-scale potato farmers now douse their land with so many fumigants, fungicides, herbicides, and insecticides that they create what are known, euphemistically, as "clean fields"—swept free of life, except for potato plants. (In addition, the crops are sprayed with artificial fertilizer, usually once a week during growing season.) If rain doesn't fall for a few days, the powders and solutions can build up on the surface of the soil, creating a residue that resembles the aftermath of a chemical-warfare test. In my area, the Northeast, I have met farmers who claimed not to allow their children to walk around their fields. One doesn't have to be an organic fanatic to wonder about the prospects of a system that turns the production of food into a toxic act.

Worse still, many researchers believe that the chemical assault is counterproductive. Strong pesticides kill not only target species but their insect enemies as well. When the target species develop resistance, they often find their prospects better than before—everything that had previously kept them in check is gone. In this way, paradoxically, insecticides can end up *increasing* the number of harmful insects—unless farmers control them with yet more chemical weapons. "Secondary pests," insects that previously were controlled by some of the species killed off by insecticides, also profit. Here, too, industry has a solution: more pesticides. "A number of new chemistries are expected to appear on the market in the near future," one research team announced in the *American Journal of Potato Research* in 2008. But

> there is no reason to believe that any of them will break the seemingly endless insecticide–resistance–new-insecticide cycle that is so characteristic of Colorado potato beetle management. . . . Despite all the scientific and technological advances, the Colorado potato beetle continues to be a major threat to potato production.

Blight, too, has returned. Swiss researchers were dismayed in 1981 to discover that the second type of *P. infestans* oomycete, previously known only in Mexico, had found its way to Europe. Because the blight was now capable of "sexual" reproduction, it had greater genetic diversity—more resources, that is, to adapt to chemical control. Similar introductions occurred in the United States. In both cases the new strains were more virulent, and more resistant to metalaxyl, the chief current anti-blight treatment. No good substitute has yet appeared. In 2009, as I was writing this book, potato blight wiped out most of the tomatoes and potatoes on the East Coast of the United States. Driven by an unusually wet summer, it turned gardens all around me into slime. It destroyed the few tomatoes in my garden that hadn't been drowned by rain. Accurately or not, one of my neighbors blamed the attack on the Columbian Exchange. More specifically, he charged that blight had arrived on tomato seedlings sold in big-box stores. "Those tomatoes come from China," he said.

7

Black Gold

NO BIRDS OR INSECTS

It looked like a forest but ecologists probably wouldn't call it one. It sprawled over miles of low hills outside the village of Longyin Le, at the southern tip of China, less than forty miles from the border with Laos. Prosperous by the standards of rural China, Longyin Le had houses with curtained windows and painted walls. Solar hot-water heaters and satellite dishes sprouted from the roofs on the houses beside the road. At the edge of the village the cab drove past barns and animal pens and then I was among the trees.

They were perhaps fifty feet tall and graceful to my eye, with mottled gray-green limbs and leaves that were pale on one side and glossy dark green on the other. All were of one species and all were the same age—forty-five years old, I had been told, give or take a year. That was when the government put them in the ground. With impressive thoroughness every other plant species that grew higher than my ankles had been cleared away. The effect was park-like, except that the trees, planted in rows about eight feet apart, created an almost unbroken canopy overhead. Spiraling down each trunk was a shallow incision the width of a knife blade. Stuck to the lower edge of the incision, following it down the tree, was a flexible plastic strip perhaps three inches wide. At the bottom of each spiral was a small ceramic bowl or a place to mount one.

The trees were *Hevea brasiliensis,* the Pará rubber tree. Villagers in Longyin Le had cut the bark and attached the strips as guides. A milky, sap-like goo—*latex,* from the Latin for "liquid"— emerged from the fissure and slowly dripped along the strip until it ran into the bowl. Depending on the tree and season, latex is as much as 90 percent water. Some of the remainder consists of tiny grains of natural rubber. At first hearing, "natural rubber" may sound like something sold in pricey New Age boutiques. In fact it is a major industrial product, highly desired by high-tech manufacturers. The natural rubber in *H. brasiliensis* had lifted Longyin Le and scores of neighboring communities from destitution.

After ten or fifteen minutes of driving I left the cab and wandered about. I had come to a slope ridged by low terraces, each bearing a line of rubber trees. Beyond the crest of the slope was

Guide strips for latex and collection cups mark this rubber plantation in Xishuangbanna, China, an autonomous southern area near the Laotian border.

a sharp drop-off, and beyond that were hills, irregular as the wrinkles of a sheet thrown to the floor, their colors fading with distance in the hazy afternoon. Every living thing that I could discern was a rubber tree.

The driver was walking with me. He said he had not been to this area since he was young. The hills had been full of mammals and birds then. All had been replaced by rubber. Even the insects were still. It may have been the quietest forest I had ever walked in. Every now and then there was a quick breath of wind and the leaves rippled like tiny flags, momentarily exposing their satin tops. "There's nothing left," the driver said, visibly upset. "People want to cut and cut and plant and plant—damn them."

More than a century ago, a handful of rubber trees had come to Asia from their home in Brazil. Now the descendants of these trees carpeted sections of the Philippines, Indonesia, Malaysia, Thailand, and this part of China. Across the border *H. brasiliensis* was marching into Laos and Vietnam. A plant that before 1492 had never existed outside of the Amazon basin now dominated Southeast Asian ecosystems. Indeed, rubber reigned over such a wide area that botanists had long warned that a single potato blight–style epidemic could precipitate an ecological calamity—and, just possibly, a global economic breakdown.

In Longyin Le I wandered from house to house, talking to farmers about rubber. To a person they were thankful for the opportunities it provided. Rubber was putting food on the table, paying for children's education, building and repairing roads. Just as the potato played a critical part in helping Europe escape the Malthusian trap (though perhaps only for a time), rubber had helped bring about the Industrial Revolution, the transition from an economy based on manual labor and draft animals to one based on mechanized manufacturing. The people in Longyin Le were its latest beneficiaries. As I looked over the lush miles of birdless trees I could still hear their grateful voices. And rising like vapor were other voices, the countless men and women whose lives, for better and worse, had become entwined with this plant:

hapless slaves, visionary engineers, hungry merchants, obsessed scientists, imperial politicians. This landscape of alien trees was the creation of countless different hands in many places, and it was much older than forty-five.

"GREASE CHEMISTRY"

In May of 1526 Andrea Navagero, the Venetian ambassador to Spain, attended an entertainment in Seville staged for the royal court. Seven years earlier, Hernán Cortés, acting without the authorization of the Spanish throne, had invaded Mexico and toppled the Triple Alliance (Aztec empire). The king and queen had to decide what to do with their millions of new subjects. Some argued that they should be enslaved, because they were naturally inferior; others, that they should be converted to Christianity and made full citizens of Spain. To demonstrate the intelligence, skills, and noble demeanor of the peoples of the Triple Alliance, the antislavery faction of the Spanish church had imported a group of them to Seville. The Indians divided into teams and played a showcase version of the Mesoamerican sport of *ullamaliztli*, which the Venetian ambassador attended.

Navagero was an insatiably curious man who translated poetic and scientific classics, wrote a history of Venice, and performed biological experiments—he created a private botanical garden in 1522, among the first on the continent. He was mesmerized by *ullamaliztli*, which he seems to have thought was a performance akin to a juggling act (team sports had been played in the Roman empire but were then almost unknown in Europe). In *ullamaliztli* two squads vied to drive a ball through hoops on the opposite ends of a field—an early version of soccer, one might say, except that the ball was never supposed to touch the ground and players could hit it only with their hips, chests, and thighs. Dressed in padded breechcloths and wrist protectors like thick fingerless mittens, the players knocked a fist-sized ball back and forth "with so much dexterity that it was marvelous to see," Navagero reported, "sometimes throwing themselves completely on the ground to return the ball, and all of this done with great speed."

As fascinating to Navagero as the ball game was the ball itself. European balls were typically made of leather and stuffed with wool or feathers. These were something different. They "bounded copiously," Navagero said, ricocheting in a headlong way unlike anything he had seen before. The Indian balls, he guessed, were somehow made "from the pith of a wood that was very light." Equally puzzled was Navagero's friend Pietro Martire d'Anghiera, who saw a game at about the same time. When the Indian balls "touch the ground, even though lightly thrown, they spring into the air with the most incredible leaps," d'Anghiera wrote. "I do not understand how these heavy balls are so elastic."

The royal chronicler Gonzalo Fernández de Oviedo y Valdés fared little better. In his *General and Natural History of the Indies* (1535), the first official account of Spain's foray into the Ameri-

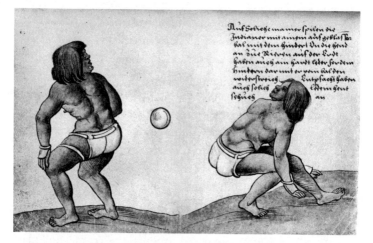

Europeans like German artist Christoph Weiditz were fascinated by the native ballplayers who toured Spain in the 1520s—and by the rubber ball, which was unlike anything ever seen in Europe.

cas, he tried to describe bouncing, a term not then in the Spanish language: "These balls jump much more than our hollow balls—by far—because even if they are only let slip from the hand to the ground, they rise much further than they started, and they make a jump, and then another and another, and many more, decreasing in height by itself, like hollow balls but more so." Indians made the strange, springy material of the balls, he wrote, by combining "tree roots and herbs and juices and a combination of things. . . . [A]fter [the mixture] is dried, it becomes rather spongy, not because it has holes or voids like a sponge, but because it becomes lighter, as if it were flabby and rather heavy." Wait a minute, one wants to say: how could something "become lighter" yet be "rather heavy"?

Navagero, d'Anghiera, and Oviedo had a right to be confounded: they were encountering a novel form of matter. The balls were made of rubber. In chemical terms, rubber is an *elastomer*, so named because many elastomers can stretch and bounce. No Europeans had ever seen one before.

To engineers, elastomers are hugely useful. They have tucked rubber and rubber-like substances into every nook and crack of the home and workplace: tapes, insulation, raingear, adhesives, footwear, engine belts and O-rings, medical gloves and hoses, balloons and life preservers, tires on bicycles, automobiles, trucks, and airplanes, and thousands of other products. This didn't happen immediately: careful studies of rubber didn't occur until the 1740s. The first simple laboratory experiments, in 1805, gave little hint that rubber might be useful—although the scientist, John Gough, did discover the fact, key to later understanding, that rubber heats up when stretched.* Only in the 1820s did rubber take off, with the invention of rubber galoshes.

---

* Gough, blind since birth, demonstrated this by touch: he pulled apart the ends of a wide rubber strip and touched it with "the edges of the lips," which are highly sensitive to heat. He also discovered that rubber shrinks when it is heated up—unlike most other substances, which increase in volume when they get hotter.

Take off for Europeans and Americans, that is; South American Indians had been using rubber for centuries. They milked rubber trees by slashing thin, V-shaped cuts on the trunk; latex dripped from the point of the V into a cup, usually a hollowed-out gourd, mounted on the bark. In a process reminiscent of making taffy, Indians extracted rubber from the latex by slowly boiling and stretching it over an intensely smoky fire of palm nuts. When the rubber was ready, they worked it into stiff pipes, dishes, and other implements. Susanna Hecht, a UCLA geographer who has worked extensively in Amazonia, believes that native people also waterproofed their hats and cloaks by impregnating the cloth with rubber. European colonists in Amazonia were manufacturing rubberized garments by the late eighteenth century, including boots made by dipping foot-shaped molds into bubbling pots of latex. A few pairs of boots made their way to the United States. Cities like Boston, Philadelphia, and Washington, D.C., were built on swamps; their streets were thick with mud and had no sidewalks. Rubber boots there were a big hit.

The epicenter of what became known as "rubber fever" was Salem, Massachusetts, north of Boston. In 1825 a young Salem entrepreneur imported five hundred pairs of rubber shoes from Brazil. Ten years later, the number of imported shoes had grown to more than 400,000, about one for every forty Americans. Villagers in tiny hamlets at the mouth of the Amazon molded thousands of shoes to the dictates of Boston merchants. Garments impregnated with rubber were modern, high-tech, exciting—a perfect urban accessory. People flocked to stores.

The crash was inevitable. The idea of impermeable rubber boots and clothes was more exciting than the fact. Rubber simply didn't work very well. In cold weather, the shoes became brittle; in hot weather, they melted. Boots placed in closets at the end of winter turned into black puddles by fall. The results smelled so bad that people found themselves burying their footgear in the garden. Daniel Webster, the senator and secretary of state, liked to tell the story of how he received a rubber cloak and hat as a gift. He wore them on a cold evening. By the time he reached his destination the cloak had become so rigid that he stood it in the street by the front door. Supposedly he propped the hat on top. "Some decorous gentlemen among us can also remember," one critic wrote later, "that, in the nocturnal combats of their college days, a flinty India-rubber shoe, in cold weather, was a missive weapon of highly effective character." Returned goods inundated rubber dealers. Public opinion swung violently against rubber.

Just before the collapse, in 1833, a bankrupt businessman named Charles Goodyear became interested in—and obsessed by—rubber. It was typical of Goodyear's entrepreneurial acumen that he began to seek financial backing for a rubber venture just at the time investors were planning their exits from the field. A few weeks after Goodyear announced his intent to produce temperature-stable rubber he was thrown into debtor's prison. In his cell he began work, mashing bits of rubber with a rolling pin. He was untroubled by any knowledge of chemistry but boundlessly determined. For years Goodyear wandered about the northeastern United States in a cloud of penury, trailed by his hungry wife and children, dodging bailiffs and pawning heirlooms. All the while he was mixing toxic chemicals, more or less randomly, in the hope that they would make rubber more stable. The Goodyears lived in an abandoned rubber factory in Staten Island. They lived in an abandoned rubber factory in Massachusetts. They lived in a shack in a Connecticut neighborhood called Sodom Hill (the name indicated its wholesomeness). They lived in a second abandoned rubber factory in Massachusetts. Sometimes the houses had no heat or food. Two of Goodyear's children died.

Taking his cue from a dream told to him by another rubber obsessive, Goodyear began mixing rubber with sulfur. Nothing happened, he said later, until he accidentally dropped a lump of sulfur-treated rubber onto a wood stove. To his amazement, the rubber didn't melt. The surface charred, but the inner material changed into a new kind of rubber that retained its shape and elasticity at high temperatures. Goodyear threw himself into reproducing the accident, a task impeded by his inability to afford any laboratory apparatus—he had to traipse from neighbor to neighbor, asking to use their wood stoves. Sometimes the sulfur process worked, sometimes it didn't. Goodyear kept working, frustrated, hungry, haunted. When he was again thrown into debtor's prison, he wrote to acquaintances from his cell, asking for supplies "to establish an India rubber factory for myself on the spot." Eventually he borrowed money and paid the debt. A month later he was in another jail.

Along the way he befriended a young Englishman. Goodyear gave him a few of his successful samples and asked him to seek investors in Britain. By a circuitous path two thin, inch-and-a-half-long strips of Goodyear's processed rubber ended up in the fall of 1842 at the laboratory of Thomas Hancock, a Manchester engineer who had developed processes for manipulating rubber. Hancock had no idea where these bits of rubber had originated. But he quickly realized that they didn't melt in hot weather or become stiff in cold weather. The question was whether he could duplicate the accomplishment. It is unclear how much he was able to learn from Goodyear's samples. Later he claimed to have "made no analysis of these little bits" from the other man—a remarkable demonstration of incuriosity, if true. In any case Hancock was more organized and knowledgeable than Goodyear and had better equipment. For a year and a half he systematically performed hundreds of small experiments. Eventually he, too, learned that immersing rubber in melted sulfur would transform it into something that would stay stretchy in cold weather and solid in hot weather. Later he called the process "vulcanization," after the Roman god of fire. The British government granted Hancock a patent on May 21, 1844.

Three weeks later, the U.S. government awarded Goodyear *his* vulcanization patent. A glance at the patent shows that Goodyear never fully understood the process: a key ingredient, he claimed, was white lead, a metal-based pigment whose effect

Identifying the inventor of the process of vulcanization, which makes rubber usable for industrial purposes, is complex. Charles Goodyear (left) had the basic idea first, but never fully understood the process; Thomas Hancock (right) patented the process before Goodyear and understood it better, but likely derived inspiration from seeing Goodyear's initial samples.

on rubber's stability is "secondary, if anything," according to E. Bryan Coughlin, of the Silvio O. Conte National Center for Polymer Research at the University of Massachusetts. "I'm not sure, because it's not a standard treatment—maybe it has some catalytic effect." By contrast, Coughlin told me, Hancock's patent was "pretty straightforward." Hancock stirred softened rubber into sulfur heated to 240°–250° F, just above its melting point. The longer he subjected it to heat, the more elasticity it lost. "That's pretty much what I teach my students," Coughlin said.

Goodyear didn't understand the recipe for vulcanization, but he did understand that at last he had a business opportunity. Showing a previously unsuspected knack for publicity stunts, he spent $30,000 he did not have to create an entire room made of rubber for the Great Exhibition of 1851 at the Crystal Palace in London, the first world's fair. Four years later he borrowed $50,000 more to display an even more lavish rubber room at the second world's fair, the Exposition Universelle in Paris. Parisians lost their urban hauteur and gawped like rubes at Goodyear's rubber vanity table, complete with rubber-framed mirror; arranged on the top was a battalion of rubber combs and rubber-handled brushes. In the center of the rubber floor was a hard rubber desk with a rubber inkwell and rubber pens. Rubber umbrellas stood at attention in a rubber umbrella-stand in the corner of two rubber walls, each decorated with paintings on rubber canvases. For weapons fans, there was a stand of knives in rubber sheaths, swords in rubber scabbards, and rifles with rubber stocks. Except for the unpleasant rubber smell, Goodyear's exhibit was a triumph. "Napoleon III invested him with the Legion of Honor," wrote the diplomat and historian Austin Coates, "and a Paris court sent him to prison for debt." He received the medal in his cell. Goodyear was forced to sell some of his wife's possessions to pay for their trip home. He died four years later, still awash in debt.

Afterward, Americans lionized Goodyear as a visionary. Books extolled him to children as an exemplar of the can-do spirit; a major tire company named itself after him. Meanwhile, Coates noted, "Hancock received English treatment: due respect while living, fading notice when dead, and on some suitable centenary thereafter, a postage stamp."

Neither Goodyear nor Hancock had any idea *why* sulfur stabilized rubber—or why, for that matter, unadulterated rubber bounced and stretched. Nineteenth-century scientists found bouncing balls exactly as mystifying as sixteenth-century Spaniards. Stretch a thin hoop of iron: it will elongate slightly, then snap in two. A rubber band, by contrast, can stretch to three times its ordinary length, then return to its original shape. Why? And why did sulfur stop rubber from melting in the summer? "Nobody knew," Coughlin told me. "It was a huge puzzle. And it was made harder by the fact that a lot of chemists didn't really want to study it."

The last half of the nineteenth century was a heady time for chemistry. Researchers were deciphering the underlying order of the physical world. They were placing the chemical elements into the periodic table, discovering the rules by which atoms combine into molecules, and learning that molecules could form regular crystals with structures that could be precisely identified.

Nowhere in these tidy intellectual schemes was a place for rubber. Chemists couldn't make it form crystals. Worse, many standard chemical tests on rubber produced nonsensical answers. The analyses demonstrated that each rubber molecule was made up of carbon and hydrogen atoms. No problem there. But they also indicated that the carbon and hydrogen were piled up into jumbo-sized molecules made up of tens of thousands of atoms. To most chemists, this was absurd—molecules are the fundamental building blocks of chemical compounds, and no fundamental building block should be that big.

The obvious conclusion, chemists said, is that rubber must be a *colloid:* one or more compounds finely ground up and dispersed throughout other compounds. Glue is a colloid; so are peanut butter, bacon fat, and mud. Because colloids aren't one substance but a mishmash of many different substances, they have no fundamental constituents. Looking for one would be like trying to find the molecular building blocks of a garbage heap. The chemistry of rubber was, one German researcher scoffed, *Schmierenchemie.* Literally, *Schmierenchemie* means "grease chemistry," though Coughlin told me it might be better translated as "the chemistry of the gunk on the bottom of a test tube."

Nonetheless, a few chemists ignored the disdain of their colleagues for rubber, prominent among them Hermann Staudinger, then at the Swiss Federal Institute of Technology in Zurich. A well-known researcher, he had already derived the chemical formulae for the basic flavors in coffee and pepper. (It is not unfair to charge Staudinger with inflicting instant coffee on the world.) Sometime during the First World War, he jumped into the entirely different field of rubber because of an intuitive belief that "high molecular compounds," as he called them, *did* have basic building blocks, which were fantastically large molecules. Readers familiar with stories of successful scientific mavericks will not be surprised to learn that Staudinger attracted vehement opposition, that he kept piling up evidence for his hypothesis, and that the resistance grew irrational and vituperative. When he left Zurich to work at the University of Freiburg in 1925 he was denounced by colleagues during his farewell lecture. Presumably the antagonism was heightened by Staudinger's penchant for picking fights. He once greeted the arrival of a rival's book by gluing a denunciation—"This book is not a scientific work but propaganda"—onto the cover of the copy in his university library. In the end, though, Staudinger's tale reached its denouement in the customary location: Stockholm, where he won the Nobel Prize for Chemistry in 1953.

Rubber and other elastomers, Staudinger showed, have molecules shaped like long chains.* "Long" is an accurate adjective: if a rubber molecule were as thick as a pencil, it would be as long as a football field. "Chain," too, is accurate: all rubber molecules are made up of tens of thousands of identical, repeating links, each consisting of five carbon atoms and eight hydrogen atoms. The molecules of ordinary solid substances—the copper in a wire, say—are usually distributed in orderly arrays. Rubber molecules, by contrast, are higgledy-piggledy, the chains scrambled around each other in no discernible pattern. "The classic analogy is a bowl of spaghetti," Coughlin explained to me. "But the analogy doesn't really work unless you're willing to say the noodles are a hundred yards long." Stretching a rubber band pulls the tangled molecules into alignment, lining them up in parallel like strands of spaghetti in a box. As they unkink, the molecules go from a clumped snarl to their full length, which is why rubber can

---

* In general, long-molecule substances are called *polymers*. Many types of polymers are familiar: fibers like silk and wool, for instance, and proteins like the gluten in bread or the albumin in egg whites. Elastomers, with their puzzling behavior, are a special type of polymer.

stretch. By contrast, the copper molecules in a wire are *already* lined up in an array, making it much harder for the material to lengthen—the difference is the difference between pulling the end of a loose, tangled string and trying to tug at a fully extended string. (The energy required to pull the chains straight is why rubber heats up when stretched.) As soon as the pressure is relaxed, the rubber molecules begin moving randomly, which naturally ensnarls them again; the rubber shrinks back to its original size.

When a lump of pure rubber is heated up, the rubber chains vibrate and slither around each other every which way and get even more chaotically disordered; the rubber loses whatever shape it has and turns into a puddle. Vulcanization prevents this. Immersing rubber in sulfur causes a chemical reaction in which rubber molecules link themselves together with chemical "bridges" formed of sulfur atoms. So ubiquitous are the bonds that a rubber band—a loop of vulcanized rubber—is actually a single, enormous, cross-linked molecule. With the molecules anchored together, they are more resistant to change: harder to align, harder to entangle, more resistant to extremes of temperature. Rubber suddenly becomes a stable material.

The impact of vulcanization was profound, the inflatable rubber tire—key to the adoption of both the bicycle and the automobile—being the most celebrated example. But rubber also made electrification possible: try to imagine a modern building without insulation on its wiring. Or imagine dishwashers, washing machines, and clothes dryers without the belts that transmit the motion of their engines to the appliance itself. Equally important but less visible, every internal combustion engine contains many pipes and valves that channel, usually under pressure, water, oil, gasoline, and exhaust vapor. Unless the parts are manufactured perfectly, engine vibrations will cause liquids or gases to vent dangerously from the joints. Flexible rubber gaskets, washers, and O-rings almost invisibly fill the gaps. Without them, every home furnace would be at constant risk of leaking natural gas, heating oil, or coal exhaust—a potential death trap.

"Three fundamental materials were required for the Industrial Revolution," Hecht, the UCLA geographer, told me. "Steel, fossil fuels, and rubber." The rapidly industrializing nations of Europe and North America had more than adequate access to steel and fossil fuels. Which made it all the more imperative to secure a supply of rubber.

"THE BATHER IN THE BUBBLY"

In my living room hangs a portrait of either my grandmother's uncle or her great-great-uncle. Both men were named Neville Burgoyne Craig. My grandfather, who found the painting in a thrift shop, thought that the subject was the older Craig (1787–1863), founding editor of the first daily newspaper in Pittsburgh. But the late-nineteenth-century style of the painting suggests that it was the younger Craig (1847–1926), an engineer who took ship for the Amazon a week after his thirty-first birthday. He intended to make his fortune in rubber.

Craig was not planning to work directly with rubber. Instead he intended to help build a railroad to transport it. Then as now the primary source of natural rubber was latex from *Hevea brasiliensis*. Native to the Amazon basin, the tree is most abundant on the borderlands between Brazil and Bolivia. The ports nearest to this area are on the Pacific coast, across the Andes. Sending rubber to those ports would mean carrying it across the high, icy mountains. After doing that, shipping the latex to England would involve dispatching ships around the stormy southern tip of South America, a long and dangerous trip of almost twelve thousand miles. The entire route was so difficult, in fact, that the secretary of the Royal Geographical Society calculated in 1871 that it would be four times faster to ship rubber to London from the western Amazon by transporting it down the Madeira River to the Amazon itself, and thence to the Atlantic. The problem was that waterfalls and violent rapids blocked a 229-mile section of the lower Madeira. West of this stretch were three thousand miles of navigable river in Bolivia and vast supplies of rubber and other valuable goods; east of it was the wide Amazon, and then the Atlantic. The downstream end of the impassable stretch was the Brazilian hamlet of Santo Antônio. My ancestor went to Santo Antônio to build a railroad around the rapids.

Born in Pittsburgh, Craig took his undergraduate and engineering degrees at Yale. He was a fine student who won two university mathematics prizes and was hired by the U.S. Coast and Geodetic Survey before graduation. Five years later, seeking excitement, he joined P. & T. Collins, a Philadelphia railway-construction firm, which had obtained the contract, secured by the Bolivian government, to build the Madeira railroad. The two Collins brothers seem to have believed that their considerable experience with railroads trumped their utter lack of experience with the Amazon. In January 1878 they sent out two shiploads of eager engineers and laborers from Philadelphia. Craig went in the first vessel.

As he later recounted in a memoir, winter gales plagued the journey to Amazonia. The storms wrecked the second—and, alas, much less seaworthy—ship about one hundred miles south of Jamestown, Virginia. More than eighty people drowned. Com-

Neville B. Craig

pany officers had trouble replacing the lost men—Philadelphians, shocked by the disaster, had lost their enthusiasm for the venture. Eventually Collins hired a new workforce from "the slums of several of our large eastern cities," to quote Craig's book, people "exhibiting in shape, countenance and gesture, striking evidence of the soundness of Darwin's theory." Most were immigrants from southern Italy; many had been pushed out of their homes for their anarchist beliefs. As my ancestor's snarky put-down suggests, anti-Italian prejudice was then widespread; these newly arrived Americans in consequence were desperate for work. The Collins brothers took advantage of their desperation to sign them up for lower wages than they paid the laborers on the first ship— $1.50 per day, instead of $2.00. Apparently it did not occur to the brothers that the anarchists would discover this arrangement, or that they would find it unacceptable.

Meanwhile Craig steamed up the Amazon and the Madeira to the proposed railway terminus at Santo Antônio and set to work surveying the route. He learned of the fate of the men on the second ship only when the Italians arrived as replacements. At the same time the Italians found out that they were being paid less than everyone else. Within days they went on strike. The engineers, Craig among them, constructed a cage from the steel rails for the railway and forced the strikers into it at gunpoint. Reading the memoir, I waited in vain for any recognition from Craig that imprisoning the workforce could have a negative impact on the construction schedule. Ultimately the strikers went back to work, sullenly hacking at the forest. A few weeks later "seventy-five or more" took off for Bolivia. None made it—perhaps, Craig luridly speculated, because they had "served as food to gratify the none too dainty appetites of the anthropophagous Parentintins." (The Parentintins, a nearby native group, kept potential colonists at bay by cultivating a reputation for ferocity.)

In one way the workers' flight may have been a boon: the expedition was running out of food. Like the Jamestown colonists, my ancestor's party was starving in the midst of plenty. Ten years before, the German engineer Franz Keller had surveyed the Madeira rapids with a party of Mojos Indians who so regularly feasted on turtle that he groused about the monotony; Keller preferred the pirarucu, an armored fish so big that Amazonians regularly toss huge pirarucu steaks on the barbecue, and the Amazonian manatee, a bulbous aquatic mammal whose meat, "when properly prepared, decidedly reminds one of pork."

Western Amazonia offered more—much more—than edible reptiles, fish, and mammals. Agricultural geneticists have long argued that the area around the railroad route—the Brazil-Bolivia border—was the development ground for peanuts, Brazilian broad beans (*Canavalia plagiosperma*), and two species of chili pepper (*Capsicum baccatum* and *C. pubescens*). But in recent years evidence has accumulated that the area was also the domestication site for tobacco, chocolate, peach palm (*Bactris gasipaes,* a major Amazonian tree crop), and, most important, the worldwide staple manioc (*Manihot esculenta,* also known as cassava or yuca). My ancestor nearly died from lack of food in one of the world's agricultural heartlands.

Only after five famished months did Craig learn from a local resident to fish not in the main channel, as the Americans had been doing, but in the smaller tributaries. Rather than using hooks and lines, to which Amazonian fish rarely respond, Indians sprinkled over the water a paralyzing elixir made from the tree genus *Strychnos* (the name suggests the poison). Temporarily unable to breathe, fish floated to the surface and were scooped into baskets. Craig's crew put down their fishing rods and learned to make poison. They stopped trying to grow peas and carrots in their gardens and began eating palm fruits and manioc.

What finally capsized the venture was malaria. Introduced to the coast by African slaves, probably in the seventeenth century, *Plasmodium* slowly transformed the Amazon basin into a collection of depopulated fever valleys that few foreigners wanted to enter. (I am picking up a story I began in Chapter 3.) Vulcanization brought people back. At a stroke European and American industries found themselves hungering for huge amounts of rubber. Most of it initially came from the mouth of the Amazon, near the port city of Belém do Pará. Each rubber tree produced perhaps an ounce of rubber per day, could only be milked 100 to 140 days a year, and needed to recuperate every few years. As demand grew, Belém's rubber tappers worked their trees too hard, killing many of them. Then the entire northeast coast was hit by a terrible drought in 1877–79. As many as half a million people died. Abandoning their stricken fields and tapped-out trees, burning with cholera, smallpox, tuberculosis, malaria, yellow fever, and beriberi, the starving backlanders—*flagelados,* they were called, the scourged—fled upstream on the Amazon's new steamships by the tens of thousands, hoping to make a living from rubber. Those with a little money or political clout obtained land grants or concessions from local officials; those with only ambition or ruthlessness just looked for untapped *H. brasiliensis* and set up shop. Ultimately they created about twenty-five thousand rubber estates, the Brazilian historian Roberto Santos has estimated, most of them small, employing more than 150,000 laborers overall. The throngs of migrants provided new targets for malaria. Keller, the German engineer, traveled the Madeira in 1867 and saw little malaria. Neville Craig arrived there a decade later and saw little else.

The toll was appalling. Craig landed at Santo Antônio on February 19, 1878. On March 23 the second ship arrived and the number of workers swelled to about seven hundred. Malaria had incapacitated almost half of them by the end of May. By the end of July, two-thirds of the crew were too sick to work; three weeks later, the proportion had risen to three-fourths. Some thirty-five people had died, the first of many. Only about 120 Americans, more than half of them sick, were left by January 1879. The next month, my ancestor wrote, the enterprise achieved "complete collapse." As a capstone, the railroad's British bankers, perhaps anticipating legal action, refused to pay the survivors' accumulated wages. Sick and broke, shoeless and ragged, Craig and a hundred or so others straggled down the Amazon into Belém, where they had to beg passage home. But even as they haunted the docks, financiers in Europe and the United States were already planning another shot at building the railroad—there was too much money in rubber to let the idea go.

Even in a time of crazy boom-and-bust cycles the rubber boom stood out. Brazil's rubber exports grew more than tenfold between 1856 and 1896, then quadrupled again by 1912. Ordinarily such an enormous increase would drive down prices. But instead they kept climbing. Attracted by tales of fortunes gained, speculators leaped into the market—"even the widow and parson are in for all they are worth," the *New York Times* observed—and pushed up prices higher still. How high? Meaningful figures are hard to provide, because speculation caused markets to shoot up and down erratically; in 1910, to pick an extreme example, New York rubber oscillated between $1.34 and $3.06 a pound. On top of that, the inflation, financial panics, and political instability of the era caused the currencies of Brazil, Britain, and the United States to gyrate wildly in value. Still, rubber kept going up. Its "soaring price is turning rubber manufacturers gray," the *Times* claimed on March 20, 1910. "One ounce of rubber, washed and prepared for manufacture, is worth nearly its weight in pure silver."

The newspaper was hyperventilating, but not entirely wrong. One economist recently calculated that the average London price of rubber roughly tripled between 1870 and 1910. The statistic is more remarkable than it may seem. Compare what happened to the price of rubber to what happened to the price of oil after a huge strike was discovered in Texas in 1900. World oil production doubled—and prices crashed. Crude didn't reach its 1900 price for twenty years. That rubber production went up by an order of magnitude while prices tripled is the kind of thing that makes natural-resource economists rub their eyes in bemusement. "It's pretty amazing," said Michael C. Lynch, president of Strategic Energy and Economics Research, of Winchester, Massachusetts. "No wonder people were going crazy."

The financial center of the trade was Belém. Founded in 1616 at the entrance to the world's greatest river, it had a strategic location—but little ability to take advantage of it. So much sediment washed in from the Amazon that the harbor was shallow and treacherous. Worse, the currents and winds generated by the river's vast outflow isolated the city from the rest of Brazil—incredibly, from Belém it was faster to sail to Lisbon, a distance of 3,700 miles, than to Rio de Janeiro, a distance of 2,500 miles. In consequence the city's population had never risen much above twenty-five thousand. The rubber boom allowed it to become, at last, what Amazonian dreamers had long hoped: the economic capital of a vibrantly growing realm.

Convinced they were building the Paris of the Americas, Belém's newly rich rubber elite filled the cobbled streets with sidewalk cafés, European-style strolling parks, and Beaux-Arts mansions with (a concession to the tropics) the exceptionally tall, narrow windows that promote air circulation. Social life centered around the neoclassical Teatro de Paz, where rubber barons in box seats smoked cigars and drank cachaça, the distilled sugar-cane liquor that is Brazil's preferred high-alcohol beverage. Tall mango trees shaded the avenues that led to the harbor, where gangs of laborers sliced open the rubber lumps sent from upriver, looking for adulterants like stones or chunks of wood. After inspection, the rubber went into a series of immense warehouses that lined the shore like sleeping beasts. Rubber was everywhere, one visitor wrote in 1911, "on the sidewalks, in the streets, on trucks, in the great storehouses and in the air—that is, the smell of it." Indeed, the city's rubber district had such a powerful aroma that people claimed they could tell what part of the city they were in by the intensity of the odor.

Belém was the bank and insurance house of the rubber trade, but the center of rubber collection was the city of Manaus. Located almost a thousand miles inland, where two big rivers join to form the Amazon proper, it was one of the most remote urban places on earth. It was also one of the richest. Brash, sybaritic, and imposing, the city sprawled across four hills on the north bank of the great current. Atop one hill was the cathedral, a Jesuit-built structure with a design so austere that it looks like a rebuke to the monstrosity that dominated the next hill over—the Teatro Amazonas, a preposterous fantasia of Carrara marble, Venetian chandeliers, Strasbourg tiles, Parisian mirrors, and Glasgow ironwork. Finished in 1897 and intended as an opera house, it was a financial folly: the auditorium had only 658 seats, not enough to offset the cost of importing musicians, let alone the expense of construction. Wide stone sidewalks with undulating black-and-white patterns led downhill from the theater through a jumble of brothels, rubber warehouses, and nouveau-riche mansions to the docks: two enormous platforms that rode up and down with the river on hundreds of wooden pillars. State governor Eduardo Ribeiro aggressively boosted the city, laying out streets in a modern grid, paving them with cobblestones from Portugal (the Amazon has little stone), overseeing the installation of what was then one of the globe's most advanced streetcar networks (fifteen miles of track), and directing the construction of three hospitals (one for Europeans, one for the insane, and one for everybody else). A celebrant of urban life, Ribeiro took part in everything his city had to offer, including its sybaritic whorehouses, in one of which he died amid what historian John Hemming delicately referred to as "a sexual romp."

The city's many brothels were largely for the rubber tappers and field operatives who staggered into Manaus after months of labor on remote tributaries. The owners and managers had mistresses, with whom they sported in the decadent style then fashionable. "Guests once knelt to lap champagne from the bathtub of the naked beauty Sarah Lubousk from Trieste," Hemming wrote in his prodigious history of the region, *Tree of Rivers*. "The bather in the bubbly," as Hemming called her, was the mistress of Waldemar Scholtz, a recent migrant who had become the city's dominant rubber shipper—and the honorary consul from Austria. A few blocks away lived Aria Ramos, who led a celebrated double life as a carnival performer and a call girl; when she was slain in a hunting accident, her wealthy clients erected a life-size statue in the cemetery. Teeming bordellos, liquor-soaked cafés, cowboy-style barroom brawls—Manaus was the very model of a turn-of-the-century boomtown, from the warnings against the discharge of weapons on the street to the obligatory lighting of cigars with high-denomination banknotes.

So much wealth—wealth that literally *grew on trees*—in such a strategic material naturally attracted huge interest, domestic and foreign, economic and political. On the domestic side, the rubber trade came to be controlled by a baker's dozen of export houses, which in turn were dominated by Scholtz & Co., owned by the man who owned the woman in the bathtub. Like Scholtz & Co., the export houses were usually run by Europeans—intense, pallid men whose waxed moustaches and pomaded beards helped them stand apart from the beardless Indian population. Classic middlemen, they unloaded and stored the rubber that came from the interior before sending it to the mouth of the Amazon, where other European-run merchant houses shipped it to Europe and North America. The rubber itself was obtained by yet another group of corporate entities. These controlled the most critical resource in the interior: human beings.

Because latex coagulates after it is exposed to air, tappers constantly had to recut trees, tending them daily through the four-to-five-month tapping season. And they had to process the latex into crude rubber before it dried and became difficult to work with. Both tapping and processing required large amounts of care and attention. And that care and attention had to occur in remote, malarial camps—the trees couldn't be moved to more convenient, salubrious locations and the latex was too heavy to transport in its liquid state. Disease and European raids had harshly cut back the original indigenous population. Europeans had not replaced them. The ever-greater hunger for rubber was accompanied by an ever-more-desperate shortage of workers. Solutions to the labor problem emerged, many of them bestial.

At first the rubber boom seemed like a godsend—an arboreal jobs program—to the region's perpetually impoverished people. Needing workers, rubber estates hired local Indians, shipped in penniless farmers from downriver, or shanghaied hands from Bolivia. Economic theory suggests that in a labor shortage the estates would have to promise high wages and comfortable working conditions. They often did, but the promised wages were offset by stiff charges for transportation, supplies, and board. Many supposedly well-paid men were never able to work themselves out of debt; malaria, yellow fever, or beriberi felled others. To keep the labor force from finding better offers—or running away—owners stored workers onsite in barren dormitories with armed guards. Neville Craig's boss, the chief railroad engineer, visited the concessionaires who controlled the middle reaches of the Madeira. Living in three-story houses with sweeping verandas, the engineer wrote, the concessionaires were "surrounded, like medieval barons, with a retinue of Bolivian servants and their families. . . . These men are absolute masters of their peons."

In the 1890s the boom went still further upstream, into the Andean foothills—areas that until then had been regarded as useless, and so left largely to their original inhabitants, most of whom had minimal contact with Europeans. Because *H. brasiliensis* can't tolerate the cooler temperatures on the slopes, entrepreneurs focused on another species, *Castilla elastica,* which provided a less-valuable grade of rubber known as *caucho.* Although Indians tapped *Castilla* trees in Mesoamerica—the latex "emerges from *sajaduras* [the shallow cuts made when marinating meat] on the tree," one Spanish eyewitness wrote in 1574—they did not in Amazonia. The incisions, *caucheiros* (*caucho* collectors) believed, let in diseases and insects that quickly wiped out *Castilla.* Rather than futilely try to protect the trees, *caucheiros* simply cut them down, gouged off the bark, and let the latex drain into holes dug beneath the fallen trunk. Sometimes collectors could obtain several hundred pounds of latex from a single tree, thus making up in volume for *caucho*'s lower price.

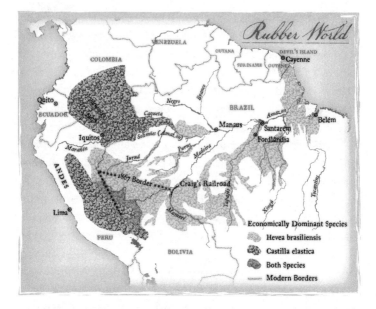

Because *caucheiros* killed the trees they harvested, they naturally put a premium on being the first into a new area. The goal was to extract the most rubber in the least amount of time; every minute not at the ax was a minute when someone else was taking down irreplaceable trees. Work crews spent weeks or months trekking from tree to tree through steep, muddy, forested hills, carrying heavy loads of *caucho* from the areas they had just looted. Few people from outside the area were willing to come into the forest for this. *Caucheiros* thus turned to the people who already lived there: Indians. The situation invited abuse—and there are always people ready to take up such invitations.

Among them was Carlos Fitzcarrald, son of an immigrant to Peru who had changed his name from the hard-to-pronounce "Fitzgerald." Beginning in the late 1880s Fitzcarrald forced thousands of Indians to work the *caucho* circuit. Brazilian writer and engineer Euclides da Cunha, who surveyed part of the western Amazon at this time, learned that at one point Fitzcarrald invaded a *Castilla*-rich area that was home to the Mashco Indians. Leading a squad of gunmen, da Cunha recounted, the *caucheiro* presented himself to the Mashco leader

> and showed him his weapons and equipment, as well as his small army, in which were mingled the varied physiognomies of the tribes he had already subdued. Then he tried to demonstrate the advantageous alternatives to the inconvenience of a disastrous battle. The sole response of the Mashco was to inquire what arrows Fitzcarrald carried. Smiling, the explorer passed him a bullet from his Winchester. The native examined it for a long time, absorbed by the small projectile. He tried to wound himself with it, dragging the bullet across his chest. Then he took one of his own arrows and, breaking it, thrust it into his own arm. Smiling and indifferent to the pain, he proudly contemplated the flowing blood which covered the point. Without another word he turned his back on the startled adventurer, returning to his village with the illusion of a superiority which in a short time would be entirely discounted.
>
> And indeed, half an hour later roughly one hundred Mashcos, including their recalcitrant chief, lay murdered, stretched out on the riverbank which to this day bears the name Playa Mashco in memory of that bloody episode.
>
> Thus they mastered this wild region. The *caucheiros* acted with feverish haste. They ransacked the surroundings, killing or enslaving everyone for a radius of several leagues. . . . The *caucheiros* would stay until the last *caucho* tree fell. They came, they ravaged and they left.

More brutal still was Julio César Arana. The son of a Peruvian hatmaker, Arana came to exert near-total command over more than twenty-two thousand square miles on the upper Putumayo River, then claimed by both Peru and Colombia. Colombia had a heavier presence on the ground but was then convulsed by civil war. The Peruvian Arana took advantage of its inattention to push into the region, shoving aside rival *caucheiros*. Not wanting to lure laborers from other areas with high wages, he turned to indigenous people. At first they were willing to do some rubber collecting in exchange for knives, hatchets, and other trade goods. But when Arana asked for more they balked. So he enslaved them. By 1902 he had five Indian nations under his thumb. *Caucho* flowed from his land in ever-larger amounts.

Arana moved with his family to Manaus and established a reputation for quiet probity—he had the biggest library in town. Meanwhile his minions expanded his realm on the Putumayo, bribing government officials and killing his competitors. He controlled his slave force with a goon squad led by more than a hundred toughs imported from Barbados. Isolated in the forest and utterly dependent on Arana, the Barbadans executed every command they were given. No one other than Arana's agents was allowed to enter the Putumayo from outside. Twenty-three custom-built cruise boats enforced his rule.

In December 1907 two U.S. travelers stumbled into the region. Encountering a *caucheiro* whose wife had been abducted by Arana's thugs, the young men impulsively decided to help him confront the wrongdoers. Arana's private police force beat and imprisoned them in one of the company's bases, a charnel house where their guards, one of the travelers later recounted, amused themselves with "some thirteen young girls, who varied in age from nine to sixteen." Outside, the "sick and dying" lay in untended heaps "about the house and out in the adjacent woods . . . until death released them from their sufferings. Then their companions carried their cold corpses—many of them in an almost complete state of putrefaction—to the river." By claiming they were representatives of "a huge American syndicate," the tourists managed to talk their way free.

One of them vowed to expose the situation. His name was Walter Hardenburg. The son of a farmer in upstate New York, he

Julio César Arana

was a clever, restless man, self-taught as an engineer and surveyor. He had gone to the Amazon with a friend in the vague hope of seeking employment on the Madeira railroad, which a new group of Americans was trying to build. Hardenburg was not a crusader by temperament, as Hemming notes in *Tree of Rivers,* but what he saw enraged him. To document the abuses he traveled to Iquitos, Peru, on the headwaters of the Amazon. Located almost two thousand miles from the river's mouth, it is often described today as the biggest city in the world that cannot be reached by road. It was then a boomtown port like Manaus, the main difference being that it was much smaller and completely dominated by Julio César Arana. At great personal risk Hardenburg spent a year and five months in Iquitos, finding witnesses to atrocities and obtaining their notarized testimony. With the last of his money he went to England in June 1908 to stir up public opinion. The first newspaper article appeared fifteen months later.

Arana had incorporated his company in London in an attempt to go public and cash out, as software entrepreneurs would do a century later. It had a placidly respectable British board of directors whose members apparently believed Arana's lies about having clear title to the rubber land and using company profits to educate tens of thousands of Indians. The slavery was therefore a British matter. Eventually there was a parliamentary investigation and a years-long public furor. London sent an investigatory team that included Roger Casement, an Irish-born British diplomat who was a pioneering human-rights activist—he had exposed atrocities committed in the Congo by agents of Belgian king Leopold II. Casement shuttled about the Putumayo, confirming Hardenburg's charges by obtaining detailed confessions of murder and torture. In a misguided fit of nationalism, Peru defended its citizen against foreign meddling. Nonetheless Arana's empire disintegrated. He died penniless in 1952.*

Arana was by no means the only force trying to build a rubber empire in this area of unsettled borders. Political and business leaders in Europe and the United States were infuriated that a material so vital to their economies was completely controlled by foreigners. The result was what Hecht has dubbed the "scramble for the Amazon." Arguing that the southern border of its colony in Guyane actually extended into rubber country, France sent troops into the forest. Brazil did the same. A standoff ensued. King Leopold II offered to settle the dispute by taking control of the rubber himself, an offer that pleased neither

---

* Casement was rewarded with a knighthood. Soon after, Sir Roger quit the Foreign Office to devote himself to the cause of Irish independence. He traveled to Germany to persuade the kaiser to provide arms for an uprising. The plot was discovered and Casement arrested as a German submarine deposited him on the Irish coast. Convicted of treason, he was sentenced to death. Influential friends begged the court for mercy. Casement was unlucky enough to be gay and unwise enough to detail his sex life in diaries. Their discovery after the trial sealed his fate. He was stripped of his honors and hanged on August 3, 1916.

Julio César Arana controlled his private rubber domain in the upper Amazon with guards imported from Barbados (left). Unfamiliar with local people and utterly dependent on him, they enforced his every rule with immediate brutality. Laborers who failed to perform were given the "mark of Arana"— whipped until the skin fell off (right).

side. France, unable to maintain its force in the forest, gave up in 1900. Britain was more successful in claiming that its colony reached into rubber territory. Rather than resorting to force of arms, it deployed the Royal Geographic Society, which produced a scientific-looking survey—proof enough for the Italian foreign minister, who had been selected to mediate the dispute. British Guiana acquired some rubber land.

From Brazil's point of view, the greatest threat to its dominance of the rubber trade was the United States. The U.S. interest in Amazonia dated back to Matthew Fontaine Maury (1806–73), founder of both the U.S. Naval Observatory and modern oceanography. An ardent advocate of slavery, Maury became possessed in the 1850s by the fear that the South would lose its political clout because it was not big enough to withstand the North. In a widely circulated pamphlet, he proposed a solution: the United States should annex the Amazon basin. Ocean currents push the river's outflow into the Caribbean, where it meets the outflow from the Mississippi—proof, to Maury's mind, that the Amazon was, oceanographically speaking, part of North America, not South America. For this reason, he argued, the Amazon valley was a natural "safety valve for our Southern States." He sent two cartographers to map Amazonia for the future day when U.S. slaveholders would go "with their goods and chattels to settle and to revolutionize and to republicanize and Anglo Saxonize that valley." Southern plantation owners should resettle there, Maury argued, converting the river basin into the biggest U.S. slave state. Few planters paid attention until the South lost the Civil War. Hoping to re-create slave society in the forest, ten thousand Confederates fled to the Amazon. All but a few hundred quickly fled back. The remaining die-hards formed a sort of micro-satellite of the Confederacy in the town of Santarém, in the lower Amazon.

With Maury, Washington gave up any idea of directly annexing Amazonia. But it was willing to try to control the rubber country through a proxy: Bolivia. Bolivia and Brazil had long contested their borders. After a short war in the 1870s, Bolivia ceded part of its territory in the south, receiving as compensation title to land to its north, around the Acre River, one of the richest areas, it later turned out, for *H. brasiliensis*. Unfortunately, all the rivers in the area—the main conduits for traffic—flowed into Brazil. It was thus vastly easier to reach Acre from Brazil than from La Paz, the Bolivian capital, up eleven thousand feet in the Andes. Taking advantage of these geographical circumstances, Brazilian tappers moved illegally across the border into Acre. Bolivia, too poor to mount an effective military response, sold the rights to Acre's rubber to a U.S. syndicate. Now the Brazilian squatters would be taking money not from powerless Bolivians, but from wealthy, politically connected U.S. businessmen. The syndicate persuaded the U.S. government to send a gunboat up the Amazon. It was turned back near Manaus.

Angered by the move, the Brazilians in Acre attacked the Bolivian regional capital of Cobija on August 6, 1902: Bolivia's national day. Asleep in its barracks after a drunken holiday feast, the garrison in Cobija was captured without a shot. The Bolivian army took three months to descend from the heights of La Paz, by which time the fight was over—Acre was Brazilian, the U.S. syndicate was routed, and Cobija, formerly in the center of Acre, was now a Bolivian border town. Today almost the only trace of the battle is at the airport in Cobija, where a monument at the entrance extols the "heroes of Acre."

Victory in Acre sealed Brazil's triumph. Having beat back almost all challenges to its control over rubber, it was producing ever more of this vital elastomer and controlling most of the trade in the rubber it didn't produce. Hundreds of thousands of people were making a living from the forest. The situation was in many ways much like what environmentalists hoped for in the 1990s and 2000s when they argued that Brazilians should sustainably gather rubber and other forest products in the Amazon, rather than set up short-lived cattle ranches. But instead Brazil showed how these schemes can go awry.

WHAT WICKHAM WROUGHT

When the man from the rubber company came to the village of Ban Namma, men drifted from their homes to meet him. They hunkered down in their sandals and worn T-shirts on the bare ground in front of the village headquarters. Surrounding them was an asteroid belt of silent women and almost-silent children. The company agent had a sports coat and a glad-handing manner. He distributed cigarettes, snapping them from the pack with the expert flick of a prestidigitator. Villagers tucked them in shirt pockets or behind ears. The man from the rubber company told a joke and the men laughed. A moment later the women laughed.

Ban Namma straggles up a hill next to the two-lane track that is the main road—often the only road—in the northwest corner of the Lao People's Democratic Republic. It is at the edge of the Golden Triangle, the intersection of the borders of Laos, Myanmar (formerly Burma), and China, a region long infamous for its opium and heroin. Some of the biggest producers were the brutish descendants of the Nationalist military officers who fled Mao Zedong's takeover of Beijing in 1949. They were joined and to some extent replaced in the 1960s by guerrillas from Communist uprisings in Myanmar. Because Beijing was subsidizing these guerrillas, its simultaneous efforts to shut down the Golden Triangle drug trade were, not surprisingly, less than successful. Eventually China tired of having criminal gangs on its border. In the 1990s it attacked them with a new weapon: corporate capitalism. Tax and tariff subsidies, some from United Nations anti-drug funds, pushed Chinese firms to create rubber plantations in the tiny, impoverished villages across the Laotian border. One of these villages was Ban Namma. The man with the cigarettes had persuaded its inhabitants to plant 1,325 acres of their land with *Hevea brasiliensis*.

The rubber man introduced himself as Mr. Chen. The venture had not been entirely successful, he told me. Rubber trees need to be planted on warm, sunny slopes that are not exposed to wind or cold and must be carefully tended for seven years before they can be tapped. In Ban Namma, Mr. Chen said, the villagers had no experience with *H. brasiliensis* and had made beginners' mistakes. They cleared land at the wrong elevation and failed to water abundantly. The promised 1,325 acres of thriving trees had become less than 500 acres of hard-pressed trees.

Despite this kind of setback Laotian rubber was booming. For miles around Ban Namma forestland had been shaved clean at the direction of Chinese rubber firms. Young rubber trees rose like morning stubble in the cleared patches. To the far west, near the border with Burma, a big Chinese holding company, China-Lao Ruifeng Rubber, was cutting and planting almost 1,200 square miles; a second firm, Yunnan Natural Rubber, planned to convert another 650 square miles. Much more was projected, according to a 2008 report by economist Weiyi Shi for the German development agency GTZ. The area was being transformed into an organic factory, primed to pump out latex for the trucks that were already beginning to thunder down the narrow roads.

If this ecological tumult could be laid at the door of a single person, it would be Henry Alexander Wickham. Wickham's life is difficult to assess: he has been called a thief and a patriot, a major figure in industrial history and a hapless dolt whose main accomplishment was failing in business ventures on three continents. Perhaps the most accurate way to describe his role was that he was a conscious human agent of the Columbian Exchange. He was born in 1846 to a respectable London solicitor and a milliner's daughter from Wales. When the boy was four, cholera took his father's life and the family he left behind slid slowly down the social ladder. Wickham spent the rest of his life trying to climb back up. In this quest he traveled the world, wrecking his marriage and alienating his family as he tried with blind tenacity to found great plantations of tropical species. Manioc in Brazil, tobacco in Australia, bananas in Honduras, coconuts in the Conflict Islands off New Guinea—Henry Wickham failed at them all. His adventure in Brazil cost the life of his mother and his sister, who had accompanied him. The coconut plantation, on an otherwise uninhabited island, was so lonely and barren that Wickham's wife, who had endured years of privation without complaint, at last demanded that he choose between the coconuts and her. Wickham chose coconuts. They never spoke again. Nonetheless at the end of his days he was a respected man. Crowds applauded as he walked onto testimonial stages wearing a silver-buttoned coat and a nautilus-shell tie clip. His waxed moustache curved ferociously beneath his jaw like the moustache of an anime character. He was knighted at the age of seventy-four.

Wickham won the honor for smuggling seventy thousand rubber-tree seeds to England in 1876. He was acting at the behest of Clements R. Markham, a scholar-adventurer with considerable experience in tree bootlegging. As a young man, Markham had directed a British quest in the Andes for cinchona trees. Cinchona

Henry Wickham

bark was the sole source of quinine, the only effective antimalaria drug then known. Peru, Bolivia, and Ecuador, which had a monopoly, zealously guarded the supply, forbidding foreigners to take cinchona trees. Markham dispatched three near-simultaneous covert missions to the Andes, leading one himself. Hiding from the police, almost without food, he descended the mountains on foot with thousands of seedlings in special cases. All three teams obtained cinchona, which was soon thriving in India. Markham's project saved thousands of lives, not least because Ecuador, Peru, and Bolivia were running out of cinchona trees—they had killed them by stripping the bark. Riding the success to the position of director of the India Office's Geographical Department, Markham decided to repeat with rubber trees "what had already been done with such happy results for the cinchona trees." British industry's dependence on rubber was leaving the nation's prosperity in the hands of foreigners, he believed. "When it is considered that every steam vessel afloat, every railway train, and every factory on shore employing steam power, must of necessity use india-rubber," Markham argued, "it is hardly possible to over-rate the importance of securing a permanent supply." Glory would attach to those who secured that supply. In the early

1870s Markham let it be known that Britain would pay for rubber seeds. When the seeds arrived, they would be sown at the Royal Botanic Gardens, at Kew in southwest London, and the successful seedlings dispatched to Britain's Asian colonies. Two separate hopeful adventurers sent batches of rubber seed. Neither batch would sprout. Wickham became the third to try.

Rubber was Wickham's exit from his failing manioc plantation in Brazil. Cannily eliciting Markham's promise that the India Office would buy every rubber seed he sent, Wickham sought the help of his neighbors in collecting them. His plantation was located in Santarém, four hundred miles from the river's mouth, a rubber town built atop a Jesuit mission built atop a native city. It was also the biggest center of ex-Confederates in the Amazon. With the aid of Confederate families, Wickham gathered seventy thousand seeds, enough to pay for passage back to Britain for himself and his wife. (He left behind, apparently without warning, his brother and his family, as well as his widowed brother-in-law.) To judge by the frigid reception he received in London, the India Office had not expected to be billed for three-quarters of a ton of rubber seeds. Nor were they overly happy that only 2,700 germinated—evidence, suggested the environmental historian Warren Dean, that Wickham and his associates scrambled through the forest in a hot-brained hurry, grabbing seeds from the ground without consideration for their viability.

Today Wickham is reviled in Brazil. Tourist guides refer to him as the "prince of thieves," a pioneer of what has come to be called "bio-piracy"; the leading economic history of Amazonia denounces his actions as "hardly defensible in the light of international law." At a literal level this claim is untrue; Brazil then had no bio-piracy laws. Nor is there any evidence that anyone tried to stop Wickham. The British were hardly secretive—London newspapers trumpeted Markham's quest for rubber. And authorities in Santarém surely were aware that an English madman was packing up cases of rubber seeds. In any case Brazilians themselves have not hesitated to import exotic species. The nation's primary agricultural exports today are soybeans, beef, sugar, and coffee. Not one is native to the Americas.*

More important, the transport of useful species out of their home environments has been a boon to humankind. The quinine supply in the Andes was far too small for the world's needs, even if collectors had hunted down every cinchona tree. Markham's "bio-piracy" saved countless thousands in Asia and Africa from premature death. Transplanting the potato to Europe and the sweet potato to China created catastrophic social and environmental problems, as I have been at pains to argue. But it also kept millions of Europeans and Chinese from malnutrition and famine. The huge benefits of moving species outweigh the huge harms, though the balance can be closer than free-exchange advocates tend to admit. As Dean put it, "The transfer of seeds, even across national borders, even for the sake of crass profit, even in behalf of imperialism, may be counted as a foremost means of the aggrandizement of the human species."

Two months after Wickham appeared in London, Kew shipped out the seedlings, most of them to Sri Lanka. Irritated with Wickham, the gardeners paid no attention to his recommendation that the trees be planted in open slopes away from marshes and riverbanks—the roots wouldn't grow properly in soggy ground. Instead they planted the seedlings in forest wetlands. Even if the plants had flourished, Sri Lanka's British colonists in 1876 were not interested in creating a new plantation industry. Two decades before, they had installed almost eight hundred square miles of coffee trees in the island's hills and imported a quarter of a million Indians to tend them. A previously unknown

---

* In 1727 the Brazilian diplomat Francisco de Melo Palheta visited Cayenne, in French Guiana, to negotiate a border dispute. Somehow he obtained coffee seeds—he is said to have received them as a farewell gift from the governor's wife, whom he had seduced. Under French colonial law coffee seeds were strictly forbidden to foreigners. Melo Palheta smuggled them to Brazil, the rubber historian Warren Dean wrote, launching "a plantation industry that was the mainstay of the Brazilian economy for a century and a half."

fungus affected "two or three acres" of coffee in 1869. Three years later the director of the Sri Lanka botanical gardens was reporting that "not a single estate has quite escaped it." Wickham's seedlings arrived just as unhappy colonists were ripping out stricken coffee trees and planting tea bushes. (The coffee plague is sometimes claimed to be why the British hot beverage of choice is tea, rather than coffee.) Few were interested in replacing their new tea bushes with rubber. The same coffee disease struck Malaysia and Indonesia in the 1890s. Forced to restart there, planters tried the rubber trees that had been languishing in Sri Lanka. The fortunes quickly made in Malaysia—and Indonesia, a Dutch colony that also took some of Wickham's trees—convinced Sri Lanka to take another look. Malaysia and Sri Lanka had a thousand acres of rubber plantations in 1897. Fifteen years later, the figure had grown to more than 650,000. For the first time more rubber came from Asia than the Americas. Prices fell, and the Brazilian rubber industry was reduced to dust.

Few in Manaus saw it coming—more evidence, if any were needed, of the human propensity to believe that flukes of good fortune will never come to an end. The city sank into lassitude, its opera house shuttered, its mansions abandoned. Rubber executives realized to their shock that laborers scattered across a forest the size of a continent could not produce latex nearly as efficiently as workers who moved down rows of closely packed trees. In their dismay few Amazonian businesses even tried to develop plantations. The first real chance the region had to recoup occurred in 1922, when British colonies in Asia, which had overplanted rubber, sought to control prices by forming a cartel. Among those enraged by this action were Harvey Firestone, the world's biggest tire maker, and Henry Ford, the world's biggest car maker. Firestone responded by creating a huge rubber plantation in Liberia. Ford planned one of equal size in the Amazon.

As a site he chose the Tapajós River, near Santarém, close to where Wickham had acquired his seeds. In an inauspicious debut for the project, Ford hired a Brazilian go-between who in

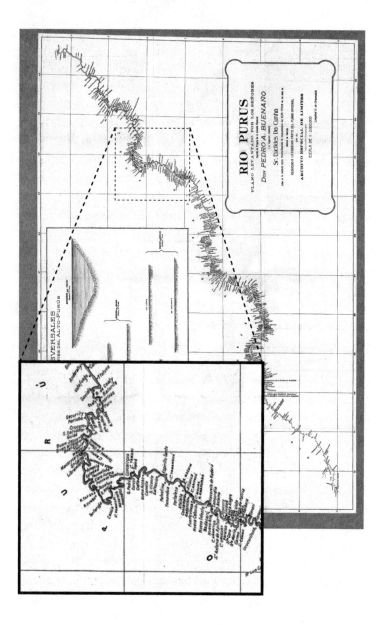

At the peak of the rubber boom, Brazil sent engineer and writer Euclides da Cunha to survey its disputed western border. Lining the banks of the Purus River, an upper Amazon tributary, da Cunha found hundreds of rubber-processing facilities (opposite). To stoke the fires that boiled down the latex and to fuel the steamboats that took it downstream, each plant consumed huge quantities of wood (above)—an early example of tropical forest destruction.

1927 sold him almost four thousand square miles of land up the Tapajós that happened to be owned by the go-between. To house his workers Ford built a replica of a middle-class Michigan town, complete with a hospital, schools, stores, movie theaters, Methodist churches, and wooden bungalows on tree-lined streets. On a hill was the Amazon basin's only eighteen-hole golf course. Orderly and straitlaced as Ford himself, the town was the opposite of boomtown Manaus. Wags immediately dubbed the project Fordlândia. Because Fordlândia was hilly, removing the vegetation "caused massive erosion and drainage problems," explained William I. Woods, a soil scientist and geographer at the University of Kansas. To prevent erosion, he told me, the company had to terrace the land, a "hideously expensive" process. In any case, Woods said, the soil was too sandy. Because the land was 135 miles up the Tapajós, oceangoing ships couldn't dock there during the dry season. "Even if they got rubber, they couldn't ship it out."

For Ford, the next few years were a series of unhappy surprises. Only after the first season's rubber trees died did the company find out that *H. brasiliensis* must be established at particular times of the year to thrive. Only after paying steamship bills did the company realize that it would not be possible to offset the cost of clearing all the hardwood trees on its land by selling the timber in the United States. And only after planting thousands of acres did the company learn that the Amazon has a fungus, *Microcyclus ulei,* that is partial to rubber trees. This last sentence is imprecise. The company did know that *M. ulei* existed. What it didn't grasp was that there was no way to stop it.

Microcyclus ulei causes South American leaf blight. Leaf blight begins when a spore lands on a *Hevea* leaf. Somewhat like the potato-blight spore, the minute, two-celled leaf blight spore grows a thin, rootlike tube that extends sideways, along the top of the leaf. Usually the tube is tipped by a structure called the appressorium. Executing a right-angle turn, the appressorium drills into the inner cells of the leaf. Depending on the leaf's defenses, the details of the infection process vary. In any case the fungus almost always wins, penetrating the leaf. Inside it produces spores—many, many spores—which emerge from new tubes on the bottom of the leaf. They are knocked free by raindrops or brushed off by rubbing against other leaves. Left behind are ruined, blackened leaves, which fall off the tree. Leaf blight defoliates *H. brasiliensis*. The blighted trees I have seen, with their sparse black foliage, looked as if someone had gone after them with a blowtorch. Many trees survive a bout with *M. ulei,* but their growth is stunted; a second or third episode will kill them.

M. ulei spores do not survive long after parting from their natal leaf. Thus *Hevea* trees in the wild are usually spaced widely apart; if one succumbs to leaf blight, the others are too distant to be attacked. In plantations, by contrast, trees are so close together that their upper branches are entangled. Spores hop from tree to tree like so many squirrels. Or the fungus can travel on the clothes and fingernails of plantation workers. That is what happened in Fordlândia.

Ironists will appreciate that *M. ulei* attacked just as Ford finally hired its first actual rubber expert, James R. Weir, a plant pathologist who was the ex-director of the U.S. National Fungus Collections. Weir's first action for Ford was to travel to the Indonesian island of Sumatra, home to many rubber estates. Its rubber planters had found especially productive trees and learned how to propagate them by grafting wood from these trees onto sturdy rootstock. In thirty years they had created prodigious groves of high-yielding clones. Weir purchased 2,046 grafted buds in December 1933. Like the Brazilians who failed to block Wickham, the Sumatrans who didn't stop Weir were upset about it later. Five months after his departure, Asian rubber producers formed a second, stronger cartel—and explicitly prohibited the removal of "leaves, flowers, seeds, buds, twigs, branches, roots or any living portion of the rubber plant." By then Weir had carried his precious sprouts to Brazil, where they were about to be wiped out by *M. ulei*.

M. ulei exists in many different strains; if a fungicide wipes out one, the others move in. Weir launched an emergency testing program to look for resistant trees. Meanwhile he tried to establish a new, fungus-free plantation eighty miles away on better land that was closer to the mouth of the Tapajós. He filled it with the high-producing clones from Sumatra. The fungus overran the new plantation even faster than the old. By selecting their trees exclusively for latex yield, Asian farmers had inadvertently produced varieties with even less resistance to blight. The disaster effectively ended Fordlândia, though it wasn't formally abandoned until 1945. Its fate made most Brazilians conclude that rubber plantations are not viable in the Amazon. When Ford bought land in Brazil, 92 percent of the world's natural rubber came from Asia. Five years after Fordlândia ended the figure was 95 percent.

The advent of synthetic rubber during the First World War

failed to drive the Asians out of business. Despite the brilliance of industrial chemists, there is still no synthetic able to match natural rubber's resistance to fatigue and vibration. Natural rubber still claims more than 40 percent of the market, a figure that has been slowly rising. Only natural rubber can be steam-cleaned in a medical sterilizer, then thrust into a freezer—and still adhere flexibly to glass and steel. Big airplane and truck tires are almost entirely natural rubber; radial tires use natural rubber in their sidewalls, whereas the earlier bias-ply tires were entirely synthetic. High-tech manufacturers and utilities use high-performance natural-rubber hoses, gaskets, and O-rings. So do condom manufacturers—one of Brazil's few remaining natural-rubber enterprises is a condom factory in the western Amazon. With its need for materials that can withstand battle conditions the military is a major consumer—which is why the United States imposed a rubber blockade on China during the Korean War.

The blockade helped convince the Chinese of the need to grow their own *H. brasiliensis*. Alas, the nation had only a few areas warm enough for this tropical species. The biggest was Xishuangbanna (syee-schwong-ban-na, more or less), at the extreme southern tip of Yunnan Province, bordering Laos and Burma. A homeland for the Dai and Akha (Hani), two of China's minority ethnic groups, Xishuangbanna Prefecture is China's most tropical place. Although it comprises just 0.2 percent of the nation's land, it contains 25 percent of its higher plant species, 36 percent of its birds, and 22 percent of its mammals, as well as significant numbers of amphibians and freshwater fish.

A few people had dabbled in rubber there as early as 1904, but the efforts had not been sustained. In the 1960s the People's Liberation Army worked to turn the prefecture into a rubber haven. Xishuangbanna plantations were, in effect, army bases; entry was forbidden to outsiders. Outsiders included the Dai and Akha who lived nearby. As suspicious of the minorities in the mountains as the Qing, the Communists imported more than 100,000 Han workers, many of them urban students from faraway provinces, and put them into labor gangs charged with revolutionary fervor. "China needs rubber!" they were told. "This is your chance to use your hands to help your country!" Workers were awakened every day at 3:00 a.m. and sent to clear the forest, one former Xishuangbanna laborer told anthropologist Judith Shapiro, author of *Mao's War Against Nature*.

> Every day we cut until 7:00 or 8:00 a.m., then ate a breakfast of rice gruel sent by the [Yunnan Army] Corps kitchen. We recited and studied Chairman Mao's "Three Articles" and struggled against capitalism and revisionism. Then it was back to work until lunch break, then more work until 6:00. After we washed and ate, there were more hours of study and criticism meetings.

Sneering at botanists' admonitions as counterrevolutionary, the government repeatedly planted rubber trees at altitudes where they were killed by storms and frost. Then it planted them again in the exact same place—socialism would master nature, it insisted. The frenzy laid waste to hillsides, exacerbated erosion, and destroyed streams. But it didn't actually yield much rubber. In the late 1970s the nation began its economic reforms. The educated young people fled back to their home cities, precipitating a labor shortage. Local Dai and Akha villagers were finally permitted to establish rubber farms. They were effective and efficient. Between 1976 and 2003 the area devoted to rubber expanded by a factor of ten, shrinking tropical montane forest in that time from 50.8 percent of the prefecture to 10.3 percent. The prefecture was a sea of *Hevea brasiliensis*.

Unlike the flat Amazon basin, Xishuangbanna is a mass of hills. Planting on slopes exposed the trees to sunlight and ensured that they didn't grow in pools of water, a constant risk in the Amazon because it damages the roots. In addition, according to Hu Zhouyong of the Tropical Crops Research Institute in Jing-hong, the prefecture capital, the relative extremes of temperature let growers select for exceptionally robust trees that would produce more rubber in every circumstance. "Xishuangbanna is ahead of everywhere else in the world in terms of productivity," Hu said to me.

Even as burgeoning China became the world's biggest rubber consumer, its rubber producers were running out of space in Xishuangbanna—every inch of land was already taken. They looked enviously over the border at Laos; with about six million people in an area the size of the United Kingdom, it is the emptiest country in Asia. A few villages in northern Laos had begun planting as early as 1994. But the real push didn't begin until the end of the decade, when China announced its "Go Out" strategy, which pushed Chinese companies to invest abroad. The country had already changed the old military farms into private enterprises—corporations with abundant political clout. As part of Go Out, Beijing announced that it would treat rubber growing in Laos and Myanmar as an opium-replacement program, making the former military farms eligible for subsidies: up to 80 percent of the initial costs for companies to grow rubber across the border, as well as the interest on loans. In addition, it would exempt incoming rubber from most tariffs. (Which companies are receiving the money is unclear; "the subsidy distribution process," economist Weiyi Shi told me, "lacks transparency and appears to be plagued with cronyism.")

Duly incentivized, companies and smallholders flowed over the border. They hired Dai and Akha who lived in China to work with their distant relatives in Laos. Most Laotians lived in hamlets without electricity or running water; schools and hospitals were a distant dream. Seeing a chance to improve their material conditions, villagers jumped starry-eyed on the rubber bandwagon, cutting deals with firms and farms in China. "In China, they were as poor as us," one village head told me. "Now they are rich—they have motorcycles and cars—because they planted rubber. We want to have the same."

No one knows exactly how much *H. brasiliensis* is now in Laos; the government doesn't have the resources for surveys. According to anthropologist Yayoi Fujita of the University of Chicago, in 2003 rubber covered about a third of a square mile in Sing District, next to the border. Three years later it covered seventeen square miles there. Similar growth has occurred in many other districts. The Laotian government estimated that rubber covered seven hundred square miles of the nation by 2010, eight times more than it covered just four years before. And the pace of clearing will only accelerate, along with the effects of that clearing.*

"To harvest a couple thousand square miles of rubber, you need a couple hundred thousand workers," Klaus Goldnick, a regional planner in the northern provincial capital of Luang Namtha, said to me. "The whole province has only 120,000 people. The only solution is to bring in Chinese workers." He said, "Many people here live off the forest. When the forest is gone, it will be difficult for them to survive." He said, "Foreign companies are paying a concession fee"—about $1.50 per tree—"to the government. The more trees, the more fees."

Most of the first plantations were created by villagers who knocked down a few acres on their own or worked with equally small plantations in China. Later the bigger Chinese operations moved in, among them the former state farms. Because rubber trees take seven years to mature, companies naturally want to make long-term arrangements with the people who plant and tend them. I was allowed to look at one of the resultant contracts, between the Chinese firm Huipeng Rubber and three hamlets in Luang Namtha Province.

---

* Jefferson Fox of the East-West Center in Hawaii, who is working with colleagues to evaluate rubber's impact in Southeast Asia, notes that Vietnam plans to increase its companies' rubber area by 1,500 square miles—a quarter of that in southern Laos. In January 2009 Fox visited big plantations in southern Laos, he told me, "from which smallholders had been removed from their land in order to grant land concessions to Vietnamese investors."

Written in both Chinese and Lao, the contract consisted of twenty-four numbered paragraphs. Three were boilerplate: legal descriptions of Huipeng and the villages. Eighteen explained the rights and privileges of the company. One listed the villagers' rights and privileges. In the confusion of the moment, I may have got the numbers slightly wrong—the papers were shown to me while a village head and the rubber agent were telling me their views, each in a different language. But it was impossible to miss that Huipeng's executives had affirmed the contract with their signatures whereas the villagers had affirmed it by rolling their thumbprints onto the page. Each village would plant a certain amount of its land with rubber, the contract explained. Huipeng would in return improve both the roads within the village and the highway to it. But the firm could then sell its rights to the land at its own discretion and hire anybody it wanted to tend the trees, including people from China. Some 70 percent of the proceeds from rubber would go to the villages, "depending on the effective results of the planting"—a big loophole, it seemed to me. Contracts of this sort between companies and villages are common in China (the tobacco plots I visited in the Fujianese hill village in Chapter 5 were governed by one). But such arrangements seemed less benign in Luang Namtha. To my eye the contract looked like the kind of document that emerges when one party has a lawyer looking after its interests and the other party doesn't know what a lawyer is.

In Ban Songma, the next settlement down the road, the village leader who negotiated the contract was about thirty years old. On the day we met he wore a white T-shirt and soccer shorts with a Munich logo. Beside him stood his wife, holding a baby girl wrapped in a faded Hello Kitty blanket. I asked him the name of the rubber company, how much land the village was supposed to provide to it, and the split of the proceeds. He couldn't answer these questions. This was not because he was stupid—he was obviously a shrewd, sparky man—but because the questions were beyond his frame of reference. To be a modern economic agent requires a huge set of habits, assumptions, and expectations. Few of them had been needed in Ban Songma even ten years before. Indeed, they may have been counterproductive. Venturing onto the clawed ground of global capitalism, the village head was as out of his element as Neville Craig was on the Madeira River. That he wanted the fruits of capitalism—Chinese motorbikes and Japanese televisions and nylon shorts emblazoned with the logos of European sports teams—didn't make the likelihood of a happy outcome any greater.

Already Huipeng had imported Chinese workers to plant the seedlings. The village head didn't know whether he and his neighbors would be taught how to graft trees themselves, or to tap latex, or to do the initial processing for rubber. But he did know that people who worked for the Chinese ended up with motorbikes, which liberated them from hours of walking up and down the steep hills. The baby in the Hello Kitty blanket will grow up knowing more than her father about the whirling new world Ban Songma is entering. Huipeng's contract will be in force for forty years. It will be interesting to see at its end how that child regards the deal that her father signed.

THE END OF THE WORLD

The morning had been clear and bright, an ominous sign. On the pedestrian bridge that leads to the Xishuangbanna Tropical Botanical Garden I could see the faintest swirl of fog on the hills. Researchers had drawn their office curtains on the building's sunny side. Founded in 1959, the garden grew up with Xishuangbanna's rubber industry. Its scores of scientists monitored the impact of the refashioning of the regional ecosystem and didn't like what they saw. "We all hate rubber," one researcher told me. "But then we're all ecologists here."

Although the Golden Triangle receives as much as one hundred inches of rain a year, three-quarters of it falls between May

Almost every bit of Xishuangbanna that can support rubber trees has been cleared and planted (top), a change that is profoundly altering the environment—the region's morning mists are vanishing, along with its water supply. With China's rubber companies running out of suitable land in China itself, they have moved across the border to northern Laos (above, a freshly logged hillside).

and October. The rest of the year the forest survives largely on dew from morning fog. "Back in the 1980s and 1990s there was still fog at lunchtime," XTBG ecologist Tang Jianwei told me. "Now it's gone by eleven." The "very obvious" change, he said, is a symptom of a profoundly altered hydrological regime.

Rubber is to blame, Tang said. *H. brasiliensis* usually sheds its leaves in January and new leaves begin budding in late March. The absence of leaves means that the forest has fewer surfaces to retain dew, which reduces water absorption during the dry season. Surface runoff rises by a factor of three—which in turn jacks up soil erosion by a remarkable factor of forty-five. Worse, the new leaves' most intense growth occurs in April, at the dry season's hottest, driest point. To propel growth, the roots suck up water from three to six feet below the surface. Tapping begins as the new leaves appear and continues until they fall. To replace the lost latex, the roots suck up still more water from the ground. How much water? Tang did some rough estimates with pen and paper. Half a kilogram of latex a day, twenty days a month tapping, 180 trees to the acre . . . good latex is 60 to 70 percent water . . . 4,400 pounds of water a year per acre. Rubber producers are effectively putting all the water in the hills into trucks and driving it away. "A lot of smaller streams are drying up," he said. "Villages have had to move because there's no drinking water." Now spread this impact across Laos and Thailand, he said. It would be a slow-motion remaking of a huge area. "It's not easy to tell what the effects would be," Tang said.

Beginning to heed ecologists' worries, Xishuangbanna effectively banned new rubber planting in 2006 by freezing all land rotation. The scheme is unlikely to have much effect. To begin with, as Shi notes, it seems to violate China's newly reformed land laws. But even if Xishuangbanna farmers were to stop planting *H. brasiliensis* tomorrow, its area would keep rising—on their own, rubber trees are invading the remaining forest.

Hillside rubber plantations surround Tang's office in the botanical garden. Because trees are grafted from the wood of high-yielding specimens, the great majority of the rubber trees in Southeast Asia are clones. And the majority of the trees used to create those clones descended from the few sprouts that survived from Henry Wickham's original expedition—a slice of a slice of a slice. These are the trees that Weir brought to Fordlândia, the varieties so highly susceptible to *M. ulei*. The trees make a canopy of green so unbroken that Beijing legally describes rubber plantations as "forests"; locals can fill fallow farmland with rubber and fulfill government conservation dictates. As the area of rubber increases, it becomes an increasingly inviting target for pests. "That's the lesson of biology," Tang said. "Diseases always come in. Sooner or later, they find a way."

For a century, isolation—the isolation of Southeast Asia from Brazil, of Southeast Asian nations from each other—has spared the rubber plantations. But the world is knitting itself together ever more closely. There are still no direct flights between Amazonia and Southeast Asia, but they will come. And in April 2008 the governments of Cambodia, China, Laos, Myanmar, and Thailand opened a brand-new highway that for the first time links all of these nations and connects them to Malaysia and Singapore. Trucks will be able to zoom in three days from Singapore to Kunming, the capital of Yunnan Province. If and when *M. ulei* arrives from Brazil, this will provide transportation. "In ten or twenty years, Xishuangbanna's trees could be wiped out," Tang said. "So would everyone else's trees, probably."

The disaster would take a long time to repair. The industrial revolution, one recalls, depends on three raw materials: steel, fossil fuels, and rubber. If one member of that triad suddenly vanished, it would have unwelcome effects. Imagine transportation networks without tires, electric power plants without gaskets and seals, hospitals without sterile rubber hoses and gloves. Industrial civilization could face such disruption worldwide that organizations from the United Nations to the U.S. Department of Defense list *Microcyclus ulei* as a potential biological weapon. Synthetic rubber will be deployed to replace it, but only as an imperfect replacement. "I sure as hell wouldn't want to be in a 747 about to land on synthetic tires," the director of the U.S. National Defense Stockpile Center has said.

Breeders are working on new, resistant plants, but progress is slow. "All control measures against this disease have been unsuccessful," stated the *Annals of Botany* in 2007. Even the most modern techniques "have failed to prevent large losses and dieback of trees." Asian scientists pulled some more trees from Brazil in 1981 to increase plantations' genetic diversity. These are being evaluated and cross-bred with more productive plants. Researchers in France announced in 2006 that they had fully resistant clones. But few plantation owners want to take up these varieties, which are new and therefore risky. Every ecologist I spoke with in Brazil, China, and Laos believed that Asia was almost as unprepared for leaf blight today as it was fifty years ago.

When I visited Xishuangbanna, I wore the same shoes that I had worn a few months before in Brazil. Because the spores are fragile, I was pretty sure I wouldn't cause an epidemic. Still, I sprayed my shoes with fungicide. At the border neither the Chinese nor the Laotian customs officials batted an eye at the two Brazilian visas in my passport, or the entry stamps that said I had passed through Manaus, epicenter of leaf blight. I wanted to do my work, so I didn't say anything.

Someday, though, there will be a problem. The cycle of the Columbian Exchange will be complete, taking away what it once gave. Trees will die fast. The epidemic will cover an area large enough to be visible from space: black-leaved splotches scattered from the tip of China to the end of Indonesia. There will be a major international mobilization of resources to fight the outbreak. And planters will suddenly be aware that they are living in the Homogenocene, an era in which Asia and the Americas are increasingly alike.

PART FOUR

Africa in the World

8

Crazy Soup

JOHNNY GOOD-LOOKING

In the 1520s a solitary man constructed a small chapel on the western highway out of Mexico City, just beyond the causeway that led to the city's western gate. No description of the chapel survives, but it was probably just two whitewashed adobe rooms: one for the shrine itself, with an altar and cross; one for the man who built and maintained it. Nearby were a few small fields on which he grew crops. The structure was known as the Chapel of the Martyrs or, more impressively, the Chapel of the Eleven Thousand Martyrs. It may have been the first Christian church in mainland America.

The man in the chapel was named Juan Garrido. Almost the only thing we know about the background of this Spanish man is that he was not Spanish and his name was not Garrido. The first hint of his existence in the historical record comes from 1477, when João II, the future king and current regent of Portugal, granted freedom to an African slave who called himself João Garrido (João is the Portuguese equivalent of Juan, or John). Enslaved as a boy and taught Portuguese, João had served as an interpreter on several slaving ventures before jumping ship in Guinea or southern Mauritania, "from a desire to be free and not to return again to being a slave." Yet Garrido didn't want to be free in Africa. Instead he boldly sought to return to Portugal, possibly because he was a Berber, a lighter-skinned North African ethnic group that was often at odds with darker African groups to the south. (I owe this suggestion to Alastair Saunders, the historian who found Garrido's freedom papers.) Garrido must have been a valuable interpreter; the future king made him "absolutely free . . . like any other Christian Portuguese," on the condition that he continue working in the slave trade.

Was this Joaõ Garrido the Juan Garrido who later built the chapel in Mexico? Most likely not, according to Matthew Restall, a Pennsylvania State University historian who has also studied Garrido's life—it would have meant that he was almost sixty years old when he came to Mexico. But the name of Garrido was uncommon among the tens of thousands of slaves in the Iberian Peninsula; it is likely that the two men were connected. The Spanish Juan could have been Joaõ's son or cousin, Restall suggests. Or, Saunders asks, "Was the Juan Garrido in Mexico one

An African man, possibly Juan Garrido, holds Hernán Cortés's horse as the conquistador, feathered hat in hand, approaches Motecuhzoma, paramount leader of the Triple Alliance. The drawing is from Diego Durán's account of the conquest of Mexico, *The History of the Indies of New Spain* (c. 1581).

of the black slaves brought back from Africa by the newly freed Berber or mulatto João Garrido, who also received the name João and adopted his master's surname when he in turn was freed?"

Like his father, uncle, or former master, this Juan Garrido had an eye for the main chance. Rather than remaining in Portugal, he crossed the Spanish border, arriving just after Colón revealed the existence of the Americas to Europe. He spent seven years in Seville, the center of Spain's growing American trade. Something of his personality is hinted at by the name he and his predecessor had chosen to live under: Juan Garrido, which means, more or less, Johnny Good-looking.

Johnny Good-looking crossed the Atlantic early in the sixteenth century, according to his biographer, Ricardo E. Alegría, an anthropologist in Puerto Rico. He landed in Hispaniola.

As aggressive and ambitious as any other conquistador, a young man with his blood aboil, he quickly attached himself to a local sub-governor, Juan Ponce de León y Figueroa, accompanying him on a mission to take over the island of Puerto Rico. When Ponce de León sank his fortune into an off-kilter hunt for the Fountain of Youth, Garrido joined the futile quest. (Along the way, they became the first people from the opposite shore of the Atlantic to touch down on Florida.) When Spain launched punitive expeditions against Caribe Indians on half a dozen Caribbean islands, Garrido brought his gun. And when Hernán Cortés seized the Triple Alliance, Johnny Good-looking was at his side.

The alliance is more commonly known as the Aztec empire, but the term is a nineteenth-century invention, and historians increasingly avoid it. It was a consortium of three militarized city-states in the middle of Mexico: Texcoco, Tlacopan, and Tenochtitlan, the last by far the most powerful partner. When the Spaniards arrived, this Triple Alliance ruled central Mexico from ocean to ocean and Tenochtitlan was bigger and richer than any city in Spain.

As canny a politician as he was a fighter, Cortés was able to foment an assault on the empire by its many enemies and place himself at its head. But despite taking the Triple Alliance emperor hostage in his own palace—a paralyzing surprise to the enemy—the initial assault failed calamitously. Indeed, the Spaniards barely escaped from Tenochtitlan. When all seemed lost, Cortés had a stroke of luck: the accidental introduction of the smallpox virus. Never before seen in the Americas, transmittable with horrific ease, the virus swept through densely packed central Mexico, killing a third or more of its population in a few months.*

As the Triple Alliance reeled from the epidemic, the Spanish-Indian army attacked the capital a second time in May 1521, with as many as 200,000 troops. Tenochtitlan occupied a Venice-like clump of islands, many of them human-made, on the west side of an eighty-mile-long, artificially recontoured lake. Spiderwebbing from the metropolis was an intricate network of causeways, dikes, dams, baffles, and channels that both kept back floods during the wet season and funneled water around the city during the dry season.

Cortés's strategy was in part to avoid the heavily defended causeways into the city by draining and filling the moat-like channels around them, thus creating dry land from which he could assault less-protected areas of the perimeter. During the siege, the attackers repeatedly tore out dikes and piled up stones and earth during the day, and the Triple Alliance repeatedly reassembled the dikes and reflooded the channels at night. On June 30, the Alliance set a trap at the shore entrance to Tenochtitlan's western causeway, undermining a bridge that crossed a shallow, reed-thick waterway. When the attackers charged across the bridge, wrote the sixteenth-century chronicler Diego Durán,

* This direst instance of the Columbian Exchange is often said to have been introduced in the body of an African slave named Francisco de Eguía or Baguía. Other reports contend that the carriers were Cuban Indians brought as auxiliaries by the Spaniards. Restall suspects that "granting the role of patient zero" to Africans or Indians is "classic Spanish scapegoating." So horrific was the epidemic, he suggests, that Spaniards did not want to be seen as the cause.

Tenochtitlan, seen in a present-day artist's reconstruction, dazzled the Spaniards when they saw it—the city was grander than any in Spain. Protecting the city was an irregular, ten-mile-long dike (far right in image) that separated the brackish water of the main lake from a new, human-made freshwater lake that surrounded the city and provided water for a network of artificial wetland farms known as *chinampas*.

"the entire thing collapsed, together with the Spaniards and Indians who stood upon it." From hiding places in the reeds shot canoes loaded with men wielding bows, spears, and stolen Spanish swords. Flailing in the brackish water, the Spaniards and their horses were easy prey; Cortés himself was wounded and almost captured.

As the surviving attackers fled to safety, they heard the boom of an enormous drum—"so vast in its dimensions," the conquistador Bernal Díaz del Castillo later recalled, "that it could be heard from eight to twelve miles distance." The Spaniards spun on their heels. Across the water they could see Triple Alliance soldiers dragging Spanish prisoners, still dripping from the watery ambush, to the summit of a great, pyramidical temple. In an act meant to terrify and demoralize, Alliance soldiers and priests ripped open the captives' chests, tore out their hearts, and kicked the bodies down the temple steps. The next morning they marched another prisoner—"a handsome Sevillian," Durán wrote—to the edge of the channel and in full view of his friends "ripped him to bits then and there." When Tenochtitlan fell, Cortés had his revenge. He stood by as his troops and their native allies despoiled the shattered city, slaughtering the men and raping the women.

Juan Garrido may have been at the ambush or known the sacrificed Spaniards or both. In any case, he was asked by Cortés to build the Chapel of the Martyrs, a monument and graveyard for fallen conquistadors, on the spot where the ambush took place. The assignment was but one of many, for Garrido soon became one of the conqueror's go-to men as he erected Spanish Mexico City literally atop the wreckage of Indian Tenochtitlan. Johnny Good-looking became a kind of majordomo for the new municipal government; protector of the trees that shaded the highways into town (the records give no reason for the position, but one can guess the trees were being cut for fuel); guardian of the main city water supply (Tenochtitlan, which had no water of its own, was supplied via aqueduct from mountain springs); and town crier—a position, Restall says, that could include the duties of a "constable, auctioneer, executioner, piper, master of weights [responsible for assaying silver and gold], and doorkeeper or guard." As lagniappe, Garrido accompanied Cortés in 1535 on the latter's ill-fated attempt to cross Mexico and sail to China—the ultimate goal of Spanish adventurers.

Garrido's biggest contribution occurred after Cortés found three kernels of bread wheat (*Triticum aestivum*) in a sack of rice that had been sent from Spain. The conqueror asked his go-to man to plant them in a plot near the chapel that served as a kind of experimental farm. "Two of them grew," the historian Francisco López de Gomara reported in 1552, and one of them produced 180 kernels. They later turned around and planted those kernels, and little by little there was boundless wheat: one [kernel] yields a hundred, three hundred, and even more with irrigation and sowing by hand. . . . To a black man and slave is owed so much benefit!

Wheat was not only desired by roll-eating, cake-munching, beer-guzzling conquistadors, it was a necessity for the politically powerful clergy, who needed bread to celebrate Mass properly. Repeatedly Spaniards had tried to grow *T. aestivum* in Hispaniola, and repeatedly it had failed in the hot, humid climate. Garrido's wheat was greeted with joy—in a strange land, it was the taste of home. Soon the golden herringbone tassels of wheat spikelets waved across central Mexico, replacing thousands of acres of maize and woodland. More than that, Mexican smallholders say, Spaniards carried Garrido's *T. aestivum* to Texas, from where it spread up the Mississippi. If this is accurate, much or most of the wheat that by the nineteenth century had transformed the Midwest into an agricultural powerhouse came from an African roadside chapel in Mexico City.

In planting Cortés's wheat, Garrido was acting as an agent of the Columbian Exchange. More important, though, he himself was part of the exchange, as were Cortés and the other foreigners.

Previously in this book, I described researchers' evolving view of the Columbian Exchange. I first looked at the Atlantic (Chapters 2 and 3), where the most important effects were caused by microscopic imports to the Americas (initially the diseases that depopulated Indian societies, then malaria and yellow fever, which encouraged plantation slavery). Next I treated the Pacific (Chapters 4 and 5), where the major introductions were American food crops, which both helped sustain a population boom and led indirectly to massive environmental problems. In the next section (Chapters 6 and 7), I showed how environmental historians have increasingly come to believe that the Columbian Exchange played a role in the agricultural revolution of the eighteenth century and the industrial revolution of the nineteenth. Both occurred first in Europe, and so this ecological phenomenon had large-scale political and economic implications—it fostered the rise of the West. In all this discussion, I have acted as if humankind were in the director's chair, distributing other species at will, sometimes being surprised by the results. But to biologists *Homo sapiens* is a species that like any other has its own distribution and range. Not only did human beings cause the Columbian Exchange, they were buffeted by its currents—a convulsion within our own species that is the subject of this section of the book.

For millennia, almost all Europeans were found in Europe, few Africans existed outside Africa, and Asians lived, nearly without exception, in Asia alone. No one in the Eastern Hemisphere in 1492, so far as is known, had ever seen an American native. (Some researchers believe that English fishing vessels crossed the Atlantic a few decades before Colón, but the principle holds—one didn't find communities of Europeans or Africans in Asia or the Americas.) Colón's voyages inaugurated an unprecedented reshuffling of *Homo sapiens*: the human wing of the Columbian Exchange. People shot around the world like dice flung on a gaming table. Europeans became the majority in Argentina and Australia, Africans were found from São Paulo to Seattle, and Chinatowns sprang up all over the globe.

The movement was dominated by the African slave trade—dominated by Garrido, so to speak, rather than by Cortés. For a long time the scale of slavery in the Americas was not fully grasped. The first systematic attempt at a count, Philip Curtin's *The Atlantic Slave Trade: A Census,* did not appear until 1969, more than a century after its subject's extirpation. Partly stimulated by Curtin's study, David Eltis and Martin Halbert of Emory University, in Atlanta, led a remarkable effort in which scholars from a dozen nations pooled their work to create an online database of records from almost 35,000 separate slave voyages. Its most recent iteration, released in 2009, estimates that between 1500 and

1840, the heyday of the slave trade, 11.7 million captive Africans left for the Americas—a massive transfer of human flesh unlike anything before it. In that period, perhaps 3.4 million Europeans emigrated. Roughly speaking, for every European who came to the Americas, three Africans made the trip.

The implications of these figures are as staggering as their size. Textbooks commonly present American history in terms of Europeans moving into a lightly settled hemisphere. In fact, the hemisphere was full of Indians—tens of millions of them. And most of the movement into the Americas was by Africans, who soon became the majority population in almost every place that wasn't controlled by Indians. Demographically speaking, Eltis has written, "America was an extension of Africa rather than Europe until late in the nineteenth century."*

In the three centuries after Colón, migrants from across the Atlantic created new cities and filled them with houses, churches, taverns, warehouses, and stables. They cleared forests, planted fields, laid out roads, and tended horses, cattle, and sheep—animals that had not walked the Americas before. They stripped forests to build boats and powered mills with rivers and waged war on other newcomers. Along the way, they collectively reworked and reshaped the American landscape, creating a new world that was an ecological and cultural mix of old and new and something else besides.

This great transformation, a turning point in the story of our species, was wrought largely by African hands. The crowds thronging the streets in the new cities were mainly African crowds. The farmers growing rice and wheat in the new farms were mainly African farmers. The people rowing boats on rivers, then the most important highways, were mainly African people.

---

* New England was an exception, but it was only a small fraction of English migration—the colonies to its south were much bigger. Until the end of the eighteenth century, African slaves outnumbered Europeans in England's American holdings by about two to one.

The men and women on the ships and in the battles and around the mills were mainly African men and women. Slavery was the foundational institution of the modern Americas.

The nineteenth century saw another, even larger, wave of migration, this one dominated by Europeans. It changed the demographic balance a second time, so that descendants of Europeans became the majority in most of the hemisphere. Surrounded by people like themselves, this second group of immigrants was rarely aware that it was following trails that had been set for more than three hundred years by Africans.

Two migrations from Africa were turning points in the spread of *Homo sapiens* around the globe. The first was humankind's original departure, seventy thousand years ago or more, from its homeland in Africa's eastern plains. The second was the transatlantic slave trade, the main focus of this section of the book. The first wave of the human Columbian Exchange, the slave trade was the biggest impetus to the migratory flood that broke through the long-standing geographic barriers that kept apart Africans, Americans, Asians, and Europeans. In this chapter, I focus on two related topics: first, the rise of plantation slavery, which largely drove the forced migration of Africans; and second, the extraordinary cultural mix that slavery inadvertently promoted. The next chapter focuses on the interactions of what became the Americas' two biggest populations, Indian and African. Largely conducted out of sight of Europeans, the meeting of red and black centered on their common resistance to the European presence in their lives—a rebellion that simmered across the hemisphere, and that had consequences that are felt to this day.

Johnny Good-looking lived with his family in the center of the whirlpool: teeming, polyethnic Mexico City. A giddy buzz and snarl of African slaves, Asian shopkeepers, Indian farmers and laborers, and European clerics, mercenaries, and second-tier aristocrats, it was a city of exiles and travelers, the first urban complex in which a majority of the inhabitants had their ancestry across an ocean. This was the social world created by the human wing of the Columbian Exchange; Garrido, an African turned European turned American, was a prototypical citizen.

He was married to a Spanish woman (a sign of high status, because not many had come to Mexico), and had three children, a home near his chapel in one of the city's poshest neighborhoods, and the knowledge that he had participated in a pivotal moment in history. Nonetheless, he was a disappointed, unsatisfied man, so hard to live with that at one point his wife tried to pay an African woman to use witchcraft to make him go on another faraway adventure. In 1538, probably in his fifties, he petitioned the court, begging the king to "recompense me for my services and for the little favor your governors have done me, having served as I have served." When he went to Spain to deliver his request, he made a little money on the side by selling his Indian servant into slavery. (The servant sued, and the sale was revoked.) Despite all the effort, Garrido's plea apparently went unheard. It says something about that chaotic time and place that this remarkable figure—a slave-descendant-turned-conquistador, an African who became a confidant of Cortés, a man from a Muslim land who married a woman from a Christian land in an animist land—was able to drop from sight. After the petition, no trace of his life has been found.

According to Alegría, Garrido's biographer, he probably died in the next decade, forgotten in the hubbub and tumult of the new world he had helped to bring into existence.

BAD BEGINNINGS

It seems fair to observe that the planners of the war did not prepare for its consequences. Scholars argue over its origins, but the goal of the war as fought was to eject a Middle Eastern dictator whom many Western leaders viewed as a threat to civilization. After impassioned speeches, they formed a multinational coalition that marched toward the ancient city that was their central objective. After a surprisingly brief battle the allied forces seized control. Unfortunately, they had made no plans for what to do next. The coalition's military leadership simply declared the mission accomplished and left for home. Only a skeleton military crew was left to face a growing Muslim insurgency in the countryside.

This was in 1096 A.D., during the First Crusade. Godfrey of Bouillon, appointed to rule newly conquered Jerusalem, had to find some way to support his remaining army, the swarm of monks, priests, deacons, and bishops who had accompanied it, the pilgrims / cannon fodder who had accompanied the religious leaders, and the Venetian merchants who had provided invaluable logistical support. An obvious answer, from the Crusaders' point of view, was to seize Muslim property. European entities took ownership of entire urban neighborhoods and even cities; Venice fastened onto the port of Tyre, for example, and the Knights of Malta (as they are now known) acquired as much as a fifth of Jerusalem. In the countryside, Crusaders ultimately assembled more than two hundred grand estates, growing olives, wine, oranges, dates, figs, wheat, and barley. Most important in the long run, though, was a sticky, grainy product that the farms' new masters had never before encountered: *al-zucar,* as the locals called it, or sugar.

Sugarcane was initially domesticated in New Guinea about ten thousand years ago. As much as half of the plant by weight consists of sucrose, a white, powdery substance known to ordinary people as "table sugar" and to scientists as $C_{12}H_{22}O_{11}$. In a chemist's lexicon, "sugar" refers to a few dozen types of carbohydrate with similar chemical structures and properties. Sucrose is among the simpler members of the group: one molecule of glucose (the type of sugar that provides energy for most animal bodies) joined to one molecule of fructose (the main sugar in honey and fruit juice). Culturally, historically, psychologically, and perhaps even genetically, though, sucrose is anything but simple. A sweet tooth, unlike a taste for salt or spice, seems to be present in all cultures and places, as fundamental a part of the human condition as the search for love or spiritual transcendence. Scientists debate among themselves whether $C_{12}H_{22}O_{11}$ is actually an addictive substance, or if people just act like it is. Either way, it has been an amazingly powerful force in human affairs.

Sugarcane is easy to grow in tropical places but hard to transport far because the stalks ferment rapidly, turning into a smelly brown mass. People who wanted something sweet thus had to grow it themselves. The crop marched steadily north and west, infiltrating China and India. *Crops,* rather—the sugarcane in farm fields is a hodgepodge of hybrids from two species in the grass genus *Saccharum.* The spoilage problem was solved in India around 500 B.C. when unknown innovators discovered how to use simple horse- or cattle-powered mills to extract the juice from the stalks, then boil down the juice to produce a hard golden-brown cake of relatively pure $C_{12}H_{22}O_{11}$. In cake form, sugar could be stored in warehouses, shipped in chests and jars, and sold in faraway places. An industry of sweetness was born.

Almost all of the Middle East is too dry to grow sugarcane. Nonetheless, people figured out how to do it anyway, irrigating river valleys in Iran, Iraq, and Syria. By about 800 A.D. cane had become particularly common on the Mediterranean coast of what are now Lebanon and Israel, which is where the Crusaders for the first time encountered "reeds filled with a kind of honey known as *Zucar*"—the description comes from the twelfth-century chronicler Albert of Aachen.

The writer Michael Pollan has recounted his son's inaugural experience of sugar: the icing on his first birthday cake.

[H]e was beside himself with the pleasure of it, no longer here with me in space and time in quite the same way he had been just a moment before. Between bites Isaac gazed up at me in amazement (he was on my lap, and I was delivering the ambrosial forkfuls to his gaping mouth) as if to exclaim, "Your world contains *this*? From this day forward I shall dedicate my life to it."

Much the same thing happened to the Crusaders' army in Lebanon. Clerics, knights, and common soldiers alike drank *al-zucar* juice "with extreme pleasure," Albert of Aachen reported; the chance to sample sugar was, in and of itself, "some compensation for the sufferings they had endured." As with Pollan's son, a single, heavenly taste was enough to ensure a lifelong craving: "the pilgrims could not get enough of its sweetness."

In their new sugar estates the Crusaders saw an opportunity: exporting to Europe large quantities of $C_{12}H_{22}O_{11}$—"a most precious product," said the archbishop of Tyre, the new rulers' first sugar center, "very necessary for the use and health of mankind." Sugar was then a rarity in Europe; regarded as an exotic Asian spice like pepper or ginger, it was found only in the kitchens of a few princes and nobles. The Crusaders proceeded to stoke a hunger for sweetness in the continent's wealthy, and to make money by temporarily satisfying it.

As important as sugar itself was its manner of production: plantation agriculture. A plantation is a big farm that sells its harvest in faraway places. To maximize output, plantations usually plant a single crop on big expanses of land. The big expanse requires a big labor force, especially during planting and harvesting. Because agricultural products spoil, plantations typically ship their crops in processed form: cured tobacco, pressed olive oil, heat-solidified latex rubber, fermented tea, and dried coffee. They must also have some way to transport their products. Thus plantations as a rule consist of a lot of land near a port or highway with an attached industrial facility and a pool of laborers.

Sugar is the plantation product *par excellence*. Even the most sugar-mad grower cannot consume the entire harvest at home; some must always be sold off the farm. Once refined, sugar can be easily packaged and shipped for long distances. And there is always a market abroad: nobody has ever overestimated humankind's appetite for sweetness. The main pitfall is labor: without workers, fields, mills, and boilers will sit idle. To avoid this calamity, plantation owners must take steps to ensure an adequate supply of employees. In an exhaustive study published in 2008, the University of Provence historian Mohamed Ouerfelli has shown that Islamic sugar plantations kept their workers by paying relatively high wages. European-owned plantations initially adopted the same strategy—in Sicily, Ouerfelli showed, people actually migrated from other parts of Europe to work on sugar plantations. But over the course of time Europe's sugar producers reconsidered.

After the First Crusade, European Catholics in later anti-Muslim missions seized sugar plantations from their Muslim and Byzantine creators in Cyprus, Crete, Sicily, Majorca, and southern Spain (Islamic empires later took some of them back). But no matter how much sugar they produced, Europeans wanted more. Eventually they ran out of areas warm and wet enough to grow sugar in the Mediterranean. Portugal looked overseas, to the Atlantic island chains: Madeira, the Azores, the Cape Verde Islands, and São Tomé (St. Thomas) and Príncipe. Spain went to another set, the Canary Islands.

Madeira was first and in some ways most important. It set precedents and established patterns. Located about six hundred miles off the Moroccan coast, the archipelago consists of more than a dozen islands, of which two are by far the largest: Porto

Santo and Madeira itself. Both are the summits of extinct volcanoes, but Porto Santo is low and partly ringed by beaches whereas
Madeira is high and bristles with cliffs.

Both islands were uninhabited until 1420, when they were visited by an expedition led by two squires in the Portuguese court
and a Genoese navigator resident in Lisbon, Bartolomeu Perestrello. Two decades after his death, Perestrello became a footnote to history: his daughter married Colón, who may have lived
on the islands and inherited his private navigational charts. While
Perestrello was alive, he was best known as the man who brought
rabbits to Madeira—or, more precisely, to Porto Santo, where the
party initially disembarked. In Perestrello's luggage was a pregnant rabbit, which gave birth aboard ship. Upon arrival he released
mother and offspring, presumably intending to hunt them later
for stew. To the colonists' horror the animals "multiplied so much
as to overspread the land," Gomes Eanes de Zurara, Portugal's
royal archivist, recounted in 1453. Being rabbits, they ate everything in sight, including the colonists' gardens. The Portuguese
"killed a very great quantity of these rabbits," Zurara reported, but
"there yet remained no lack of them . . . our men could sow nothing that was not destroyed by them." Starved out of Porto Santo
by its own fecklessness, the expedition retreated to Madeira island.

So delicious is this ecological parable that one naturally doubts
its veracity. But Zurara, generally regarded as a careful writer,
had visited the island; rabbits still plagued it at the time he wrote.
Adding to the plausibility of the tale, much the same occurred
after Spain took the Canary Islands. Colonists brought donkeys to
Fuerteventura, the chain's second-biggest island. Inevitably, the
animals escaped. So many bands of donkeys rampaged through
grain fields, reported a historian who lived there at the time, that
the government "assemble[d] all the inhabitants and dogs in the
island, to endeavor to destroy them." An orgy of asinine slaughter ensued.

Even if the Portuguese did not cause rabbit chaos on Porto
Santo, they wreaked still worse ecological mayhem on Madeira.

The island, unlike relatively open Porto Santo, was covered with deep forest (its name is the Portuguese word for "wood"). To plant crops, some portion of that forest had to be removed. The settlers chose the simplest method: fire. Predictably, the burn escaped control and engulfed much of the island. The settlers fled into the sea, where they stayed for two days in neck-deep water as flames roared overhead. Supposedly the fire continued, burning in roots underground, for seven years. The settlers planted wheat on the burned land, exporting the harvest to Portugal. Not until the 1440s did they learn that the island's warm climate was better suited to another, more profitable crop: sugarcane.

Meteorologically, Madeira was a fine place to grow sugar. Geographically, it was a challenge. The island has little land level enough for agriculture, and most of that little is on three high, inaccessible "shoulders" around the island's two main volcanic peaks, the tallest more than six thousand feet. Elsewhere the terrain is so steep that in some parts cattle are kept in small, shed-like byres for their entire lives for fear they will tumble fatally down the slopes. (Tourist guides extol Madeira as "the island of sad cows.")

The first settlers parceled out most of the land among themselves. Late arrivals either had to lease fields as sharecroppers or hack terraces from unused land. In either case, they had to channel water from the wet peaks to their plots, which involved creating an octopoid network of tunnels and conduits that twisted every which way through the stony hills. Despite the obstacles, sugar boomed. According to Alberto Vieira, the islands' most prominent contemporary historian, between 1472 and 1493 production grew by a factor of more than a thousand. Prices fell, as one would expect. Planters who had been making huge profits suddenly saw those profits threatened. The only way to keep the money rolling in was to ramp up production: build new terraces, carve out new waterways, and construct new mills. They clamored for workers— wanted them *now*—to slash cane, extract juice, boil down sugar, and ship the crystallized results. With little evident reflection, some colonists made a fateful decision: they bought slaves.

In some sense this was nothing new; slavery had existed in the Iberian peninsula since at least Roman times. At first many slaves had been taken from Slavic countries (the origin of the word "slave") but in the intervening centuries the main source of bondsmen had become captured Muslim soldiers. As a rule, they worked as domestic servants and were treated in much the same way as other domestic servants; their main purpose, according to Antonio Domínguez Ortiz, a historian at the University of Granada, was to serve as "sumptuary articles"—status symbols. Slaves were living, breathing testaments to the wealth and rank of their owners. Being able to summon a captive Muslim or African to pour wine was proof that one was important enough to possess an exotic foreign human being. The system was not benevolent, but it had escape hatches big enough to avoid murder, insurrection, mutiny, and the other problems with slave labor identified by Adam Smith. Slaves often were allowed to earn their

Sugar mills were smoky, steamy places that required many workers, as shown in this engraving of a Sicilian mill around 1600 (it is based on a painting by Jan Van der Straet, a Flemish artist active in Florence).

own money, for instance, with which they could rent their freedom on a monthly basis. Often enough, this led to emancipation. Domínguez Ortiz has speculated that Iberian slavery, left to itself, would have evolved into a system in which owners had the right to extract money from slaves, rather than labor, and only for designated periods of time.

In Madeira, Iberian slavery was transformed. True, most of the Europeans there had only a little land and couldn't afford bondsmen. And even those who did buy slaves rarely had more than two or three, and often they didn't grow sugarcane. Initially the slaves themselves were not from the Gulf of Guinea, the great indentation along the west-central African coast that was the origin of the great majority of slaves in the Americas. Instead the first captive workers were a luckless, scattershot collection of convicts, Guanches (the original inhabitants of the Canary Islands), Berbers (the people of northwest Africa, long-term adversaries of the Portuguese), and, probably, conversos (Iberian Jews and Muslims who had been more or less forcibly converted to Christianity—many Portuguese and Spaniards viewed them as potential traitors). Nevertheless Madeira was where plantation agriculture was joined, however shakily, to African slavery. In time, Vieira says, the convicts, Guanches, Berbers, and conversos were replaced by west-central Africans. Africans grew and processed sugar, and their numbers rose and fell with the fortunes of the sugar industry. The world of plantation slavery was coming, terribly, into existence. And Madeira was, in Vieira's phrase, its "social, political and economic starting point."

Two key elements, though, were missing: the organisms responsible for malaria and yellow fever. Both were abundant in São Tomé and Príncipe, two islands in the Gulf of Guinea that Portugal took over in 1486. Like Madeira, they were uninhabited, thickly forested places with a warm climate, good volcanic soil, and plenty of water—perfect for producing $C_{12}H_{22}O_{11}$. Like Madeira, they were settled by entrepreneurially minded petty nobles who hoped to cash in on Europe's sweet tooth. Unlike

Madeira, though, São Tomé and Príncipe swarmed with *Anopheles gambiae,* Africa's worst malaria carrier; and *Aedes aegypti,* which transmits yellow fever. It was a bit like a natural scientific experiment: change one variable and see what happens.

The first two, small movements into São Tomé failed—killed off by disease. A third, larger attempt in 1493 succeeded, partly because it was accompanied by a mass of slave labor: convicts and undesirables, notable among the latter some two thousand Jewish children who had been taken forcibly from their parents. Sugar planters and sugar processors, criminals and children—all died in droves. After six years, only six hundred of the children were alive. Nonetheless the colony somehow kept going. A Dutch force landed in 1599 on Príncipe, the second island, intending to transform it, too, into a sugar center; the invaders departed after just four months, leaving more than 80 percent of their men beneath the ground. A year later the Dutch tried a different tactic: occupying São Tomé itself. Two weeks and 1,200 dead Dutchmen later, they bolted. Europeans perished with such routine dispatch on the islands that the Portuguese government took to exiling troublesome priests there, thus ensuring their deaths while technically avoiding the Vatican's ban on executing its functionaries. In 1554, six decades after colonization began, São Tomé had but 1,200 Europeans. By 1600 the figure had shrunk to about two hundred—slaves outnumbered masters by more than a hundred to one. In 1785 an official report claimed that just four people—four people!—of pure European stock lived on the island. Trying to build up the colonial population, the monarchy ordered that female African slaves be awarded to every new male European arrival, along with exhortations to breed. The ploy failed to boost the number of migrants—the risk was not worth it. Even the bishops the Vatican appointed to the island refused to come. After the post had been vacant for forty-three years, a new bishop finally had the courage to land in São Tomé in 1675. He was dead in two months. "In São Tomé, there's a door to come in," the Portuguese sang, "but there's no door to go out."

Despite the lack of colonists the colony flourished—for a while. At the height of the boom São Tomé exported four times as much sugar as Madeira. About a third of the island's surface had been converted to sugarcane; much of the forest had vanished to fuel sugar mills. Because few Europeans ventured there, the land was not sliced into small parcels, like Madeira. Instead São Tomé was divided into a few dozen big plantations, each with several hundred slaves. From a distance, the plantations looked like tiny cities, with the slave huts clustering like suburbs around the high-timbered "big house" for the plantation manager and his family, many of whom were the mixed-race results of the free-concubine system (the owners themselves remained in Portugal if they could). With their tiny, fever-ridden European populations brutally overseeing thousands of enchained workers, São Tomé and Príncipe were the progenitors of the extractive state.

An onslaught of sugar from big new plantations in Brazil knocked both Madeira and São Tomé out of the sugar market in the 1560s and 1570s. But what happened to the two islands was entirely different. Madeira's lack of malaria and yellow fever had long been noted, though only in the last century did scientists discover the cause: the island does not host the mosquito vectors for the diseases. In the absence of disease, wealthy Europeans, many of them not Portuguese, had moved to the warm island. Around their manors and palaces they erected cathedrals, hospitals, convents, schools, and customs houses—tourist attractions today, valuable investments then. And the farms themselves weren't monocultures, entirely devoted to a single crop, because they had to feed their owners and their owners' neighbors. When the sugar market crashed, sugar squires were reluctant to abandon the homes, fields, and neighborhoods into which they had sunk so much effort. Instead they switched to a newly invented product: the fortified, heat-treated wine today called Madeira.

Wine making, which typically emphasizes quality rather than quantity, is not well suited to plantation slavery. In 1552, the apex of the island's sugar era, three out of ten of its inhabitants were slaves; four decades later, with Brazilian sugar washing across the Atlantic like a white tide, the figure was one out of twenty. By and large, Madeirans freed their slaves; because they weren't working sugarcane anymore, it was cheaper than feeding them. The ex-slaves, having no way off the island, became tenant farmers and sharecroppers for their former masters, who were now building wine presses and cellars. Constantly eyeing famine, the freed slaves survived, like shack people in the Chinese mountains half a world away, mainly on sweet potatoes. But they did survive; Madeira remained a crowded place. At the end of the nineteenth century the island became a tourist destination, touted in guidebooks as a mecca "for those convalescent and requiring rest after dangerous illnesses, malarial fever, etc."

No one has ever advertised São Tomé as a place to rest and recuperate from malaria. Its economy, too, crashed before the onslaught of Brazilian sugar. But São Tomé, unlike Madeira, did not adapt and recover—it simply marched on, though in ever-more-degraded form. Not having neighborhoods to protect, many of the island's offshore landowners contented themselves with watching from afar as their Afro-European managers in their rotting haciendas half-heartedly tried to continue operations by growing food to provision European slave ships. Other planters simply transferred their interests to Brazil, walking away from their property in São Tomé. Some former overseers acquired their own land and bought slaves to tend it. So did some former slaves. By the mid-eighteenth century, São Tomé's colonial masters had been replaced by a new elite of "Creoles" who traced their ancestry (or said they did) to the mixed-race children of the Portuguese and the first emancipated slaves. But the new management changed nothing about the plantations themselves. Even though there was little to sell and few customers, these zombie enterprises struggled on, slaves planting under the lash as the forest overran former sugar fields and colonial buildings crumbled into the harbor.

Resistance was a constant presence. It didn't matter to slaves whether they were owned by Portuguese, Afro-Portuguese, or Africans; they escaped when they could. Runaways joined together to form armed bands in the forest. To guard against their attacks, landowners built wooden forts staffed by gun-toting slaves. Judging by the frequency of successful assaults, the guards were rarely diligent. In a revolt in 1595 as many as five thousand slaves destroyed thirty sugar mills. The destruction was as understandable as it was pointless; the mills were going silent anyway. In a violent stasis, guerrilla warfare between plantations and runaways continued for almost two hundred years.

São Tomé's plantations eventually did switch to other crops: cocoa (from Brazil) and coffee (from the other side of Africa). These became profitable enough to lure back several hundred Portuguese, who dispossessed the Creoles, taking their land and slaves. Cocoa and coffee covered almost every square inch of arable land by the beginning of the twentieth century. Slavery had long been abolished legally, but Portugal kept it going as a practical matter by instituting special taxes in its African colonies. People unable to pay the levies were shipped to São Tomé to work off their debts, de facto slaves locked at night into dilapidated barracks on the plantation. As other nations joined the chocolate industry and improved manufacturing methods, the

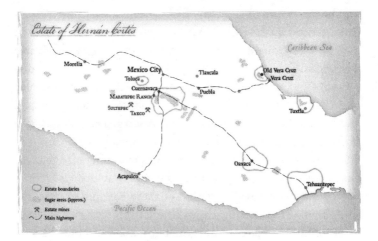

island's antique cocoa plantations became less and less viable. An independence movement sprang up in the 1950s, its primary goal to end the plantation system. When Portugal left in 1975, the country was one of the poorest on earth. The new government nationalized the plantations. It combined them into fifteen super-plantations, then ran them almost exactly as before.

This was the system that crossed the Atlantic to the Americas.

NEW WORLD BORN

Like Juan Garrido, Hernán Cortés died a disappointed man. After subjugating the Triple Alliance, he was awarded a title—Marquis of the Valley of Oaxaca—and given his choice of real estate in the lands he had conquered. He chose six spreads in central and southern Mexico: 7,700 square miles in total, an expanse the size of El Salvador. The biggest chunk, 2,200 square miles of temperate plains south of Mexico City, was where he built his thick-walled, castle-like home. An opulent place, it had no less than twenty-two tapestries, each at least fifteen feet wide; the conqueror, something of a dandy, liked to roam about his tapestries in brocaded velvet jackets and pearl-studded dressing gowns.

Having acquired his property, Cortés threw himself with characteristic energy into a series of entrepreneurial ventures: digging silver mines; establishing cattle ranches and hog farms; panning for gold; opening a shipyard on the Pacific coast; creating a kind of shopping mall in central Mexico City; growing maize, beans, and Garrido's wheat; lending money, goods, livestock, and slaves to entrepreneurs and adventurers in return for a share of the profits; importing silkworms (and mulberry trees to feed them); and raising big stone structures as monuments to himself. Sugarcane, which he began growing in 1523, was high on his list.

Cortés might have succeeded at these enterprises if he had paid attention to them. Instead he kept looking for new kingdoms to vanquish. He marched into Guatemala. He schemed to send ships to Peru. He went to the Pacific and nearly killed himself looking for a route to China. All the while, he flagrantly disobeyed orders. Eventually he ran out of his own money and other people's patience. He returned to Spain in 1540, hoping to obtain more royal favors and positions for himself and his friends. Cortés followed the king from place to place, seeking an audience. Carlos V refused to see him. The heartbroken conquistador was unable to fathom why the sovereign might worry about creating a powerful new aristocracy of unreliable, impulsive men of action. The story, told by Voltaire but surely apocryphal, is that at one point Cortés bulled his way onto the emperor's carriage. Carlos V, annoyed, asked who he was. "It is he," Cortés supposedly said, "who has given you more states than your ancestors left you cities."

His timing was dreadful. As he followed the court, the king was talking with Bartolomé de las Casas, a fiery Dominican priest who had just completed *Brief Account of the Destruction of the Indies,* an indictment of Spanish conduct that remains a landmark both in the history of human-rights activism and in the literature of sustained invective. Reading his first draft before the shocked court, Las Casas branded the conquest of Mexico as "the climax of injustice and violence and tyranny committed against the Indians." He denounced Indian slavery as "torments even harder to endure and longer lasting than the torments of those who are put to the sword." Troubled by Las Casas's lurid descriptions of cruelties committed in the name of Spain, Carlos V had asked his council of advisors to investigate the nation's policies toward Indians.

As the king surely knew, the Spanish monarchy had been struggling to define its Indian policy since before he was born. His grandparents, King Fernando and Queen Isabel, had been stunned when Colón informed them that they now ruled over multitudes of people whose very existence had been previously unsuspected. The monarchs, devout Christians, worried that the conquest could not be justified in the eyes of God. Colón's new lands had the potential of enriching Spain, an outcome they of course viewed as highly desirable. But obtaining the wealth of the

Americas would involve subjugating people who had committed no offense against Spain.

As Fernando and Isabel saw it, Indian lands were not like the Islamic empires whom they and their royal ancestors had fought for centuries. Muslim troops, in their view, could be legitimately enslaved—they had conquered most of Spain, exploited Spanish people, and, by embracing Islam, rejected Christianity. (For similar reasons, the Islamic empires freely enslaved Spanish POWs.) Most Indians, by contrast, had done no wrong to Spaniards. Because American natives had never heard of Christianity, they could not have turned away from it. In 1493, Pope Alexander VI resolved this dilemma of conscience. He awarded the sovereigns "full, free and complete power, authority, and jurisdiction" over the Taino of Hispaniola if they sent "prudent and God-fearing men, learned, skilled, and proven, to instruct [them] in the Catholic faith." Conquest was acceptable if done for the purpose of bringing the conquered to salvation.

The Spaniards who actually went to the new lands, though, had little interest in evangelization. Although often personally pious, they were more concerned with Indian labor than Indian souls. Colón was an example. Despite being fervently, passionately devout, he had appalled Isabel in 1495 by sending 550 captured Taino to Spain to sell as galley slaves. (Galleys were still common on the Mediterranean.) Colón argued that enslaving prisoners of war was justified—he was treating the Indians who had attacked La Isabela as Spaniards had long treated their military enemies. In addition, he said, the Indians' fate would deter further rebellions. Isabel didn't agree. Slowly growing angry, she watched shackled Taino trickle into the slave markets of Seville. In an outburst of fury in 1499 she ordered all Spaniards who had acquired Indians to send them back to the Americas. Death was the penalty for noncompliance.

The queen seems mainly to have been outraged by the presumption of the colonists—they were disobeying instructions and enslaving the wrong people. But she also must have known that the monarchs hadn't addressed the fundamental problem. On the one hand, the pope had justified Spain's conquest because it would allow missionaries to convert the Indians—a goal unlikely to be accomplished if they were enslaved in large numbers. On the other hand, the colonies were supposed to contribute to the glory of Spain, a task that could not be accomplished without acquiring a labor force. Spain, unlike England, did not have a well-developed system of indentured servitude. And unlike England it did not have mobs of unemployed to lure over the ocean. To profit from its colonies, the monarchs believed, Spain would have to rely on Indian labor.

In 1503 the monarchs provided their answer to the dilemma: the *encomienda* system. Individual Spaniards became trustees of indigenous groups, promising to ensure their safety, freedom, and religious instruction. In fine protection-racket style, Indians paid for Spanish "security" with their labor. The *encomienda* can be thought of as an attempt to answer the objections to slavery raised by Adam Smith. By restricting the demands on Indians, the monarchs sought to reduce the incentive for revolt—a benefit to the Spaniards who employed them.

It didn't work. Both Indians and conquistadors disliked the *encomienda* system. Legally, Hispaniola's Indians were free people, their towns and villages still governed by their native leaders. In practice the rulers had little power and workers were often treated as slaves. *Encomenderos* (trustees) loathed negotiating with Taino leaders, which required more tact and delicacy than they typically wished to muster. When native workers didn't feel like showing up—why *would* they, if they could avoid it?—they vanished into the countryside, where their whereabouts were concealed by relatives, friends, and sympathetic Indian leaders. For their part, the Taino came to view the system as little but a legal justification for slavery. Under the law, Indian Christians were entitled after baptism to be treated exactly like Spanish Christians, who could not be enslaved. But colonists argued the contrary; Indians were, in effect, less human than Europeans, and thus could be forced to work even after they converted.

Cortés, conqueror of Mexico, may have had more unfree Indians than anyone else in the world. In addition to owning three thousand or more indigenous slaves outright, his estate forced as many as twenty-four thousand laborers a year to work as tribute (they were sent by their home villages for a week at a time). Indian hands had unwillingly planted thousands of acres of sugarcane on his land and cut wood for the great boilers that crystallized the sugar in his cane juice and constructed his water-driven sugar mill, a two-story edifice made of stone and adobe bricks mortared with sand and lime. Always keenly aware of political currents, Cortés surely would have been following the regal hand-wringing over Indian policy. The council of deputies issued a memorandum in April 1542 begging Carlos V "to remedy the cruelties that are happening to the Indians in the Indies." Seven months later, the king responded: he issued the so-called New Laws, which banned Indian slavery.

The New Laws had big loopholes. Indians still could be enslaved if they were captured while resisting Spanish authority. Because one could always claim that a given person or group was resisting authority, the loophole amounted to a license to enslave. Nonetheless the New Laws so angered the conquistadors that they decapitated the new viceroy of Peru when he tried to enforce them. The viceroy of New Spain (the empire's holdings north of Panama) prudently suspended the laws before they came into effect. Nonetheless, the trend was clear: it was going to be harder for people like Cortés to force Indians to work for them.

A few weeks after the deputies' memorandum, the conqueror cut a deal with two Genoese merchants to bring in five hundred African slaves—the first big contract for Africans on the mainland, and one of the biggest to date. Two years later the initial shipment of a hundred captives arrived at Veracruz, on the Gulf of Mexico. It marked the arrival of the Atlantic slave trade.

Africans had been trickling into the Americas almost as long as Europeans. A U.S.-Mexican archaeological team announced in 2009 that three men in La Isabela's cemetery were probably of African descent (their teeth had the biochemical signatures of a diet rich in African plants). By 1501, seven years after La Isabela's founding, so many Africans had come to Hispaniola that the alarmed Spanish king and queen instructed the island's governor not to allow any more to land. (Also on the no-entry list: Jews and Jews who had converted to Christianity, "heretics" and heretics who had converted to orthodox Christianity.) The instructions made an exception for people of African descent born in Christendom. Slavers claimed their "pieces" were Spanish or Portuguese and sent them over anyway. Within a few months the governor was begging the king and queen to ban all Africans of any sort from Hispaniola. "They flee to the Indians, and they learn bad customs from them, and they cannot be captured." Nobody listened. The colonists saw that Africans appeared immune to disease, didn't have local social networks that would help them escape, and possessed useful skills—many African societies were well known for their ironworking and horsemanship. Slave ships bellied up to the docks of Santo Domingo in ever-greater numbers.

The slaves were not as easily controlled as the colonists had hoped. Exactly as Adam Smith would have predicted, they were dreadful employees. Faking sickness, working with deliberate lassitude, losing supplies, sabotaging equipment, pilfering valuables, maiming the animals that hauled the cane, purposefully ruining the finished sugar—all were part of the furniture of plantation slavery. "Weapons of the weak," political scientist James Scott called them in a classic study of the same name. The slaves were not so weak when they escaped to the heights. Hidden by the forest from European eyes, they made it their business to wreck the industry that had enchained them. For more than a century, African irregulars ranged unhindered over most of Hispaniola, funding their activities by covertly exchanging gold panned from mountain rivers with Spanish merchants for clothing, liquor, and iron (ex-slave blacksmiths made arrow points and swords). Little wonder that the island's sugar producers moved to the mainland! Not only did Mexico have more land and lots of Indian labor, it was not plagued by thousands of anti-sugar guerrillas. (I will further discuss slave rebellions in the next chapter.)

Among the sugar men who relocated was Hernán Cortés, who as a teenage newcomer to Hispaniola had watched the industry rise in the settlement of Azúa de Compostela. Sugar mills were a primary focus at his new estates in Mexico, though his penchant for adventuring delayed their completion for a decade. Other mills at other *encomiendas* came online too, as sugar plantations spread along the Gulf coast, clustering around the warm, wet port of Veracruz.

Between 1550 and 1600, production soared even as the price tripled. Economists would say this phenomenon—rising prices despite increasing supplies—indicates surging demand. They would be correct. Spain's conquest of the Triple Alliance had introduced its citizens to the delights of $C_{12}H_{22}O_{11}$. Like Europeans, the peoples of central Mexico turned out to have an insatiable yen for sweetness. "It is a crazy thing how much sugar and conserves are consumed in the Indies," marveled historian José de Acosta in the 1580s.

No longer were Africans slipped into the Americas by the handful. The rise of sugar production in Mexico and the concurrent rise in Brazil opened the floodgates. Between 1550 and 1650—the century after Cortés's contract, roughly speaking—slave ships ferried across about 650,000 Africans, with the total split more or less equally between Spanish and Portuguese America. (England, France, and other European nations as yet played little role in the slave trade.) In these places, the number of African immigrants outnumbered European immigrants by more than two to one. Everywhere Spaniards and Portuguese went, Africans accompanied them. Soon they were more ubiquitous in the Americas than Europeans, with results the latter never expected.

Africans walked with Spanish conquistadors—some as soldiers, some as servants and slaves—as they assailed Guatemala and Panama. They poured by the thousands into Peru and Ecuador—Francisco Pizarro, conqueror of the Inka, and his family received more than 250 licenses to import slaves in the first years of conquest. On the Rio Grande, Africans assimilated into Indian groups, even participating in attacks on their former masters. Luring them to native life, according to one appalled report, was peyote, "which stirs up the reason in the manner of drunkenness." (Some Spaniards joined the Indians, too.) Juan Valiente, born in Africa, enslaved in Mexico, joined conquistador Pedro de Valdivia's foray into Chile in 1540 as a full partner and was rewarded after its success with an estate and his own Indian slaves. He was in the midst of buying his freedom from his owner in Mexico when he died alongside Valdivia in the native uprising of 1553. African slaves were part of the first European colony in what is now the United States, San Miguel de Gualdape, established by Spain in 1526, probably on the coast of Georgia. First colony, first slaves—San Miguel de Gualdape was also the site of the first slave revolt north of the Rio Grande. The insurrection burned down the colony within a few months of its founding, putting it to an end. It is widely thought that the slaves ran away and made their homes with the local Guale Indians. If so, they were the first long-term residents of what is now the United States from across the Atlantic since the Vikings.

By the seventeenth century, Africans were everywhere in the Spanish world. Six companies in Argentina were sending slaves up to the Andean silver town of Potosí; slightly more than half of the people in Lima, Peru, were African or of African descent; and African slaves were building boats on the Pacific coast of Panama. All the while more Africans were pouring into Cartagena, in what is now Colombia—ten to twelve thousand a year, the Jesuit Josef Fernández claimed in 1633. At the time the city held fewer than two thousand Europeans. Most of those people's livelihood depended on the slave trade. Bribes paid to land Africans illegally were a major source of income. Portuguese Brazil turned to Africans more slowly. Indians were so plentiful there that slaves weren't imported in any number until the end of the sixteenth century and slowly for a few decades after that. The colony's

powerful Jesuit priests were partly responsible for the turn to Africans; enslaving Indians was a sin, they explained, whereas Africans were fair game. (The Jesuits practiced what they preached:
in their sugar mills, Africans alone were in bondage.)

Cortés established what may have been the first cattle ranch in
Mexico. To tend the animals, he did not select native workers—
they had no experience with cows or horses. Africa has been a
center of cattle-herding and horse-riding for thousands of years.
Cortés's first ranch hand, possibly the first cowboy in the mainland
Americas, was an African slave. Thousands of others followed.
In Argentina Africans fled the restrictions of the cities and plantations to the grasslands of the pampas. Driving herds of stolen
cattle with stolen horses, these roaming vagabonds reproduced a
pastoral way of life that was familiar in the West African plains—
"liv[ing] free / and without depending on anyone," as the classic
Argentine poem *Martín Fierro* put it in the 1870s. Later called *gau-
chos,* they became symbols of Argentina in much the same way
that North American cowboys became symbols of the U.S. West.

The paradigmatic example of the African diaspora may be
the man known variously as Esteban, Estevan, Estevanico, or
Estebanico de Dorantes, an Arabic-speaking Muslim/Christian
raised in Azemmour, Morocco. Plagued by drought and civil war
in the sixteenth century, Moroccans fled by the desperate tens
of thousands to the Iberian Peninsula, glumly accepting slavery
and Christianity as the price of survival. Many came from Azemmour, which Portugal, taking advantage of the region's instability,
occupied during Esteban's childhood. He was bought, probably in
Lisbon, by a minor Spanish noble named Andrés Dorantes de Car-
ranza. Dreaming of repeating Cortés's feats of conquest, Dorantes,
with Esteban in tow, joined an overseas expedition led by Pánfilo
de Narváez, a rich, fiercely ambitious Castilian duke with every
quality required of a leader except good judgment and good luck.

More than four hundred men, an unknown number of them
African, landed under Narváez's command in southern Florida on
April 14, 1528. One catastrophe followed another as they moved

up Florida's Gulf coast in search of gold. Narváez vanished at sea; Indians, disease, and starvation picked off most of the rest. After about a year, the survivors built ragtag boats and tried to escape for Hispaniola. They ran aground off the coast of Texas, losing most of their remaining supplies. Of the original four hundred men, just fourteen were still alive. Soon the tally was down to four, one of whom was Esteban. Another was Esteban's owner, Dorantes.

The four men trekked west, toward Mexico, in a passage of stunning hardship. They ate spiders, ant eggs, and prickly pear. They lost all their possessions and walked naked. They were enslaved and tortured and humiliated. As they passed from one Indian realm to the next, they began to be taken for spirit healers—as if native people believed their horrific journey of itself must have brought these strange, naked, bearded people close to the numinous. Perhaps the Indians were right, for Esteban and the Spaniards began curing diseases by chant and the sign of the cross. One of the Spaniards brought back a man from the dead, or said he did. They wore shells on their arms and feathers on their legs and carried flint scalpels. As wandering healers they acquired an entourage of followers, hundreds strong. Grateful patients handed them gifts: bountiful meals, precious stones, six hundred dried deer hearts.

Esteban was the scout and ambassador, the front man who contacted each new culture in turn as they walked thousands of miles across the Southwest, along the Gulf of California and into the mountains of central Mexico. By some measures, Esteban was the leader of the group. He certainly held the Spaniards' lives in his hands every time he encountered a new group and, rattling his shaman's gourd, explained who they were.

Eight years after their departure, the four Narváez survivors entered Mexico City. The three Spaniards were feted and honored. Esteban was re-enslaved and sold. His new owner was Antonio de Mendoza, viceroy of New Spain. Mendoza soon assigned him as the guide to a reconnaissance party going north—Esteban was back on the road. The party was searching for the Seven Cities of Gold. Supposedly these had been established in the eighth century by Portuguese clerics escaping from Muslim invasions. For decades, people from Spain and Portugal had been hunting for them—the Seven Cities were an Iberian version of the Sasquatch or Yeti. Why anyone should imagine these cities were in the U.S. Southwest is unexplained and perhaps unexplainable. Somehow the tales of the Narváez survivors reignited this passion, and Mendoza had succumbed.

Leading the expedition was Marcos de Niza, a Franciscan missionary who has never been charged with insufficient zeal. Mendoza's instructions took pains to command Esteban to obey him. But Esteban had no interest in following orders. As they moved north he encountered Indians who recalled him from his previous journey. He shed his Spanish garb, wore bells, feathers, and chunks of turquoise, and shook a rattle in a spiritual fashion. He again acquired several hundred followers. He ignored Niza's demand that he stop performing ritual cures and refuse his patients' gifts of alcohol and women.

In a decision that the missionary claimed was his own, Esteban and his followers went ahead of the rest of the party after crossing what is now the U.S. border. Quickly they gained a lead of many miles. Once again, Esteban was moving into an area never before seen by someone from across the ocean. Days after the separation, Niza encountered some of Esteban's entourage, wounded and bleeding. In the mountains at the Arizona–New Mexico border, they told him, the group had come across the Zuni town of Hawikuh, a collection of two- and three-story sandstone homes that climbed like white steps up a hill. Its ruler angrily refused entrance. They barricaded Esteban and his cohort into a big hut outside town without food or water. Esteban was slain when he tried to escape Hawikuh the next day, along with most of the people accompanying him.

The Zuni themselves have a different story—*stories,* I should say, because many have been recounted. In one version told to me, Esteban is not refused entry, but welcomed into Hawikuh. The people have heard of this man and his extraordinary journey.

They want to keep him there—want this very badly, at least in the story. He is a man like no other they have encountered, an incredible physical specimen with his skin and hair, a man whose spirit holds a great wealth of knowledge and perhaps more, a valuable possession they have no desire to lose.

To prevent his departure, they cut off his lower legs, lay him gently on his back, and bathe themselves in his supernatural presence. Esteban lives in this way for many years, the story goes, always treated with the respect due to such uncommon figures, always on his back, legs stretched out, with the wrappings on his stumps carefully tended.

All versions of his end are based on stories that people have told to themselves. His actual fate may never be known with certainty. What seems clear is that in the end this man who crossed so many bridges fell into the same delusion that possessed so many Spaniards. He thought that he understood the shook-up world he was creating and that he was in control. He forgot that under bridges is only air.

FAMILY VALUES

Tenochtitlan fell on August 13, 1521, in a welter of massacre and chaos. In the waterways outside the disintegrating city Spanish troops discovered a small flotilla of canoes. Spanish writings say their occupants were hiding in the reeds and found only by determined search. Native accounts say they sought out the invaders to surrender. Historians today tend toward the latter interpretation. In the tumult of the disintegrating city, concealment would have been so easy that it seems likely that the people in the canoes were not even trying to avoid discovery.

In one boat was Cuauhtemoc, last leader of the Triple Alliance; others contained his wife and family. Tenochtitlan rulers, like their European counterparts, had long consolidated power by marrying within a select group of other elite families. As in

Europe, men in authority had children by multiple women. The imperial family tree hence was complicated. It was about to become even more complicated.

Cuauhtemoc, then in his early twenties, was the nephew of Motecuhzoma II, the famous "Montezuma," who had been held hostage by Cortés in his own palace during the Spaniards' first assault on the capital city. Motecuhzoma was killed—exactly how is in dispute—during the counterattack that drove Cortés's force from the city. His successor reigned for barely two months before dying of smallpox. To bolster his legitimacy, the successor had married Motecuhzoma's daughter, Tecuichpotzin, who had been widowed during the first assault. The successor died as the Spanish-Indian alliance began its second assault on Tenochtitlan. Cuauhtemoc, then eighteen, took the throne. He quickly married Tecuichpotzin for the same reason as his predecessor. She was in the canoes with him.

As a captive, Motecuhzoma had asked Cortés to protect his family. This was a big job: the emperor had nineteen children. The conquistador failed—smallpox and war killed all but three of the nineteen. One of the survivors was Tecuichpotzin. (The Spaniards gave her a European name that they could pronounce: Isabel.) Tecuichpotzin was the daughter of the emperor's principal wife, whereas the other two surviving children were from wives of lesser value. All were then adolescents. Tecuichpotzin, twice a widow, was about twelve.

Cortés regarded them as the legitimate rulers of the Triple Alliance, Tecuichpotzin the most important. The conqueror's task, as he saw it, was to graft Spanish authority onto native roots. Europeans would rule through Indian institutions. To do this, he made the straight-faced claim that while held hostage Motecuhzoma had voluntarily given sovereignty over the Alliance to Carlos V. Because Indian elites therefore were now good Spanish subjects, they had to be treated as equivalent to Spanish elites. The two groups would have to mingle on equal terms. Cortés gently nudged this accommodation forward by impregnating Tecuichpotzin.

He didn't do this immediately—she was still married to Cuauhtemoc. Claiming that the Triple Alliance leader was plotting against Spain, Cortés executed him in 1525. He then arranged for Tecuichpotzin to marry her fourth husband, a conquistador he regarded with especial fondness. This man died a few months later. Cortés considerately moved the widow, now sixteen or seventeen, into his own spacious home, which is where she became pregnant, and where he arranged for her fifth marriage, to another favored conquistador. Leonor Cortés Moctezuma was born in 1528, four or five months after the wedding.*

Leonor was not the conqueror's only illegitimate child—he had at least four others. Nor was she his only half-Indian child. Throughout the assault on the Triple Alliance, Cortés traveled with a guide and interpreter: a woman whose name has come down to the present as, variously, Malinche, Marina, or Malintzin. Born to a noble family in a neutral zone between the Triple Alliance and the Maya, she was sold to the Maya after she became an impediment to her stepfather's family. Because Malinche had learned the language of the Triple Alliance as a child, the Maya gave her to Cortés, who was bound in that direction. A sexual relationship began quickly. The conqueror's son Martín came into the world in May or June 1522, which means he was conceived in August or September, in the celebratory aftermath of the empire's fall. (Another half-native daughter, María, is referred to in Cortés's will, but nothing else is known about her except that her mother, too, was one of Motecuhzoma's daughters. One assumes María was conceived during the months when Cortés held Motecuhzoma hostage and that her mother died in the war.)

Cortés did not hide his illegitimate, hybrid children. Leonor was raised by her father's cousin, the administrator of his vast estate. Sugar profits provided a dowry big enough for her to

---

* "Motecuhzoma" is the most common scholarly Romanization of the emperor's name today. At the time, Spaniards usually called him "Moctezuma," which became the name of his grandchildren.

AMERICAN IMPERIAL FAMILIES
OF THE 15TH AND 16TH CENTURY

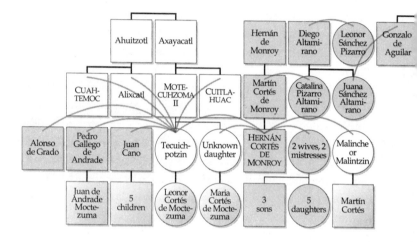

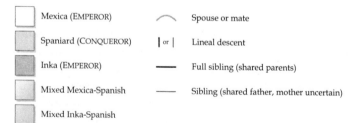

Mexica (EMPEROR)

Spaniard (CONQUEROR)

Inka (EMPEROR)

Mixed Mexica-Spanish

Mixed Inka-Spanish

Spouse or mate

Lineal descent

Full sibling (shared parents)

Sibling (shared father, mother uncertain)

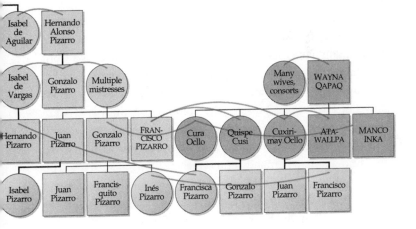

To bolster the legitimacy of their rule, conquistadors often married into or took consorts from the elite of the peoples they conquered, Cortés and Pizarro being among the leading examples. They created a generation of mixed-culture children who became some of the new colonies' most powerful citizens. Because many of the conquistadors were from Extremadura, a mountainous region dominated by a few interrelated families, they were often as tightly related as Indian nobility. The result was a multicultural family web unlike any other.

attract the hand of Juan de Tolosa, discoverer of Mexico's biggest silver mine. Cortés took more dramatic action for Martín: he sent the boy to the Spanish court to serve as a page and hired a Roman lawyer to petition Pope Clement VII to legitimize him. The pope, born as Giulio de' Medici, had every reason to sympathize. Not only was he himself illegitimate, he had his own illegitimate, hybrid child—Alessandro de' Medici, whose mother was a freed African slave—and had tried to ensure his future by appointing him duke of Florence. The pope did indeed legitimize Martín Cortés. Along with Cortés's oldest legitimate son, also named Martín Cortés, he was a principal heir in the conqueror's will. Both were full members of Spanish society—and proved it by spending five years in a court battle over their bequests from their father. Naturally, they fought over Indian slaves.

Europeans and Indians had been mixing since Colón touched down at Hispaniola. Most of the colonists on the island were young, single men; in a census of Hispaniola in 1514, only a third of its *encomenderos* were married. Of these, a third were married to Taino women. Fernando and Isabel encouraged such intercultural coupling, though they believed it should lead to Christian marriage. Christian marriage, perhaps surprisingly, was also the goal of some natives: by marrying their daughters to Spaniards in a Christian ceremony, elite Indians could reinforce their status. For many Spaniards, though, a Taino ceremony was more useful than a Christian wedding—only through marrying a native woman could a low-ranking Spaniard gain access to the goods and workers controlled by high-status Indians. As a result, many of the Spaniards whom the clergy viewed as living in sin thought of themselves as married.

A hybrid society was coming into existence, first in the Caribbean, then everywhere else in the Americas. The mixing began at the top—Cortés was an example. Like many members of the first generation of conquistadors, Cortés came from Extremadura, a poor, mountainous area controlled by powerful families who had been marrying into each other for generations. His distant cousin was Francisco Pizarro, conqueror of the Inka empire—Pizarro's great-uncle was married to Cortés's aunt. When the intertwined conquistador families married into the equally intertwined families of noble native societies, they produced the kind of baroque, multibranched family trees that wake up genealogists at 3:00 a.m.—Cortés's relations with the Mexica (Tenochtitlan's people) were prototypical.

Cortés was only the beginning. Like his Extremaduran cousin, Pizarro set up shop with a noble native woman: Quispe Cusi, the sister or half sister of Atawallpa, the Inka emperor whom Pizarro overthrew. Quispe Cusi bore Pizarro two children, Francisca and Gonzalo, whom he asked the king to legitimize by royal decree. Pizarro often said that Quispe Cusi was his wife, but he didn't actually marry her. Nor did he let this "marriage" interfere with his liaisons with two other royal Inka sisters, one of whom bore him another two children. An illegitimate child himself, Pizarro did not turn his back on his half-Inka offspring. Francisca, his daughter by Quispe Cusi, became his principal heir. (Her brother Gonzalo died at the age of nine.)

The conqueror came to Peru with three brothers. One took an Inka princess as a mistress. Another took an actual Inka queen—he stole the wife of the puppet emperor whom Francisco Pizarro had installed after killing Atawallpa. The remaining Pizarro brother, Hernando, was the only one to return to Spain alive. The wary Carlos V put him under house arrest—Hernando, after all, had a history of impulsively overthrowing kings. Besides, he had murdered a lot of Spaniards in battles over the spoils of Peru. When the king abdicated, his successor, Felipe (Philip) II, continued the imprisonment. Altogether Hernando was confined for twenty-one years. "His confinement was gentle enough," John Hemming observed in *The Conquest of the Incas* (1970), his marvelous account of the Pizarro brothers' assault on Peru. "He was in the same prison and apartments that had harbored [French] King Francis I after his capture [in a battle with Spain] in 1525." Rising at noon, Hernando ate and drank lavishly in his sumptuous quarters, then entertained Spain's elite far into the night. He had a mistress who bore him a daughter in prison.

Hernando met Francisca for the first time since infancy when she was seventeen and had just inherited her father's vast fortune. The fifty-year-old Pizarro married her almost on the spot, Hemming wrote, "unperturbed by consanguinity, the thirty-three-year difference in age, or his own imprisonment." When Hernando was at last released from house arrest, the couple built a massive Renaissance-style palace on the main plaza of Trujillo, the city where the Pizarros had been born. In a kind of colonial fantasy, they dined on gold plates with Peruvian food and imported a squadron of Inka servants to wait on them.

The Pizarros were wealthier than their fellow conquistadors but in other ways not exceptional. Historians have tracked the lives of ninety-seven of the 150 men who founded Santiago, Chile, in 1541. They had 392 children and grandchildren, of whom 226 (57 percent) were of Indian descent. One conquistador in Chile proudly told the Inquisition in 1569 he had produced fifty children with non-European mothers.*

---

* Such mingled relationships were not restricted to Spanish and Portuguese America. As time passed, the Princeton historian Linda Colley has written, Britain "evolved a more hybrid construction of its empire" as a balance of different, rapidly mixing groups. The conception was embraced by some early U.S. leaders, including President Thomas Jefferson, who argued that Europeans and Indians should "meet and blend together, to intermix, and become one people." A classic example of this mixing was Sam Houston, first president of Texas and later its governor, who ran away from his childhood home and was adopted by a Cherokee family. He returned to the society of his birth and launched a violent, alcohol-fueled political career. At thirty-six, his marriage having ended, he returned to the Cherokee, married a half-Cherokee woman, became the Cherokee ambassador to Washington, and took to wearing native garb. Angered by his constant drinking, the Cherokee ejected him from his job and threw him out of the group. Houston became president of Texas after it seceded from Mexico. In office, he tried to forge an alliance with local Cherokees to invade northern Mexico and create a bicultural state. Jefferson, too, helped create a mixed society. As demonstrated by DNA tests in 1998, he was the likely father of one or more children by his part-African slave, Sally Hemings, who may have been his wife's half sister. Jefferson freed all six

Few of those children had African blood. That would change—rapidly. As plantation slavery spread, the percentage of Africans in the hemisphere rose, and with it the number of Afro-Indians, Afro-Europeans, and Afro-Euro-Indians. By 1570 there were three times as many Africans as Europeans in Mexico and twice as many people of mixed parentage. (Both were outnumbered by Indians, of course.) Seventy years later there were still three times as many Africans as Europeans—and *twenty-eight* times as many mixed people, most of them free Afro-Europeans.

On the one hand, Spaniards in some ways easily accepted the hybrid world they were making. Europeans then did not have the same concept of "race" as later generations and thus did not see themselves as being different from Africans or Indians on a biological level. They did not fear what today would be called genetic contamination. On the other hand, the blending of native and newcomer led to enormous fear about *moral* contamination.

Spain justified its conquest, one recalls, by promising to convert the Indians. Spaniards' consistent mistreatment of native people impeded this mission. The Franciscan friars who held sway over New Spain's religious life proposed an apartheid solution: turning the colony into two "republics," one for Indians, one for Europeans. Untroubled by European demands, Indians would focus on conversion in all-Indian neighborhoods and towns; Spaniards would focus on growing rich from the fruits of conquest in an all-Spanish setting. Thus in 1538 Bishop Vasco de Quiroga began to group thirty thousand Indians in the mountains west of Mexico City into reservation towns that he intended to make into an American utopia—literally, for Quiroga laid out the settlements according to the prescription in Thomas More's novel *Utopia*, which had been published twenty-one years before.

Exemplifying the pitfalls of the two-republic scheme was the Franciscans' near-simultaneous founding of an all-European town,

---

of Hemings's children—the only slaves he emancipated—and three went on as adults to live as "whites."

The cultural and ethnic jumble in the streets of Spain's American colonies was often reflected in its art—as in this anonymous eighteenth-century oil, which depicts the Virgin Mary embedded in the great silver mountain of Potosí, visually uniting Christianity with the Andean tradition that mountains are the embodiment of a deity.

Puebla de los Ángeles, on the road between Mexico City and Veracruz. The goal was to solve a problem: Spanish lowlifes were living as parasites in native villages, their constant demands for food, shelter, and women interfering with the crucial work of orderly conversion. The Franciscans' solution: forcibly collecting the vagabonds and planting them in a city of their own under church supervision. Half of Puebla's first inhabitants abandoned the project when they discovered that they were not going to be presented with their own personal contingent of Indian laborers. To build the city, its architects ended up relying on *encomienda* labor. Spaniards kept leaving, and the clerics had to sweeten the pot. Ultimately, each Pueblan household received the services of forty to fifty Indian workers every week. The city created to protect Indians from Spanish calls for forced Indian labor thus was wholly based on forced Indian labor. And Indians and Spaniards were again completely intermingled. Even when the authorities were able to keep them separate, free Africans acted as arbitrageurs, taking advantage of price differences between native and Spanish neighborhoods to buy goods in one and sell them in the other.

The inexorably rising number of people with mixed descent made a mockery of the two republics—what group did they belong to? Mexican churches kept separate baptismal, marriage, and death registers for Indians and Spaniards. Did they have to begin a third set of records? Worse, the growing number of mixed people aroused fears for the purity of the colonists' blood.

At the time many Spaniards believed that parents passed on their ideas and moral characters to their children, with the effect amplified by the atmosphere of the home. A mother who was born Jewish or Muslim somehow would instill the essence of Judaism or Islam in her offspring, even if she never exposed them to the religion. If the children lived in a family with Jewish or Muslim customs like not eating pork or frequent bathing, the inner stain would be darker and more ineradicable. Conversely, the stain was reduced, though not eliminated, if the child had a Christian parent and ate Christian food and learned Christian habits. In this view, Africans were to be feared not because of their African genes, but because their ancestors had embraced the immoral heresy of Islam, which would lodge in their descendants' hearts.

Initially Indians were not seen as dangerous in this way. Because the Gospel had never come to the Americas before Colón, their ancestors had never rejected the Savior. Their heathen beliefs were mistakes born of ignorance, not of evil. As innocents, they could not pass the brand of heresy to their children. Over time, though, it became clear that many Indians were resisting full evangelization, and they became suspect as a class. Meanwhile, the number of Africans and mixed people rose inexorably. Surrounded in the seventeenth century by an ever-larger population of untrustworthy groups, the elites who had embraced mixed unions in the sixteenth century felt themselves losing control. With this loss of control went their previous tolerance for the populace's freewheeling ways.

Views on race are a complicated subject that scholars have devoted careers to elucidating. The issue has a charged history that prompts suspicion and defensiveness. As one might imagine, there is considerable dispute. My brief discussion above is an attempt to summarize part of what seems to me a persuasive analysis by María Elena Martínez, a historian at the University of Southern California. Some scholars surely would roll their eyes at her views, or at least my truncated version of them. But few doubt that as colonial society grew more diverse, colonial authorities tried to put the genie back into the bottle.

In the second half of the sixteenth century, Spanish governments began restricting mixed people, forbidding them from carrying weapons, becoming priests, practicing prestigious trades (silk making, glove making, needle making), and serving in most governmental positions. A Spanish butcher who used fraudulent weights on scales to cheat customers was to be fined twenty pesos. A butcher with Indian blood who did the same thing was to receive a hundred lashes. Men and women with African blood could not be seen in public after 8:00 p.m. or congregate in groups of more than four. In addition, they had to pay special assessments every year—a sort of original-sin tax. Indo-European women were not allowed to wear Indian clothes. Afro-European women were not allowed to wear Spanish-style gold jewelry or the elegant embroidered cloaks called *mantas*. And so on—scores of petty regulations, issued in uncoordinated bouts of malice and anxiety, a quibbling, bureaucratic assault by Spain against its unruly offspring.

As the restrictions increased, so did the fear of the restricted, which led to more restrictions and more fear. Clerics took to arguing that Indians were not innocent—that, like Jews, they bore the stain of their previous un-Christian beliefs. Maybe they actually *had* descended from Jews—the lost tribes of Israel! Maybe some of them, like some of the ex-Jews in the Iberian Peninsula, hadn't *really* converted to Christianity. Maybe they were jointly plotting with Africans to attack Christians. New Spain, the Augustinian friar Nicolás de Witte stated in 1552,

> is full of mestizos, who are [born] badly inclined. It is full of black men and women who are descended from slaves. It is full of black men who marry Indian women, from which derive mulattoes. And it is full of mestizos who marry Indian women, from which derive a diverse casta [caste] of infinite number, and from all of these mixtures derive other diverse and not very good mixtures.

"Mestizo" and "mulatto" later became key parts of the elaborate classificatory scheme known as the *casta* system. Never formally codified on an empire-wide level but recognized in hundreds of separate local, ecclesiastical, and trade-guild rules, the *casta* system was an attempt to categorize the peoples of New Spain according to moral and spiritual worth, which was linked to descent. Each group had a fundamental, unalterable nature that combined in distinct, predictable ways with people outside that group. A mulatto (Afro-European) was different from a mestizo (Indo-European) was different from a zambo (Afro-Indian—the term comes, unflatteringly, from *zambaigo,* knock-kneed). When a Spaniard produced a child with a mestizo, the offspring was a *castizo;* with a mulatto, a morisco (the name, oddly, means "Moor"). Over time the classifications grew more baroque, refined, and absurd: coyote, *lobo* (wolf), albino, *cambujo* (swarthy), *albarazado* (white-spotted), *barcino* (the opposite—

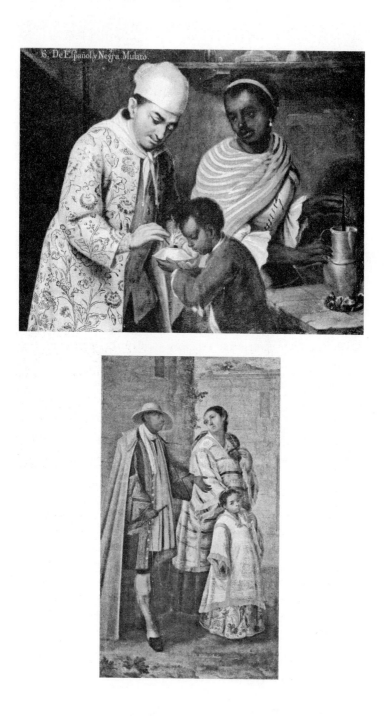

S. De Español, y Negra, Mulato.

Casta paintings, bizarre versions of the natural-history paintings then becoming popular in Europe, were intended to instruct outsiders about the cultural mixes in Spain's American colonies. Laying out a complex racial scheme that parceled out European, Indian, and African ancestry, they were pictorial anthropological treatises, the racial types neatly labeled. Sometimes they warned of the dire consequences of certain ethnic combinations: spouses murdering each other, children who don't resemble their parents. Clockwise from lower left: *Negro and Indian Makes Wolf* (José de Ibarra, ca. 1725); *Spaniard and Negro Makes Mulato* (attrib. José de Alcibar, 1760–70); *Chamizo and Indian Makes Cambuja* (Unknown, ca. 1780); *Spaniard and Albino Makes Return-Backwards* (Ramón Torres, 1770–80).

color-spotted, so to speak), *tente en el aire* (suspended in air), *no te entiendo* (I don't understand you).*

None of it worked quite as the government intended. Rather than being confined to their allocated social slots, people used the categories as tools to better their condition, shopping for the identity that most suited them. The half-Indian son of the conquistador Diego Muñoz married a native noblewoman; his son, who would theoretically be classed a a coyote, was declared an Indian, and this grandson of a Spaniard became the "Indian governor" in Tlaxcala, east of Mexico City. Meanwhile, other Indians claimed to be Africans—slaves paid fewer taxes, and the Indians didn't see why they should pay them, either. Local officials were supposed to police the categories; strapped for cash, they were in fact ready to sell people whatever identity they wanted to assume. When Spaniards in the Caribbean died before producing legitimate offspring, their mestizo and mulatto offspring were promoted to "Spaniards" and pressed into duty as heirs—a transformation that occurred so often that the bishop of Puerto Rico sniffed in 1738 that the islands had "very few white families without mixture of all the bad races." Later that century a traveler sardonically noted that although "many whites are listed" in Hispaniola's official census, local parish registers listed the same people as "mixtures of whites and Indians and these with zambos, mulattos, and blacks."

The New Laws that banned indigenous slavery added to the ethnic jumble. Because the Spanish legal code known as the Siete Partidas declared that children inherited the status of their mother, the offspring of European and Indian women had to be free, at least in theory. In consequence, African men sought out

---

* Spaniards weren't alone in this preoccupation. The eighteenth-century French polymath Louis-Élie Moreau de Saint-Méry tried to split Haiti's jumbled population into 128 minutely differentiated groups ("the twelve combinations of Mulatto range from 56 to 70 parts white").

non-African women (in any case, the colonies didn't have enough African women for them—three-quarters of the slaves were male). Madrid demanded that Africans only marry Africans, but the colony's powerful clergy pushed slaves in illicit relationships to get church marriages—it was a way of bringing pagan Africans into the Christian fold. Half or more of all Africans ended up with non-African spouses. Colonial authorities proclaimed that the Siete Partidas didn't apply and tried to enslave Afro-Indian and Afro-European children anyway. In an act of collective defiance, many of them simply moved away and with their relatively light complexions told their new neighbors that they were Indians or Spaniards.

Human beliefs and actions about ethnic and racial differences rarely withstand logical scrutiny, and Mexico was no exception. From a geneticist's point of view, the population blended increasingly over time. By the end of the eighteenth century "pure" Africans were disappearing, disease and intermarriage were reducing the number of "pure" Indians at a tremendous rate, and even the remaining "pure" Spaniards—a tiny group that in Mexico City consisted of less than 5 percent of the populace—were marrying outside their category at such a rate that they soon would no longer exist as a separate entity. Yet as it became ever more difficult to distinguish one individual from another, colonial authorities tried ever harder to separate them, a peculiar dynamic perhaps best exemplified in one of the world's odder artistic genres: *casta* paintings.

Casta paintings are sets of images, usually but not always sixteen in number, that purport to depict the categories of people in New Spain. Painted or engraved in the colony itself, they portray the mestizos, mulattos, coyotes, *lobos,* and *tente en el aires* of Spanish America with the frozen exactitude of Audubon birds. Several sets, in fact, were displayed at Madrid's natural history museum, the strains of *Homo sapiens* in Spain's American colonies side by side with exhibits of fossils and exotic plants. Almost all the paint-

ings present the viewer with a family group: a man of one category, a woman of another, and their offspring. Gilt labels, painted
directly on the canvas, act as explanatory captions:

From Black Man and Indian Woman, Wolf
From Spanish Man and Moorish Woman, Albino
From Mulatto Man and Mestiza Woman, Wolf-Suspended-in-Air
From Indian Woman and Male Wolf-Return-Backwards, Wolf Again

More than a hundred sets of *casta* paintings are known. Many
are beautifully crafted. Some were painted by mixed people
themselves.

Looking at these images today, it is hard to imagine what
their creators were thinking at the time. They must have known
that Europeans were fascinated and repulsed by New Spain's
exotic inhabitants. The portraits were intended to parade their
fellows like specimens in a zoo. Yet at the same time most show
the *castizos,* mestizos, and mulattos dressed sumptuously, moving happily about their daily business, tall and robustly healthy
each and every one. Looking at the smooth, smiling faces now,
one would never know that on the streets of the cities where they
were painted these people were scorned for their very diversity.
One would also never know that the *casta* paintings were not
diverse *enough*—not a single one portrayed New Spain's Asian
population, by far the biggest outside Asia.

SHOOK-UP CITY

In January 1688 crowds of the faithful forced their way into the
chapel of Holy Innocents in the Jesuit church of the Holy Spirit,
in Puebla de los Ángeles. Resting inside was the body of Catarina de San Juan, a renowned local holy woman who had died in
her eighties. Officers at the city cathedral and the local religious
orders had taken turns carrying her elaborately carved coffin

into the chapel, where it rested on a bier bedecked with art and manuscript poetry. In an ecstasy of faith, worshippers tore at the shroud covering the body, trying to cut off fingers, ears, or other gobbets of flesh as relics. To protect Catarina's corpse from her fans, ecclesiastical authorities emplaced a team of armed soldiers.

Leading members of the city council and Puebla's religious establishment attended the interment, then walked to the cathedral for a memorial mass. The sermon was given by the Jesuit Francisco de Aguilera, who recounted Catarina's life in elaborate, fanciful detail. Although Catarina spent most of her day praying, Aguilera told the assembled dignitaries, she was in fact voyaging spiritually across the planet. Indeed, she was responsible for Christian victories over Muslim armadas in the Mediterranean. Later supporters would learn that she had joined the Virgin Mary to save the Spanish treasure fleet from a demonic hurricane; helped Spanish ships beat back English and French pirates; flown over Japan and China to spread Christianity; and personally witnessed the martyrdom of Franciscan missionaries in New Mexico.

These feats were unusual in number but not in type for people who went on to become saints in that period. Not unusual, too, were the hagiographic biographies written after her death by churchmen who knew Catarina, though it was remarkable for three to appear, one of them almost a thousand pages long. What was peculiar was Aguilera's claim about her birth: Catarina de San Juan, an obscure visionary in the mountains of Mexico, was the granddaughter of an Asian emperor. More peculiar still, this claim was probably accurate, or mostly so.

Named Mirra at birth, she was born around 1605 into an aristocratic family in a city in the Mughal empire, probably Lahore, in modern Pakistan, or Agra, later famous for the Taj Mahal. The Mughal empire was a Muslim dynasty, and Mirra's family, which seems to have been distantly related to the imperial family, was Muslim as well. Mirra/Catarina's biographers claim that she lived in a palace beside a river with the rest of the emperor's extended family and that her family had Christian sympathies. The latter

claim is not preposterous. Akbar, emperor at the time, was celebrated for tolerance; Jesuits, welcomed at his court, converted
some high-ranking courtiers to Christ. Images of Christian saints
were common in courtly gardens, statuary, and tombs—they
were taken as symbols of Akbar's divinely guided reign.

Everything changed when Mirra was seven. Portuguese
pirates seized a ship of Mughal pilgrims on their way to Mecca.
Interpreting the attack as a deliberate religious insult, Akbar
booted out the Jesuits and turned to persecuting Christians. Mirra's parents were implicated in the crackdown and moved to the
coast—possibly Surat, on the Arabian Sea, which had a big European community. Surat also had, alas, a big piracy problem. As
recounted by one of Mirra's biographers, who claimed to have
heard the story from her own lips, pirates disguised as Portuguese merchants abducted her on the beach and transported her
to Kochi (Cochin), near the southern tip of India. Jesuits there
baptized her. Christians were not supposed to be enslaved by
other Christians, but the pirates took the young girl back from
the Jesuits. On the seas she was repeatedly degraded before being
deposited in Manila, where she was acquired by a ship captain
from Puebla.

In Mexico the girl now known as Catarina became ever
more fervent and ascetic in her beliefs, retreating to a cell, drinking little and eating less, twisting straps with sharp metal studs
around her limbs, and rejecting any hint of sexual contact—she
once told the naked Christ, whom she saw in a vision, to put his
clothes on. Shuttered in a bare, tiny room, she battled the assault
of devils on a nightly basis, arming herself with holy water, reli-
quaries, and the cross. Visions possessed her, according to her
most determined chronicler, Alonso Ramos, a Jesuit whom
Catarina had selected as her confessor. She saw the commu-
nion host transfigured into a star that shot magical rays into her
mouth, Ramos said. She saw the soul of the Queen of Heaven
rise in a splash of brilliance and fire, twelve lights upon her head

like a diadem. She saw the vaults of the church explode and the roof crack open to reveal a floating, magical table covered with flowers and sparkling gold and a great feast attended by the Savior himself. She saw a staircase made of "delicate and shimmering clouds" with souls climbing it to heaven and her own prayers turning into angels and flowers raining down on everything and everyone.

Ramos narrated these events in three jumbo-sized volumes released in 1689, 1690, and 1692—the longest work ever published in New Spain. Four years later the Inquisition condemned all three as "useless, improbable, full of contradictions and . . . rash doctrines." Ramos was removed from his position as rector of the Puebla Jesuit college and confined to a cell. Already an alcoholic, he seems to have gone mad in captivity. He escaped, tried to murder his successor as rector, and died a forgotten man.

Catarina de San Juan, too, was almost forgotten. Forgotten as well were the Asians who preceded and followed her to the Americas—fifty to a hundred thousand of them, according to Edward R. Slack, a historian at Eastern Washington University. They came via the galleon trade: sailors, servants, and slaves disembarking in Acapulco and scattering across New Spain. By the early seventeenth century, Asians—Filipinos, Fujianese, and Filipino-Fujianese—were building Spanish ships in Manila Bay. When Spaniards proved reluctant to make the long and arduous trip across the ocean, Asians took their place. Some may have shipped to Mexico as early as 1565, when Urdaneta made the first successful crossing of the Pacific from west to east. (On that voyage, Legazpi sent Asian slaves to his hacienda in Coyuca, northwest of Acapulco.) Slack estimates that 60 to 80 percent of the crew on the great ships and their accompanying vessels were Asian. Many never went back to Manila. One example is the seventy-five Asian sailors known to have landed in Acapulco in 1618 on the galleon *Espíritu Sancto*. Only five were aboard for the return trip. Over the decades thousands of sailors jumped ship in the Americas, taking jobs in the city's shipyards or building forts and other public works.*

Sometimes Asian sailors worked side by side with Asian slaves like Catarina de San Juan, who trickled in despite the disapproval of the colonial government. They came from India, Malaysia, Burma, and Sri Lanka to Manila, transported by Portuguese slavers; Chinese junks brought others from Vietnam and Borneo. From Manila they were shipped in the great galleons with the silk and porcelain. In 1672 Manila banned Asian slavery. The ban was rarely effective. Almost a century later, the municipal council of Veracruz forced a company of Jesuits from Manila to get rid of the twenty Asian servants whom they were taking to Madrid. They were too much like slaves.

Known collectively as *chinos,* Asian migrants spread slowly along the silver highway from Acapulco to Mexico City, Puebla, and Veracruz. Indeed, the road was patrolled by them—Japanese samurai perhaps in particular. Katana-swinging Japanese had helped suppress Chinese rebellions in Manila in 1603 and 1609. When Japan closed its borders to foreigners in the 1630s, Japanese expatriates were stranded wherever they were. Scores, perhaps hundreds, migrated to Mexico. Initially the viceroy had forbidden mestizos, mullatos, negros, *zambaigos,* and *chinos* to carry weapons. The Spaniards made an exception for samurai, allowing them to wield their katanas and tantos to protect the silver shipments against the escaped-slaves-turned-highwaymen in the hills. The results were so encouraging that the authorities reversed course and drafted mixed-race people into the militias. By the eighteenth century Afro-Indo-Asian paramilitary units on Mexico's Pacific coast were protecting mail deliveries, patrolling for bandits, and repelling attacks by British ships. Acapulco, terminus of the silver

---

* Not all went to Mexico. A census of Lima, Peru, in 1613 found 114 Asians living there, almost half of them women. Presumably the actual tally was bigger, because Asians would have tried to avoid the census takers. Many were "ruff openers" (*abridores de cuellos*), fixing the mechanisms on the stiff ruffs wealthy men then wore about their necks.

Carried across the Pacific from Manila by the galleon trade, the Chinese artist Esteban Sampzon became one of Buenos Aires's leading sculptors at the end of the eighteenth century. The sensitively rendered features of his *Christ of Humility and Patience* (ca. 1790) still adorn the city's Basílica de Nuestra Señora de la Merced.

trade, was guarded by a force of *morenos, pardos,* Spaniards, and *chinos,* the latter mostly Filipinos and Fujianese. When the British admiral/pirate George Anson invaded western Mexico in 1741, the multicultural force played a major role in his defeat.

Puebla was bigger than Acapulco and had a more tight-knit Asian community. Indeed, Catarina's owner found another Asian slave there for her to marry. (The marriage did not take. It may have been doomed from the wedding night, when Catarina told her new spouse that St. Peter and St. Paul had appeared at the bedside to deny him from exercising his conjugal rights.) One of the city's most important industries was ceramics—Puebla clay is of exceptional quality. Working with eye-straining attention to detail, skilled potters created pieces that imitated blue-and-white Ming dynasty porcelain. Guild regulations specified that "the coloring should be in imitation of Chinese ware, very blue, finished in the same style." Edward Slack, the Eastern Washington historian, points out that the manufacturers would hardly have ignored the skilled Asian craftspeople in their midst. More than likely, Puebla's fake Chinese pottery was created in part by real Chinese potters. If so, they did a splendid job: talavera ware, as it is known today, is now so highly prized that when I visited Puebla shopkeepers complained that the country was fighting an invasion of counterfeits from China—a Chinese imitation of a Chinese-made Mexican imitation of a Chinese original.

Larger still was the Asian community in Mexico City. The first real Chinatown in the Americas, it was centered around an outdoor Asian marketplace under a tent-like roof in the Plaza Mayor, the city's grand central square, built atop the city center of old Tenochtitlan. The marketplace was called the Parián, after the Asian ghetto in Manila. In a cacophony of languages, Chinese tailors, cobblers, butchers, embroiderers, musicians, and scribes competed with African, Indian, and Spanish shopkeepers for business. Alarming to colonial authorities, Chinese goldsmiths drove European goldsmiths out of business—"the people of China that have been made Christians and every year come thither, have perfected the Spaniards at that trade," a Dominican friar lamented in the 1620s.

Spanish goldsmiths evidently took the loss of business calmly. Spanish barbers did not. In those days a barber was both a hair and beard trimmer and a low-ranking medical provider who performed dental surgery. About two hundred *chino* barbers set up shop in the Plaza Mayor, treating maladies with a combination of Eastern and Western techniques: cauterization and acupuncture, bloodletting and Chinese herbal medicine. Wealthy women flocked to their kiosks. It was not just a New Age fad—Chinese dentistry was then the most sophisticated in the world. In the Tang dynasty the savants of Beijing had realized that periodontal disease could be prevented by scraping away dental plaque. They treated the bleeding with pastes made with roots and herbs that recent research has shown to have antibacterial and anti-inflammatory properties.

In 1635 the city's Spanish barbers petitioned the municipal council to stop the *chinos'* "excesses" and "inconveniences." The complaint was artfully worded, but one detects the real cause of grievance: the Chinese were willing to pay higher rents for space in the center of town, even at the risk of lowering their profits, because that brought them closer to their customers. And they spent long hours on the job, forcing European barbers to work equally hard to compete. To Spaniards, the solution was obvious: expel the Chinese from the city center and restrict hair-cutting hours so that they wouldn't have to work so hard and accept such low profits. Six months later the viceroy banned Asian barbers from the Plaza Mayor. Twisting the knife, he restricted the number of razors they could possess, thus ensuring that their shops couldn't grow too large.

Despite the ban, the government kept approving applications for *chino* barbershops in the Playa Mayor—perhaps, one is tempted to speculate, because influential customers didn't want to have to travel long distances to have their hair cut and their teeth cleaned. European businesses again complained about the competition. In 1650 the government created a barbershop czar, empowered to extract hefty fines from bootleg hair salons. The post was ineffective: Chinese barbers proliferated by the score. An especially zealous Spanish barber won the czarship in 1670. Slack, whose account I am following here, found no indication of success.

The city's raucous mix of peoples was nowhere better expressed than its festivals, such as the Easter processions. Orga-

nized by the lay religious groups called confraternities, they were
ostensibly intended as public acts of penitence but functioned
as ethnically based civic celebrations. Asians helped found the
Confraternity of the Holy Christ in the mid-sixteenth century;
aligned with the Franciscans, its members were allowed to construct a chapel in the monastery and decorate it with imported
ivory gewgaws. The Italian traveler Giovanni Francesco Gemelli
Careri watched them march in an Easter parade in Mexico City in
April 1697. Carrying statues and torches, three costumed confraternities went out from city hall that day: the brotherhood of the
Holy Trinity, the Jesuits of the Church of San Gregorio, and the
Franciscans. The march of the Franciscans, Gemelli Careri noted,
was called "the Procession of the Chinese," because the march-
ers were all from the Philippines. Each procession, he wrote, was
walked with

> a company of soldiers . . . on horseback, and was preceded
> by mournful horn-players. When the procession came to
> the royal palace, the Chinese and the [Franciscans] fought
> to be at the head of the line; they beat each other over the
> shoulders with clubs, and with their Crosses; and many
> were wounded.

The big Chinese population reflected the city's status as
the clearinghouse for information about the East. In 1585 Juan
González de Mendoza, a Dominican there, compiled sources
from the galleon trade into a *History of the Most Notable Things,
Rituals and Customs of the Great Kingdom of China*. Published in
dozens of editions in many languages, it became the standard text
on China for educated Europeans. Not only did the China trade
fascinate Mexico City's civil government, it preoccupied many of
the clerics in the city's many churches, who begged their superiors for the chance to get on a galleon and save Chinese souls.
Much of their fascination was fueled by a miscalculation—they
believed Mexico to be much closer to China than it actually is.

(In fact, as the Canadian historian Luke Clossey has pointed out, Beijing is closer to Rome than to Mexico City.) The Dominican Martín de Valencia spent months on Mexico's west coast waiting for Cortés's ships to take him to China on the conqueror's failed expedition to the Pacific. The ships never appeared. Lying on his deathbed in Mexico City, Valencia said, "I have been cheated of my desire."

Scuffling in the streets, struggling to pull strings in the government, uneasily cooperating in the military, Mexico City's multitude of poorly defined ethnic groups from Africa, Asia, Europe, and the Americas made it the world's first truly global city—the Homogenocene for *Homo sapiens*. A showpiece for the human branch of the Columbian Exchange, it was the place where East met West under an African and Indian gaze. Its inhabitants were ashamed of the genetic mix even as they were proud of their cosmopolitan culture, perhaps none more so than the poet Bernardo de Balbuena, whose *Grandeza Mexicana* is a two-hundred-page love letter to his adopted home. "In thee," he wrote, addressing Mexico City,

> *Spain is joined with China,*
> *Italy with Japan, and finally*
> *an entire world in trade and order.*
> *In thee, we enjoy the best of the treasures*
> *of the West; in thee, the cream*
> *of all luster created in the East.*

Balbuena wrote his panegyric while the city he extolled was under water. Cortés's siege wrecked the intricate network of dikes and baffles that kept the island from flooding every spring; now the city was inundated for months at a time. (Repairing the damage took almost four centuries and in some ways left the city worse off than before.) Balbuena seemed not to mind. It was evidently worth the inconvenience of wading through the flood to live in an urban dream of chanting religious processions, swishing silk dresses, groaning carriages of silver and gold, and great clanging church bells, a city where people drifted in canoes down canals lined with flowers as sunlight gleamed from the mountains. But it was both more and less than that. Menaced by environmental problems, torn by struggles between the tiny coterie of wealthy Spaniards at the center and a teeming, fractious polyglot periphery, battered by a corrupt and inept civic and religious establishment, troubled by a past that it barely understood—to the contemporary eye, sixteenth- and seventeenth-century Mexico City looks oddly familiar. In its dystopic way, it was an amazingly contemporary place, unlike any other then on the planet. It was the first twenty-first-century city, the first of today's modern, globalized megalopolises.

It may seem foolish to use terms like *modern* and *globalized* to describe a time and place in which there were no means of mass communication and most people had no way of buying goods or services from overseas. But even today billions of people on our networked planet have no telephones. Even today the reach of goods and services from high-tech places like the United States, Europe, and Japan is limited. Modernity is a patchy thing, a matter of shifting light and dark upon the globe. Here was one of the spots where it touched first.

Forest of Fugitives

IN CALABAR

Christian de Jesus Santana could see the secret city from his window. Known as Calabar, it was at the edge of Salvador da Bahia, in northeast Brazil, on the inland side of a ridge that paralleled the coastline. On the shore side of the ridge, invisible from Calabar, was the great Bay of All Saints, the second-biggest slave harbor in the world, the first glimpse of the Americas for more than 1.5 million captive Africans. The slaves were supposed to spend the rest of their days in Brazil's sugar plantations and mills. Most did, but countless thousands escaped their bondage, and many of these established fugitive communities—*quilombos*, as Brazilians called them—in the nation's forests. Almost always they were joined by Indians, who were also targeted by European slavers. Protected by steep terrain, thickly packed trees, treacherous rivers, and lethal booby traps, these illicit hybrid settlements endured for decades, even centuries. The great majority were small, but some grew to amazing size. Calabar, where Christian grew up, swelled to as many as twenty thousand inhabitants. (The name Calabar comes from a slave port in what is now Nigeria.) A few miles away, another Salvador *quilombo*, Liberdade (Liberty), today has a population of 600,000 and is said to be the biggest Afro-American community in the Western Hemisphere.

Good records do not exist, but Calabar and Liberdade were certainly going concerns by 1650. In Liberdade I met a local historian who told me the city actually originated decades earlier, when slaves had escaped from Salvador down a native path in the forest. The Bay of All Saints is bordered by high, forested bluffs; escapees climbed the bluffs and took over land on the other side, creating a ring of encampments between the colonial port and the indigenous interior. Sometimes their homes were just a few hundred yards away from European farms as the crow flies, but the forest and hills were impenetrable enough to conceal their location. The Portuguese constantly hunted the runaways, but they also traded with them—Calabar's residents, four miles from the center of Salvador, exchanged dried fish, manioc (cassava), rice, and palm oil for knives, guns, and cloth. In 1888 Brazil finally

Tucked behind a wall of high-rise apartments in one of the wealthiest neighborhoods in Salvador, Brazil, the hidden city of Calabar was founded four centuries ago by escaped slaves and still is only weakly attached to the larger urban complex.

abolished slavery, yet life in its *quilombos* showed little improvement. They were still regarded as illegal squatters' settlements. But the government was too weak to do much about them.

In the 1950s and 1960s Salvador grew enormously. Urban pseudopods reached over the ridges, engulfing Calabar, Liberdade, and half a dozen other *quilombos*. But these fugitive settlements never fully became part of the city—nobody had legal title to the land. Few roads entered Calabar. Sewer lines were routed around its borders. People had to steal electric power with jury-rigged hookups. By 1985, when Christian was born, the former hideaway was completely surrounded by high-rise apartments.

When I met Christian, he was kind enough to take Susanna Hecht—the UCLA geographer, who was generously sharing her linguistic and historical expertise—and me around his childhood home. The entry was a narrow, unmarked stairway. Bootleg electrical connections made snarls of wire along the walls. Houses staggered up the ridge, linked by crumbling concrete paths. There were almost no cars. At the bottom of the hill the streets were crowded with promenading people and music was in the air as in other Salvadoran neighborhoods. Teenagers in white clothing were practicing capoeira, the Afro-Brazilian dance that is also a martial art. Banners touting neighborhood programs hung over the street. Here and there new streetlights gleamed. It was a living community, or so it seemed to me, a city within a city.

Calabar and Liberdade are not unique. Thousands of fugitive communities dotted Brazil, much of the rest of South America, most of the Caribbean and Central America, and even parts of North America—more than fifty existed in the United States. Some covered huge areas and fought colonial governments for decades. Others hid in wet forests in the lower Amazon, central Mexico, and the U.S. Southeast. All were scrambling to create free domains for themselves—"inventing liberty," in the phrase of the Brazilian historian João José Reis. They have been called by a host of names: *quilombos*, yes, but also *mocambos, palenques,* and *cumbes*. In English they are usually called "maroon" communities—the term apparently comes, poignantly, from *símaran,* the Taino word for the flight of an arrow.

American history is often described in terms of Europeans entering a nearly empty wilderness. For centuries, though, most of the newcomers were African and the land was not empty, but filled with millions of indigenous people. Much of the great encounter between the two separate halves of the world thus was less a meeting of Europe and America than a meeting of Africans and Indians—a relationship forged both in the cage of slavery and in the uprisings against it. Largely conducted out of sight of Europeans, the complex interplay between red and black is a hidden history that researchers are only now beginning to unravel.

Even when schoolbooks do acknowledge the hemisphere's majority populations, they are all too often portrayed solely as helpless victims of European expansion: Indians melting away before the colonists' onslaught, Africans chained in plantations, working under the lash. In both roles, they have little volition of their own—no *agency,* as social scientists say. To be sure, slavery forced millions of Africans and Indians into lives of misery and pain. Often those lives were short: a third to a half of Brazil's slaves died within four to five years. More still died on the journey within Africa to the slave port, and on the passage across the Atlantic. Yet people always seek ways to exert their will, even in the most terrible circumstances. Africans and Indians fought with each other, claimed to be each other, and allied together for common goals, sometimes all at the same time. Whatever their tactics, the goal was constant: freedom.

More often than is commonly realized they won it. Slaves vanished from the ken of their masters by the tens or even hundreds of thousands in Brazil, Peru, and the Caribbean. Spain recognized autonomous maroon communities in Ecuador, Colombia, Panama, and Mexico and used them as buffers against its adversaries. In Suriname, "Bush Negros" fought a century-long war with the proud Dutch colonial government and in 1762 pushed it into a humiliating peace treaty—the European negotiators, following African custom, had to endorse the pact by drinking their own blood. A maroon-Indian alliance in Florida forced the U.S. government after two wars to grant liberty to its population of escaped slaves. It was the only time that Washington freed a class of slaves before the Emancipation Proclamation (to save face, the government called the pact a "capitulation"). Most important, slaves in Haiti created an entire maroon nation by driving out the French in 1804—a revolution that terrified slave owners across Europe and the Americas.

These struggles are not confined to the past. African populations in Colombia, Central America, and Mexico are increasingly climbing out of the shadows and demanding an end to discrimination. In the United States the descendants of maroons are at the center of legal battles from Florida to California. The greatest impact may be in Brazil, though, where recent laws have given maroon communities a key role in determining the future of Amazonia.

AFRICANS IN CHARGE

Back in Africa, or so the tale goes, Aqualtune was a princess and a general. It is said that she ruled one of the Imbangala states that rose in central Angola as the previously dominant Kingdom of Kongo declined. In about 1605, according to the story, she was captured in a battle against the Kongolese and sold with other POWs to Portuguese slavers. On the passage across she was raped and impregnated. Aqualtune landed in the sugar port of Recife, at the tip of Brazil's "bulge" into the Atlantic. A military strategist, she naturally began to plan an escape. Within months she was in the hinterland with about forty of her troops. Twenty-five miles from the coast, a series of abrupt basaltic extrusions dominates the plain like a line of watchtowers. Their sheer, cliff-like walls reach hundreds of feet up to flat summits with dizzying views of the surrounding plain. One of these tall hills was the Serra da Barriga—Potbelly Hill. On its peak was a pool of cool water, sheltered by trees, perhaps fifty yards across, with an indigenous community around it. Here Aqualtune founded Palmares.

Today Aqualtune's peak is a national park. A plaque by the pond proudly recounts her story—doubtless to the distress of historians, because nobody knows how much of it is true. What is known is that thirty thousand or more Africans fled to the Serra da Barriga and the nearby hills in the 1620s and 1630s, taking advantage of the disorder caused when the Dutch attacked and occupied the Portuguese coastal sugar towns during that time. Free of European control, the escapees built up as many as twenty tightly knit settlements centered on the Serra da Barriga, a haven for African, native, and European runaways. At its height in the 1650s, according to the Boston University historian John K. Thornton, the maroon state of Palmares "ruled over a vast area in the coastal mountains of Brazil, constituting a rival power unlike any other group outside Europe." It had close to as many inhabitants at the time as all of English North America. It was as if an African army had been scooped up and deposited in the Americas to control an area of more than ten thousand square miles.

Palmares's capital was Macaco, Aqualtune's springside resting place. Spread along a wide street half a mile long, it had a church, a council house, four small-scale iron foundries, and several hundred homes, the whole surrounded by irrigated fields. The head of state was Aqualtune's son, Ganga Zumba, who lived in what one European visitor described as a "palace," complete with an entourage of flattering courtiers. Other members of the royal family ruled other villages. Ganga Zumba may have been a title, rather than a name; *nganga a nzumbi* was a priestly rank in many Angolan societies. In any case, the visitor reported, he was treated with the deference due a king. His subjects had to approach him on their knees, clapping their hands in an African gesture of obeisance.

Knowing that his people were always subject to attack, Ganga Zumba organized the towns more like military camps than farming villages—strict discipline, constant guard duty, frequent drill sessions. Each major settlement was ringed by a double-walled wooden palisade with high walkways along the top and watchtowers at the corners. In turn the palisades were surrounded by protective snarls of timber, hidden deadfalls, pits lined with poisoned stakes, and fields of caltrops (antipersonnel weapons made from iron spikes welded together in such a way that one always points upward, ready to injure anyone who steps on it). Every single person who had fled slavery to live there had risked life and limb for liberty in a way that is difficult to imagine today. Palmares fairly bristled with determination to maintain command over its own destiny.

One of the most persistent myths about the slave trade is also one of the most pernicious: that Africans' role was wholly that of hapless pawns. Except for the trade's last few decades—and arguably not even then—Africans themselves controlled the supply of African slaves, selling them to Europeans in the numbers they chose at prices they negotiated as equals. To be sure, Europeans tried to play off slavers against each other to get lower prices. But Africans played off European buyers against each other, too, captain against captain, nation against nation.

If Africans were not forced by Europeans to sell other Africans, why did they do it? In some sense, the question is an example of "presentism"—the projection of contemporary beliefs onto the past. Few Europeans or Africans at this time viewed slavery as an institution that needed to be explained, still less as an evil to be decried. Slavery was part of the furniture of everyday life; in both Europe and Africa, depriving others of their liberty wasn't morally problematic, though it was bad to enslave the wrong person. Christians, for example, were generally not supposed to enslave fellow Christians, though breaking this rule was sometimes permitted. Africans sold other Africans into slavery more often than Europeans less because of their different attitudes toward liberty than because of their different economic systems.

Broadly speaking, says to Thornton, the Boston University historian, "slaves were the only form of private, revenue-producing property recognized in African law." In western and central Europe, the most important form of property was land, and the aristocracy consisted mainly of large landowners who could buy or sell property with little legal restriction. In western and central Africa, by contrast, land was effectively owned by the government—sometimes personally by the king, sometimes by a kinship or religious group, most often by the state itself, with the sovereign exercising authority in the manner of a chief executive officer. No matter which arrangement held true in a given polity, though, the land could not be readily sold or taxed. What *could* be sold and taxed was labor. Kings and emperors who wanted to enrich themselves thus didn't think in terms of occupying land but of controlling people. Napoleon sent his army to seize Egypt. An African Napoleon would have sent his army to seize *Egyptians*.

As was the case in much of Europe, Africans could be sentenced to slavery if they forfeited their membership in society by committing a crime. People could be enslaved, too, to repay a debt, whether incurred by themselves, their families, or their lineages. In times of drought or flood they pawned family members to other members of their extended families or clans. Sometimes they pawned *themselves*. But the most common way to acquire slaves was by sending troops across the border—that is, by war. Seventeenth-century West Africa was even more politically fragmented than Europe. A map prepared by Thornton shows more than sixty different states of wildly varying size. When leaders in one state wanted to aggrandize their status, a border was always nearby; it was easy to send out raiders. Captives would be taken by the king or given for sale to middlemen, who would take them to customers in North Africa or Europe.

In the beginning of the transatlantic slave trade, when European ships first became a constant presence on African shores, the difference between the two systems, European and African, was more a matter of culture than economics. Europeans could buy and sell labor—that was the purpose, to cite one example, of indentured-service contracts. And Africans could effectively own land by controlling the labor from the people who used that land. In both cases the owners ended up profiting from the fruits of the land and labor, even if the route to those profits was different. In economic terms, Europeans could own one of the factors of production (land), whereas Africans could own another (labor). Both systems gave owners the right to claim part or all of the products of that labor. Still, they were far from identical. One big distinction is that labor can be taken from one place to another in a way that land cannot. Labor is portable—a key factor for the later development of the slave trade.

Because labor was the main form of property in West Africa, rich West Africans almost by definition owned a lot of slaves. Plantations were rare in that part of the world—coastal West Africa's soil and climate typically won't support them—so big groups of slaves rarely were found working in fields as was common in American sugar or tobacco plantations. Instead slaves were soldiers, servants, or construction workers, building roads and fences and barns. Often enough they did almost nothing; wealthy, powerful slave owners kept more slaves than they needed, in the way that wealthy, powerful landowners in Europe would pile up unused land. In addition, much slave labor consisted of occasional work performed as a tax or tribute.

Foreign observers noticed that the surplus and tribute slaves didn't always have to work hard or for long periods of time, and they often concluded that African slavery was inherently less brutal than slavery in the Americas. In terms of long-term survival, this seems to have been true. On a tobacco plantation in the Americas, slaves who couldn't work had no value, and were treated that way. The same slaves would have some value in Africa—they were adornments to the owner, in somewhat the same manner that diamond necklaces are valuable despite their lack of practical use. Even the oldest, most infirm slaves could wear fine clothes and walk in a procession, chanting praises of their masters. Or they could simply be interesting to their owners. For several years the king of Dahomey had an utterly useless palace slave who had been seized as a debt repayment: a hapless Briton named Bulfinch Lamb, whom the monarch enjoyed talking to. Moreover, African slaves were more likely to be granted liberty after a period of service than they were in the Americas, both because captives often had a kin connection to their captors and because as subjects they were still valuable to the monarch (freed slaves were a total loss to plantation owners if they were still capable of work). The two factors mitigated the callousness of the institution, helping to satisfy Adam Smith's economic objections to slavery. Still, one suspects the Africans wrested from their homes in military raids would not have celebrated the humanity of the system.

When Europeans arrived, they easily tapped into the existing slave trade. African governments and merchants who were already shipping human beings could increase production to satisfy the foreigners' demands. Sometimes political leaders would hike criminal penalties to obtain slaves. Scofflaws, tax cheats, political exiles, unwanted immigrants—all went in the hopper. Usually, though, armies were sent to raid other nations. Or soldiers could abduct an important person in a neighboring polity and demand a ransom of slaves. If demand increased still further, private traders might seize captives without approval, angering the state. If no other source was available, Africans bought slaves from Europeans. In the seventeenth century, the Yale historian Robert Harms has estimated, Europeans sold forty to eighty thousand slaves to Africans in what is now Ghana.

African demand was as important as European demand in the growth of the trade. When the flintlock replaced the undependable matchlock at the end of the seventeenth century, Africans were as keen to acquire the new guns as the Indians in Georgia and Carolina. In April 1732, traders from the rapidly growing Asante empire appeared at the Dutch fort of Elmina, in Ghana. They had a convoy of captives which they demanded to exchange for guns. Frightened by the threatening tone of the conversation, Harms wrote, Elmina's "governor-general sent a desperate circular to all the other forts ordering that all flintlocks be sent to Elmina at once." Asante had become the dominant regional power by a calculated exchange of slaves for guns and gunpowder. The waves of slavery that fueled Asante's arms buildup, Harms remarked, "account for much of the rise in Dutch slave exports in the 1720s."

African merchants bought slaves from African armies, raiders, and pirates and paid Africans to convey them to African-run holding tanks. Once the contract was arranged, Africans loaded the slaves aboard the ships, which often had crews with significant numbers of Africans. Other Africans supplied the slave ships with food, rope, water, and timber for the voyage out. Europeans naturally played a role: they were customers, the demand side of the basic economic equation. A few even braved the African coast, marrying Africans; their children frequently became negotiators and middlemen in the African slave trade. A combination of disease and watchful African armies otherwise kept them confined to outposts on the edge of the continent.*

Tiny outposts, for the most part. The Dutch West Indies Company long held a legal monopoly on the Dutch slave trade, shipping out about 220,000 captives by 1800. Elmina, its African headquarters, had a European population that rarely exceeded four hundred, and was usually smaller. Three miles away was Cape Coast, the biggest base of the English Royal African Company, which had an equivalent legal monopoly on the English slave trade. From its docks left tens of thousands of enchained men, women, and children. Yet Cape Coast had fewer than a hundred foreign inhabitants. Seventeenth- and eighteenth-century European maps proudly depicted Africa's Atlantic coast as bristling with Danish, Dutch, English, French, Portuguese, Spanish, and Swedish forts,

---

* Huge numbers have watched the opening hour of the miniseries *Roots*, in which U.S. slavers raid villages in the Gambia. In fact, such forays were rare. African states didn't like trespassers, especially when the trespassers were slaving companies trying to cut them out of the supply chain—and the captives were their own subjects.

garrisons, and trading posts. But most of the stars on the maps had fewer than ten expatriate residents and many had fewer than five. The principality of Whydah, in today's Benin, exported 400,000 people in the first quarter of the eighteenth century—it was the most important depot in the Atlantic slave trade in that time. Not one hundred Europeans lived there permanently. The largest groups of foreigners were the slavers who camped on the beach as they waited to fill their ships with human cargo.

Yet these minute stations were the catalytic points for an enormous change. In the past, most African slaveholders had known something about their slaves' previous lives. Sometimes they were related to their bondsmen, distant cousins or in-laws; other times they understood exactly what familial, lineage, or tribal obligation had resulted in their enslavement. Even prisoners of war had been obtained in a known location, in a known conflict. Chattel slavery on colonial plantations, by contrast, made slaves anonymous—they were, so to speak, something bought in a store, selected purely on physical characteristics, like so many cans of soup. (In account books, slavers called their human cargo "pieces," a revealing term.) European slaveholders usually didn't even see their human property; they were thousands of miles away, safe from disease in London, Paris, and Lisbon. When they wanted to expand production of sugar or tobacco, they borrowed money from equally distant financiers and dispatched written instructions to acquire so many pieces at such-and-such a price. This transformation was not understood as it occurred. But it removed a bond, however tenuous, between slave and owner. No longer were captives an owner's relatives or vanquished enemies. Instead they were anonymous units of labor, production inputs on a balance sheet, to be disposed of purely according to an estimate of their future economic value.

Hovering in their vessels along the coast, Dutch, Portuguese, and English slavers thus had little knowledge about the origins of the unhappy men and women on their ships. The colonists who rushed to buy their cargo on the quays of Jamestown,

Cartagena, and Salvador had even less. According to Thornton, "only a handful of American slave owners seem to have actually known . . . that many thousands of them were prisoners of war." When captive soldiers organized escapes and rebellions, some owners learned the import of their military backgrounds. From the beginning, American slave owners were dogged by the problem that their army of slaves could be an enslaved army.

The first bondsmen in Hispaniola came mainly from the civil war–torn Jolof empire in what is now Senegal and Gambia. It seems likely that many of the slaves sent to the Caribbean were POWs—military men. In any case Spanish records note that the first large-scale slave revolt in the Americas was led by Jolofs. It occurred on Christmas Day, 1521, at a sugar mill owned by Diego Colón, son and heir of the admiral. About forty slaves raided a cattle ranch, killed several celebrating Spaniards, burned down a few buildings, and took numerous prisoners, including a dozen Indian slaves. Colón assembled a cavalry force that charged the renegades. The classic response for foot soldiers facing horses is to bunch together tightly, spears facing out from a defensive wall—the tactic used by Greek infantry to win the battles of Marathon and Plataea. Despite their lack of weapons, the slaves did exactly that, their line holding together until the third charge. Eventually the renegade captains fell. Survivors were hunted down and hanged along the road to deter other would-be troublemakers.

The Spaniards' troubles were not over. Even as the bodies dangled along the highway, a Taino leader called Enriquillo was setting up a European-free village in the southwestern mountains. Enriquillo, a devout Christian who had been taught by Franciscan friars, was initially co-opted by the *encomienda* system. Exactly as its designers had hoped, he sent out his people to work in exchange for status and trade goods. But Enriquillo's trustee—his *encomendero*—didn't like having to negotiate with him for workers. In a fit of anger the *encomendero* assaulted Enriquillo's wife and stole his horse. The Taino man furiously confronted him. As the Indian advocate Bartolomé de las Casas tells the story, the *encomendero* reacted to Enriquillo's protests by threatening to beat him with a club. The beating, he mocked, would complete the proverb: it would add injury to insult.

Enriquillo decamped for the hills with the rest of his family and a handful of followers. Escaped Africans and other Taino joined the revolt, swelling its numbers to perhaps five hundred. The maroons built a covert village in the hills that the Spaniards hunted in vain for more than a decade. Tired of the escapees' raids, the colonists finally negotiated a treaty in 1533. The Spaniards promised to obey the *encomendero* law and respect Enriquillo's status if his rebels would return to their homes. Enriquillo and other Taino accepted the deal—but their African allies did not. Led by one Sebastian Lemba, they refused to come back.

"Lemba" was a kind of spiritual association of wealthy merchants—a mix, perhaps, of a church and the Rotary club—based in Kongo. Lemba's name hints that he was a businessman caught in an Imbangala slave raid. If so, his organizational skills may have played a role in his leadership, which the Spaniards admitted was "extremely able." Far more vengeful than Enriquillo had been, Lemba broke his troops into small, mobile bands that pillaged sugar plantations and mills for sixteen years. So many slaves were inspired to revolt by his actions that the archdeacon of Santo Domingo claimed in 1542 that the number of guerrillas in the hills was greater than the number of Spaniards in Hispaniola. Only ten of Hispaniola's thirty-four sugar mills were open; the rest had been shut down by rebel slaves. Five years after the archdeacon's lament, Lemba was betrayed by another slave—a man who sought his own liberty sold out by a man who was rewarded with freedom. The colonists mounted Lemba's head on a pike near the main gate to Santo Domingo.

Again the insurrection didn't stop. Why would it? The colonial government had utterly lost control. Within months of Lemba's death its officials were complaining to the court that rebels were "killing and robbing Spaniards" just nine miles outside of

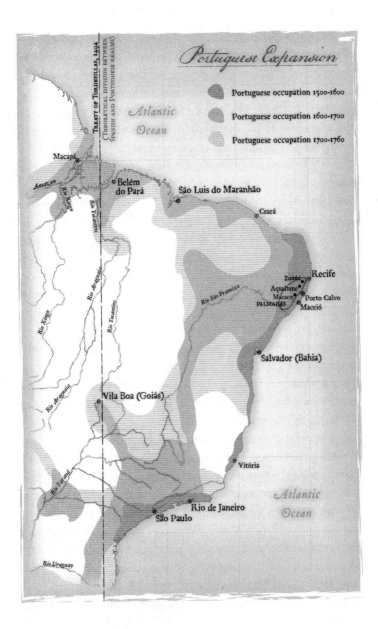

Portuguese Expansion

Portuguese occupation 1500–1600

Portuguese occupation 1600–1700

Portuguese occupation 1700–1760

TREATY OF TORDESILLAS, 1494
(THEORETICAL DIVISION BETWEEN
SPANISH AND PORTUGUESE REALMS)

Atlantic
Ocean

Macapá

Amazon

Rio Xingu

Rio Tapajós

Belém
do Pará

São Luis do Maranhão

Ceará

Rio Araguaia

Rio Xingu

Rio Tocantins

Rio São Francisco

Zumbi
Recife
Aqualtune
Mácaco
PALMARES
Porto Calvo
Maceió

Salvador (Bahia)

Rio Araguaia

Vila Boa (Goiás)

Vitória

Rio Paraná

Rio de Janeiro
São Paulo

Atlantic
Ocean

Rio Uruguay

Santo Domingo. Not that this caused them to rethink their commitment to slavery. As the Dominican historian Lynne Guitar has noted, the same letter asked the king for permission to ship five to six thousand more slaves to open more sugar plantations.

In their hunger for labor, European sugar growers were importing people who would have liked nothing better than to destroy them—people like Aqualtune and Ganga Zumba in Palmares. Maroons in Hispaniola ultimately helped drive the sugar industry from the island. Portuguese officials feared that Palmares would do the same in Brazil. The rebel confederacy was a direct military and political challenge to the colonial enterprise. Not only did its troops raid Portuguese settlements, its strategic location atop hills like the Serra da Barriga blocked further European expansion into that part of the interior. If such rebellions spread to other ports of Brazil, Lisbon feared that its colonists would become a kind of marine froth on the coast, while the interior turned into a mosaic of Afro-Indian states.

Palmares was led by Angolans, but it was not an Angolan society, or even an African society. Many of its people were Tupi-speaking Indians. Some were Europeans with uneasy relations to their own societies; Jews and converted Jews, heretics and ex-heretics, suspected witches and escaped criminals, and a salting of suspicious ethnic minorities. In the main Palmares's people looked like Africans, but they lived in Indian homes made from woven thatch and steam-bent arches and harvested Indian-style multicrop fields of maize (American), rice (African), and manioc (American). (African rice, *Oryza glaberrima*, was domesticated in West Africa and is a different species than Asian rice, *Oryza sativa*.) Palmares's blacksmiths used African-style forges to make plows, scythes, spears, and swords; colonial reports claimed they could even make guns and bullets. Their religious ceremonies, from what one can gather today, mixed Christianity with Indian and African elements. But their military organization was African, with Aqualtune's children and grandchildren in strict control of villages so battle-hardened that they should perhaps be thought of as bases.

Between 1643 and 1677 Portugal and the Netherlands (which occupied part of Brazil for some of this time) attacked Palmares more than twenty times, always unsuccessfully. When the armies approached the maroon state's outlying settlements, their inhabitants would flee to the hilltops, where fertile soils, artesian water, and storehouses of food made it possible to outlast any siege. The attackers would find empty villages stripped of food and valuables. Then they would blunder about the forest, trying to find the people. Soon they would run out of supplies. All the while they would be watched—and ambushed. Arrows flew from the trees to pick off stragglers. Advance scouts fell into hidden pits. Men woke up to find their comrades missing and their food stolen. Infuriating to the soldiers, the region's planters bought their slaves' food from Palmares. In exchange for maize and manioc, the planters had provided the maroons with the guns and knives now trained on the soldiers.

A central figure in the story of Palmares was Zumbi, who became its military commander. The nephew of Ganga Zumba, the king, Zumbi was taken as an infant by Dutch forces during an otherwise unsuccessful attack. He was raised under a European name by a priest in the small coastal town of Porto Calvo, learning Portuguese, Latin, theology, and sciences like navigation and metallurgy. In 1670 the teenaged Zumbi ran back to Palmares and resumed his maroon name and life, though he returned to the priest for sentimental visits. Charismatic, well educated, and knowledgeable about the enemy, he rose quickly to command despite a severe limp from a wound suffered in an early battle. Zumbi, too, may have been a title, rather than a name. It means something like "ancestral spirit"—a reference, perhaps, to his return from the death of colonial life.

A Portuguese assault in 1677 wounded Ganga Zumba and captured some of his children and grandchildren. Weary and saddened, the king negotiated a peace treaty the next year with the Portuguese. He promised to stop accepting new escapees and move out of the mountains if the Portuguese would stop attacking Palmares. Zumbi viewed the pact as a sellout of everything the maroons stood for. Angered beyond measure, he poisoned the king, seized the throne, and tore up the treaty. The war was on again. Colonial militias attacked every year for the next six years, achieving little.

Appalled by the meager results of the forty-year campaign against Palmares, the newly appointed governor-general of the region decided to try a different tack. He had received a request from a man named Domingos Jorge Velho for a license to conquer more Indians. Reluctantly, the governor agreed to meet him.

Jorge Velho was a *bandeirante,* a backwoodsman. Often the product of a union between a Portuguese man and an Indian woman, *bandeirantes* used their mothers' connections to advance the agenda of their fathers—indeed, the term *bandeirante* means "flag-bearer," and refers to their role in claiming land for Portugal. Jorge Velho was an exemplary case. A Kiplingesque adventurer, he had assembled a private army and created a kind of private kingdom in southern Amazonia. Hundreds of Indians served him as fieldworkers and soldiers, controlled partly by his promise to protect them from other, worse *bandeirantes.* Jorge Velho had the gangster's predilection for boasting of his magnanimity. He seized Indians and their land, he later proclaimed in a letter to the Portuguese court, for the natives' own good, not merely for profit. By taking natives from the forest, he

> domesticate[d] them to the knowledge of civilized life and human society and to association and rational dealings. . . . If afterward we use them in our fields we do them no injustice, for this is to support them and their children as much as to support us and ours.

The letter's flowery phrases, as well as the letter itself, were doubtless written for him by someone else; Jorge Velho was illiterate.

As the governor discovered at the meeting, the *bandeirante* had more in common with the maroons than with other Europeans. He spoke Portuguese so badly that he had to use an interpreter to speak with colonial officials. "This man is one of the worst savages I have ever encountered," reported the appalled bishop of Pernambuco, who reserved especial ire for the *bandeirante*'s penchant for traveling with seven Indian concubines "to exercise his lusts." (The concubines, more than sexual partners, were Jorge Velho's links into native communities. For the same reason, he also had a Portuguese wife.)

The colonial administration knew that Jorge Velho might be able to break Zumbi, but its officials were reluctant in the extreme to hire him. Only after almost seven years of dithering did the authorities finally cave in. By that time Jorge Velho had them over a barrel—he was their last chance. If he would move on taking care of the Palmares problem, the governor-general promised, the administration would provide his men with gunpowder, bullets, food, a tax-free hand with any booty, a reward for every captured African, and, perhaps most important, full pardons for any previous crimes.

Accompanied by about a thousand native troops and almost a hundred Portuguese, Indo-Portuguese, and Afro-Portuguese, Jorge Velho marched out from his estate in 1692. The journey to Palmares, almost five hundred miles long, occurred in what he modestly described as "the worst conditions of toil, hunger, thirst and destitution that have yet been known and perhaps ever will be known." Two hundred of his troops died; another two hundred deserted. They ran out of food and ammunition and had to wait for ten starving months in the forest for supplies promised by the colonial authorities in Recife. Reduced to "six hundred native soldiers and forty-five whites," Jorge Velho's force returned to the assault in December 1693.

Zumbi's headquarters in Macaco was next to impossible to approach. I got a hint of what it had been like when I visited the park atop Serra da Barriga. Ruts in the muddy, unmarked route tore out the exhaust in the rental car; local teenagers kindly tied it back on with wire scavenged from a telephone pole. From the summit, everything for miles around was visible, cars and tractors picked out by the sun with dizzying clarity. I could imagine maroons watching Jorge Velho's men below like a line of ants on a tablecloth. Attackers and defenders both were mainly Indians and Africans with a sprinkling of Europeans. The difference was that in Palmares the Europeans were not running the show. Scrambling up the hill to Macaco, the *bandeirantes* had to twist through a maze of defenses, caltrops slicing at their feet and hands, maroon troops shooting at them from the palisade towers. The attackers formed a ring around the peak in an attempt to starve out the town. It was like a medieval siege in the tropical forest.

After several weeks of stalemate, the besiegers apparently realized that the maroons had more supplies than they did. Jorge Velho instructed his forces to construct a series of stout, movable barricades. Crouching behind them, his men shoved and wedged the walls up the hill a few feet at a time, scanning the coming ground for caltrops, snares, drop-traps, and poisoned stakes, heedless of the arrows and bullets thunking into the other side of the wood. Although the *bandeirantes* had timed their assault for the dry season, rain fell for days on end, turning every inch of ground into thick mud. Realizing that the movable barricades were blocking their shots, maroon archers and gunmen slipped out of the palisade and climbed high into trees. When the attackers' walls moved beneath them, they shot the *bandeirantes* in the back.

Zumbi paced the walkways atop the palisades, rallying his wet, exhausted forces. On the moonless night of February 5, 1694, he discovered that *bandeirantes* had killed two sentries. (The story comes from maroon testimony afterward.) In the darkness and rain the rest of the guard had not noticed the gap in the defenses—or that the attackers closest to it had taken advantage of that inattention to bring their barricades within a few feet of the walls. Squinting through the downpour at the barely visible

From the summit of Serra da Barriga, the maroons of Palmares could see every movement below.

attackers, Zumbi apparently realized that it now would be impossible to stop the assault from breaching the palisade. News of the imminent attack radiated through Macaco like terror itself. As Zumbi tried to rally his force for a final defense, some of his men realized that the attackers, too, had a gap in their line. They tore down part of the palisade and fled through it. The *bandeirantes*, caught by surprise, let most of the maroons pass, firing only a single volley at their heels. Then they poured into Macaco through the fallen wall.

Neither side had expected the final assault to occur when and where it did. In the darkness and confusion and rain, Indians, Africans, and Europeans on both sides smashed clumsily at each other with sticks and blades. Guns were useless in an hour when fighters could barely see and weapons slipped from muddy hands. Covered in a thick impasto of blood and earth, shouting and sobbing, the two forces assailed each other without compunction. Half the six hundred *bandeirantes* died within minutes, as did an equal number of maroons. Perhaps two hundred more maroons were forced off the cliff or threw themselves off rather than face captivity—no one is sure. When dawn at last shone on the sodden Serra da Barriga, Macaco was in ruins.

Somehow Zumbi escaped. The surviving *bandeirantes* thought at first that he had flung himself off the cliff. Instead he continued to skirmish with the Portuguese for more than a year, until one of his aides revealed his location. Zumbi and a small band of followers were ambushed and killed on November 20, 1695. His body was taken to Porto Calvo and identified by people who had known him as a child. All along the coast colonists celebrated the victory, parading through the streets with torches night after night in an improvised festival of joy. Zumbi's decapitated head was taken to Recife, where it was displayed on a spike to forestall any claims that he had somehow survived. Ninety years after Aqualtune arrived in the Americas, her city had at last been destroyed. But it was anything but the end of *quilombos* and maroons in Brazil or anywhere else in the Americas.

IN THE ISTHMUS

Vasco Núñez de Balboa, like Cortés and Pizarro, was from the remote Spanish region of Extremadura. Like them, he was a bold, ruthless, antically ambitious man. Economical with the truth and recklessly impulsive, he was, according to an acquaintance, "tall and well-built, with good, strong limbs and the refined gestures of an educated man." As the younger son of a down-at-heels noble family, he had prospects bleak enough to encourage him to ship across the ocean in 1500, when he was about twenty-five. He established himself as a farmer in Salvatierra de la Sabana, a remote hamlet in southwestern Hispaniola.

In retrospect, it was a terrible career choice. "The calm and tranquil life of the farmer wasn't a match for his great aspirations and adventurous, energetic spirit," explained one admiring Spanish biographer. Indeed, Núñez de Balboa's great aspirations and adventurous, energetic spirit led him to pile up debts at such a clip that he fled his creditors by stuffing himself into a barrel and having himself rolled aboard a ship bound with supplies for a new colony on the mainland, Spain's first attempt to establish a base there. (According to some reports, he stowed away in the barrel *with his dog.*)

The settlement, located in what is now Colombia, near the border with Panama, had been established to find gold mines. Labor was to be provided by enslaving local Indians, some of whom would also be sold in Hispaniola. The Indians saw no reason to participate in this scheme and expressed their lack of enthusiasm by riddling the invaders with poisoned arrows. With the colony near collapse, its founder sailed for help in Hispaniola in July 1510. His ship ran aground off the coast of Cuba and he staggered half-starved across the island. After being rescued, he immediately retired from the discovery-and-conquest business. Meanwhile, another ship had left from Santo Domingo in September to aid the settlement. This was the vessel that contained Núñez de Balboa and his barrel.

He was quickly discovered. Charismatic and clever, he managed to talk the irate captain out of stranding him on a desert island. Within weeks Núñez de Balboa was one of the captain's most valued lieutenants. Within months he had persuaded the captain to relocate the colony to what he thought would be a better location. Within a year he had deposed the captain and was leading an expedition up the coast of Panama, looking for gold.

In Panama, Núñez de Balboa became the first European to see the Pacific Ocean from the American side, an exploit that won him enduring fame. Today, five centuries later, a cursory online search for "Núñez de Balboa" will find countless pictures of the conquistador standing athwart a crag or striding into the waves, sometimes in full armor, gazing in wonder at the endless sea ahead. But the heroic images have not kept up with his reputation among historians. Núñez de Balboa was unquestionably bold and brave, but he also committed actions that are difficult to justify in any current ethical scheme. And he may well not have been the first person from the other side of the Atlantic to see the Pacific from the American side.

The newly moved colony, Santa María la Antigua del Darién (Antigua), was legally under the jurisdiction of another conquistador. When this conquistador came to Antigua to demand control, Núñez de Balboa put him onto a leaky brigantine and told him to sail away. He was never seen again. Now feeling more secure about his command, Núñez de Balboa turned his attention to the resident Kuna and Choco peoples, whose penchant for draping themselves with gold jewelry made them fascinating in Spanish eyes. He began asking around for the source of the gold.

About fifty miles north of Antigua reigned a man named Comagre, who lived with his many wives and children in what the historian Pietro Martire d'Anghiera described as "a house made of big, interwoven timbers, with a hall 80 paces wide and 150 long and what looked like a coffered ceiling." His domain—the Spaniards called it a "seigneury"—had about ten thousand inhabitants. When Núñez de Balboa paid a visit, Comagre plied the expedition with "wine made from grain and fruit," assigned the visitors seventy slaves for the duration of their visit, and gave them "four thousand ounces of gold in jewelry and finely worked pieces." The Spaniards whipped out scales and weighed out shares of the booty amid much quarreling. Laughing at their cartoonish greed, Comagre's son told them of the existence of another seigneury with even more gold, on the shores of "another sea which has never been sailed by your little boats."

Another sea! More gold! Núñez de Balboa was beside himself with excitement. He returned to Antigua, put together an expedition of about eight hundred—two hundred Spaniards and six hundred Indians—and set off on September 1, 1513. (Along for the

ride were at least one mixed-race man and an African, both probably bondsmen; the African would later be given his freedom,
land in Nicaragua, and 150 Indian slaves.) The journey began in
the steep, wet, thickly forested hills of eastern Panama, which
rise up almost directly from the coast. It was the height of the
rainy season—annual precipitation there is as much as sixteen
feet. Staggering under the weight of armor, plagued by insects
and snakes, covered in mud, the Spaniards soon began falling to
illness and injury. Núñez de Balboa led his increasingly ragged
force from one native group to the next, asking questions and
seeking food, leaving his weak and sick behind at every stop. The
coastal ridges descend vertiginously into the hot, mucky valley
of the Chucunaque River, so close to the Pacific that tides cause
daily floods far upstream. From the river's other bank ascend a
jumble of craggy low peaks atoss with palms. The exhausted men
reached these slopes on September 24, having traveled about
forty miles in three weeks.

Near the summit they encountered Quarequa, lord of a small
seigneury of the same name. Backed by hundreds of men with
bows and spears, he refused to let the foreigners enter his land.
The Indians, who had never seen firearms and swords, confronted the Spaniards in a mass. Without warning, Núñez de Balboa ordered his men to fire at point-blank range. Into the smoke
the Spaniards ran with naked swords. Hundreds died, including
Quarequa, the bodies piled atop one another. The Spaniards
chased the survivors into their main village, where they found
all the gold and food stores gone. The next day, September 25,
Núñez de Balboa and his tattered band climbed to the summit
and saw the dizzying vastness of the Pacific before them. In a
gesture that now seems touchingly absurd, he claimed all of the
ocean and attendant lands it touched for Spain.*

---

* Less touching from today's perspective were Núñez de Balboa's actions
in Quarequa's village. In it he had found forty members of Quarequa's family and court dressed as women. The story is that he had them torn apart by

Left behind in Quarequa's village were women, children, and some African slaves—"black men with big bodies and big bellies, and long beards and crooked hair," as one report described them a year and a half later. The Spaniards had been stunned to see them, and stunned again when they were told that an entire community of escaped African slaves existed just two days' walk away. Indians and Africans had been fighting for years, each side forcing captives from the other into slavery.

The Spaniards' identification of the slaves as Africans is unlikely to be mistaken—they were traveling with at least two. Nor does the story seem to be apocryphal; half a dozen Spanish sources attest to it. Not one of these sources, however, drew out the implications. First, the existence of escaped slaves in the mountains likely meant that Africans, not Europeans, were the first people from across the Atlantic to settle on the mainland—and to see the Pacific from the American side. Second, it meant that the isthmus was a good place for maroons to evade capture. The latter fact would come to preoccupy the Spanish crown.

Finding a route to the Pacific electrified the Spaniards in Antigua. They soon abandoned the colony, which became a ghost town.* Most of its former inhabitants went on to found two new settlements: Panamá, on the Pacific side of the isthmus, and Nombre de Dios, on the Atlantic. The idea was that spices from the Maluku Islands, which Spain intended to capture, would be

---

dogs (one of whom, supposedly, was the dog in his barrel). Other villagers then pointed out more transvestites and persuaded him to kill them, too. The sequence of events is hard to credit as presented. Although Panamanian native groups were reputedly tolerant of homosexuals, their presence in big, cohesive groups is unlikely. One can speculate that the Spaniards mistook some form of courtly attire for women's clothing. In the political vacuum caused by Quarequa's death, the courtiers' enemies may have used this misapprehension to get the Spaniards to eliminate rivals.

* Santa María la Antigua del Darién is often said to be the first permanent European settlement on the mainland. "Permanent" is a stretch; the colonists abandoned it after 9 years. About 170 years later, Scotland tried to establish a colony just a few miles away, with results that I described in Chapter 3.

transported to the Americas, carried on a new road between the two towns, then loaded onto ships for Europe. When Spain failed to seize the Malukus, both would-be ports shrank.

Neither Panamá nor Nombre de Dios had more than forty European residents in 1533, when the unexpected news arrived that Francisco Pizarro, one of Núñez de Balboa's companions on his journey across the isthmus, had conquered a great Indian empire in the Andes, and was sending gold and silver to Panamá. (Núñez de Balboa did not participate in the subjugation of the Inka. His flagrant machinations had caught up with him, and he had been executed in 1519.) Twelve years later, in 1545, silver was discovered at Potosí. Half or more of the silver—including most of the king's taxes and fees from the mines and mint—was shipped to Panamá.

The road between Panamá and Nombre de Dios thus became a critical chokepoint for the empire, a single passage down which flowed much of the monarchy's financial lifeblood. From an engineering point of view, it was not ideal. Knee-deep in mud and choked by debris, barely wide enough for two mules to pass, the road plunged in a tangle of switchbacks between crag and swamp and back again. Traversing it terrified the Spaniards—the forest, one chronicler complained, swarmed with "lions, tigers, bears, and jaguars." Screaming monkeys threw rocks from trees. Fer-de-lances and bushmasters, two of the planet's most deadly snakes, were active at night. Travelers could paint themselves from head to foot with oil and mud to ward off mosquitoes but were helpless against the bats—"biting so delicately on the tips of [sleepers'] toes, and the hands, and the end of the nose, and the ears," an Italian moaned, "that one is never the wiser, and gnawing that little mouthful of meat, and sucking the blood that comes from it." They couldn't be warded off, he said, because the heat made it necessary "to sleep naked atop the covers." Even in the dry season it was sweaty going for men in European armor, a necessary precaution against Indian attack. During the rainy season, the road was flat-out impassable; travelers had to pole barges up or down the Chagres River, navigable when swelled by rain but dangerous for the same reason. Sixteenth- and seventeenth-century Europe simply did not have the means and technology to maintain an adequate highway in these conditions. It remained "an extremely bad road, the worst that I have seen on my travels," one annoyed voyager wrote in 1640, 120 years after its initial construction.

To convey the king's silver across the isthmus required many hands. As ever, labor was in short supply. Few Spaniards would leave their homes to toil in a remote forest. To would-be silver transporters, there was an obvious recourse: Indian slaves. At the time that Núñez de Balboa saw the Pacific, the isthmus of Panama was filled by perhaps a hundred small, fractious polities, honeycombed so tightly together that the sixteenth-century historian Gonzalo Fernández de Oviedo y Valdés claimed the native population "surpassed two million, or they were uncountable." Modern estimates are much lower: because most seigneuries (as I am calling them) had no more than three thousand people, researchers say, the total population must have been at most a quarter of a million. The exact figure almost doesn't matter, though, because the isthmus was rapidly depopulated. By the time Potosí began exporting silver, historians estimate that fewer than twenty thousand people lived there. Even if the remaining Indians had allowed themselves to be captured, there simply weren't enough hands to satisfy European demand. In consequence, the empire imported slaves from the Andes, Venezuela, and Nicaragua—so many that in Spanish areas they quickly became more numerous than locals.

After Spain banned Indian slavery, the colonists turned to Africa—beginning with Núñez de Balboa, who before his death took thirty African captives to the Pacific to build ships. Soon Africans were pushing barges on the Chagres River, eighteen or twenty straining men on each one, twenty or more vessels in a row. Mule trains, scores of animals tied nose-to-tail, were crossing between the oceans, driven by dozens of whip-toting Africans, themselves driven by gun-toting Spaniards. Sometimes the journey took as long as a month. The path, the bat-hating Italian said, was lined with the corpses of mules and men.

Africans outnumbered Europeans seven to one by 1565. Unsurprisingly, Europeans found it hard to control their human property. Runaways grouped hundreds strong into multiethnic villages that were joined by escaped Indian slaves from the Andes and Venezuela and the remnants of free Indian groups from the isthmus. United by their loathing of Spaniards, they liberated slaves, slew colonists, and stole mules and cattle. Sometimes they abducted women. Losses mounted. Spain had a dreadful maroon problem.

The issue was noticed as early as 1521, but the first serious effort to eliminate a maroon settlement in the isthmus didn't occur for another thirty years, after a young slave known as Felipillo, a pearl harvester in the islands outside Panamá, led a group of fugitive Africans and Indians into the mangrove thickets of the Gulf of San Miguel. Their entire village was wiped out in 1551, after two years of freedom. Other maroons learned a lesson from Felipillo's fate: don't hide out in the lowlands, which were too accessible.

That same year, the municipal government of Nombre de Dios complained to the crown that 600 maroons were robbing and killing travelers on the road to Panamá. Two years later the havoc was worse and the number had risen to 800. Two years after that it was 1,200. In the isthmus, not only slaves but escaped slaves outnumbered Europeans. Maroons wiped out the first two Spanish expeditions against them, in 1554 and 1555. In Nombre de Dios, they stole so many captive Africans and Indians that surviving colonists feared to send their slaves outside to fetch water. Most residents fled to Panamá, returning to Nombre de Dios only when the silver fleet came into view.

Leading the maroons was a man whose name has come down to us as, variously, Bayano, Bayamo, Vallano, Vayamo, and Ballano. Like Aqualtune, he seems to have been a captured military leader. "Burly and fierce, coarse and stalwart, rudely dressed and roughly witty," the poet Juan de Miramontes described him, Bayano was "agile, bold, sudden, and sharp"—a man with a "warrior spirit." He oversaw the construction of a palisaded fortress atop a cliff-ringed hill in the ridges overlooking the Caribbean. Guards stood ready to roll stones down into the marshy ravines that were its only entrances. Located far enough from Nombre de Dios that Spaniards would be unlikely to discover it, the stronghold was mainly populated by young men whom Bayano ordered about with soldierly dispatch. Farther away was a second village for the community's women, children, and elderly. Mixing Indians from Peru to Nicaragua and a dozen African ethnicities, Bayano's mini-kingdom was an extraordinary cultural potpourri, one sixteenth-century priest remarked, with "every different mixture of people, all dissimilar in color to their fathers and mothers." Their religion, too, was an equally various jumble of Christian, Islamic, and indigenous traditions, according to Jean-Pierre Tardieu, a historian at the University of La Réunion whose work I am relying upon here. Nobody knows what language they spoke together.

A new viceroy of Peru traveled to Nombre de Dios in 1556 en route to Lima. Infuriated by Bayano's depradations, he established a fund to hire an anti-maroon force. Nobody accepted the offer. Finally the viceroy filled the roster by visiting the prison in Nombre de Dios and telling the inmates that they could either wage war against the ex-slaves or effectively become slaves themselves and be sent to the galleys. The response was positive. Seventy armed ex-convicts went out in October 1556, led by Pedro de Ursúa, an experienced soldier whom the viceroy had persuaded to take on Bayano.

Guided by a captured maroon who had become an informer, Ursúa's troops hiked through the forest for twenty-five days to reach Bayano's hilltop. Realizing that he could not successfully lay siege to the place, Ursúa instead persuaded the maroon leader to negotiate. He offered to split the isthmus into two kingdoms, one ruled by Felipe II of Spain, one ruled by Bayano I of Panamá. Bayano accepted the flattering offer and the Spaniards hung around for weeks, hunting and fishing with the former slaves and amusing themselves with contests of strength and skill. Just before leaving, Ursúa threw a celebratory feast. Bayano and forty of his court attended. The Spaniards drugged their wine, incapacitating them. The maroons were hauled back to Nombre de Dios and returned to slavery. Ursúa took Bayano in chains to Lima as a trophy for the viceroy. Other maroons learned a lesson from Bayano's fate: Spaniards cannot be trusted.

The maroon problem did not go away. Not only did the remnants of Bayano's community regroup, but others sprang up in its wake. Eradicating them, the colonists realized, would require a long-term military campaign with as many as a thousand soldiers, most of whom would have to be sent from Europe. To obtain a thousand soldiers, the government would have to import as many as two thousand, because new European arrivals (one part of the Columbian Exchange) fell at horrific rates to malaria and yellow fever (another part of the Columbian Exchange). Nombre de Dios in particular became so unhealthy that European visitors gave it a bleakly rhyming nickname: "Nombre de Dios, Sepultura de Vivos"—Buried Alive. The king, appalled at the dying, ordered the populace moved entire to a new location, Portobelo, in 1584. It was scarcely less deadly. Visiting the new city in 1625, the English priest Thomas Gage noted that the silver fleet, once landed in Portobelo, "made great haste to be gone"; nonetheless, the ships' two-week stay in the "open grave" of Portobelo was enough to kill "about five hundred of the soldiers, merchants, and mariners." Such losses would ensure that importing an anti-maroon force from Europe would be hugely expensive.

Nobody could agree on who should pay for it. Europeans in the isthmus were mainly agents for Seville merchants. Unlike the Portuguese sugar growers who fought Palmares, few of the Spaniards in Nombre de Dios and Panamá town intended to create permanent establishments; instead the goal was to make a quick killing and leave. Naturally enough, these people did not want to spend much of their potential profit on a project—expunging maroons—that would accrue most of its benefits after their departure. Instead they asked Madrid to ship in and maintain the soldiers. As the king stood to lose most from the attacks, the merchants reasoned, he stood to gain most from their suppression, and therefore he should foot the bills. The crown, for its part, was too far away to monitor expenditures closely. With no way to ensure that the isthmus's short-termers wouldn't pocket funds designated for the anti-maroon campaign, the king was reluctant to, so to speak, sign the check. The conflict was a version of what economists call the "principal agent" problem: when one party pays another to act on its behalf but can't readily measure its performance. And it was enough to stall large-scale action against the maroons, even though the stakes for Spain kept rising.

From the colonists' point of view, it was bad enough when nude, grease-smeared ex-slaves and Indians swept into Panamá town with their "very big and strong bows" and iron-tipped arrows, as one colonial official wrote in 1575, stealing cattle, carrying off slaves, and "usually killing the [Europeans] they meet." Worse, the maroons, out of spite, threw whole shipments of silver and gold into the river. But then the maroons joined forces with the man who would become Spain's most hated enemy: Francis Drake, the English pirate/privateer.

Drake, then on his first major independent voyage, came in July 1572 to the isthmus, looking to loot Spanish treasure. Finding African slaves loading wood on an island outside Nombre de Dios, he asked them about the town's defenses. (The slaves had been left by their owners, who presumably intended to return for them; Drake set them ashore, so that they could run away.) The English attacked at 3:00 a.m. on July 29 in a flurry of gunfire. The exchange wounded Drake badly enough that his men pulled back, regretfully leaving behind, according to his authorized biography, "a pile of barres of silver, of (as neere as we could guesse) seventie foot in length, of ten foot in breadth, and twelve foot in hight." Drake was not discouraged. Just after he set off for Nombre de Dios, the men whom he had left behind to guard his ships were hailed by an African—a maroon offering the assistance of his fellows.

After some fumbling about, Drake met in September with a maroon captain, Pedro Mandinga. To the dismay of the English, Mandinga told them that the flow of silver from Peru had stopped for the year. The next shipments would not occur until March, when the rainy season ended. Drake decided to wait. With Mandinga, he devised a plan to steal silver not on the coast, but in Venta de Cruces, a transshipment area on the Chagres River where mule trains were unloaded onto barges. Mandinga sent spies into Panamá to find out when the silver ships would arrive. Meanwhile, the English hid from Spanish eyes in a cove west of Nombre de Dios, their victuals largely provided by maroon bows and fish hooks. Waiting was riskier than the English anticipated; yellow fever killed half their number in December. Among the victims was Drake's younger brother, Joseph. (Another brother had died a few weeks before.)

Early in February 1573 Mandinga and twenty-nine other maroons led Drake and eighteen surviving buccaneers through the forest toward the Pacific. They moved in total silence, military style, maroons deploying ahead of the English, to mark the trail, and behind, to cover their tracks. After reaching Venta de Cruces in the morning of February 14, the party waited for the silver in the long grass by the side of the highway. Because the first stretch of the road on the Pacific side passed through low, open grassland, the mule trains traveled by night, to avoid the sun. (Later, in the deep forest, they traveled by day.) Within a few hours of Drake's arrival one of Mandinga's spies in Panamá delivered some news. The treasurer of the regional government in Lima was leaving town with fourteen mules, nine of them laden with gold and jewels. Behind him would follow two mule trains, each of fifty to seventy animals, carrying silver.

The pirates and maroons split into two groups, one led by

Drake, the other by Mandinga, about fifty yards apart from each other on the road. Drake's group would let the mule train pass until it could be ambushed by Mandinga's group. Then Drake and his men would close in from the rear, trapping the convoy fore and aft. Late in the evening the attackers heard the bells on the harnesses of the approaching mules. As soon as they came into view, an English sailor in Drake's group charged drunkenly out of hiding, waving his weapon. One of the maroons yanked him back into the grass, but the damage was done—a Spanish advance scout had spotted the sailor's white shirt in the moon-light. The scout wheeled about his horse, galloped back to the mule train, and told the treasurer to turn back to Panamá. The chagrined English rampaged through Venta de Cruces, wreck-ing warehouses and spoiling stores. But they found little and so fled to the coast, led by Mandinga. The maroons learned a lesson: Europeans were unreliable allies.

While Drake pondered his next move, his men spotted a ship belonging to a French pirate named Guillaume le Testu, who had learned that the English were on the isthmus and had been try-ing to find them for weeks. A fine cartographer who had helped found a short-lived French colony near Rio de Janeiro, Testu had been jailed for four years in France because of his Protestant faith. Freed after protests to the king, he had accepted a privateering commission, probably from Italian merchants. Now he hoped to join with Drake in swiping Spanish treasure. Drake, Testu, and Mandinga agreed to work together and take a silver convoy as it descended the hills in the outskirts of Nombre de Dios.

Again maroons led Europeans in a silent march through the forest, arriving at the ambush site on April 1. Again they split into two groups fifty yards apart along the road. In midmorning the waiting pirates and maroons heard bells—120 mules, the biog-raphy said, "every [one] of which caryed 300. Pound weight of silver, which in all amounted to neere thirty Tun." This time the scheme succeeded. The guards fled, leaving the convoy in the hands of the pirates. Giddy but too weary to lug all the silver through the hills, the Anglo-Franco-Afro-Indian force stripped the mules of their glittering burden and in true pirate fashion buried the booty at the bottom of a nearby stream. They carried away a few silver bars as trophies. Not until they were miles from the ambush did they realize that a Frenchman was missing. Later they learned that he had gotten drunk while burying silver and missed their departure. He was caught by Spanish troops and revealed, under torture, the location of the silver. From Nombre de Dios, the biography reported, "Neere 2000. Spaniards and Negroes [went out] to dig and search for it." They tore apart the area, found the precious metal, and transported it to Nombre de Dios. Drake's men, returning, were only able to find "thirteen bars of silver, and some few quoits of Gold"—less than 2 percent of the shipment.

Decades later, Philip Nichols, who had served as Drake's chaplain and become a friend, compiled surviving sailors' reminiscences of the expedition, passed the manuscript by Drake for editorial approval, and published the result—the authorized biography I have been quoting—under the curious title of *Sir Francis Drake Revived*. The book portrays Drake's sojourn in the isthmus—a time when he failed three times to seize large quantities of silver and lost half his men to disease and battle, including two of his brothers—as a rousing success. This view is not entirely wrong. The assaults on Nombre de Dios and Venta de Cruces *were* a triumph—for the maroons.

"CAPITULATIONS"

Reports of the maroon-pirate alliance appalled the Spanish crown, especially given that the Nombre de Dios merchants who reported the seizure of the silver shipment neglected to inform the government that they in fact had recovered almost all of the stolen money. (Much of the silver was tax payments for the court, so its disappearance truly stung.) Colonial officials used the incident to demand that the king send the fleet to clean out the maroons. "What grieves us most is to see with our own eyes the ruin of this realm imminent unless your majesty remedy the situation promptly," the governors of Nombre de Dios claimed a month after the attack. The court, justifiably fearful of being cheated, dragged its feet. While colonial officials dithered, sometimes trying to negotiate with Afro-Indian communities, sometimes seeking to raze them, maroons continued to steal cattle, free slaves, and kill Spaniards. Some of the dead Spaniards were priests; in their hatred of Catholic Spain, the maroons had happily let Drake convert them to Protestantism. (No evidence exists that they actually changed their previous religious practice.) Even when the two sides finally committed to negotiating, their mutual suspicion and hostility made progress agonizingly slow.

All the while, English, French, and Dutch pirates were coming to the isthmus, asking the maroons to help them as they had helped Drake. Most didn't get any assistance—the maroons seem to have acquired a low opinion of European competence. Nonetheless, Spanish fears of a maroon-pirate alliance continued to grow, reaching a kind of frenzy in 1578 and 1579, as the now-infamous Drake sailed up the Pacific coast of South America on another voyage, wrecking Spanish possessions along the way. Colonial officials approached Domingo Congo, leader of the regrouped maroons in Bayano's territory, with a deal: if his maroons promised to be loyal to the king, they would be given good farmland, cattle, and pigs, tilling and harvesting equipment, a year's worth of maize seed, and—most important—their liberty. As lagniappe, the colonists promised to exempt them from the taxes paid by Spanish residents. The terms were attractive, but Domingo Congo hesitated to accept—every maroon knew what had happened to Bayano when he negotiated with Spaniards. The colonists, for their part, were leery of rewarding people whom they viewed as thieves, murderers, and stolen property. Despite their distaste, though, they issued similar offers to the scatter of runaway groups in the hills outside Panamá town and the bigger, more centralized maroon "kingdom" near the planned location of Portobelo.

Portobelo's "king" put his mark on the treaty on September 15, 1579. The action delighted Felipe II, king of Spain. Four months later, when Domingo Congo's maroons in Bayano hadn't followed suit, the king urged the colonial government to close the deal:

> Because of the great importance of subduing the maroon blacks for the peace and quiet of these lands, we took great contentment in learning from your letter of the good state you have reached with them in Portobelo and we expect that their example can make those of Bayano understand the great favor that they will have from pardoning their crimes and the safe places they will live in and the other benefits that will follow the capitulation that you will send to our Council of the Indies.

"Capitulation"? The term *capitulación* means both "contract" and "military surrender." That is, the king described giving the maroons almost everything they wanted in exchange for ending a notional alliance with foreign pirates as a surrender—*by the maroons.* True, the maroons did not get to return to their African homes. But that would have been next to impossible; even had the colonists not reenslaved the maroons once they were confined on a ship, they wouldn't have known where to return them. Moreover, many maroons by this point had wives from other parts of Africa and the Americas. For better and worse, the isthmus had become their home. By "capitulating," they won the lasting, if uneasy, freedom to live as they wished, tax-free, in their own communities.

Two years later, Domingo Congo signed the treaty, as did the maroons outside Panamá. These agreements did not stop future escapes, as Tardieu, the University of La Réunion historian, has noted. Indeed, runaways continued to disappear into the forest until the end of the slave trade. Many escapees filtered into free maroon villages. By 1819, when the isthmus won its freedom from Spain, these communities' origin had been almost forgotten. Maroons had won the highest kind of liberty—they were ordinary citizens.*

The story is not exceptional. Although governments throughout the Americas wiped out many maroon groups, others won their freedom—along with the later anonymity that was its concomitant. A few examples are worth listing, if only because slaves' prospects for autonomy are all too often portrayed as completely dependent on the goodwill of their masters.

Mexico

Even as Spain was giving in to Africans who menaced the silver road in Panamá, it was facing Africans who menaced the silver road in Mexico. Sporadic, small-scale violence in the sugarlands of Veracruz flared into full-scale revolt after about 1570, with the escape of Gaspar Yanga or Nyanga, said to be a prince and general in what is now Ghana. Like Aqualtune in Palmares, he may actually have been one. Yanga, by all accounts a compelling, canny figure, united hundreds of Africans into a confederation in the mountains outside Veracruz. Driven by a kind of serene fury toward the people who had taken him in chains across the ocean, he led countless raids of sugar plantations, gleefully snatching slaves and provisions. Most important to New Spain, the maroons attacked convoys carrying silk and silver on the Veracruz–Mexico City road. Horrified colonists spread rumors that the maroons killed anyone who saw their faces and drank their victims' blood in Satanic ceremonies.

The colonial government, confounded by the rugged ter-

---

* They were not spared racial discrimination, of course. The ex-maroons were free to be treated just the same—just as badly, that is—as other free citizens of African descent.

rain, did little about the assaults until Yanga's forces committed the unforgivable sin of destroying a shipment of the most recent fashions from Europe. A military expedition of a hundred soldiers, an equal number of Indians, and two hundred colonists and their slaves charged into the mountains in January 1609. Six weeks later they occupied Yanga's base—and accomplished nothing, because the maroons had evacuated to a second, more remote base. Yanga dispatched a Spanish prisoner with eleven nonnegotiable demands, chief among them "that all those who escaped before last September will be free." The discouraged colonists accepted all eleven. Like the maroons of Bayano and Portobelo, Yanga's people were presented with their own domain: San Lorenzo de los Negros. Later renamed Yanga, honoring its founder, it was the Americas' first sunset town: Europeans were legally prohibited from staying the night there. Yanga and his descendants prospered so much that local Spaniards eventually paid them the ultimate compliment and moved in, ignoring the ban on whites. As a result, the town of Yanga is now almost completely "Mexican."

Two other, legally free African towns are known in Mexico proper, one in the mountains west of Veracruz and one on Mexico's west coast. But the maroons' greatest success may have occurred in the eighteenth century, on the Pacific coast of Guatemala. A hotbed of maroon activity, it was assaulted by Spain until its militia ran out of soldiers—a problem the government solved by replacing the militia with the Afro-Indian groups they were attacking. Once they controlled the army, the maroons used subtle threats to persuade officials to remove the last vestiges of slavery.

Nicaragua

English Pilgrims launched two colonies: the famous Plimoth, the first successful colony in New England, in 1620; and a short-lived effort in Providence Island, 140 miles off the coast of Nicaragua, in 1631. Unlike their brethren in non-malarial New

England, the Providence Pilgrims imported African slaves in numbers and with enthusiasm. As many as six hundred escaped when Spain drove out the Pilgrims in 1641. Landing in what is now Nicaragua by either shipwreck or design, they ended up mixing with Miskitu-speaking Indians and a small number of Europeans. More African and Indian refugees kept trickling in, swelling the ranks of the Miskitu, as these hybrid people came to be called. Viewing Spain as the biggest potential threat, they allied with the English who had previously enslaved some of their number. Riding with English buccaneers, armed with English swords and English guns, they raided Spanish plantations from Costa Rica to Panama, capturing Indian and African slaves and selling them to English sugar plantations; once the Miskitu even sent troops to Jamaica to help the English put down a maroon rebellion. London sealed the alliance by staging coronation ceremonies for Miskitu kings in Jamaica, Belize, or, occasionally, England. "King" was the word used at the time

Francisco de Arobe (middle) led Esmeraldas, an independent maroon society on the north coast of Ecuador. In 1599, two years after signing a treaty in which de Arobe accepted nominal Spanish sovereignty in return for a free hand in Esmeraldas, the colonial governor commissioned Andrés Sánchez Gallque, an Indian trained in Quito, to make this portrait of the leader, his twenty-two-year-old son, and a friend.

but is perhaps misleading; the Miskitu "kingdom" was a collection of four allied polities along the coast ruled by (from north to south) a "general," a "king," a "governor," and an "admiral."

As European diseases took their toll on Miskitu with native-American ancestry, all four areas became more African, genetically speaking. Culturally speaking, though, they increasingly claimed to be "pure" Indian—a claim that seems strangely at odds with their kings' habits of performing their functions in gold-spangled military uniforms with white satin or cotton vests, breeches, and stockings, leaning on the gold- and silver-headed walking canes that had become a symbol of their office. Thousands of Britons moved into the area in the nineteenth century, paying taxes to Miskitu governments and promising to obey Miskitu laws. If they began to throw their weight around, the Miskitu would remind the British of the usefulness of having an ally on the otherwise solidly Spanish expanse of Central America. The kingdom thrived, controlling its own destiny, for more than three centuries. Only in 1894 did the now-independent nation of Nicaragua formally incorporate it.

The United States

Maroons were fewer in the United States than farther south, because slaves could escape bondage altogether if they traveled north of the Mason-Dixon line. In addition, they found it harder to survive on their own in unfamiliar temperate ecosystems. Nonetheless, maroon encampments were common in places like the valley of the Savannah River, the Mississippi River delta, and, especially, the Great Dismal Swamp, a peat bog that then sprawled across more than two thousand square miles of Virginia and North Carolina. (It is now smaller, because much of the swamp was drained in the nineteenth century.) To escape European incursions, Indians moved there in numbers after about 1630, living in scattered, small settlements of ten to fifty houses. Africans soon followed. Thousands eventually made their base there, according to the historians John Hope Franklin and Loren Schweninger, building villages on raised "islands" in the rarely seen heart of the swamp. Hidden from slaveholding society, some maroons had children who reportedly went their entire lives without encountering a European. This happy isolation ended at the end of the seventeenth century, when Virginia initiated big swamp-drainage projects, sending thousands of slaves to dig drainage canals in wretched conditions. Would-be maroons and would-be maroon-hunters alike used the canals to penetrate the marsh, setting off low-intensity guerrilla warfare that did not truly let up until the end of U.S. slavery. (Harriet Beecher Stowe, author of *Uncle Tom's Cabin*, wrote her second novel, *Dred*, about the Great Dismal Swamp in that time of conflict.) By that time, though, the establishment of the "underground railroad" to freedom in the north had robbed the swamp of much of its allure.

Farther south, the best hope for slaves who wished to rid themselves of their bonds was the Spanish colony of Florida. Carolina was founded in 1670 (I described this in Chapter 3). Large numbers of slaves began to arrive a few years after. Quickly they began to escape, also in large numbers, crossing the border into Spanish Florida. A few Europeans, fleeing for one reason or another from their colonial governments, took refuge there as well. Seeing the military potential in these England-hating maroons, the Spanish king promised in 1693 to grant automatic liberty to all Africans who came to Florida from the Carolinas and Georgia, provided that they (1) agreed to convert to Christianity; and (2) promised to stand by Spain and fight any English invasion. Near the Spanish capital of St. Augustine the colonial government in 1739 established a new town, Gracia Real de Santa Teresa de Mosé, to house what amounted to a militia of ex-bondsmen—the first legally recognized free African-American community north of the Rio Grande. (Other free maroon communities surely existed, but were not officially viewed as legitimate.) Most Florida maroons, though, went deep into the interior of the peninsula, territory dominated by Seminole Indians, a group that had split off from the Creeks decades before, taking over land that had been depopulated by disease. In this low, sandy area, a savannah that had been annually burned for hundreds of years, the two groups formed a strong but carefully delineated alliance.

That any two groups of Indians and Africans would cooperate was not a given—just north of Florida, the main body of the Creek enthusiastically hunted maroons and sold them to the English. Ultimately the Seminole established more than thirty towns, some with thousands of inhabitants, all surrounded by farmland, polycropped in the indigenous mode. Four of those towns were mainly inhabited by Africans—Black Seminole, as they are often called. The relationship between "red" and "black" Seminole was complex, beginning with the fact that some Africans were "red" and some European refugees were "black." Under Seminole law, most Africans in those towns had the legal status of slaves, but native bondage resembled European feudalism more than European slavery. Seminole slaves owed little work; instead they were supposed to provide native villages with tribute, usually in the form of crops. The burden, though of course unwelcome and resented, usually was not onerous. Many of the slaves were African soldiers, disciplined and organized as one would expect from prisoners of war in wartime. Determined to establish themselves, maroons opened up trade with the Spanish and as a group became more prosperous than their Indian owners. For the most part they lived adjacent to but carefully separate from the Seminole, unincorporated into the big kinship-linked clans that were a principal aspect of Indian social networks. Yet they willingly joined their owners in common fights, of which there were, alas, all too many.

The Seminole faced a parade of adversaries. England took over Florida in 1763; the Seminole resisted all efforts at incorporation. Twenty years later, the United States came into existence; the English stopped seeking to dominate the Seminole and instead asked them to ally with them against the new nation (England had held on to Florida after the revolution). In 1812, the Seminole violently opposed U.S. efforts to annex Florida. Another flareup

European societies invariably portrayed their conflicts with maroons as victories. The Battle of Okeechobee, fought on Christmas Day, 1837, during the Second Seminole War, ended with the U.S. forces being driven back with twice as many dead and many more wounded than the Seminoles. Much of the blame for the disaster belongs to Col. Zachary Taylor, the commanding officer and future president, who foolishly insisted that the Seminoles would flee if attacked directly. Yet this typical engraving from 1878 depicts the Seminoles melting away before Taylor's heroic, bayonet-wielding charge.

occurred in 1816–18; many Seminole, black and red, were driven south to new settlements, the biggest of which, Angola, was at the mouth of the Manatee River in Tampa Bay. Some fled to the Bahamas. In both cases the Seminole received covert support from British guerrillas. Conflict grew more intense still when the United States took over Florida in 1821 and the government, responding to popular pressure, planned to "remove" the native peoples of the Southeast, the Seminole among them, to Indian Territory, a big reservation in what is now Oklahoma. Overt war began in 1835. Maroons joined in, fighting as allies but under their own command.

The Seminole strategy was twofold: First, they destroyed the plantations that supplied U.S. troops, capturing their slaves to bolster the native army. Second, they waited for yellow fever and malaria to kill northern soldiers. If they got in a jam, they pretended to negotiate until the onset of the "sickly season" forced U.S. forces to withdraw. It was so brilliantly successful that in 1839 Thomas Sidney Jesup, commander of the U.S. army in Florida, wrote to Washington, D.C., to ask permission to give the Seminole everything they wanted if they would simply stop wrecking plantations. The idea was indignantly rejected, but Jesup did come up with what would eventually become a winning strategy: he promised that any Africans who gave up fighting and consented to settle in the West would be given their liberty. Slowly the offer pried apart the Seminole-maroon alliance. Its success was understandable, as the abolitionist Joshua Gibbons recognized, for it gave the maroons "that security for which they had contended for a century and a half." After seven years of increasingly brutal war, the conflict ground to a halt with a cease-fire. Several hundred Seminole remained, unconquered, on the land they had fought to keep; the rest had accepted offers of land and liberty, establishing communities that still exist in Texas, Oklahoma, and Mexico.

Haiti

A French possession with about eight thousand plantations rich with sugar, coffee, and yellow fever, eighteenth-century Haiti was a true extractive state: forty thousand fabulously rich European colonists atop half a million seething African slaves. St. Domingue, as the colony was then called, was shaken by the advent of the French Revolution in 1789. *Liberté, egalité, fraternité!*—the resonance, for an island of French slaves, was obvious. Paradoxically, though, the loudest local supporters of the revolution were French sugar growers, slaveholders who had long chafed at royal restrictions on the slave trade. (Freedom, to

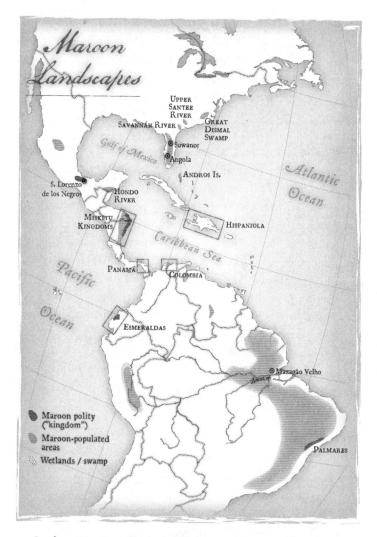

Maroon Landscapes

UPPER
SANTEE
RIVER

SAVANNAH RIVER

GREAT
DISMAL
SWAMP

Suwance

Angola

Gulf of Mexico

ANDROS IS.

S. Lorenzo
de los Negros

HONDO
RIVER

MISKITU
KINGDOMS

Atlantic
Ocean

HISPANIOLA

Caribbean Sea

Pacific

Ocean

PANAMÁ

COLOMBIA

ESMERALDAS

Mazagão Velho

Amazon

Maroon polity
("kingdom")

Maroon-populated
areas

Wetlands / swamp

PALMARES

In the centuries of the slave trade, flight was frequent and often successful. Mixing with native groups, escaped Africans and their descendants scattered across the hemisphere. Many formed Afro-Indian polities, microstates that often won de facto independence from Spain—a tenacious struggle for liberty that created large free areas in the Americas decades and even centuries before the U.S. Declaration of Independence.

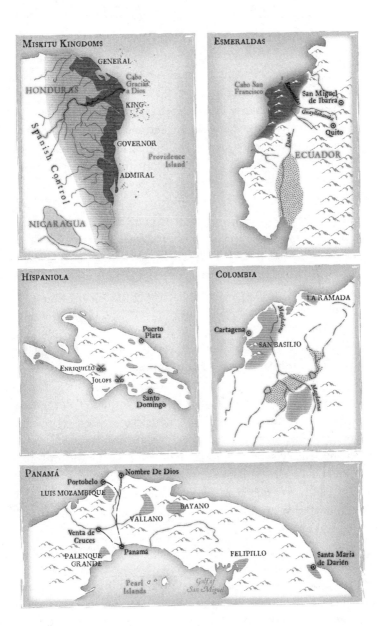

them, meant the freedom to enslave.) Fearing the consequences of planter rule, Africans opposed the forces chanting *"Liberté, egalité, fraternité!"* Seizing the moment, they launched a revolution against the revolution.

The new republic in France, ensnarled in internecine battles, became involved in a war with England and its allies. Wanting to deny sugar revenues to France, England seized Haiti's main cities in 1793. Its troops proved welcome hosts to that malign participant in the Columbian Exchange, the yellow fever virus. According to J. R. McNeill, the Georgetown historian of mosquito-borne disease, the British army lost roughly 10 percent of its troops every month between June and November of 1794. Survivors of yellow fever were prostrated by malaria. The army hung on, helped by reinforcements, until the next summer, when the monthly death rate rose to as high as 22 percent. "The newly arrived died with astonishing quickness," McNeill wrote, "seemingly disembarking from ships straight to their graves." Again they were reinforced: 13,000 more troops arrived in February 1796. In weeks 6,000 were dead. The British abandoned Haiti in 1798.

All the while the slave revolt continued, led by the brilliant, charismatic, and dictatorial Toussaint Louverture. Toussaint, as he is known, had little time to savor Britain's defeat. Napoleon Bonaparte had staged a coup in France and determined to retain the immensely profitable sugar and coffee plantations of Haiti. A French force of perhaps 65,000 landed in February 1802. Toussaint had barely half as many men and so little equipment and weaponry that his army was, he said, "naked as earthworms." He ordered his rebels to retreat to the hills and await the fever season. Toussaint was captured and imprisoned but his strategy prevailed. By September some 28,000 French were dead; another 4,400 were hospitalized. Two months later the French commander died. His army struggled on, but it was trying to conquer its own cemetery. The effort collapsed in November 1803, having lost 50,000 of its 65,000 troops. As McNeill noted, the same malaria and yellow fever that had done so much to promote African slavery here helped Africans to destroy it. Napoleon, his hopes for a Caribbean empire in ruins, sold the United States all of France's North American territories: the Louisiana Purchase.

Much of the United States' present territory is thus owed indirectly to maroons—not that the newly expanded nation showed much gratitude. Independent Haiti, an entire maroon nation, became a global symbol that terrified slaveholders throughout the world, including the United States. All of Europe and the United States put a punishing economic embargo on Haiti for decades. Deprived of the trade in sugar and coffee that had been its economic lifeblood, the nation's economy collapsed, impoverishing what had been the wealthiest society in the Caribbean.

Suriname

A few Dutch and English adventurers showed up in coastal Suriname, north of Brazil, in the early seventeenth century, intending to grow coffee, cocoa, tobacco, and sugarcane. Because the Europeans had valuable trade goods, indigenous rulers initially tolerated their presence—they could be expelled at any time. Indeed, one imagines the Indians watching with amusement as the tiny Dutch and English colonies promptly went to war with each other over their notional possession of the area. The struggle was part of a worldwide battle between the English and Dutch over the part of global trade not dominated by Spain. In 1667 a treaty was hammered out on terms favorable to the Dutch. The Netherlands won Suriname, with its rich potential. As a kind of booby prize, the English received official title to a cold, thin-soiled island known to its original inhabitants as Mannahatta.

Quickly the Dutch set to work. Ships full of imprisoned Africans docked at the minute port of Paramaribo, at the mouth of the Suriname River. Slave-rowed barges conducted them thirty miles upstream, to sugar plantations centered on the village of Jodensavanne (Jews' Savanna), founded by Jews fleeing the Spanish Inquisition.* There the Indians' managed forest was replaced by waving expanses of Dutch sugarcane. Interspersed with the cane were fields of African rice. As in the Caribbean, logging and farming benefitted mosquitoes, especially *Anopheles darlingi,* which I noted in Chapter 3 was South America's most important malaria vector. Slave ships introduced *Aedes aegypti,* the yellow fever mosquito. The slaves themselves brought falciparum malaria and yellow fever. All went upstream to Jodensavanne. *A. darlingi* likes to breed in recently cleared land, where it can dash back and forth between the edge of the forest and human houses. As the colonists forced slaves to cut trees, European death rates soared. Dutch landowners responded by staying home and hiring overseers to manage their properties. "Managing properties" mainly meant importing Africans. About 300,000 landed on Suriname's shores. Another way of saying this is that a colony about the size of Wisconsin absorbed nearly as many slaves as the entire United States. For each European, the colony had more than twenty-five Africans.

As one would expect, the few malarial Dutch were unable to prevent their captives from escaping. Africans ran away by the thousands, mixing with native groups and establishing outlaw hybrid societies in the boondocks. Guerrilla war broke out in the 1670s and continued for almost a century, the Dutch slowly losing. In 1762 the colonial government signed a humiliating peace treaty—the Dutch signatories, following African custom, reluctantly guaranteed the peace by cutting themselves and drinking their blood. The maroons' main concession was to promise that they would give back new escapees. As a result, runaways went to other parts of the forest and established new communities. Efforts to pursue them ignited a second guerrilla war. Suriname's planters begged for help.

---

* Sephardic Jews were prominent landowners and slaveholders in Suriname. Elsewhere in the Americas, though, they were not especially important slaveowners.

More than a thousand soldiers came across the Atlantic in 1772, among them John Gabriel Stedman, born in the Netherlands to a father who had fled Scotland's famines. Stedman kept a diary that is an encyclopedia of medico-military calamity. Soon after landing he "became so ill by a fever—that I was not expected more to recover." None of the other soldiers helped him: "Seekness being so common in this Country, and every one having so much ado to mind themselves, that neglect takes place betwixt the nearest acquaintances."

Stedman was lucky enough to survive his seasoning and go upstream. The once carefully managed Indian landscape was now a nightmare of pests. Stedman's diary fairly pulses with complaint about the "inconceivable numerous" mosquitoes—insects in such thick, buzzing clouds that they smothered candles and made it impossible to see or hear people a hundred feet away. Stedman once clapped his hands together and killed thirty-eight.

Sick, miserable, insect-bitten, dressed in tatters, Stedman's force futilely chased runaway slaves through the forest for three years. They fought exactly one battle. They won that battle, as the adage goes, but lost the war. "Out of a number of near twelve hundred Able bodied men, now not one hundred did return to theyr Friends at home," Stedman wrote sadly, "Amongst whom Perhaps not 20 were to be found in perfect health." All the others, he said, were "sick; discharged, past all Remedy; Lost; kill'd; & murdered by the Climate, while no less than 10 or 12 were drown'd & Snapt away by the Alligators."

Eventually the Dutch and the maroons reached a kind of accommodation. The Europeans kept shipping in Africans and growing cane, accepting that a certain number of slaves would escape each year. Meanwhile, most of the Dutch colonists stayed as little as they could; in 1850, after two centuries of colonization, Suriname had perhaps eight thousand European residents, most of them agents for sugar planters who lived safely in the Netherlands. Not residing in the colony, the growers had little interest in creating the institutions that underlie a productive society.

Every scrap of profit went back to the home country; education, innovation, and investment in Suriname were almost entirely ignored. When Suriname became independent in 1975, it was one of the poorest countries in the world.

Naturally, the new nation sought development. Suriname has large deposits of bauxite, gold, diamond, and oil and more tropical forest per capita than any other nation. The cash-strapped government—both the military dictatorship that seized power in 1980 and its civilian successor, which began in 1992—awarded mining and timber rights to foreign companies. In the 1960s, the colonial government had let Alcoa, the big aluminum company, build a six-hundred-square-mile lake to feed a hydroelectric dam for aluminum refining. Now the independent government awarded China International Marine Containers, the world's biggest container-manufacturing firm, the rights to log almost eight hundred square miles to make wooden shipping pallets. Other firms followed suit. By 2007 some 40 percent of the country's surface area had been leased for logging.

All the while the government was fending off environmentalists' criticism by creating parks. At a joint press conference in 1998 with Conservation International, the nation announced that it had set aside six thousand square miles—10 percent of its territory—to create the Central Suriname Nature Reserve, the world's biggest protected tropical forest. "Suriname's example," *The New York Times* editorialized, is "a small ray of hope." UNESCO named the park a World Heritage Site in 2000, lauding it as "one of the very few undisturbed forest areas in the Amazonian region with no inhabitants and no human use."

Beginning with the blood-drinking treaty of 1762, the Dutch had recognized the autonomy of six maroon groups, of which the biggest today are the Saramaka and Ndyuka, with about fifty thousand people each. None had been apprised beforehand about the logging and mining concessions, though many were on their land. None had been consulted about the dam, which inundated maroon villages (in a further insult, the turbines silted up and are now useless). Nor had they been asked about the park, which includes part of the homeland of the Kwinti, the smallest of the six maroon groups, who have been in that area since about 1750. (It also houses an Indian group called the Trio.) The government's actions led a coalition of Saramaka leaders to file a complaint with the Inter-American Commission on Human Rights in October 2000. Angered, Suriname's president charged that the Saramaka petition showed that they wanted to ally with Colombian narco-guerrillas to foment civil war. The government vowed to continue opening land to logging and mining, a stance it reiterated when the commission ordered that the process be suspended, and reiterated again in November 2007, when the Inter-American Court of Human Rights demanded that Suriname give the Saramaka control over their resources.

The nation has not complied, as of the time of this writing. Indeed, the jousting among maroons, governments, and large corporations seems likely to last for years. The stakes are nothing less than the future of the tropical forest itself, and the maroons are not fighting only in Suriname.

SHAKE IT, OX!

In 1991 Maria do Rosario Costa Cabral and her siblings bought twenty-five acres on the banks of Igarapé Espinel (Espinel Creek), a sub-sub-tributary of the Amazon in Amapá, Brazil's northeasternmost province. A wiry, watchful woman of sixty-two, Dona Rosario was born into a maroon community called Ipanema—a place so poor, she told me, that families cut their matches in half lengthwise to make a box last twice as long. Her father spent his days as a rubber tapper, toting the latex to one of the small natural-rubber distributors that still hang on in the area. If he and his friends showed up with a lot of rubber, wealthier people would realize they had found an especially productive group of trees. They would figure out the location, force out the rubber tappers, and take over. The same thing happened with their farms. They would acquire abandoned land—a plantation that had failed twenty or thirty years before—and pull out a few harvests. Just as the family was settling in, men with guns would show up. You are squatters, they would say. If they had a contract, they would say the title was invalid. Leave now, they would say, touching their weapons. Little changed when Dona Rosario reached adulthood. Repeatedly she set up farms and repeatedly she was pushed off them. Still, she jumped at the opportunity to buy the land on Igarapé Espinel.

To non-Amazonians, the property wouldn't have seemed worth troubling about. It is located about two hundred miles from the river's mouth, where the Amazon is so large that it acts like a tidal body—tides flood the area twice a day. The force is so large that deep within the forest nameless streams well over their banks and march inland, sometimes for miles. People build their homes on stilts and paddle their canoes between the trees. Even when the surface is exposed, it is thick with gooey mud. I visited Dona Rosario's farm recently with Susanna Hecht, the UCLA geographer. The mud soon covered us to our knees and practically ripped the boots from our feet.

Dona Rosario told us that she got the property cheap, because it had been ravaged by the heart of palm craze of the late 1980s, when every fashionable menu from London to Los Angeles had to feature heart of palm salad. Heart of palm is the growing tip and inner core of young palm trees, particularly South American species like açai (*Euterpe oleracea*), jucura (*Euterpe edulis*), and pupunha (peach palm, *Bactris gasipae*). Determined to wring every penny out of the forest that they could, palm hunters scoured the lower Amazon with the implacability of paid assassins. Barges discharged crews with axes and winches who chopped down entire palm groves to obtain the edible tips (the hearts can be removed without killing the tree, but this takes more time). If they spotted anything else that looked valuable, they took that, too. "The land was looted," Dona Rosario told us. "It was a mass of vines and scrub."

She set out to bring it back with techniques she had learned from her father in the region of her birth. With help from her sisters and brothers, she planted fast-growing timber trees for sawmills upriver. For the market, they put in fruit trees: limes, coconut, cupuaçu (a relative of cacao prized for its fragrant pulp, rather than its seeds), and açaí (formerly used for heart of palm, the tree has purple fruit that produce a yogurt-like pulp). With woven shrimp traps—identical to those in West Africa, Hecht told me—the family caught shrimp and kept them alive in cages that drifted in the creek. At the river's edge they encouraged shrubs that made habitat for fish and fry and planted trees with seeds and fruit that would attract them into the flooded forest. To an outside visitor, the result looked like a wild tropical landscape. The difference was that almost every species in it had been selected and tended by Dona Rosario and her family.

Dona Rosario lives on the fringes of a sprawling *quilombo* complex centered on Mazagão Velho (Old Mazagão), founded in 1770 by transplanting almost entire the last Portuguese colony in North Africa. The year before, the inhabitants had fled before a Muslim army, arriving as a body in Lisbon. Treating defeat as opportunity, the Portuguese court ordered the community to resettle en masse in Amapá, where its presence was supposed to thwart potential incursions by French Guiana, Amapá's northern neighbor. A Genoese engineer designed the new town as a graceful Enlightenment-era city, complete with public squares and gridded streets. Slaves actually built more than two hundred houses in what was then called Vila Nova Mazagão (Mazagão New Town); the Portuguese may have moved as many as 1,900 people into them. The transition was eased by grants of cash, livestock, and several hundred slaves. Soon the newcomers made the unhappy discovery that the lower Amazon, unlike the dry, breezy Moroccan coast, is hot and humid—it is located almost exactly on the equator. Within a decade of arrival the colonists—malarial, famished, living in wretched huts they were too poor to repair—were begging the crown to relocate them. Ultimately, almost all of the surviving Europeans slipped away. The remainder soon died. Through no act of their own, the slaves found themselves at liberty. Vila Nova Mazagão had become a *quilombo*.

They were free as long as they pretended they weren't. The Portuguese administration wanted to be able to report to the king that his subjects were guarding Brazil's northern flank. The slaves were willing to say they were doing it, if that meant they would be left alone. Everyone was happy: the maroons pretended they were Portuguese subjects in a Portuguese colony and the Portuguese pretended the maroons were guarding the frontier. As the decades went by, the descendants of the colony's Africans spread out along the riverbanks, living much like their Indian neighbors. The river supplied fish and shrimp, the small-scale garden cultivation yielded manioc, the trees provided everything else. Two centuries of constant tending and harvesting structured the forest. Mixing together native and African techniques, maroons created landscapes lush enough to be mistaken for pristine wilderness.

So did others. Portuguese euphoria from the destruction of Palmares had been short-lived. Slaves continued to escape and to live in the forest. But they didn't repeat the mistake of forming big, centralized communities like Palmares. Instead they created ten thousand or more small villages in a flexible, shifting network that spread across much of eastern Brazil and the lower Amazon. They mixed with extant native settlements, collected Indian slave escapees, threw open their doors to Portuguese misfits and criminals. Many Africans had lived in tropical environments before being shipped across the ocean. They were comfortable in hot, wet places where people farmed palms and kept trapfuls of shrimp in the stream. They were happy to learn when Indians showed them how to fish by scattering poison in a tributary or make protective "boots" by melting latex over their

Hundreds of *quilombos* were established in the lower Amazon, a maze of rivers that tidally spill over their banks twice a day, washing a mile or more into the interior. Because the rivers are the main transport routes, villages spread out along the banks (top, Anauerapucu, in the state of Macapá); houses are built on stilts (bottom, in Mazagão Velho) to let the tidewater pass beneath the floorboards.

feet or squeeze the bitter compounds out of manioc with long, tubular baskets. Ideologically opposed to "going native," the Portuguese were much less willing to adjust. In consequence, the forest seemed dangerous to them, a place to be ventured into only with an army. Ceding the field to *quilombos,* the colonists were only partially aware that the escaped slaves were living within a short walk of the plantations, as in Calabar or Liberdade. In consequence, the *quilombos* were left largely alone—unless they were unlucky enough to be in the path of gold miners, rubber tappers, or other people who sought quick wealth in the forest.

Brazil has a host of hybrid spiritual regimes—Candomblé, Umbanda, Macumba, Santería—often focused on special areas where Afro-Brazilians drum, dance, and practice the ritualized martial art of capoeira. In their isolation, Brazil's *quilombos* built their own pageants and festivals atop these spiritual traditions, binding together communities in steel hoops of shared memory. Consider the satirical *bumba-meu-boi* (loosely, "shake it, ox"), celebrated in *quilombos* across northeastern Brazil. In the version celebrated by the *quilombo* of Soledade (Solitude) in the eastern state of Maranhão, villagers pay festive homage to the fable of Pai Francisco, a henpecked African slave whose pregnant wife hankers for a taste of ox tongue. Alas, the only nearby ox is the pride and joy of Francisco's brutal master. Worse still, Francisco has been entrusted with its care. Nonetheless, he leads the beast into the forest and puts his knife into it. Quickly apprehended, Francisco is threatened with death unless the ox can be resurrected. Dancers representing authorities from the local mayor to the national president haplessly struggle to bring the beast back to life, giving spectators a chance to hoot at their failures. Ultimately, native priests revive the animal with blasts of tobacco, perfumed waters, and the shaking of special rattles: the indigenous arsenal of cure. Crowds cheer as the ox staggers to its feet and exhort it to dance lively—*bumba, meu boi!* A cheerful mashup of America (tobacco, priests, and forest creatures) and Africa (cows, slaves), *bumba-meu-boi* is the tale of the *quilombo* itself:

Seringa (*Hevea brasiliensis*) rubber—many uses

Acasuzeira: timber (poles, firewood)

Pão Mulato (*Mirtaceoe facies*) timber (needs fire, light to grow, so disturbance indicator)

Guava (*Psidium spp.*): fruit, fish food

Açaí (*Euterpe oleracea*): see above

Mango (*Mangifera spp.*) fruit (intro. from India)

Andiroba (*Carapa guianensis*): lamp oil, bug repellent, medicines

Inga (*Inga edulis*) fruit, bird and monkey attractant

Açaí (*Euterpe oleracea*): fruit, heart of palm, timber

Aninga (*Montrichardia linifera*): Erosion control, fish/shrimp habitat

Seringa (*Hevea brasiliensis*) rubber, bird attractant, fish food

Acasuzeira: timber (poles, firewood)

Lemon (*Citrus limon*): fruit

To inexpert eyes, the riverbank across from Maria do Rosario's home looks like a typical tropical hodgepodge. But almost every plant in this image was sown and tended by Rosario and her family, creating an environment as ecologically rich as it is artificial.

slaves escaping their fate with the help of Brazil's original inhabitants.

Five hundred miles southwest, the *quilombo* struggle for freedom is revisited even more overtly at the rite of *lambe-sujo* (an insulting reference to the red African cloth used for turbans—the equivalent, perhaps, of "towelhead"). Covering their entire bodies with a shimmering, tarry coat of charcoal and oil, *quilombo* dwellers in the state of Alagoas reenact their ancestors' lives in an annual pageant. The day begins with men and women playing runaway slaves gathered in a protective circle around a king and queen—African nobility, like Aqualtune and Yanga. Some of the slaves suck on baby pacifiers, symbolizing the cruel circular plugs strapped into the mouths of recalcitrant slaves. Ominously lurking at the edges are *caboclinhos* (another pejorative term, perhaps translatable as "redskins")—Indian trackers who were agents of the Portuguese. Their bodies dyed red with plant oil, brilliantly colored feathers exploding from their heads, the trackers meet the Africans in their protected circle. After ritualized struggle, the *caboclinhos* win; as the *lambe-sujos* are dragged through the streets, they beseech bystanders for money in a final attempt to buy their freedom.

In these Afro-Indian communities, the context is head-spinning: people with African ancestors in what amounts to blackface, people with native ancestors who allied with Africans playing other natives who fought with them. Somehow stepping across the centuries, the eighteenth- and nineteenth-century Africans beg contemporary Brazilians for the means to attain liberty.

Legally, Brazil's *quilombos* had had nothing to fear after the nation abolished slavery in 1888—nobody was going to return runaway slaves to captivity. But the end of slavery did not mean an end to discrimination, poverty, and anti-maroon violence. The nation's maroon communities continued to conceal themselves, staying so far out of official sight that by the middle of last century most Brazilians believed that *quilombos* no longer existed. In the

1960s, the generals who then ruled Brazil looked on their maps and observed to their displeasure that about 60 percent of the country was blank (actually, it was filled with Indians, peasant farmers, and *quilombos,* but the government dismissed them). To the generals' way of thinking, filling the emptiness was a matter of national security. In a breathtakingly ambitious program, they linked the brand-new, ultramodernist capital, Brasília, the western frontier, and the ports of the Amazon by slashing a network of highways across the interior.

In the 1970s and 1980s hundreds of thousands of migrants from central and southern Brazil thronged up the highways, believing the generals' promises that they could begin new lives

Constantly hunted by slavers, the escaped slaves and natives who coalesced into Brazil's *quilombos* naturally sought spiritual comfort—and found it in an extraordinary variety of religious observances that mixed African, Indian, and Christian elements. These limbs hang in the Room of Miracles in Salvador's Igreja de Bonfim, votive offerings given as thanks for miraculous cures in a church that is a holy place for both Catholicism and the Afro-Indian religion Candomblé.

in new agricultural settlements. Instead, they encountered bad roads, poor land, and lawless violence: *Deadwood* with malaria. Many smallholders abandoned their farms soon after clearing them—few conventional annual crops would grow in Amazonia's aluminum-saturated soil. In the long run, the big ranches didn't do much better, even though many received subsidies from the military government. In the short run, they deemed all people found on their property to be squatters and removed them, often at gunpoint. In this way countless *quilombos* were expunged, their inhabitants scattered—Dona Rosario's family was probably among them.

The onslaught of ranches was greeted by worldwide protest. Chico Mendes, a kind of Brazilian Martin Luther King, led an international campaign to recognize the rights of the Amazon's inhabitants to their land. Meanwhile, the dictatorship's hold on power unraveled as Brazil plunged into economic crisis. The nation enacted a new, democratic constitution in October 1988. Two months later a rancher-paid hit man killed Mendes. But the assassination was too late to stop his cause. Among other things, the new constitution already declared that "quilombo communities" are "the legitimate owners of the lands they occupy, for which the State shall issue the respective title deeds."

"Nobody understood the implications of this," said Alberto Lorenço Pereira, undersecretary for sustainable development in the Brazilian ministry of long-term planning, which formulates the nation's land-use policy. When the new constitution was enacted, he told Hecht and me, its drafters imagined "a few remnant *quilombos* somewhere in the forest" whose elderly members would be rewarded with their fields. Now many researchers believe that as many as five thousand may survive in Brazil, most of them in the Amazon basin, occupying perhaps 30 million hectares—115,000 square miles, an area the size of Italy. Not only did the *quilombos* occupy an enormous territory, much of it spread out along riverbanks, which meant that they controlled access into a still-larger expanse in the interior. Conflict was inevitable, Pereira said. "A lot of people want that land."

I saw what he meant when I visited the *quilombo* of Mojú, four hours of bone-jarring muddy road from Belém, the city at the mouth of the Amazon. Its twelve linked settlements had been founded by runaways sometime in the late eighteenth century. It had existed in hiding for almost two hundred years, Manuel Almeida, head of the village *quilombo* association, told me. The end of slavery had brought no relief, Almeida said. The rubber tappers had come first, grabbing Mojú's rubber trees. Then came the timber companies, stripping the forest of mahogany and dye-wood. Cattle ranches had seized land in the 1960s and 1970s—the properties, though little used, were still fenced off. A company punched through roads to a bauxite mine upstream. Two other firms that mine kaolin, a special white clay used in porcelain-making and paper-making, had jammed pipelines through the middle of the village. Now the bauxite firm—a subsidiary of Companhia Vale do Rio Doce, the biggest mining company in the Americas—wanted to put a pipeline for crushed bauxite through Mojú on the way to a big refinery west of Belém. All of this had occurred without permission or consultation, Almeida said. The government had granted the firms concessions that gave them the right to build these things because the *quilombo* had no legal existence.

Almeida was talking in his home, in a room that was bare except for a hammock and a crucifix on the wall. Now and then his wife and brother walked in and offered glasses of water. He said that he had heard that Brazilian companies were prospecting in the region for natural gas. He said that he had heard that American companies wanted to put in resorts at the mouth of the Amazon. He said that a man had come by with some papers that he said gave him the right to put in a farm for oil palms. He said that Mojú's twelve communities had existed for two centuries and that this ought to count for something.

THE VIEW FROM DONA ROSARIO'S FARM

Two years after relocating Mazagão from North Africa to the northern Amazon, the Portuguese feted their own bravery by honoring St. James, the patron saint of Iberian anti-Muslim activities. For the colonists, isolated on the equator, it must have been an apprehensive time; according to Laurent Vidal, a historian at the University of La Rochelle who is the author of a study of Mazagão, the clergy, too, were gloomy, fearful that civilization itself was under assault. Perhaps that is why they jointly chose to honor a golden moment in Mazagão's history: the day, two centuries before, when the favor of St. James had allowed them to repel an attack by Sultan Abdallah al-Ghalib Billah, the powerful ruler of much of what is now Morocco. Something about the occasion took hold in the celebrants' imaginations—not just those of the colonists, but also their slaves. As the Portuguese left Vila Nova Mazagão, their slaves stepped in to take their places in the ritual. Decades after the last European had departed, its African and Indian inhabitants were still reenacting a faraway battle between Islam and Christendom. They still do today.

Over time, the celebration has grown ever more elaborate, ever more encrusted with ritual—and ever more disconnected from actual events. The battle that maroon descendants celebrate today is entirely different from the battle commemorated by the founders of Vila Nova Mazagão. Sultan Abdallah has vanished, replaced by a Muslim leader named, mysteriously, Caldeira (Boiler). When Caldeira's siege does not breach the walls of Mazagão, Caldeira tries a Trojan Horse–like ruse. Admitting the failure of his attack, he proposes rewarding the Christians' courage with a masked ball, at which he will serve platters of delicacies, a treat for hungry soldiers. In fact, the sultan plans to use the masked ball as a cover to persuade Portuguese soldiers to defect. Those who remain loyal will be given the sweets, which are poisoned. The Portuguese wisely suspect the gifts. They slip some of the food to Caldeira's horses, which expire promptly. At the ball, they give some to his men, killing them. Then they feed Caldeira, killing him. By morning, the dance floor is littered with corpses.

Enraged by his father's death, Caldeira's son Caldeirinha (Little Boiler) attacks the fort. The weary Christians are overwhelmed by the vengeful Muslims. To demoralize them further, Little Boiler orders his men to kidnap all the children in the city. Now enraged and vengeful themselves, the Christians counterattack. The tide of battle turns as the day draws to an end. Realizing that night will give the Muslims time to retreat and regroup, the Portuguese pray for more time. In the heavens, St. James hears their pleas. His holy fingers reach into the sky and stop the sun from setting. With the extra hours of daylight the Christians drive away Little Boiler's army, capturing him along the way.

An epidemic in 1915 forced many of Vila Nova Mazagão's people to move the town again, to an area about an hour down the river. They called its third incarnation Mazagão Nova; the second one was changed to Mazagão Velho, Old Mazagão. Ultimately many of the maroons didn't like the new city, which was more accessible. They returned to Mazagão Velho. Again the festival proved to be a way of knitting together a community spread over dozens of rivers. It grew into a full-fledged theatrical reenactment, complete with a delivery of "poisoned" sweets, an all-male masked ball, a "stoning" of a Muslim spy with tomatoes and oranges, an "abduction" of children, and a stylized battle on horseback in orange and green costumes.

I took a boat one morning to visit Mazagão Velho. The rivers were crowded with vessels taking children to school—one of them held an entire soccer team, exuberant in handmade uniforms. The town was getting ready for the festival. Somebody was testing the loudspeakers on the main church with carimbó, the dance music of the lower Amazon. Children ran from the boats to their classrooms under displays of flags and bunting.

The laughter belied a division in the town. Newcomers, we were told, were trying to make the festival into a tourist attraction. They were throwing out the old costumes and masks and bringing in new ones with more international appeal. The old costumes had been hidden away. A woman named Joseane Jacarandá showed me the old costumes in a back room lined with flags bearing Christian crosses and Muslim scimitars. Her grandson strutted around the living room with a gigantic bishop's hat. Jacarandá's eyes glittered with angry tears. For more than two centuries the maroons had been left largely alone. Now the world was coming in and wrecking something she held dear.

Dona Rosario had entirely different feelings about coming out of the shadows. Three years before my visit, men had laid electric wire along Igarapé Espinel. I had seen it on the boat to her home, a thin, fragile link, draped from tree to tree along the water. The power had allowed her to buy a cell-phone charger—which is to say, she now had a telephone. If somebody in her family was hurt or sick, she could call for help. For people who have always lived a phone call away from an ambulance or police car, the magnitude of this change is difficult to grasp. As is the magnitude of the change represented by her second big purchase: a chest freezer. Until buying the freezer, she had always had to sell açai immediately after harvest, to avoid spoilage—she couldn't wait for a better deal. Without a phone, she couldn't call around to find the best price. Knowing her circumstances, buyers had always offered her the worst terms—she couldn't walk away from the deal. Now she could process the fruit into pulp and stick it in the freezer until she was ready to sell. Açai had become faddishly popular in the United States and Europe for its purportedly high levels of antioxidants. Now she could take advantage of the fad.

In January 2009 Dona Rosario stumbled across a surveying party on her farm. Planting stakes and tying ribbons around trees, they were slicing up her holding into smaller parcels. "They were saying, 'What a great açaí place—let's divide it up and sell it,'" she told me. The buyers would then use the courts to boot out the helpless occupants—a common practice in Amazonia, as Dona Rosario knew all too well.

"I had a fit," she said. "I said, 'I *own* this land, I *planted* this land.' " The surveyors ignored her. After she bought the land she had been told the title was invalid—the previous owners had walked away from their taxes. She had spent a decade paying the back taxes and acquiring the title even as she restored the land. She had grown up seeing her parents lose one piece of land after the next. Here the same thing was happening to her.

One difference between Dona Rosario and her parents was that she had a phone. Another was that she had some capital—a freezerful of açai and a bank account with a little in it. With the phone she called government inspectors and showed her documents to them, all the while threatening to use her money to hire a lawyer. "They looked it up and said, 'Wait a minute, you can't steal this land.' " The surveyors backed down.

Similar stories are being repeated throughout Amazonia. Six months after Dona Rosario saw the surveyors on her land, Brazil-

Vendors in the main market of Belém, the port at the mouth of the Amazon, sell tree seeds for the region's farmers, many of whom replant the forest with useful tree species like açai (celebrated for its fruit juice), bacuri (a fruit somewhat like a sweet-and-sour papaya), and bacaba (another popular fruit).

ian president Ignacio Lula da Silva signed Provisional Law 458, a remarkably ambitious attempt to straighten out land tenure in Amazonia—a root cause of the violence and ecological destruction of the past forty years. It gave title to maroon communities whose members already occupied the land and had less than two hundred acres apiece, effectively bringing a struggle that has lasted for centuries to a victorious close. Pulling these thousands of settlements out of the shadows, Pereira said, will allow the state to invest in schools and clinics, something it can't legally do while their existence is contested.

Provisional Law 458 was immediately challenged in court on behalf of industrial and environmental groups, both of whom argued that it would reward squatters for taking land illegally. Their alarm was easy to understand. The law would give control of a substantial part of the Amazon to its residents, and nobody was sure what they would do.

I happened to visit Dona Rosario not long after Lula's signature. In her isolated area, she had heard little about the new law. As Hecht told her about it, she nodded in forceful agreement. Her ancestors had come from Africa and blended with American natives and created something new. In their mixed way they had taken care of the forest; it was no accident, she believed, that all of the most valuable and beautiful areas of the Amazon were full of *quilombos*.

"Forest" was perhaps the wrong word. Outsiders saw the region as a forest—impenetrable, dark, full of threats. People like Dona Rosario saw it another way, as a place that their ancestors had tended and shaped, mixing old traditions with something of their own. They had been forced to live covert, hidden lives, always worried about dispossession. Now they would be free to live in their creation, the world's richest garden.

CODA

Currents of Life

10

In Bulalacao

FRACTURED CEREBRATION

In the Philippines, children learn a folk song called "Bahay Kubo"—the title refers to the single-room house made of palm leaves that was long traditional on the islands. Built on stilts to avoid flooding, open to the cooling breeze, the *bahay kubo* was surrounded by a generous plot of fruits and vegetables. Sitting on the high doorstep, householders could luxuriate in the sights and smells of their family garden. Like "Home on the Range," "Bahay Kubo" nostalgically evokes the values of those simpler, perhaps better days before cell phones and computers, stock-market gyrations and stressed-out commutes—except that unlike "Home on the Range," which celebrates the beauties of unmarked wilderness, "Bahay Kubo" extols an entirely human-ized landscape.

Bahay kubo, kahit munti, the children sing (in Tagalog, the islands' main language). *Ang halaman doon, ay sari-sari*. Even though my palm-leaf house is small, it has many different plants. And the song continues by enumerating the contents of an ideal-ized Filipino garden:

Jicama and eggplant, winged bean and peanut,
String bean, lima bean, hyacinth bean,
Winter melon, sponge gourd, wax gourd and winter squash,
And there is also radish, mustard,
Onion, tomato, garlic, and ginger!
And all around are sesame seeds.

The botanists in Manila who told me about this song chuckled as they wrote down the lyrics. Every single one of these age-old traditional garden plants, they said, is in fact an introduced species, native to Africa, the Americas, or East Asia. Like my own tomato patch, the garden extolled in "Bahay Kubo" is an exotic modern object. Far from being an exemplar of age-old custom, it is a polyglot, cosmopolitan, thoroughly contemporary artifact.

The botanists told me this in the local office of Conservation International, an environmental-activism organization based outside Washington, D.C. The office halls and doors were covered with wanted-style posters and flyers proclaiming the dangers of invasive species. Hundreds of exotic creatures have made the Philippines their home since Legazpi arrived in the 1560s. Introduced fish like tilapia and Thai catfish have wiped out almost all the local species of fish in Filipino lakes. South American shrubs have driven out local palms and bushes in Filipino parks. Water hyacinth from Brazil chokes the rivers in Manila; weeds from Africa grow over rice paddies. Seven of the immigrants are on a hit list of the one hundred worst invasive species compiled by the International Union for the Conservation of Nature.

A small minority of the newcomers were environmentally or economically damaging and only a very few harmed the ecosystem itself, impairing its ability to filter water or grow plant matter or process nutrients into the soil. But to the scientists in the room almost all the exotics were problematic, because they were helping, in ways large and small, to turn the Philippines from what it had been before Spain into something else—a homogenized, internationalized, airport-shopping-mall version of itself, a vest-pocket version of the Homogenocene. The island landscape, they said with some heat, was less and less what it had been before. Like too many places around the world, it was becoming a nursery of canny opportunists—the sort of species equally at home in an abandoned pasture and at the edge of the big-box parking lot that replaced the pasture. It was no longer the Philippines.

Not until I had left the building did it occur to me to wonder: Why are the species in "Bahay Kubo" not foreign invaders? Surely Filipino gardens must have grown something before Legazpi. Why didn't Conservation International print wanted posters of tomatoes, peanuts, and string beans? How could this dog's breakfast of recent international arrivals become a symbol of home and tradition, sung about by school kids before nostalgic parents?

Then it occurred to me: I, too, thought of my garden as a kind of home. Futzing around with the plants was my refuge from e-mail, deadlines, and my office desk. Much like the biologists, I wished more of the local nurseries sold local plants—I had complained in one of them that there was nothing in the entire space that was from anywhere within hundreds of miles. Embarrassing in retrospect, I issued this gripe as I was at the nursery cash register, paying for seedlings of bell pepper (origin: Mesoamerica), eggplant (origin: South Asia), and carrot (origin: Europe). I was simultaneously promoting and denouncing the Columbian Exchange, and the globalization that trailed in its wake. I, too, was an example of fractured cerebration.

STAIRCASES IN THE HILLS

The way I like to put it, my family is partly responsible for the worms. The worms—two species in the genus *Pheretima* and three in the genus *Polypheretima*—first appeared about forty years ago in the mountain rice terraces three hundred miles north of Manila. My family in this context means my grandfather, who in

1959 became headmaster of a small private school near New York City. One of the perks of the job was an imposing house on the school grounds. When I visited for the first time my grandfather told me that he had instituted a policy of having breakfast every day with half-a-dozen students. By careful scheduling, he could invite everyone in the school at least once a year. To accommodate his guests, he asked the school to provide him with a bigger breakfast table. The table that arrived was made of Philippine mahogany.

Philippine mahogany is not true mahogany—it comes from two tree species in a wholly different genus. But because it looks like mahogany, especially when stained, importers dubbed it "Philippine mahogany"—much to the anger of the Mahogany Association, a Chicago-based association of furniture manufacturers who used real mahogany, which originated in the Caribbean, and wanted to protect the name. Decades of litigation produced a Federal Trade Commission ruling in 1957 that Philippine mahogany could not be marketed as "mahogany," without the qualifier. More properly known as "lauan" or "luan," the tree was extremely common in the Philippines. Exports soared in the 1950s, most of the wood going to Japan and the United States, where it was made into furniture, decking, and trim. The first place timber companies paid a call was the interior of Luzon, the Philippines' biggest island, because it was close to Manila, where the logs would be put on ships.

For visitors the most notable feature in Luzon's mountainous interior are the rice terraces. Long, skinny rice paddies, the terraces ladder up hills for miles in every direction. Tourist brochures say they were built two thousand years ago by refugees, Miao people from southwest China fleeing an ethnic purge. The Miao built terraces like those in their homeland, but even more spectacular. When the sun stabs through the clouds the young rice shimmers in a grass-green band along the stony edges of the terrace walls—the sort of impossibly beautiful sight that makes visitors clutch reflexively at their cameras. So many tourists have clutched at their cameras that UNESCO selected Ifugao, the most photographed area, as a World Heritage site. Some Ifugao terraces wrap completely around hills, making them look like wedding cakes fifty layers high. Women in ankle-deep water were weeding the paddies when I arrived. Below them the terraces fell and then gleaming fell. Two boys were fishing in a stand of rice. The terraces stepped up and down with the crazy order of an Escher drawing.

A man I had met on the bus to Ifugao walked with me for a while. The terraces were dying, he said, all four hundred square miles of them. Giant earthworms from somewhere overseas had invaded them. He spread his hands two feet apart to indicate their size, complicated tattoos weaving in chains over his upper arms as he gestured. Water sluiced out of the paddies through their huge tunnels, killing the rice plants. The worms, foreign intruders, were making the terraces porous and sponge-like. "Porous" and "sponge-like" are not adjectives that should ever modify the noun "terrace." The terraces that had lasted two millennia would disappear in less than a decade.

That wasn't the only introduced plague. The golden apple snail (*Pomacea canaliculata*) was sent from Brazil to Taiwan in 1979 to start an escargot industry. The industry never got off the ground, because would-be escargot magnates discovered that the snail was vulnerable to rat lungworm, a parasite that can infect humans. Also Taiwanese didn't like the snail's taste. Not long after their arrival, the snails escaped from their snail plantations and into the countryside. Farmers who grew other crops discovered to their dismay that golden apple snails are omnivorous, fast reproducing, surprisingly mobile, and very hungry. Proliferating along rivers and streams, they ate fish and amphibian eggs, other snails, many insects, and countless types of plants. They had a special liking for rice stalks—a big problem in an East Asian country. Despite this record the Philippines government asked U.S. Peace Corps volunteers to introduce the apple snail into the country's rice paddies in the early 1980s. Again the hope was to start an escargot industry. Again the hope proved delusory. Soon snails were eating everything in sight.

The man from the bus told me that his name was Manuel. We came to his home and sat on the pieces of striped cloth that seemed to be everywhere in people's houses. Jars and cans were stored in bamboo baskets. Rice was steaming in a pot. Manuel saw me looking at the pot and asked if I wanted some—it was his own crop. One bite would have been enough to convince even the most casual diner that Ifugao's terraces produced something special. I stuck my nose above the plate and took a deep breath. Into my nose rose an odor good enough to be called a perfume. It had more to say than any rice I had encountered before.

On the terraces grow more than five hundred different traditional varieties (landraces) of rice. Farmers constantly mix and match them in an effort to develop better-tasting, better-growing varieties. People in one area might prefer one landrace because of its texture when cooked; those in another might prefer a second because it is easier to prepare; those in a third might concentrate on landraces with higher yields, or ones that are less attractive to birds and rats. At each stage in the growing cycle village priests and landowners perform ceremonies, fueled by rice wine and often involving the sacrifice of chickens, pigs, or water buffalo, to seek the spiritual guidance of the area's hundreds of local deities. Many of the farmers are Christians, but they perform the ceremonies anyway. All the while the terraces and the irrigation channels that feed them are meticulously maintained in a web of complex actions guided by ritual. It is a way of existence that has protected the genetic diversity of rice and maintained the soil despite centuries of intensive cultivation. This entire social, cultural, and ecological world would disappear with the terraces.

Today farmers have learned how to control the snails. Worms remain the more important problem. In 2008 two biologists discovered in the terraces nine worms new to science. They weren't exotic—they were Filipino natives. They had always been living in the forest, probably in small numbers. But when the slopes around Ifugao were logged for mahogany, the environment changed around them, and they migrated to the paddies. The source of the problem thus wasn't introduced species. It was the worldwide demand for Philippine mahogany.

The problem, in short, was my grandfather. People like him, activists say, are agents, however unwitting, of globalization. His innocent wish for a new table, multiplied ten-thousand-fold, set off an island version of the Amazon's Great Heart of Palm Zap: chainsaw-wielding goons flooded Luzon's mountains, wreaking ecological havoc in their frenzied effort to tear out every lauan tree in the hills. Left unchecked, greed would destroy this beautiful, age-old arrangement, as it had destroyed many others. Corporate capitalism sweeping across oceans and frontiers, wiping out traditional livelihoods with hardly a thought! Manuel was about sixty-five—he wasn't sure of the exact figure—and thought he would live to see the end of the rice terraces. It was an object lesson in the evils of globalization.

Or was it? The first two anthropologists to study the Ifugao terraces arrived there before the First World War. Both were amazed at their apparent age. "It took a really long time to build those terraces," marveled Henry Otley Beyer, the better-known member of the pair. A chemist who moved to the islands, married the teenage daughter of an Ifugao leader, and became known as the father of Philippines archaeology, Beyer firmly stated that Ifugao's people "took between two and three thousand years to cover northern Luzon with the great terraced areas that exist there now. . . . [I]t was a thousand or 1,500 years ago when the terraced areas were at their maximum."

Beyer's estimate has long been accepted as gospel, repeated countless times in guidebooks like the one then in my bag. Unfortunately, he had no concrete evidence for this conclusion. He simply made a seat-of-the-pants guess about the time people without modern tools would need to build four hundred square miles of terraces. Not until 1962 did Felix Keesing, a Stanford anthropologist, try a different approach: he pored through Spanish records

for mentions of the terraces. Although colonial "military commanders, mission fathers, and other visitors" crisscrossed Ifugao, not one mentioned the terraces until 1801. Because Keesing
couldn't believe that visitors wouldn't marvel at this huge engineering feat, he reasoned that the terraces were a "comparatively
recent innovation"—they were not a millennia-old tradition.

Neither Beyer nor Keesing had any archaeological evidence—
they hadn't so much as taken a shovel to the terraces. To be sure,
terraces are difficult to date. Farmers constantly move around the
soil, destroying the archaeological record. And modern archaeological tools like radiocarbon dating weren't widely available
until the 1960s.

Robert F. Maher of Western Michigan University, in Kalama-
zoo, became, in the 1960s, the first archaeologist to excavate
the terraces. Surprisingly, his work was not followed up until the
early 2000s. Radiocarbon dates in both studies showed that the
heartland of the terraced area was, as Beyer had guessed, as much
as two thousand years old. But the area outside the center—the
great bulk of the terraces—was at most a few hundred years old,
as Keesing had thought. When Legazpi seized Manila, many of
its inhabitants moved into the hills to escape Spanish demands
for labor—workers to build city walls and construct great ships
for silk and porcelain. The radiocarbon dates suggested that the
Ifugao were among the refugees. They had poured into an outlying area that was hilly enough to force them to build terraces to
survive. An explosion of earth moving followed soon after, as did
a flourish of ritual and custom. The terraces thus were largely
the creation of the same great exchange that was now destroying them—they were, in their way, a monument to the galleon
trade, created by globalization like the worms that were wrecking them.

Looking around Ifugao, I was struck by the number of abandoned, crumbling terraces. People were walking away from their
farms. It was easy to understand—Ifugao is among the poorer
regions in the Philippines. More than 90 percent of its income

comes from government programs. The terraces are beautiful but small; the cool climate limits the rice harvest. A typical family's holdings, one U.N. report estimates, can feed it for just five months. In this capital of rice the crop most people actually depend on for meals is the sweet potato. Others buy rice at subsidized prices from the government's National Food Authority—a photograph of Ifugao farmers lining up for rice handouts in front of their terraces stirred a brief outcry in 2008. (The Manila government is Asia's biggest rice importer.) Below, meanwhile, is the city, great Manila athrob with lights and sound, promising jobs, education, and excitement to hungry young men in knee-deep water. So many people have left the terrace lands that the communities Manuel wanted to preserve now exist mainly to provide a fine backdrop for photographs.

More subsidies, that's what the terrace farmers need to continue! So argue pro-farmer activists and the national Department of Environment and Natural Resources. While waiting for the money to flow in, the mayor of Banaue, the most important town in the terrace zone, hired unemployed people to grow rice. To maximize returns, they planted new, hybrid varieties of rice, which grow faster than traditional varieties. All the while, the worm problem worsened. The deforestation that had let in the worms also reduced the slopes' water-retention capacity. Rising numbers of hotels and restaurants for tourists competed with farms for the remaining paddy water. The paddy soil got drier. In drier soil, worms reproduce more rapidly.

A ray of light came from Eighth Wonder, a rice-importing company founded in Ulm, Montana, by Mary Hensley, a social worker and travel agent who had been a Peace Corps volunteer in Ifugao. With a partner in Manila—Vicky Garcia of Revitalize Indigenous Cordilleran Entrepreneurs (RICE), a nonprofit organization—Hensley in 2005 launched a plan to export "heirloom" rice to the United States and Europe. It was a struggle. To get enough rice to sell abroad, the partnership had to persuade the farmers to form cooperatives (not a local tradition), teach them to dry their rice uniformly to ensure quality, build special milling equipment that could process the thick hulls of the area's ancient landraces, and push regional utilities to provide electricity to run the equipment. Landslides blocked roads, typhoons battered ships, equipment broke down, and spare parts could not be found. There was little precedent at a legal level: Eighth Wonder, according to Manila newspapers, was the only rice exporter in all of the Philippines. Sales in the United States launched in 2009. Seven varieties were available, selling for $5.75 and $6.00 a pound. When I bought a pound, shipping cost $11.75. Ifugao rice was almost sixteen times more expensive than the rice in my supermarket.

Reactions to Eighth Wonder have been mixed, as I discovered when I mentioned the company to scientists in Manila. As growing numbers of Ifugao farmers flock to join the project, a rising percentage of the area's harvest—a precious cultural artifact—is being sent out of the country to affluent foreign food snobs. Worse, the cooperatives, standardization, and mechanized processing are dramatically changing Ifugao culture—all for the benefit (as one scientist put it) of faraway people who want to pat themselves on the back for their enlightenment as they click the link to order fancy multicolored rice. The global market is not the solution, activists say, but the problem! These supposed do-gooders are just hooking Ifugao into the worldwide network of exchange, making them dependent as never before on the whims of faraway yuppies! Antipoverty activists charge the anti-trade activists with wanting to condemn the poor to backbreaking labor so that they can feel good about themselves as they sit in their air-conditioned offices in Manila. The terraces have been linked to the global network almost from the beginning—why should they experience only the harms (falling commodity prices, environmental damage) and not the benefits (communicating with people who are willing to pay sixteen times as much for rice)?

What's being lost here? What would count as saving it?

ON THE BOAT

During another trip to Manila, I decided to see the place where Legazpi had first encountered Chinese junks: the beginning of today's world-encompassing trade network. The encounter, I knew, occurred on the southern part of the island of Mindoro. But exactly where on Mindoro was unclear—the Spanish description of the meeting was confusing, at least to me. I thought a visit might dispel my confusion. Besides, I was curious.

A friend of a friend contacted one of her friends, who ran a hotel on the east coast of Mindoro. The message was conveyed to me: don't drive to southern Mindoro. Guerrillas were active there. I was surprised—Mindoro, the closest big island to Manila, has a lot of pricey resorts on its north side. An Internet search showed that Mindoro's hills indeed housed an old-style Communist insurgency, the New People's Army. They are often photographed in green shirts with arm badges: a red triangle with an AK-47. Sometimes they wear berets. Sometimes they wave red flags with the hammer and sickle. Legazpi's meeting had occurred, I knew, somewhere around the small town of Bulalacao. A year before

The *Traveller-7*

my trip, the New People's Army had visited there, blowing up a bulldozer, a dump truck, and some construction equipment.

I saw no indication that the guerrillas cared about individual American visitors. Still, taking a boat seemed prudent. Besides, I like boats.

The hotel owner found a vessel that I could charter inexpensively. I took a bus through Manila's appalling traffic to the Mindoro ferry, climbed into a tiny, cheerfully crowded jitney van after landing, and bounced to the hotel, in the village of Bongabong. At five thirty the next morning I was wading to the boat: a modern version of the traditional shallow-draft *proa*, with two sweeping wooden outriggers. The *Traveller*-7 had a tiny cabin, barely big enough to contain the engine batteries, a few liters of water, and a lighted coal brazier with a bubbling pot of rice. Flapping above the deck was a blue plastic tarp. With the Philippines' tenacious refusal to fulfill touristic fantasies of exoticism, the three-man crew was wearing baseball caps and droopy basketball shorts with NBA logos.

After four hours along the cliff-bristling coast, we anchored off Bulalacao's long concrete esplanade. The town had electricity and (intermittent) cell-phone service but was physically cut off—the road to the rest of the island was not only guerrilla-infested but unpaved and often impassable for anything but four-wheel-drive vehicles. I saw only a single car. A breeze riffled the surface of the water and the plastic tarps above the market kiosks. At the fringes of the market people were holding a cock fight. Signs of large-scale economic activity were not readily apparent.

I had no appointment or anyone to see. It was my notion that I would attract attention, and the attention would take me to the right person. After I had walked around for about fifteen minutes, a man showed up on a motor scooter. He took me up a long slope to the South Drive Bar and Grill, Bulalacao's sole restaurant. The floor was gravel. In one corner was a small, dusty stage with three guitars, an electronic drum kit, several ramshackle speakers, and a laptop playing, unbelievably, "What a Wonderful World"—the original Louis Armstrong version. As the shuffle function switched the music to Japanese pop I was greeted by Chiquita "Ching" Cabagay-Jano, proprietor of the South Drive Bar and Grill and Bulalacao's municipal planning and development coordinator and tourism administrator.

In the manner of town planners everywhere, Cabagay-Jano was enthusiastic about Bulalacao's prospects. Investors were coming in from the resorts to the north, she said. Investors were coming from China. Investors were coming from *America*. Land in Bulalacao was there to be acquired—one man had snapped up 250 acres for a golf course. The government was paving the road around southern Mindoro, which would allow regular bus service. Last year the town had held the First Bulalacao Windsurfing Invitational Cup—banners from the competition adorned the restaurant walls. A crew was coming the very next day to install a permanent webcam above the town beach. Bulalacao was poor now, but soon it would swim in the stream of global commerce. It was waiting for the world.

When I asked about Legazpi, Cabagay-Jano summoned a son, Rudmar, and instructed him to guide my boat to the place where Spain had encountered China. The site was in a shallow bay, a nick in the coast just to the south, occupied by the hamlet of Maujao. Just past the high-tide line was a spring covered by a concrete pillbox. A metal pipe dribbled water into a cement channel, which channeled it to the beach. Two kids were filling up plastic buckets with water.

For centuries Mangyan people had waited there in their embroidered bark-cloth shirts and indigo-dyed cotton loincloths for the junks from Fujian and Guangzhou. White parasols made from Chinese silk shielded them from the sun. The smoke from their beach fires must have been like a welcome signal to the ships from afar. Both the Mangyan and the Chinese had a written language. It is tempting to imagine scribes keeping track of the exchange, so many cakes of wax and bundles of cotton for so many porcelain plates, shiny bronze gongs, iron pots, and needles. The southern wing of the little bay was a sharp point like a finger into the sea. At dawn four and a half centuries ago the Spaniards had abruptly rounded that point in their strangely shaped vessels. Stand back, the Chinese had cried. Many did not live to see the sunset.

Occupying the point was a small, half-complete resort: Thelma's Paradise. Workers were building the main guest house on the shore. Thelma's Paradise was going to be a "farm resort." Visitors from Manila would stay there and "participate in the Bulalacao farming lifestyle"—the phrase comes from a handout Rudmar gave to me. I asked one of the workers what this meant. Rudmar translated, perhaps imperfectly, the response. Busy city executives would come to Maujao and weed Thelma's gardens—a refuge from e-mail, deadlines, and the office desk.

People from Manila? I asked.

Not just Manila, I was told. From Legazpi's time, poverty, colonialism, and slavery had scattered Filipinos across the world. Filipinos were nannies, nurses, and construction workers in Hong Kong, Sydney, Tokyo, San Francisco, and Paris. They had made money and wanted to visit home. Home was the sea and the beach and a cookout beneath the palms. Home was *bahay kubo*.

Rudmar stood with his back to the water, scowling at the hills. After logging companies had exhausted the hills of Luzon, they turned to the other seven thousand islands in the archipelago. Industrial ships pulled into Mindoro's lightly settled bays and unloaded bulldozers and trucks and men with saws and come-alongs. They stripped the slopes bare. Floods ensued, wiping out farms and villages. Compounds in the runoff washed over the island's white beaches, staining them permanently yellow. The government ultimately banned most logging, but the damage had been done. *They took the color from the earth*, Rudmar said. He wanted his home back.

This anger, magnified and distorted, was the well from which the New People's Army drew. Their bases were in the ravaged hills, perhaps close enough to watch me bumble around Thel-

A small resort occupies the point at Maujao, where Asia, Europe, and the Americas met for the first time.

ma's Paradise. Living amid ecological mayhem, the guerrillas saw all the costs of the great market and none of the benefits. It was no accident that their attack on Bulalacao the year before had targeted the construction equipment that built resorts. A few months after my visit, they came out of the hills again, assaulting a nearby military outpost—troops from a government that they viewed as a corrupt handmaiden to global capitalism.

Yet there were real benefits from the logging, too. My grandfather got his table. Craftsmen got paid to build it. Shipping companies got paid to carry it, giving people jobs. The students got to have breakfast with my grandfather, a wonderful raconteur. Even the men with chainsaws should be considered. These agents of destruction were just putting food on the table.

Economists have developed theoretical tools for evaluating these incommensurate costs and benefits. But the magnitude of the costs and benefits is less important than their distribution. The gains are diffuse and spread around the world, whereas the pain is intense and local. Economists say that the transactions in such cases have *externalities*: spillovers on parties who are otherwise

Great Manila, like small Bulalacao, is wrestling with the push and pull of the global market. Its outer harbor is a mass of sleek, wired-up international buildings, but the populous inner harbor is unchanged in many ways—houses still crowd the water, and people still live their lives on boats much like those in Legazpi's day.

not involved. The side effects can be positive; some Mindoro villagers are using the semi-legally cleared land to plant bigger gardens. But the worry is negative externalities: erosion, landslides, yellow sand. In theory, the solution is obvious: increase the price to take into account the full range of costs. Rather than paying, say, $100 for his table, my grandfather should have paid $125, with the extra money going either to compensate villagers for their yellow beaches or to cover the costs of logging companies adopting protective measures. In practice, making those arrangements is not easy.

Complicating all is the welter of mixed motives. On the one hand, people want the wash of goods and services that the worldwide market provides. No one forced Thelma to build a resort for foreigners. In Amapá, nobody twisted Dona Rosario's arm to buy a television and a freezer. Nobody was holding a gun to the head of the teenage Chinese villagers in Shaanxi who clamor for Nintendo games and U.S.-brand cigarettes and DVDs of Will Smith movies. Or, for that matter, their counterparts in Beijing and Shanghai, whose demand for French wine is driving Bordeaux prices to amazing heights. Smart phones, aerodynamic sneakers, beige faux-leather living-room sets—people desire these things. Absent catastrophe, they will get them. Or their children will.

On the other hand, the same people who want to satisfy their desires also resist the consequences of satisfaction. They want to have what everyone else has, but still be aggressively themselves—a contradictory enterprise. Floating in the capitalist stream, they reach down with their feet, looking for solid ground. To be a good place to stand, it must be their own, not somebody else's place. As human desires bring the Homogenocene into existence, billions of people marching through increasingly identical landscapes, that special place becomes ever harder to find. Things feel changed and scary. Some people hunker down into their local dialects or customary clothing or an imagined version of their own history or religion. Others enfold themselves in their homes and gardens. A few pick up weapons. Even as the world unifies, its constituent parts fragment into halves, and the halves into quarters. Unity or division—Thelma's Paradise or New People's Army—which will win out? Or is the conflict inevitable?

After an hour or two, the pilot hurried us back to Bulalacao. He was worried about taking a boat with no lights, charts, or navigation equipment around the rocky, island-dappled coast at night. I walked along the town esplanade with Rudmar, looking for a place to buy some water. Afternoon light was beginning to throw deep shadows. I came upon some women and children in what looked, to my inexpert eye, like a family garden around a palm-thatched home—a *bahay kubo*.

The women and children moved with enviable efficiency— they were getting things done. Towering above their heads were tall stalks of maize, now the second most important crop in the Philippines. Below it were squashes and peppers. I could see why the botanists had been amused by the song—the plants they were growing would not have been out of place in Mexico. Yet at the same time the garden was obviously something else.

Gardeners work in partnership, more or less successfully, with what nature provides. They experiment all the time, fiddling with this, trying out that. People take seeds and stick them in the ground to see what happens—that's how Ifugao villagers bred hundreds of types of rice in a few centuries. An essential factor is that gardeners experience the consequences of their own actions. They make decisions and expend labor; a few months later they discover what they have wrought. Externalities are rare. Gardens are places of constant change, but the changes are owned by the gardener—which is why they feel like home.

Despite the visible impatience of the pilot, I spent a few minutes watching the family in their garden. In this place the Columbian Exchange had been adapted and remade. Families had embraced the biological assaults of the outside world—some of them, anyway—and made them into something of their own. Other problems would be dealt with as they came. Even people trying to preserve the past by growing traditional varieties of rice are necessarily facing the future. The women were weeding around the maize. Every stalk carried its American past in its DNA, but the kernels swelling in the cobs were concerned with next season's growth.

APPENDIX A

Fighting Words

A book like this must thread a path through terminological quicksand. The problems are threefold. First, many of the names that readers are familiar with are inaccurate; sometimes they are viewed as insulting. Second, different people perceive things in different ways, so a term that may be spot-on from one point of view may seem wildly off target from another. Third, words can be used in different ways in past than present, so that one can employ a term accurately (that is, use it in the way it was used by the people under discussion at the time and place under discussion) but convey entirely the wrong thing.

Take the word "Asian." In countries like the United States, the term is a replacement for "Oriental," which is viewed as Eurocentric. In other parts of the world, though, "Oriental" and its translated equivalents seem unexceptionable. Because "Asian" is a common word in all places, substitution should be unproblematic, at least at first glance—what's the cost? The cost is that although the dictionary defines "Asian" as meaning "of, relating to, or characteristic of the continent of Asia"—the entire landmass, from Israel to Siberia—in practice the word generally refers to specific groups. In the United States it usually indicates East and Southeast Asia (China, Japan, and Vietnam, for example), whereas in Britain it is mainly used for South Asia (India and Pakistan, for example).

That is a comparatively simple distinction. Consider the Parián, the big Chinese ghetto in Manila that played an important role in the silver trade. Spanish records routinely refer to its inhabitants as *chinos* and *sangleys*. Using the latter is discourteous, to say the least—*sangley* is a pejorative, analogous in weight, perhaps, to "kraut" or "frog" for German and French people. *Chino* means "Chinese person." It isn't particularly pejorative, but it also isn't particularly accurate: more than a few Parián residents were not from China. As used in Manila, the term really meant something like "people from Asia who are not from the Philippines." (Because the Spaniards often distinguished the Japanese from other Asian peoples, it may be more accurate to say that it meant something akin to "people from Asia who are not from the Philippines or Japan.") Unsurprisingly, Parián residents didn't think of themselves in this way. Most were from Fujian, and Fujianese people typically described themselves as Hakka or Min—"Chinese," in their view, applied mainly to the Han, the dominant ethnic group.

Matters get more complex still when one considers that Spaniards in different places used *chino* to mean different things. In Mexico, the rulers of New Spain viewed anyone with "Asiatic" features as a *chino*, including people from the Philippines. Thus a Spanish word used to distinguish Filipinos from other Asians in one place was used to describe them in another. Worse still, the word *chino* in Spanish America soon lost its connection with China, and even Asia. Peculiarly, some of the mixed descendants of Indians came to be known as *chinos*. (A popular folk figure in the Mexican city of Puebla is the *china poblana*, the Chinese Pueblan woman, a racy, flirtatious sort who wears a white blouse, a colorfully patterned skirt, and a shawl. Visitors to Puebla are told that the style was originated by Catarina de San Juan, the pious, vision-filled Mughal slave whom I described in Chapter 8; the patterned skirt, one is solemnly assured, was inspired by her sari. But Muslim women like Catarina didn't wear saris; purdah was becoming popular, and they wore concealing garments. In addition, there is ample testimony that in Puebla Catarina wore black and was anything but flirtatious. The dress style, researchers say, is simply an adaptation of Indian dress.)

Similar concerns apply to "European." The idea of Europe as a geographic entity has existed for a long time. The idea that this entity was populated by people with commonalities enough to be described as a group has not. According to the *Oxford English Dictionary*, the first English use of the word to mean "a resident of Europe" occurred only in 1639. For most of the time surveyed in this book, people from the eastern Atlantic shore referred to themselves by nationality: English, French, Dutch, and so on. The people from the Iberian Peninsula who play such a large role in this book often identified themselves by region—Extremaduran, Basque, Castilian, and so on. If all these diverse people had recourse to a collective noun, it was "Christian," because Europe was part of Christendom. (Early in the writing of this book, I tried using "Christian" in that sense. I gave a few pages to a friend, who asked why I was dragging religion into a story about trade—was I writing some kind of pro- or anti-Christian tract?)

The peoples of Africa, America, and Asia quickly learned that Spaniards, Portuguese, Dutch, and English were different. Nevertheless, they also regarded them as members of a single group— people who showed up from another continent, wanting to take over. In China, Europeans were often grouped together, disparagingly, as *gweilo* or *laowai;* the terms retain some sting for those to whom they are applied.

Given these spiraling complexities, I couldn't find a consistent way to use historically accurate terminology. Instead, I refer to people *geographically,* by their place of origin, using modern terms. Thus I call the conqueror of the Philippines, Miguel López de Legazpi, a Spaniard, even though he was Basque, led an expedition composed mainly of Basques, and presumably spoke Basque at home. When the regional origin becomes important, as in my discussion of the Basque-Vicuña war in Potosí, I use more local geographic names. This scheme courts anachronism, though I have tried to avoid it. Because the United Kingdom of Great Britain didn't exist until Scotland and England merged in 1707, for instance, I don't call anyone before that date from these countries "British." At the same time, I also don't ever label anyone from Ireland as a Briton, even though Ireland was formally part of the United Kingdom between 1800 and 1921—it's too confusing. I am sure that I have made mistakes; readers who wish to tell me about them can contact me at charlesmann.org.

Despite its problems, this scheme has the virtue of allowing me to avoid another intractable issue: race. Race is part of any discussion today of the interactions of people of European, African, Asian, and Indian descent. But at the dawn of globalization modern concepts of race didn't exist. Fighting off the yoke of African Islamic empires, the inhabitants of the Iberian Peninsula didn't, as a rule, kill or enslave "blacks," they killed and enslaved "Moors" or "infidels" or "idolaters." At the beginning, slavery had little racial baggage; the question that preoccupied Spaniards was not whether "black" or "red" people could be enslaved but whether Christians could be put in bondage; heathens, heretics, and criminals of any color were fair game.

The term *negro,* a Portuguese word for "black," didn't come into wide use until the 1450s, when Portuguese ships came to what is now Senegal and dubbed it the *terra dos negros* (land of the blacks). Although "negro" referred to skin color, it was mostly an ethnic descriptor, much like "Irish" or "Malay." An analogy might be *ang mo,* the Fujianese word for "redhead." *Ang mo* was used to label the Dutch, even though most didn't have red hair. Later *negro* came to mean "slave," and was used by Africans themselves. As the historians Linda M. Heywood and John K. Thornton have noted, Central Africans insisted that European visitors use one Portuguese word for black (*negro*) to describe slaves and a second, alternative Portuguese word for black (*preto*) to describe free Africans.

From the beginning, Europeans had terrible things to say about "blacks," but the disdain wasn't as monolithic as sometimes portrayed, and hard to distinguish from the garden-variety ethnocentrism that seems to be an ineradicable part of the human condition. More important, the negative beliefs weren't racial in the modern sense—they didn't invoke an inheritable genetic makeup. Europeans criticized African *behavior,* not African racial stock; Africans were bad because they were supposedly "promiscuous," "thievish," or engaged in "devil worship," not because they were physically or mentally inferior. (I am oversimplifying a little: Europeans also believed that parents who engaged in harmful practices like devil worship would pass on a terrible moral stain to their children, who would grow up to be physically and mentally inferior. But this is still quite different from the modern conception of race.)

Races in the contemporary sense of heritable genetic patterns associated with geographic origin certainly exist, though actually identifying which genes make someone "African" or "Caucasian" remains a tall order. Are men and women "black" if they have very dark complexions and broad noses, but their hair is not twisted into corkscrew curls? Are they "white" if they have aquiline noses and flat hair but dark complexions? The complications are endless, and nobody has come close to resolving them. They are also beside the point: this type of scientific description is not what eighteenth- and nineteenth-century theorists had in mind when they evolved the social concept of "white," "yellow," "red," and "black" races. The two definitions of race, genetic and social, are only loosely connected—one reason that discussions of race are so often dialogues of the deaf. To avoid the confusion, I have always referred to people by geographic origin—African, European, Asian, and so on—except, occasionally, for rhetorical purposes.

I make one big exception to this rule. In this book, indigenous people are usually referred to by their ethnic names, not by a geographic label. In the modern context, it seemed to me forgivable to refer to people from Yuegang as "Chinese," even if they would not have used that name. But it seemed foolish to refer to, say, the Inka as "Peruvians"—the gap between the Inka empire and modern Peru is too great. I make exceptions to my exception. In Chapter 9, for instance, I refer several times to "Angolans" in Palmares, because it is not clear which ethnic group from the area that is now modern Angola they belonged to. A bigger exception, as I imagine the reader is already thinking, is my use of the term "Indian." On the simplest level, the plain sense of the word is wrong—among other things, Indians are not from India. ("Red Indians," sometimes heard in Britain, is not preferred as a way to distinguish Indians from the Americas from Indians from India.) Unfortunately, alternative terms are no better. "Native American," for instance, literally means someone who was born in the Western Hemisphere. My family and I are native Americans—yet we are not Indians. Canada has introduced the term "First Nations," an admirable term, but one that lacks useful adjectival and possessive forms. As a writer, I am reluctant to inflict terms on readers that I cannot easily say.

On a deeper level, "Indian," "Native American," and "indigenous" are remote from the way the Americas' original inhabitants thought about themselves. Just as sixteenth- and seventeenth-century Europeans did not describe themselves as "Europeans," the inhabitants of the Western Hemisphere in that same era did not think in terms of any collective entity. Today, such group nouns are important. In my experience, indigenous Americans tend to use the word "Indian" when referring to their fellows. For better or worse, I am following their example.

Globalization in Beta

Why did Fujian become the center of the silver trade, and not some other place in China? One answer is that it was the region in China most experienced with exchange across the ocean. The fabled city of Zaytun, one bay north of Yuegang, was the eastern terminus of the maritime Silk Road.

A glittering, congested metropolis, Zaytun occupied a key place in what might be called a first pass at globalization, a system of exchange across Eurasia that reached its apogee in the fourteenth century. One trade route went overland, across western China to the Middle East and Black Sea before reaching, through many middlemen, the Mediterranean. The other went by sea, touching down at Indochina and India before going up the Red Sea; it, too, finished at the Mediterranean. The overland route was dominant until the Mongol empire began falling violently apart, at which point the nautical route became safer. From Zaytun's wharfs sailed Chinese junks low in the water with chests of silk and porcelain; into them came Chinese junks laden, according to an impressed Marco Polo, with "rich assortments of jewels and pearls, upon the sale of which they obtain a considerable profit." Polo's descriptions of Fujianese trade focused obsessively on the Asian luxury goods—precious stones, silk, porcelain, spices—that fascinated Europeans. In fact, though, Fujian's traders made most of their money from items that Polo would have found mundane, such as bulk copper and iron, which temples across Southeast Asia needed for ritual objects. Zaytun was a full-service emporium, not a boutique.

The city was ringed by a twenty-foot-high wall, faced with glazed tile and brick. Outside the wall, trading prosperity paid for massive marsh-drainage projects, a network of irrigation canals and waterworks to prevent the harbor from filling up with the sediment from the Jin River. Inside the wall, shaded by the tiger's-claw trees that lined the streets, walked people of every ethnicity: Malays, Persians, Indians, Vietnamese, even a few Europeans, each group with its own neighborhood. Rising into Zaytun's coal-smoke-filled sky were seven great mosques, three churches (Eastern Orthodox and Nestorian), and a cathedral (Roman Catholic), and countless Buddhist institutions—one visitor claimed that a single monastery had three thousand monks. The Moroccan traveler Ibn Battuta, who visited in the 1340s, marveled at the scores of huge junks in the harbor; around them, he said, swarmed small vessels "past counting," buying and selling. Ibn Battuta called the port "one of the biggest in the world—I'm wrong, it is *the* biggest." The traveler was not simply exaggerating to make a good tale; Zaytun, with several hundred thousand people crammed into the littoral beneath the hills, was one of humankind's richest, most populous cities. Little wonder that Polo's account inspired people like Colón to dream of going there!

After the Song dynasty fell to the Mongol invasion in the 1270s, the last embers of resistance burned in Fujian. An opposition movement there installed a Song prince as emperor. The Mongols quickly attacked in great force, and the Song prince took refuge in Zaytun with his courtiers and troops. A well-connected Muslim Arab merchant named Pu Shougeng had long been the local superintendent of trade ships, which placed him in charge of both the local militia and the local navy. The Song prince asked Pu to give him control of Zaytun's hundreds of ships—an instant navy. The prince's sudden acquisition of naval power would pose a threat to the Mongols, who had no navy.

A Mongol general sent emissaries to Pu, asking him not to back the Song emperor. After consulting with local scholars and landlords and other foreign trading families, Pu presented Zaytun and all its ships to the Mongols in 1276. To seal the deal, he ordered the murder of some of the prince's family, who happened to live in town. The Song forces had been camped outside the city. Angered, they besieged Zaytun for three months before fleeing the advance of the Mongols.

The Mongols—who had now formed the Yuan dynasty— lavishly rewarded the conspirators, effectively giving control of the port to the Pu family and their allies in the Muslim trading families.* So powerful did Zaytun's Muslim minority become that some Fujianese converted to Islam, which allowed them to register as foreigners and enjoy foreigners' privileges. Eventually most government positions throughout Fujian were held by Chinese converts.

As one might expect, the Islam practiced by these newcomers was far from the pure faith of Arabia. Rather than making the pilgrimage to distant Mecca, Fujianese believers traveled to the hills outside the city to walk seven times around the tombs of two early Sufi missionaries. Others adopted the Chinese custom of venerating their ancestors' graves. Few learned the precepts of the Qu'ran—the book was not fully translated into Chinese until 1927. Fujianese imams, most of whom did not speak Arabic, memorized the original text, declaiming it phonetically in the mosques. As memories faded, the services descended into gibberish, meaningless recitations before uncomprehending audiences. In one way, though, this remote outpost of Islam preserved tradi-

---

* The Mongols eagerly absorbed Han Chinese culture but were leery of granting too much power to the Han themselves. (The Han, one recalls, are China's dominant ethnic group—the group Westerners refer to as "Chinese.") As a result, the Yuan often installed non-Han leaders as local rulers. Giving Arabs and Persians control over Zaytun was an extension of this stratagem.

tion most faithfully: Zaytun's Muslim families, old and new alike, were split into quarrelsome factions, Sunni, Shi'ite, and Sufi.

Each faction dominated part of the government, controlled a section of the harbor, and had its own private militias. Pu's lineage and its associates, who were apparently Sunni, had the Mongols' favor and thus the most political power. The bulk of Zaytun's foreign population, though, was Persian, and therefore Shi'ite. The Shi'ites had the biggest militias—enough to stop the Sunnis from grinding them under their heels. (Little is known about Sufis in Fujian.)

The balance of power held until the 1350s, when peasants throughout the nation rebelled against their Mongol masters. One of these revolts would eventually topple the Yuan and establish the Ming dynasty. To safeguard Fujian against the insurgencies, the Yuan emperor authorized Zaytun merchants to build up their private militias even more by recruiting and training thousands of foreign Muslim soldiers (or, perhaps more accurately, "foreign" Muslim soldiers—many were not from the Middle East but were converted Chinese). The emperor asked two Sunni militia leaders to suppress an insurrection by Chinese around Zaytun in 1357. The next year they stopped revolts in Xinghua and Fuzhou, the next two port cities to the north. Nonetheless, the Yuan were not entirely pleased. Overcome by enthusiasm, one Sunni militia had plundered Xinghua for days; the other had occupied Fuzhou, turning it into a private satrapy. The leader of the first militia was slain by a rival Sunni—a Pu family confederate who was superintendent of marine affairs in Zaytun. The second was killed by the Yuan, who didn't like it when their creatures acted too boldly.

Proclaiming his loyalty to the Mongols, the Pu confederate took over the dead man's militia and used it to stamp out peasant uprisings. But he also took advantage of the chaos to turn Zaytun into an independent fiefdom and "exterminate" the city's remaining Shi'ites (the verb comes from an account in an official city gazetteer). After three years of sporadic conflict the local Yuan commanders allied with the Shi'ite militias they had previously fought against, persuaded one of the few surviving Shi'ites in Zaytun to open the city gates secretly, and wiped out the Sunni. Then the commanders switched to the side of the incoming Ming.

It was too late to save Zaytun. Years of conflict had reduced all but one of the city's seven great mosques to rubble. (Wealthy Arabs are supposedly about to restore the surviving building, now a park, to its former glory.) Most of the foreign population was dead. The survivors fled into the hills and became farmers. They stopped identifying themselves as Muslim. The Ming were loath to restore a city that had been, in its way, a center of pro-Yuan sentiment. They allowed its waterworks to break down and fill the harbor with silt. Foreign trade did not openly resume for two centuries. The center of its revival was not Zaytun, but Yuegang, the harbor to the south. But that didn't stop many of the old Zaytun trading families from leaving the hills to participate in the birth of globalization.

Many of the Chinese merchants who filled the junks at Yuegang thus were descendants of families that had prospered from its first pass at globalization. They were doing the work of the centuries. They were agents of humankind's unending quest to enlace its most far-flung members in a single skein, a journey whose endpoints the travelers have rarely been able to anticipate.

⁀ ACKNOWLEDGMENTS ⁀

Years after reading Alfred Crosby's books, *The Columbian Exchange* and *Ecological Imperialism,* I met the author and got to know him a little bit. Almost every time we spoke, I suggested that he should update those books to take into account the enormous amount of research they had stimulated. Crosby was never interested; he was on to other, newer things. One day when I had mentioned this notion a few too many times, he growled, "Well, if you think it's such a good idea, why don't *you* do it?" Naturally, I took his offhand quip as license. The project quickly got out of hand. *1493,* the result, is scribbled in the margins of *The Columbian Exchange.*

Crosby is far from the only person to whom I owe gratitude. All the way I have benefited from the help and counsel of William Denevan, William I. Woods, and William Doolittle (the three Bills). A veritable Solecism Squad read the manuscript in part or whole: Robert C. Anderson, James Boyce, Richard Casagrande, David Christian, Robert P. Crease, Josh D'Aluisio-Guerrieri, Clark Erickson, Dan Farmer, Dennis Flynn, Susanna Hecht, John Hemming, Mike Lynch, Stephen Mann, Charles McAleese, J. R. McNeill, Edward Melillo, Nicholas Menzies, Brian Ogilvie, Mark Plummer, Kenneth Pomeranz, Matthew Restall, William Thorndale, and Bart Voorzanger. They saved me from many mistakes. Nonetheless, this book is mine, along with all its problems.

Even Isaac Newton, never a modest man, admitted that he was able to see far only because he stood on the shoulders of giants. In this way—if only in this way—all writers can claim kinship to Newton. For this book, some of these giants are mostly invisible—they are beneath the text in so many places that I found it hard to cite them anywhere in particular. Whenever I didn't understand something as I wrote *1493*, I asked, "What did David Christian say about this?" Then I would page through *Maps of Time* and find his admirably concise take on the matter. Just as dog-eared and grease-stained is my copy of Robert Marks's crisply opinionated *Origins of the Modern World*. Encountering a question about the Spanish realm, I turned to Henry Kamen's *Empire*. When I had a question about China and the West, I turned with equal alacrity to Kenneth Pomeranz's *Great Divergence*. For the galleon trade, Dennis Flynn and Arturo Giráldez have so many papers that I'm not sure which to say I pilfered most frequently. Books by Robin Blackburn, David Brion Davis, David Eltis, and John Thornton played the same role with regard to slavery. Individual chapters owe much to individual works. Chapter 3 is indebted to *Mosquito Empires* by J. R. McNeill. Countless details in Chapters 4 and 5 are from Li Jinming's *Zhangzhou Port* (漳州港). My musings on the potato in Chapter 6 are lifted shamelessly from Michael Pollan's *Botany of Desire*. Tom Standage's *Edible History of Humanity* also played a role here, as it did in all this book's discussions of food, agriculture, and other matters. John Hemming's *Tree of Rivers* and Susanna Hecht's *Scramble for the Amazon* are sturdy underpinnings for Chapter 7. John Thornton's many works rustle alluringly in the background of Chapter 8. Richard Price's *First-Time* and *Rainforest Warriors* are the foundations of my discussion of Suriname in Chapter 9. The Brazil sections of the same chapter are based on a *National Geographic* article by Susanna Hecht and myself. If *1493* brings new readers to these books, I will be more than satisfied.

A far from pro forma acknowledgment: The author is grateful to the Lannan Foundation for an award that greatly assisted in the research of this book, and thanks the John Carter Brown Library at Brown University for its support and an appointment as Invited Research Scholar.

Any project that attempts to cover a large area must contend with humankind's linguistic creativity. Lucky for me, I was accompanied in China by Josh D'Aluisio-Guerrieri, who also found a host of Chinese sources for me from his home in Taipei, read even the most ancient gazetteer with aplomb, and put up with endless lists of e-mailed questions. All translations from Chinese in 1493 are by Josh, except a very few from Devin Fitzgerald, whom I asked for help when I couldn't bear bothering Josh any more. Scott Sessions took time from the immense assembly of the African-American Religion Documentary History Project to answer many, many questions when sixteenth- and seventeenth-century Spanish proved beyond my compass. Susanna Hecht was a boon companion in Brazil, a fine translator, generous with her immense knowledge of that great nation. There is nobody I would rather have a car breakdown with in *quilombo* country. Reiko Sono's help with matters Japanese is hereby thanked and acknowledged.

This big book about many things had many friends in many places. Maria Isabel Garcia, the finest science writer in Manila, kindly did a host of favors there for me, including finding a boat in Mindoro, and people to pilot it. Clark Erickson provided a tent and sleeping bag for me in Bolivia, and told me how to hire a plane in Trinidad. Alceu Ramzi gave me amazing aerial tours of Acre and didn't laugh when my lecture was interrupted, unbelievably, by a clown act. Dennis Flynn and Arturo Giráldez put up with repeated pleas for assistance; Dennis hosted me when I arrived very late one night on a flight over the Pacific.

In the United States, Greg Garman of Virginia Commonwealth University took me on a marvelous boat tour of the James River. Caleb True obtained permissions to reproduce the images in this book and began the arduous process of straightening out the endnotes. I shuddered to see the time stamps on the e-mails from Nick Springer and Tracy Pollock, who put together the maps in the incredibly short time they were allotted. Alvy Ray Smith created the amazing family tree in Chapter 8; the color version, available at alvyray.com, is even better. Peter Dana helped me understand area calculations and cartographic software, digitized a map of Cortés's estate, and much else. Faith d'Aluisio and Peter Menzel let me use photographs, provided instruction in photo-editing software, and, again, much else. Ellis Amdur told me interesting things about Japanese swords and the people who used them. James Fallows and Richard Stone helped me get material from Beijing. Neal Stephenson, a patient traveling companion in Xiamen, opened his immense contact list on my behalf. My thanks, too, to the bloggers and other online commenters who have discussed my work, sometimes with amazing acuity. Special thanks for corrections and suggestions in this edition to Robert Bush, Jill Ferguson, Sam Gitlitz, Sandra Knapp, John Major, Alastair Saunders, Fritz Schwaller, Rob Schwaller, William Starmer, Reed Taylor, and Martin Wall.

It is a pleasure to tip my hat to the editors who published, over the years, bits and pieces of this book: Barbara Paulsen at *National Geographic;* Jennifer Sahn at *Orion;* Richard Stone and Colin Norman at *Science;* Cullen Murphy at *Vanity Fair;* and (last but far from least) Corby Kummer, Cullen Murphy (again), and William Whitworth at *The Atlantic.* At Knopf, Jon Segal was patient beyond measure with a slow and wayward author; I am grateful for his support and advice on this, the fourth (and most difficult, from my point of view) project we have worked on together. Also at Knopf, Kevin Bourke, Joey McGarvey, Amy Stackhouse, and Virginia Tan performed all of the organizing, arranging, and tidying tasks that smooth out bumps in the reader's progress and make books and their authors look good. My thanks, too, to Henk ter Borg in Amsterdam, Francis Geffard in Paris, and Sara Halloway in London. Rick Balkin, my agent, has been a good friend almost since I began writing. Many other people gave me their good offices; I can't possibly thank them all or even acknowledge them, except to say that I hope they believe the investment was worthwhile.

✎ NOTES ✐

CHAPTER I / *Two Monuments*

3 Location of La Isabela: Colón 2004:314; Léon Guerrero 2000:247–51; Las Casas 1951:vol. 1, 362–63; Anghiera 1912:87; Chanca 1494:62–64; Colón, C. 1494(?). Relation of the Second Voyage. In Varela and Gil eds. 1992:235–54 ("a very suitable area of high land . . . not a closed port, but rather a very large bay in which all the vessels in the world will fit," 247 [my thanks to Scott Sessions for helping me with the translation]). As Morison noted, though, the harbor is open to the north, "rendering the anchorage untenable" in winter storms, and potable water was about a mile away (1983:430–31).

4 Description of La Isabela: Author's visit; Deagan and Cruxent 2002a:chap. 3; 2002b:chap. 4 (esp. fig. 4.2).

4 Colón's life: Recent biographical studies include Abulafia 2008; Wey Gómez 2008; Fernández-Armesto 2001, 1991; Taviani 1996; Phillips and Phillips 1992. Useful but dated is Morison 1983. Biographies by contemporaries are Colón 2004; Las Casas 1951:vol. 1; vol. 2:1–200 (the two frequently are identical). See also the notes to p. 12.

5 Shuttle flight: I owe this simile to William Kelso.

5 First and second voyages: Abulafia 2008:10–30, 105–212; Colón 2004:chaps 13–63; Fernández-Armesto 2001:51–114; Léon Guerrero 2000; Las Casas 1951:vol. 1 ("bailiff posthaste," 170; Colón's share of budget, 175–76); Phillips and Phillips 1992:120–211 (ship lengths, 144–45); Varela and Gil eds. 1992:95–365 (Colón's and other letters); Gould 1984 (Colón's crew); Oviedo y Valdés 1851–53:bks. 1–4; Cuneo 1495:50–63. Las Casas says the second voyage had "1,500 men, all or almost all paid by Their Highnesses" (1951:vol. 1, 346); court historian Andrés Bernáldez says (1870:vol.2, 5) that the ships had "one thousand two hundred fighting men in them, or a little less," a tally that seems not to include seamen, priests, artisans, etc.

5 "magnificent walls": Scillaccio, N. 1494. The Islands Recently Discovered in the Southern and Indian Seas. In Symcox ed. 2002:162–74, at 172.

6 Worldwide spread of tobacco: See Chaps. 2, 5; Satow 1877:70–71 (Tokyo gangs).

6 Early pan-Eurasian trade: Overviews include Bernstein 2008:1–109; Abu-Lughod 1991.

7 Colón as beginning of globalization: I adopt this point from Phillips and Phillips (1992:241), who say that the admiral "placed the world on the path" to global integration.

7 Venturesome land creatures: Decker-Walters 2001 (bottle gourds); Zizumbo-Villarreal and Quero 1998 (coconuts); Montenegro et al. 2007 (sweet potatoes).

7 Torn seams of Pangaea: Crosby 1986:9–12.

7 Columbian Exchange: Crosby 2003.

7 Comparison to death of dinosaurs: Crosby 1986:271. Crosby's point (2003:xxvi) is increasingly accepted: "Even the economic historian may occasionally miss what any ecologist or geographer would find glaringly obvious after a cursory reading of the basic original sources of the sixteenth century: the most important changes brought on by the Columbian voyages were biological in nature."

8 *Santa María,* La Navidad: Abulafia 2008:168–71; Colón 2004:108–13; Morison 1983:300–07; Colón 1493:177–86. La Navidad may have been near the town of Caracol, in northern Haiti; the anchor of the *Santa María* may have been found there in the eighteenth century (Moreau de Saint-Méry 1797–98:vol. 1, 163, 189, 208).

9 Taino: Rouse 1992.

9 Destruction of La Navidad: Abulafia 2008:168–71; Las Casas 1951:vol. 1, 356–59; Chanca 1494:51–54 ("grown over them," 54—I translate *yerba* as "weeds" and "vegetation"). Las Casas (1951:vol. 1, 357) gives the number of bodies as "seven or eight"; Colón's son (2004:312) gives the figure as eleven. Michele de Cuneo (1495) says that the Spaniards feared they had been eaten.

9 Creation of La Isabela: Abulafia 2008:192–98; Las Casas 1951:vol. 1, 363–64, 376–78; Anghiera 1912:88 (gardens); Cuneo 1495:178 ("roofed with weeds").

9 Reiter's syndrome (footnote): Disease at La Isabela: Allison 1980; Aceves-Avila et al. 1998; Chanca 1494:66–67; Las Casas 1951:vol. 1, 376. Colón's sickness: Colón 2004:329; Las Casas 1951:vol. 1, 396–97; and Colón, C. 1494. Letter to the Monarchs, 26 Feb. In Varela and Gil eds. 1992:313. Rheumatologist Gerald Weissmann has written up Colón as a case study (1998:154–55). According to Las Casas (1951:vol. 1, 363–64), whose father and brother were eyewitnesses, the admiral also fell sick in January; the summer attack may have been a second, worse bout. Colón described "a sickness that deprived me of all sense and understanding, as if it were pestilence or *modorra*" (313). Reiter's is not linked to *modorra*

(swoony somnolence, then regarded as its own disease), but has been associated with high fever and confusion, which may be close enough. Colón's later symptoms, such as inflammation, match more closely.

10 Margarit's betrayal: Abulafia 2008:202–03; Phillips and Phillips 1992:207–08; Poole 1974; Las Casas 1951:vol. 1, 399–400; Oviedo y Valdés 1851–53:vol. 1, 54. Poole argues that Margarit's departure was less a betrayal than the action of a dutiful royal servant reporting the chaos in the colony, but this is a difference of nuance—his report was strongly anti-Colón.

10 War with Taino: Abulafia 2008:201–07; Colón, C. Letter to the Monarchs, 14 Oct. 1495. In Varela and Gil eds. 1992:316–30 ("the land," 318); Castellanos 1930–32:vol. 1, 45 (Elegía II [chemical warfare]).

11 Hispaniola in Columbian Exchange: E-mail to author, Bart Voorzanger (grass); Hays and Conant 2007 (mongoose); Eastwood et al. 2006 (swallowtail); Guerrero et al. 2004 (swallowtail); Martin et al. 2004 (forest understory); Rocheleau et al. 2001 (forest change); Parsons 1972 (African grasses [bedding, 14]); Hitchcock 1936 (African grasses, 161, 259).

13 Fire ant onslaught: Wilson 2006, 2005; Williams and Matile-Ferraro 1999:146 (African scale insects); Las Casas 1951:vol. 3, 271–73 (all other quotes); Oviedo y Valdés 1851–53 ("depopulated," vol. 1, 453 [bk. 15, chap. 1]; plantains, vol. 1, 291–93 [bk. 8, chap. 1]); Herrera y Tordesillas 1601–15:vol. 2, 105–06 (Dec. 2, bk. 3, chap. 14). The plantains were imported from the Canary Islands, off the West African coast. Most of the scale insects involved were mealybugs. Oviedo reported that "the ants in these parts are very good friends" of plantains (vol. 1, 291).

13 "millions of them": Colón, C. Letter to the Catholic Monarchs, Apr.-May 1494. In Varela and Gil eds. 1992:284. *Cuento de cuentos* literally means "a million millions," or one trillion. But *trillón* did not enter Spanish until the seventeenth century. Colón clearly intended "a whole lot," so I use "millions upon millions" to convey this indeterminacy with a period term. My thanks to Scott Sessions for help with this translation.

13 Estimates of Hispaniola population fall: Livi-Bacci 2003 (table of estimates, 7; "a few hundred thousand," 48; 1514 count, 25–34); Las Casas 1992:29 ("three million"). Geographer William Denevan, author of many studies of pre-contact demography, believes (pers. comm.) the figure is 500,000–750,000.

14 Fewer than 500 Taino: Oviedo y Valdés 1851–53 ("no one believes in this year of 1548 there are as many as 500 persons," vol. 1, 71 [bk. 3, chap. 6]). Las Casas claimed that "not 1000 souls could have survived or escaped this misery" by 1518–19 (1951:vol. 1, 270). Oviedo lived on the island from 1514 to 1556, Las Casas from 1502 to about 1540.

14 Santo Domingo in poverty: Bigges 1589:32. When Sir Francis Drake sacked the city, it was too poor "for lacke of people to worke in the Mines" to give him much ransom.

14 Absence of diseases in Americas, incursion of new diseases: A summary is Mann 2005:86–133.

14 Taino genes: E-mail to author, Juan Carlos Martinez-Cruzado (University of Puerto Rico). Martinez-Cruzado reported in July 2009 "finding Native American mtDNA at ~15% in the Dominican Republic," but at the time of writing was still trying to distinguish what fraction was Taino.

15 Charges against Colón: See, e.g., Sale 2006. Index entries under "Colón" indicate his general tone: "complaining," "deceptive," "exaggeration and boasting," "preoccupation with glory, with lineage," "self-pity," "self-serving." For reactions to Colón over time, see Stavans 2001.

16 Lighthouse first proposed: Del Monte y Tejada 1852–90:vol. 1, 316–19 ("divine decree," 316).

17 Colón's collected writings: Varela and Gil eds. 1992.

17 Signature: Colón, C. Entail of estate, 22 Feb. 1498. In Varela and Gils eds. 1992:356. See also, Morison 1986:356–57; Milhou 1983:55–90.

18 "San Fernando": Colón 2004:238. See also, Las Casas 1951: Vol. 1, Chap. 2. My brief biography relies on the sources cited in the notes to p. 4.

18 Ambition and faith: Here I follow Fernández-Armesto 2001. For Colón's attempts to found a dynasty, see, e.g., his instructions for dynastic succession in Colón 1498. See also the determined efforts to preserve the records of his noble privileges in what has come to be known as the *Book of Privileges* (Nader 1996:10–13). The best analysis of his religious beliefs I have seen is Milhou 1983; see also, Delaney 2006; Watts 1985.

18 "conquest of the Holy Land": Colón 1493:181. Colón amplified his hopes in a 1493 letter to the monarchs: "Divine grace permitting, . . . seven years from today, I will be able to pay to Your Highnesses the costs of five thousand cavalry and fifty thousand infantry in the war and conquest of Jerusalem, which was the reason for undertaking this enterprise" (Varela and Gil eds. 1992:227–35, at 232). See also, Delaney 2006; Rusconi ed. 1997:71–77 (unsent Colón missive exhorting the monarchs to take Jerusalem); Colón 1498:360 (instructing his heirs to help with the conquest). My thanks to Scott Sessions for helping me with this material.

19 "by divine will": Colón, C. Relation of the Third Voyage, Aug.(?). 1498. In Varela and Gil eds. 1992:377 ("put there"); 380 ("by divine will"). Reporting to the monarchs about his third voyage, Colón claimed to have found the Paradise nipple. At the mouth of the Orinoco River, in Venezuela, he had ascended the slopes that led to "this protruding part" (ibid.:377). Heaven was presumably near the river's headwaters. Such ideas were widely shared; Dante put Paradise on top of a huge projection he called Mount Purgatory (Lester 2009:292–95). My thanks to Scott Sessions for translation help.

19 Islam, Venice, Genoa as middlemen: My thanks to Dennis Flynn for

straightening me out. See, e.g., Bernstein 2008:70–76; Hourani 1995:51–87ff.

19 Eratosthenes: Crease 2003:chap. 1 ("around the globe," 3). Eratosthenes's actual figure was probably closer to 29,000 miles, which most science historians view as remarkably accurate.

20 Colón's geography: Wey Gómez 2008:65–99, 143–58; Varela and Gil eds. 1992:90–91 (quadrant); Nunn 1924:chap. 1. Colón also relied on the cosmographer Pierre d'Ailly, who argued that East Asia had to be near West Africa (Nunn 1935). D'Ailly took his ideas from Roger Bacon (1962:311), who in turn based his on a mistaken reading of Aristotle. (Aristotle in fact wrote [1924:II.14], "[M]athematicians who try to calculate the size of the earth's circumference arrive at the figure 400,000 stades"—about 45,500 miles, twice its actual size.) Colón knew nothing of such scholarly niceties, and there is no evidence that he would have cared. Old Spanish nautical miles are close enough to current terrestrial miles that I ignore the difference.

21 "in the Apocalypse": Colón, C. 1500. Letter to Dona Juana de la Torre. In Varela and Gil eds. 1992:430–37, at 430.

21 Converting emperor of China: Colón, C. 1503. Relation of the Fourth Voyage, 7 Jul. In Varela and Gil eds. 1992:485–503, at 498.

21 Inuits in Ireland (footnote): Varela and Gil eds. 1992:89 (marginalia). The story is brushed aside by Morison (1983:25), mentioned without comment by Phillips and Phillips (1992:105), and dismissed by Quinn (1992:282–85). By contrast, Jack Forbes (2007:9) calls the marginalia "solid, indisputable evidence that Columbus and others had seen Native Americans at Galway." My thanks to Scott Sessions for helping me translate Colón's note.

21 History of lighthouse project: González 2007; Roorda 1998:114–18, 283–85; Farah 1992; French 1992a, b (protests); Wilentz 1990; Gleave 1952.

23 Homogenocene: Samways 1999. It is more often called the Anthropocene.

24 Legazpi and Urdaneta before expedition: Rubio Mañé 1970, 1964; Mitchell 1964 (Urdaneta selects Legazpi, 105); De Borja 2005:chap. 3. Enriquez, M. 1561. Letter to Felipe II, 9 Feb. In B&R 3:83–84 (Urdaneta and Legazpi as relatives); idem 1573. Letter to Felipe II, 5 Dec. In B&R 3:209–22, at 216–17 (Legazpi sells property). A popular account is Sanz y Díaz 1967:3–17.

25 King's hopes for mission: Royal Audiencia of New Spain. Instructions to Legazpi, 1 Sep. 1564. In B&R 2:89–100.

25 "and Legazpi": García-Abásolo 2004:231. Legazpi, García-Abásolo says, "always thought that his final destination would be to come to China. . . . Probably, if he had lived a few more years, Legazpi would have sponsored a diplomatic mission to China" (235). See also, Cortés 2001:266–77 passim, 444–47 (hoping to explore Pacific) and the many abortive China missions chronicled in Ollé Rodríguez 2002.

26 Legazpi's first years in the Philippines: Legarda 1999:16–31; Guerrero 1966:15–18; Rubio Mañé 1970, 1964. See also, Sanz y Díaz 1967:35–52.

26 Beeswax (footnote): Cervancia 2003; Ruttner 1988:284 (bee ranges); Cowan 1908:73, 89, 105 (scale insect).

26 Maujao trading place: Author's visit, interviews with Chiquita Cabagay-Jano and Rudmar T. Cabagay (Bulalacao development office); Horsley 1950:74–75 (parasols, drums); Legazpi, M. L. d. Letter to Felipe II, 23 Jul. 1567. In B&R 2:233–43, at 238. See also, Laufer 1908:251–52; Li 2001:76–79.

27 Encounter at Maujao: Legarda 1999:23–24; Zuñiga 1814:vol. 1, 110–11; Laveçarism G. d. 1575? "Part of a Letter to the Viceroy." In B&R 3:291–94, at 291–92; Anon. (Martín de Goiti?). 1570. Relation of the Voyage to Luzon. In B&R 3:73–104, at 73–77 (all quotes). The Spaniards went on to Manila, where they protected some Chinese junks in a trade dispute and also sacked the town (ibid.:94–96, 101–04; Anon. 1571. Relation of the Conquest of the Island of Luzon. In B&R 3:141–72, at 148–57). The distances involved are in "leagues," which I take as the *legua común* of 3.46 miles (Chardon 1980). See also, Sanz y Díaz 1967:53–59.

28 Chinese visit Manila: Pacheco Maldonado, J. 1572. Carta en Relación de Juan de Maldo-nado Tocante al Viaje y Poblacion de la Isla de Luzón en Filipinas, 6 May. Quoted in Ollé Rodríguez 2006:32, 1998:227–30; Riquel H., et al. 1573. News from the Western Islands. In B&R 3:230–49, at 235. 1573:235; Zuñiga 1814:vol. 1, 125–26.

28 Initial exchange in Manila: Lavezaris, G. d. 1573. Affairs in the Philippines, after the Death of Legazpi, 29 Jun. In B&R 3:179–89, at 181–84; Riquel et al., op. cit. In ibid.: 230–49, at 243–49 ("delighted," 245); see also the letter from the viceroy of New Spain quoted on ibid.: 226, note 75.

28 New globalized era, impact of silver trade and slavery: See Chaps. 3 and 4.

29 China's superiority: Pomeranz 2000:31–107; Frank 1998: Chap. 4.

29 Spanish inability to export European goods to China: Enriquez, M. 1573. Letter to Philip II, 5 Dec. In B&R 3:209–19, at 212 ("not already possess"), 214. The Chinese were equally blunt: "China trades in Luzon solely for the purpose of obtaining *feringhee* [foreign] silver coins," wrote Provincial Governor Xu Xueju. The foreigners had nothing else China needed (Xu, X. Initial Report on Red-Haired Foreigners. In Chen et al. eds. 1962:4726–27). See also Marks 2007:60–62.

29 "revolutionary events": Song 2007:2. See Chap. 5.

30 "Asian train": Frank 1998:277.

30 History of Legazpi-Urdaneta monument: De Borja 2005:17, 128.

33 World money supply triples: Garner 2006, using updated data from TePaske and Klein; Morineau 1985:571–99. By 1650 the Americas had produced about thirty thousand tons of silver.

33 Potosí geology: Bartos 2000; Waltham 2005.

33 Potosí as biggest American city: See Chap. 4.

33 Silver processing, transportation, fleet: Craig and Richards 2003:1–12 (sixty-five pounds, table 1–1); Goodman 2002:3–5; Cobb 1949:33–36; Acarete du Biscay 1698:54–57.

34 Deserted Andean fields: Studies of the abandonment of highland agriculture are few, but Denevan sums the evidence and concludes that it was "primarily related to sixteenth-century population decline" (2001:201–10, quote at 210).

34 1600 volcano: De Silva and Zielinski 1997.

34 Proportion of silver shipped to China: The two sides of the debate are encapsulated in Garner 2006 and Flynn and Giráldez 2001. See Chap. 4.

36 Spanish and European financial woes: Standard accounts include Kamen 2005, Elliott 2002, Lynch 1991 (price revolution, 174–84), Parker 1979 (esp. "War and Economic Change: The Economic Costs of the Dutch Revolt," 178–203), and Hamilton 1934. Spain's woes are succinctly summarized in Flynn 1982:esp. 142–43. A detailed study of Spain's flight into silver-hued bankruptcy is Carande 1990, esp. vol. 3. Silver production decline: Garner 2007:figs. 6, 8 (using updated, unpub. figures from John TePaske); Garner 1988:fig. 2; Brading and Cross 1972:fig. 2. The idea of a "general crisis" apparently originated with Roland Mousnier in 1954.

36 "Spain's wars": Flynn (pers. comm.) based his summary on Parker 1979b, c.

38 Little Ice Age in Europe: Parker 2008:1065, 1073 (Ireland); Fagan 2002 (frozen sea, 137); Reiter 2000 (Greenland, 2); Lamb 1995:chap. 12; Ladurie 1971 (wine, 52–56; bishop, 180–81).

38 Sunspots, volcanos, and Little Ice Age: Eddy 1976 (Maunder Minimum); Briffa et al. 1998 (volcano impact); Jansen et al. 2007:476–78 (skepticism on sunspots, volcanos); Hegerl et al. 2007:681–83 (sunspots and volcanos). The 1641 eruption was heard throughout the Philippines—a "noise in the air of musketry, artillery, and war drums" (Anon. 1642. News from Filipinas, 1640–42. In B&R 35:114–24, at 115).

39 Ruddiman's hypotheses: Ruddiman 2003, 2005, 2007. His additional argument that deforestation and burning affected climate as much as eight thousand years ago has been the subject of attacks (e.g., Olofsson and Hickler 2008) and defenses (e.g., Müller and Pross 2007). Accepted by many is the connection between American pandemics and CO_2 decline (Dull et al. 2010; Nevle and Bird 2008; Faust et al. 2006).

39 Fire maintains prairies: Anderson 1990; Stewart 2002:113–217; Clouser 1978. For the role of fire in worldwide grasslands, see Bond et al. 2005.

39 Indian fire and eastern forest: Johnson 2005:85 ("Parkes in England"); Stewart 2002:70–113; Williams 1989:chap. 2, esp. 43–48; Cronon 1983:48–52; Day 1953.

39 Thirty-one sites: Nevle and Bird 2008.

42 Change in landscape after epidemics: Dull et al. 2010 ("carbon budget," 765); Denevan 2007, 1992:377–79; Wood 1977:38–39 ("travel through").

42 Little Ice Age in North America: Parker 2008:1067; Pederson et al. 2005 (forest composition); Anderson 2004:100 (livestock); Kupperman 1982.

43 Climate, mosquito, disease, slavery: See Chap. 3. *A. quadrimaculatus:* Reinert et al. 1997; Freeborn 1923. Paradoxically, drought also favors the mosquito, because it kills off aquatic predators on its larvae (Chase and Knight 2003); the species thrives on climatic disturbance.

43 Introduction of horses: Hämäläinen 2008 passim; Calloway 2003:chap. 6; Holder 1974.

44 Mexico City and Acapulco: Mexico City: See Chap. 8. Acapulco: Schurz 1939:371–84; Gemelli Careri 1699–1700:vol. 6, 5–16.

45 1637–41 drought, decline in silver: Garner 2007:esp. figs. 1–3 (silver); Atwell 2005 (silver); Atwell 2001:32, 36, 62–70 (volcanos).

45 Famine and unrest in China: Parker 2008:1058–60, 1063–65. See Chap. 5.

46 Greatest cities in 1500: Chandler 1987:478–79; see also, Eggiman 1999, De Vries 1984. The sole exception to the rough thirty-degree rule is Beijing, the northern capital of a country whose population was concentrated to the south. Note: I adjust Chandler's list for sub-Saharan African and indigenous American cities, which he underestimates consistently. Explanations follow. *Tenochtitlan:* Typical estimates for the whole conurbation range from 1 to 1.5 million, with Tenochtitlan assigned, somewhat arbitrarily, a fifth to a quarter of the total (see, e.g., Smith 2002:57–59; Sanders 1992). *Qosqo:* Population numbers are even more uncertain, but recent estimates are between 100,000 and 200,000, based on Spanish colonial accounts (typically higher numbers) and archaeological surveys (typically somewhat lower). See, e.g., D'Altroy 2002:114 (100–150,000); Cook 1981:217–19 (150–200,000); Agurto Calvo 1980:122–28 (125,000). *Gao:* Data are poor, but a late-sixteenth-century census of compound houses in the central city came up with a population of forty to eighty thousand; presumably as many or more surrounded the central zone. "Such a size and population may sound exaggerated. . . . But we must remember that Gao was the epicenter of an empire that extended over 1,400,000 sq. km (500,000 square miles)" (Hunwick 1999:xlix). The nineteenth-century traveler Henry Barth, who saw the ruins in relatively undisturbed condition, estimated that Gao "had a circumference of about six miles" (Barth 1857–59:vol. 3, 482). *Paris:* Bairoch, Bateau, and Chévre estimate it at 225,000 in 1500 (cited in DeLong and Shliefer 1993:678). Chandler (1987:159) estimates Paris in 1500 at 185,000 by multiplying a 1467 estimate of the number of men who could bear arms (28–30,000) by 6 to obtain 174,000, which rises for unexplained reasons (migration?) to 185,000 in 1500. The factor of 6 seems high—indeed, Chandler uses 5 for Paris on a similar estimate a century before.

46 Change in cities: Acemoglu et al. 2002.

CHAPTER 2 / *The Tobacco Coast*

51 Introduced earthworms: Author's interviews, Hale, Reynolds, Bohlen; Frelich et al. 2006; Hendrix and Bohlen 2002:esp. 805–06, table 4; Reynolds 1994; Lee 1985:156–59.

51 Rolfe: Price 2005:154–58; Townsend 2004:88–96; Haile ed. 1998:54–56; Robert 1949:6–9. Occasionally St. John's, Newfoundland, is cited as the first long-lasting English colony, but most historians believe it had no permanent population before 1610.

51 "Drinking" tobacco: Ernst 1889:141–42; Apperson 2006:6.

51 Types of tobacco: Horn 2005:233; Robert 1949:7–8; Arents 1939:125; Strachey 1625:680 ("biting taste"). *N. rustica* was often so strong as to be hallucinogenic, which some colonists clearly enjoyed. Smoking, Thomas Hariot said, led to "manie rare and wonderful experiments" (Hariot 1588:n.p.[17]).

51 Rolfe and tobacco: Arents 1939:125. See also, Hamor 1615:820, 828 ("sweet, and strong"); Velasco, A. d. 1611. Letter to King of Spain, 26 May. In Brown 1890:vol. 1, 473.

52 English tobacco mania: Laufer 1924b:3–48; "C. T." 1615:5 (silver); Rich 1614:25–26 (seven thousand tobacco houses).

52 Tobacco exports: For export figures, see below. Size of barrels: author's visit, Jamestown archaeological site.

52 Ballast: Given that ship ballast is now viewed as a prime source of biological introductions (e.g., Bright 1988:167), it is surprisingly little studied. According to one nineteenth-century nautical textbook, ballast usually consisted of "iron, stone, or gravel, or some similar material," though "in some Colonial and other ports sand only is to be had" (Stevens 1894:75–76).

52 Earthworms as engineers: Darwin 1881 ("organized creatures," 313); Edwards 2004:4 (mass of worms, turnover rate).

53 Ice Age and worms: James 1995. The Ice Age didn't kill *all* northern American worms. But all common earthworms in North America today are imports, mostly from Europe and Japan.

53 Ecological impacts of introduced earthworms: Author's interview, Hale; Frelich et al. 2006:1239 (see fig. 1), 1236, 1238 (soil density), 1237 (impact on nutrients), 1241 (litter), 1241 (understory plants); Bohlen et al. 2004a:8 (nutrients); Bohlen et al. 2004b:432 (clears understory); Migge-Kleian et al. 2006 (declines in invertebrates, mammals, birds, lizards). Specific impact of *L. terrestris:* Proulx 2003:18; Tiunov et al. 2006. Impact of *L. rubellus:* Bohlen et al. 2004b:432; Tiunov et al. 2006:1226. Earthworms may promote invasions by exotic species (Heneghan et al. 2007).

53 Goal of colony: Horn 2005:41–42, 55–56, 80–81; Price 2005:21–22, 75–76. The colonists' instructions commanded them to hunt for "any miner-

als" and to set up camp on a river flowing from the northwest, "for that way you shall soonest find the other sea." (Haile ed. 1998:19–22). The company charter (McDonald ed. 1899:1–11) concerns itself with only three things beyond survival and defense: converting the natives (¶II); obtaining "Gold, Silver, and Copper" (¶IX); and trading with "any other Foreign Country" (¶XVI). In part the English believed there was gold and silver because of the claims of a previous English visitor to North America (Ingram 1883; DeCosta 1883).

54 First representative body, first slaves: See below.

54 Jamestown landing: Bernhard 1992:600–01; Billings 1991:5; Kelso 2006:14. The number of colonists is disputed. Bernhard and Kelso, agreeing with George Percy (Haile ed. 1998:98), argue for 104 (105 sailed, 1 died en route). But Kelso and Straube (2004:18) and Kupperman (2007:217) use 108; Price (2005:15) calls it at "105 or so."

54 Jamestown settlers: Visitors to the Jamestown archaeological site find lists of "Jamestown settlers." The same language turns up in Wikipedia and newsmagazine headlines (Lord 2007).

54 Tsenacomoco: Variant spellings include Tsenacomacah, Tsenacomma-cah, and Tsenacommacoh. I use "empire" following Fausz 1977:68–70.

55 Began with six villages: Strachey 1612:615. For accounts of Powhatan's empire building, see Rountree 2005:chap. 4 and Fausz 1977:56–68.

55 Powhatan domain boundaries: Hatfield 2003:247; Rountree 2005:40; Turner 1993:77.

55 Tsenacomoco's size and population in 1607: Subject to scholarly debate since Thomas Jefferson (1993:220) made the first population estimate (8,000 mi², 8,000 people). Recent estimates include Feest 1973 (14,300–22,300 people); Turner 1973 (18,550 km² [7,160 mi²], 10,400 people); Fausz 1977:60 ("in the neighborhood of twelve thousand"); Turner 1982 (16,400 km² [6,332 mi²], 12,940 people); Rountree 1990:15 (16,500 km² [6,370 mi²], 13,000–14,300 people); Rountree and Turner 1994:359 ("slightly less than 6,500 square miles"; "some 13,000 persons"); McCord 2001 ("sparse" population); Hatfield 2003:fig. 1 (about 6,200 mi²); Turner 2004 (13,000–15,000 people); Horn 2005:16 ("perhaps 15,000 people"); and Rountree 2005:13 ("about 15,000"), 40. I follow Rountree and Horn.

55 "of western Europe": Williams 1989:33.

56 Powhatan as man and domain: Rountree 1990:7; Rountree 2005:33. Allen (2003:64–67) explains the derivation of the name. His subjects addressed him by his common name, Wahunsenacawh (Strachey 1612:614).

56 Powhatan's capital, residence, appearance: Author's visit, archaeological site; Gallivan et al. 2006 (geography, fig. 3.1); Gallivan 2007 (town map, fig. 2); Smith 2007a:17, 22 ("expresse"), 53–54; Smith 2007b:270 ("a vault"), 296–97 (pearls, divan); Strachey 1612:614–19 ("king's house" 615); Rountree 2005:29–35.

57 Lack of domesticated animals: Strachey 1612:637; Crosby 1986:172–94; Diamond 1999:160–75. The Powhatan, like other Indians in eastern North America, had only dogs and hawks, the latter of which were not domesticated so much as tamed (Anderson 2004:34–37).

57 Criteria for domestication, tally of domesticated animals: E. O. Price 2002; Mason ed. 1984. The number of birds is disputed, one issue being whether caged birds like parakeets and canaries are domesticated.

60 English landscapes: Anderson 2004:84–90.

61 Native agriculture methods: Smith 2007b:279; Strachey 1612:676–77; Spelman 1609:492.

61 Indian maize field size: Maxwell 1910:73; Smith 2007b:284 ("their fields or gardens [are] some 20 acres, some 40. some 100. some 200"). Strachey (1612:626) noted that "so much ground [in one town] is there cleared and open" that with "little labor" the colonists could plant corn "or make vineyards of two or three thousand acres." Edward Williams argued (1650:13) that colonists need not fear the labor of opening up the forest, because of the "immense quantity of Indian fields cleared already to our hand." Scholarly summaries include Rountree et al. 2007:34–35, 41–42, 153. Citing another figure by Strachey (1612:636), Rountree (1990:280, note 22) argues that most fields were one to two hundred feet on a side. Strachey also reported that plants were separated from each other by "4 or 5 foot" and "commonly" bore two small ears, which would indicate that a 150' x 150' household field would yield about three thousand ears—food for a month or two for the "six to twenty" residents of each house (1612:636, 676). (Native maize ears were less than half the size of typical modern ears.) In anthropological annals one rarely encounters people who take the trouble to clear land for cereals but not enough to use them as staples, as Strachey's second, smaller estimate suggests.

61 Palisades, absence of fencing: Rountree 2005:42; Rountree and Turner 1998:279; Rountree et al. 2007:38.

61 Meaning of fences, domestic animals in England: Anderson 2004:78–90.

61 Use of "abandoned" fields and plants on them: Rountree, Clarke, and Mountford 2007:42; Rountree 2005:9, 56; Rountree 1993a:173–74.

62 Impact of beaver: Hemenway 2002; Naiman et al. 1988. There is a European beaver, but it had been hunted to extinction in Britain.

62 Tuckahoe: Author's visits, Jamestown; Smith 2007b:276, 391; Rountree et al. 2007:43–44, 124; Rountree 2005:12, 1990:52–53; Strachey 1625:679.

63 Smoke and fire observable from sea: De Vries 1993:22 ("it is seen"); Bigges 1589:38 ("great fire[s] . . . are very ordinarie all alongst this coast," 132).

63 Indian hunting by fire: Smith 2007a:14 ("over the wood"); Mann 2005:248–52; Williams 1989:32–49; Krech 1999:104–06; Byrd 1841:80–81.

63 Effects of native burning: Miller 2001:122; Wennersten 2000:chaps. 13–15; Pyne 1999 ("into metals," 7), 1997a:301–08, 1997b, 1991 ("corridors of travel," 504); Pyne et al. 1996:235–40; Rountree 1993b:33–38 (paths); Hammett 1992; Williams 1989:32–49; Byrd 1841:61 ("all before it"); White 1634:40 ("without molestation"). Like White, John Smith insisted that "a man may gallop a horse amongst these woods" (2007b:284), as did a seventeenth-century chronicler from Maryland ("The Woods for the most part are free from underwood, so that a man may travel on horse-backe, almost any-where" [Anon. 1635:79]). So commonly was the Virginia forest understood to be open that William Bullock, before his first visit there, explained (1649:3) that in Virginia people can see "above a mile and a half in the Wood, and the Trees stand at that distance, that you may drive Carts or Coaches between the thickest of them, being clear from boughs a great height." (Bullock 1649:3). Among the first sights that greeted the Jamestown colonists was a big fire-created clearing (Percy 1625?:90–91).

64 Jumble of ecological zones: Rountree 1996:4–14.

64 Smith tales in *True Travels:* Smith 2007c (early years, 689–94; "to Rome," 693; "Stratagem," 696; "such like," 703; single combats, 704–06; slavery, 717–18; "his necke," 720; "braines," escape and flight, 730–33; African piracy, 741–43). See also, Kupperman ed. 1988:introduction.

65 Skepticism, support of Smith: Adams 1871; Fuller 1860:vol. 1, 276 ("proclaim them"). Adams's motives: Rule 1962 ("aristocracy," 179). Refutations of skeptics: Striker 1958; Fishwick 1958; Striker and Smith 1962 ("Al Limbach," 478); Barbour 1963; Kupperman ed. 1988:2–4. A popular satirical poem, *The Legend of Captaine Jones,* appeared in 1630, mocking Smith's boastfulness.

67 Smith irritates social betters: Like a modern populist, Smith mocked the milieu of "Parliaments, Plaies, Petitions, Admiralls, Recorders, Interpreters, Chronologers, Courts of Plea, [and] Justices of peace" (2007c:329) inhabited by politically connected gentlemen like the colony leaders. In return, they denounced him (Wingfield 1608?:199–200; Percy 1625?:502; Ratcliffe [in Haile ed. 1998:354]; and Archer [ibid.:352–53]). Attempts to pass new sumptuary laws are described in Kuchta 2002:37–39. Percy's trunk is described in Nicholls ed. 2005:213–14.

67 Smith's version of capture: Smith 2007b:316–23 ("from death," 321; "with hunger," 323).

68 Skepticism on Pocahontas story: The two varying accounts are from 1608 (Smith 2007a) and 1624 (Smith 2007b). Rountree (2005:76–82) argues, convincingly to my mind, that at most Pocahontas was playing a part in a ritual whereby Powhatan made Smith his vassal (Horn 2005:66–71; Kupperman 2007a:228; Allen 2003:46–51; Richter 2001:69–78). The lovelorn women who succored Smith are cataloged by Townsend (2004:52–54)

and Smith himself (2007b:203–04). Films include *The New World* (2005), *Pocahontas* (1995), and *Captain John Smith and Pocahontas* (1953). Popular accounts are divided on accepting the story (Price 2005:59–69, 241–45; Horwitz 2008:334–37).

69 Smith story obscures real story: Kupperman 2007a.

69 English monarchy's debts, forced loans: Homer and Sylla 2005:122; Croft 2003:71–82; Scott 1912:vol. 1, 16–27, 52–54, 133–40.

70 "Slave of Wickedness": Barlow 1681:2–6 (all quotes). This is the most common seventeenth-century translation of the encyclical *Regnans in Excelsis* (1570).

70 Spanish colonies: Pre-Jamestown Spanish incursions included San Miguel de Gualdape (founded in 1525, probably in South Carolina [see Chap. 8]), Santa Rosa Island (1559, off the Florida panhandle), San Agustín (1565, now the city of St. Augustine, Florida), Guatari (1566, in South Carolina), San Antonio (1567, in southwestern Florida), Tequesta (1567, in southeastern Florida), Ajacán (1570, near Jamestown), San Pedro de Mocama (1587, on an island near the present Georgia-Florida border), Santa Catalina de Guale (early 1590s, on another Georgia island), Tolomato (1595, on the Georgia coast), Santa Clara de Tipiqui (1595, on the same coast), Talapo (1595, on the same coast), Santo Domingo de Asao (1595, on the same coast), San Pedro y San Pablo de Puturiba (1595, on the same island as San Pedro de Mocama), San Buenaventura de Guadalquini (1605, on another Georgia island), and San Joseph de Sapala (1605, on yet another). This list is not complete; in some cases sources differ on the proper spelling and exact date of founding. For details on Ajacán, see Lewis and Loomie 1953. Many more were founded after Jamestown, among them Santa Fe.

70 Colonies in New France: Charlesbourg-Royal (founded 1542, on the St. Lawrence River), Charlesfort (1562), Fort Caroline (1564), Sable Island (1598), and Port-Royal (1605). Quebec was founded in 1608, a year after Jamestown.

71 Hakluyt: Hakluyt 1584:chap. 4 ("daily piracies"), chap. 1 (all other quotes).

71 Closeness of Americas and China: See below.

71 Joint-stock companies: A standard history is Scott 1912. Succinct explanations of the companies' origins as a means for spreading risk include Kohn forthcoming:chap. 14; Brouwer 2005. Importantly, joint-stock companies let investors negotiate with the crown as a group when seeking the necessary royal permission for foreign trade. As individuals, single investors had little leverage; banded together, they were less vulnerable to royal whim. I thank Mark Plummer for many useful conversations.

72 Landes and North: Landes 1999 ("patience, tenacity," 523); North and Thomas 1973 (arrangements, "phenomenon," 1). Other works in this

sometimes polemical tradition include Gress 1998, Lal 1998, and Jones 2003.

72 Ten joint-stock companies before Jamestown: The count is the companies discussed in Scott 1912:vol. 2. I do not include mining partnerships but do include Ralegh's colonial ventures (see below). Most large-scale European trade then was controlled by merchant families and royal monopolies; an example is the Merchant Guild, the state-affiliated association of Seville merchant families that long dominated Spain's America trade. A partial exception was the Dutch East India Company, a consortium of six merchant firms supervised by a board of overseers chosen by the governments of the Netherlands' five provinces. For brief accounts of the Merchant Guild and the rivalrous English and Dutch East India Companies see, respectively, Smith 1940:chap. 6 and Bernstein 2008:chap. 9.

72 Four previous colonies: Humphrey Gilbert's venture (canceled by Gilbert's ship sinking during a reconnaissance mission in 1583); the Popham colony in Maine (1607–08); and the two efforts at Roanoke (1586–1587; 1587–?). For Roanoke, Ralegh did not create a joint-stock company but raised the money through an informal but similar arrangement (Trevelyan 2004:54, 81, 114, 138). The Popham colony began soon after Jamestown, but I include it as its prime mover was also an organizer of the Virginia Company.

72 Roanoke colony: Horn 2010; Kupperman 2007b; Oberg 2008; Donegan 2002:chap. 1; Fausz 1985:231–35; Quinn and Quinn eds. 1982. Quinn 1985 remains the history on which all others are built. Popular accounts include the enjoyable Horwitz 2008:chap. 11.

73 Roanoke introduces tobacco to England (footnote): Laufer 1924b:9–11 ("smoak," 10).

73 Virginia Company view of Spain, Tsenacomoco: Billings ed. 1975:19–22 (quotes, 19–20).

74 *Tassantassas:* Rountree 2005:6. See its usage in, e.g., Hamor 1615:811.

74 Jamestown peninsula, problems with site: Author's visit; author's interviews, William Kelso, Greg Garman; Smith 2007b:389 (well); Barlow 2003:22–25 (crater) Rountree 1996:18–29 (Indians occupied best land); Earle 1979:98–103 ("salt poisoning," 99); Strachey 1625:430–31; Percy, G. 1607(?). Observations Gathered out of a Discourse of the Plantation of the Southern Colony in Virginia by the English, 1606. In Haile ed. 1998:85–100 ("filth," 100). The reputation for picking the best land lasted. "Wherever we meet an Indian old field or place where they have lived," the clergyman Hugh Jones wrote of Virginia in 1724, "we are sure of the best ground" (quoted in Maxwell 1910:81).

74 Droughts: Stahle et al. 1998. A team of archaeologists and dendrochronologists (scientists who study tree rings) examined long-lived Virginia

cypress trees. Because rainy years create wider tree rings than do dry years, the scientists could show that the 1606–12 drought was the worst in centuries.

74 Thirty-eight left alive: Smith 2007b:323, 406; Bernhard 1992:603; Earle 1979:96–97; Kupperman 1979:24.

74 Powhatan's attitude: Rountree 2005:143–47; Fausz 1985:235–54; Fausz 1990:12 ("ignorance"); Percy 1625?:505 ("fox"); Strachey 1625:419 (stragglers). Powhatan made the threat of withholding food explicit through an intermediate (Smith 2007b:388). The Council of Virginia clearly understood the peril (1609:363). See also, West et al. 1610:457.

75 Smith takes charge: Smith 2007b:314–96 ("good hope," 341); Horn 2005:59–100. As Smith (2007b:392) notes, only seven men died on his watch.

75 Colony rose to two hundred: Two groups came in 1608, the "first supply" in January (100 or 120 people, Horn 2005:75; "neare a hundred men," Smith 2007b:324); and the "second supply" in, depending on the source, September or October (seventy men, Horn 2005:104; a few more than seventy, Smith 2007b:358). The first supply brought the total to 138–58, but deaths that summer reduced the number to about 130; the second supply lifted it to about 200 (Bernhard 1992:603).

76 Smith blows self up: Smith 2007b:402; Percy 1625?:502. Horn (2005:169–70) speculates that it was attempted murder, but this seems unlikely; Smith's enemies depended on him. His lack of children is often linked to the severe powder burns on his groin.

76 Arrival of convoy, Smith's replacement: Glover and Smith 2008:chap. 4; Smith 2007c:chap. 12; Horn 2005:chap. 6; Archer, G. 1609. Letter to ———, 31 Aug. In Haile ed. 1998:350–53; Ratcliffe, J. 1609. Letter to R. Cecil, 4 Oct. In idem:354–55.

77 First Indian War: Smith 2007c:chap. 12; Morgan 2003:79 (Smith's views); Fausz 1990 ("first Indian war"); Percy 1625?:503–04 (all quotes).

78 "starving time": Glover and Smith 2008:chap. 7; Smith 2007b:411–12 (Powhatan stops providing food); Horn 2005:174–77; Price 2005:126–29 (ruffs, 127–28); Donegan 2002:144–75; Shirley 1942 (Percy's clothing, 237–38); Percy 1625?:502–08 (all quotes, 505); "Ancient Planters" 1624:894–95. The term "starving time" comes from Smith (2007b:411). Winter 1609 death toll: Kelso 2006:90; Bernhard 1992:609–13; Kupperman 1979:24. The total number of colonists dropped from 245 to eighty or ninety. See also, Governor and Council in Virginia. 1610. Letter to Virginia Company, 7 Jul. In Haile ed. 1998:456–57.

78 Chesapeake fish: Author's interviews, Kelso (sturgeon bones); Wennersten 2000:5–7, 12–13 (underwater), 23–27; Pearson 1944.

78 Gentlemen: Smith 2007b:404 (retainers); Morgan 2003:63, 83–87 ("in England," 84).

79 Rolfe's voyage, attempted abandonment of Jamestown: Glover and Smith 2008:chaps. 3–8; Horn 2005:157–64, 177–80; Price 2005:130–39; Strachey 1625:383–427 (quotes from 384, 387); "Ancient Planters" 1624:895–97 ("not less than," 897); Somers 1610; West, T. (Baron de la Warre). 1610. Letter to Earl of Salisbury, Jul. In Haile ed. 1998:465–67; West et al. 1610.

81 Total immigrants and deaths 1607–24: See sources for chart, esp. Kolb 1980, Hecht 1969, Neill 1867, Thorndale pers. comm. A summary can be found in Kupperman 1979:24, though a ship-by-ship count suggests that her figure of six thousand for the 1607–24 influx is too low. I am grateful to William Thorndale for his kindness in discussing this material with an amateur.

81 Accounts of Jamestown deaths: KB 4:148 ("extreamities"); KB 4:160 ("left alive"); KB 4:175 ("3000 p[er]sons"); KB 4:22 ("delivered"); Percy 1625?:507 ("out of his body"); KB3:121 (are dead); KB 4:238 ("servants are dead," Rowsley arrived in spring 1622 [KB 4:162, Thorndale pers. comm.] and the note reporting the deaths was written in June); KB 4:234 ("leave the Contrey"); KB 4:235 ("well out of it").

81 Berkeley Hundred: Dowdey 1962:chap. 2; KB 3:230 (date, number of arrival), 3:207 ("Almighty god"), 3:197–99 (list of dead). In general, see KB 3:195–214, 3:271–74. "Hundred" refers to the number of acres supposedly granted to each partner in the enterprise. I am grateful to Jamie Jamieson for giving me a tour of Berkeley.

82 "from the investment": Craven 1932:24.

82 Failed ventures at Jamestown: Hecht 1982:103–26.

82 Pocahontas bio, abduction, marriage: Smith 2007c:423–27; Rountree 2005 (lack of clothing, Mataoka, 37), 2001; Horn 2005:217–18; Townsend 2004:100–06; Price 2005:153–58; Dale 1615:845–46; Hamor 1615:802–09; Rolfe 1614; Argall, S. 1613. Letter to "Master Hawes," Jun. In Haile ed. 1998:754–55; Strachey 1612:630 ("all the fort over").

83 English counterattack: Kupperman 2007a:255–59; Horn 2005:180–90; Morgan 2003:79–81 (oatmeal); Fausz 1990:30–34; Percy 1625?:509–18; Strachey 1625:434–38.

83 Initial refusal to negotiate over Pocahontas, subsequent pact: Smith 2007c:424–26 ("had stolne," 424); Horn 2005:212–16; Rountree 2005:chap. 12; Fausz 1990:44–48; Hamor 1615:802–09; Dale 1615:843–44. Argall (1613:754–55) says Powhatan did negotiate, but Horn's argument (2005:213) that he would not have wanted to seem weak by negotiating seems plausible—Argall may have been inflating the success of his disagreeable tactic.

85 Pocahontas in captivity: Rountree 2005:chap. 12; Townsend 2004:chap. 6; Hamor 1615:803 ("discontented"); Rolfe 1614.

85 Pocahontas's first marriage: Rountree 2005:142–43, 166; Townsend 2004:85–88.

85 Cease-fire and Opechancanough plan: Rountree 2005:chap. 15; Fausz 1977:320–50; Fausz 1981; Fausz 1990:47–49 ("formal winner," 48). Many English thought Opechancanough had taken charge well before Powhatan's death (Hamor 1615:808; Dale 1615:843). Powhatan did not create an orderly succession plan. Lear-like, he retired to a faraway village, dividing his kingdom among his younger brothers. Initially, another brother had the most formal power (Smith 2007c:447). Infighting was inevitable (KB 3:74, 3:483). Finally Opechancanough emerged as first among equals (KB 2:52, 3:550–51, 4:117–18; Smith 2007c:437–47 passim, 478; Rolfe 1616:868–69). Notching stick: Smith 2007c:442.

86 James and tobacco: Laufer 1924b:17–19; James I 1604:112 ("braine").

86 Virginia tobacco in England: Morgan 2003:107–10 (servant pay, productivity), 192–98 (taxes); Hecht 1969:175–93, esp. table VII:4 (1,000 percent, 188); Laufer 1924b (debts); see also, Horn 2005:246–47, 280–83; Price 2005:186–87; Wennersten 2000:40–41; Gray 1927.

87 First representative body: Horn 2005:239–41; Price 2005:189–94; KB 3:482–84 (charter).

87 Jamestown slaves: Kupperman 2007a:288; Price 2005:192–97; Sluiter 1997; Rolfe, J. 1619. Letter to Sandys, E. In KB 3:243 ("20. and odd"). An intriguing investigation is Hashaw 2007; the basic source is Rolfe (KB 3:241–48).

87 Virginia tobacco mania, near starvation: Smith 2007c:443–44 ("with Tobacco"; the quotation is attributed by Smith to Rolfe and Deputy Governor Samuel Argall); Morgan 2003:111–113 (taverns); Rolfe 1616:871 (Dale's orders); KB 1:351, 1:566, 3:221, 4:179. (The colonists, the Virginia Company treasurer said in December 1619, "by this misgovernemt [sic] reduced themselves into an extremity of being ready to starve" [KB 1:266].)

87 Clergy on Virginia: Glover and Smith 2008:62–67, 221–23; Horn 2005:137–41; Donegan 2002:3–4, Fausz 1977:256–65; Crashaw 1613 ("take it?," 24–25); Symonds 1609. See also the Crashaw sermon in Brown 1890:(vol. 1) 360–75.

87 New Orleans (footnote): Powell 2012:18–32, 127–28, 132–43.

88 Later rounds of financing: Hecht 1969:279 (known first investors: six people, £209 [no complete list survives]), 280–310 (1609–10 investors); Brown 1890:vol. 1, 209–28, 466–69 (1609–10); KB 3:79, 98, 317–39 (1610–19 investment rounds). Not every listed investor actually paid (Glover and Smith 2008:115).

88 1622 attack and company finances: Rountree 2005:chap. 16; Horn 2005:255–62; Fausz 1977:chap. 5; Waterhouse, E. 1622. A Declaration of the State of the Colony and Affaires in Virginia. In KB 3:541–71; 2:19 (debt); 3:668; 4:524–25.

90 Lack of planting, onslaught of unfed new colonists: Morgan 2003: 100–02 (captains' incentives); Hecht 1982:appendix 2; KB 4:13, 41, 74, 186 (fear of planting maize), 451, 525 (abandonment of planting). In Fausz's sum-

mary: "[J]ust as in the colony's early days, the English were dependent upon the [Indians], now their implacable enemies, for the most basic, most crucial human need" (Fausz 1977:473).

90 Second "starving time": KB 4:25, 41–42, 62 ("of the Ground"), 65 (*"bury the dead"*), 71–75, 228–39, 263, 524–25 (more than one thousand dead, i.e., two out of three). Figures cannot be precise, as emigrants kept arriving and dying throughout the year.

90 Poisoning (footnote): Rountree 2005:219–20; KB 4:102, 221–22 ("ther heades"); others put the number of dead at 150 (KB 2:478). Such treachery was common (Morgan 2003:100).

90 English inability to attack successfully: KB 2:71, 4:10 ("they retyre"). Although they did not want to leave their tobacco fields (KB 4:451), they did destroy some Indian food stores (KB 3:704–07, 709).

90 "their Countries": Smith 2007b:494.

90 Company demise: Horn 2005:272–77; Morgan 2003:101–07 ("their deaths," 102); Rabb 1966:table 5 (£200,000); Craven 1932:1–23; KB 2:381–87; 4:130–51, 490–97.

91 Traditional tobacco growing: Percy 1625?:95; Archer 1607:114 (describing one farm as "bare without wood some 100 acres, where are set beans, wheat [maize], peas, tobacco, gourds, pompions [pumpkins], and other things unknown to us").

91 Tobacco and soil depletion: Morgan 2003:141–42 (and cited sources); Craven 1993:15 ("In the tobacco regions of the South, . . . the planters seldom counted on a paying fertility lasting more than three or four years"), 29–35. Colonists observed that the "ground will hold out but 3 yrs" (KB 3:92; see also 220). Some aspects of Craven's thesis (that tobacco's capacity to exhaust the soil ultimately caused an agricultural collapse) have been contested (Nelson 1994), but not the exhaustive capacity of tobacco agriculture itself.

92 English take best land and keep it: Rountree 2005:152, 188, 228 (see also 154, 187, 200, and 260, note 23); Morgan 2003:136–40; Wennersten 2000:46–47 ("centuries"). By the 1620s some English regarded this idea—taking over previously cleared land with the best soil—as a plan of action (Martin 1622:708; Waterhouse 1622:556–57).

92 Deforestation, erosion: Craven 2006:27–29, 34–36; Williams 2006:204–16, 284–308 ("spared," 294) Wennersten 2000:51–54.

93 Animals imported, eat Indian harvests: Anderson 2004:101–03, 120–23, 188–99; Morgan 2003:136–40.

93 Impact of pigs on tuckahoe: Crosby 1986:173–76; Kalm 1773:vol. 1, 225, 387–88 ("extirpated"); KB 2:348, 3:118 ("into the woods"), 221.

94 Biological imports, honeybee invasion: Crane 1999:358–59; Crosby 1986:188–90 ("in all minds," 190); Grant 1949:217 (pollination discovery); Kalm 1773:vol. 1, 225–26 ("English flies"); KB 3:532 (list of imports).

94 Fruit that needs pollination: Flowering plants are either open pollinated or biotic pollinated, which means that either they can pollinate themselves via the wind or they can't; most mix both methods. Apples and watermelon are close to the purely biotic end of the spectrum; some (but not much) pollination of peaches can occur in the absence of insects. As a practical matter, all require bees. Apples originated in Central Asia, peaches in China, watermelons in North Africa. I am grateful to the farmers in Whately, Mass., who explained this to me.

95 Nicholas Ferrar: Skipton 1907:22–25, 61–63; KB 3:83, 324, 340 (investments).

96 Ferrar reads Bullock, longing for China: Thompson 2004. Summarizing Ferrar's reaction to tobacco, the Oxford historian Peter Thompson called it "an inedible crop whose monetary value to the state could be construed as being inversely related to its detrimental effect on the nation's morals and reputation." (121). All quotes from online transcription. See also, KB 3:30; 4:109–10. Spain thought the English were building a chain of forts in Virginia to protect the China route: Maguel, F. 1610. Report to the King of Spain, 30 Sep. In Haile ed. 1998:447–53, at 451–52.

97 Worldwide spread of tobacco: Brook 2008:117–51 ("to buy tobacco," 137); Céspedes del Castillo 1992:22–48ff.; Goodrich 1938 (daimyo ban, 654); Laufer et al. 1930 (Sierra Leone, 7–8); Laufer 1924a (Japan, 2–3; Mughals, 11–14); Laufer 1924b (pope, 56; Ottoman bribes, 61). See also Chap. 5. Three years after the khan's ban the Chinese emperor also banned the foreign plant, decreeing that all tobacco vendors "shall, no matter the quantity sold, be decapitated, and their heads exposed on a pike" (Goodrich 1938:650).

CHAPTER 3 / *Evil Air*

99 Discovery of copybook: Varela and Gil eds. 1992:69–76.

100 Translation of account of second voyage: Colón, C. Letter to the Sovereigns, Feb. 1494. In Taviani et al. eds. 1997:vol. 1, 201–39 ("tertian fever," 233); Gil, J., and Varela, C. Memorandum to Centro Nacional de Conservación y Microfilmación Documental y Bibliográfica, 29 Dec. 1985. In idem:164–65 ("revelations," 164).

100 Tertian fever: A less common type of malaria is associated with a seventy-two-hour cycle: quartan fever.

100 Cook and malaria: Cook 2002:375.

100 Lack of malaria in Americas: Rich and Ayala 2006:131–35 (monkey malaria); De Castro and Singer 2005; Carter and Mendis 2002:580–81; Wood 1975; Dunn 1965.

101 "had trouble": Colón, C. 1494. Relation of the Second Voyage. In Varela and Gil eds. 1992:235–54, at 250. My thanks to Scott Sessions for helping me with this translation.

101 Definitions of *çiçiones:* Author's interviews, Sessions (Cook also discusses the issue); Covarrubias y Orozco 2006: fol. 278v; Vallejo 1944; Real Academia Española 1726–39:vol. 2, 342. See also M. Alonso, *Diccionario Medieval Español* (Salamanca: Universidad Pontificia de Salamanca, 1986), 2 vols.; J. Corominas and J. A. Pascual, *Diccionario Crítico Etimológico Castellano e Hispánico* (Madrid: Editorial Gredos, 1980–91), 6 vols.

102 "marshy emanations": Real Academia Española 1914:753.

102 Malaria toll: World Health Organization 2010 (death, morbidity estimates, 60); Gallup and Sachs 2001 (economic burden).

103 "Extractive states": Acemoglu and Robinson: forthcoming; Acemoglu et al. 2001; Conrad 1998:84 ("disease and starvation").

104 Evolution of malaria: Rich and Ayala 2006; Carter and Mendis 2002:570–76. Half a dozen other *Plasmodium* species occasionally attack people, but the vast majority of human malaria is due to *P. vivax* and *P. falciparum,* with some from *P. malariae* and *P. ovale.*

104 Malaria life cycle: Interviews and e-mail, Andrew Spielman; Baer et al. 2007; Morrow and Moss 2007 (ten billion, 1091); Sturm et al. 2006; Rich and Ayala 2006; Carter and Mendis 2002:570–76. I ignore many, many complications here.

106 Jeake's attacks: Hunter and Gregory ed. 1988:210–25 (all quotes, 215).

106 Differences between vivax and falciparum: Mueller et al. 2009; Packard 2007:23–24. The two species have different reproductive strategies. Vivax infects only very young red blood cells, about 2 percent of the total, but does so for a long time. Mosquitoes are unlikely to pick it up at any one feeding, but have a lot of time to do it. Falciparum attacks all red blood cells, but for less time. Mosquitoes are more likely to pick it up at any one feeding, but have less time to do so.

107 Temperature sensitivity: Roberts et al. 2002:81. I drastically simplify the issue; for a careful calculation, see Guerra et al. 2008:protocol S2.

108 *Anopheles maculipennis:* Ramsdale and Snow 2000; Snow 1998; White 1978; Hackett and Missiroli 1935. The *maculipennis* species in coastal England, *A. atroparvus,* seems resistant to *P. falciparum,* an additional reason that falciparum was rare there.

108 Draining wetlands and mosquitoes: Thirsk 2006:15–22, 49–78, 108–41; Dobson 1997:320–22, 343–44. Although rarer, malaria was likely present before drainage; Hasted, for instance, reports that Archbishop John Morton died of "quartan ague" in 1500 (1797–1801:vol. 12, 434).

109 Improved drainage (footnote): Dobson 1997:320–22, 343–44 ("pigsties," 321); Kukla 1986:138–39 (cattle).

109 English malaria misery: Packard 2007:44–53; Hutchinson and Lindsay 2006 (causes of death); Dobson 1997:287–367 (Aubrey, 300; burial ratio, 345); 1980 (mortality rates, 357–64); Dickens 1978:1 ("brothers"); Defoe 1928:13

("certainly true"); Wither 1880:139 ("had there"); Hasted 1797–1801:vol. 6, 144 ("twenty-one"). The 1625 death figure is from the *Collection of Yearly Bills of Mortality*. My description is based heavily on Dobson's work.

112 Emigrants from malaria zone: Author's interviews and e-mail, Robert C. Anderson (Great Migration Project), Preservation Virginia (Jamestown colonists), William Thorndale; Dobson 1997:287 (Sheerness); Kelso and Straube 2004:18–19 (Jamestown, Blackwall); Fischer 1991:31–36; Bailyn 1988:11. Anderson told me that "about 15 percent" of the English migrants to New England came from Kent alone; Thorndale (pers. comm.) is cautious about the precision of the individual Preservation Virginia biographies, which have not been formally published.

112 Vivax hiding: Mueller et al. 2009. Worse still, victims can become carriers. By fighting off the disease, people acquire immunity—of a peculiar, dispiriting sort. If they are bitten by an infected mosquito, the "immunity" greatly reduces the symptoms of malaria. But it does not stop the infection itself, which can be passed on.

113 *A. quadrimaculatus:* Reinert et al. 1997. *A. quadrimaculatus* is strikingly similar to *A. maculipennis* (Proft et al. 1999). Indeed, their ranges almost overlap—*A. maculipennis* can be found in the northern fringes of the United States (Freeborn 1923).

113 Malaria transmissibility: Author's interviews, Spielman. In August 2002, two teenagers in northern Virginia were hospitalized with malaria. The victims, near neighbors, lived less than ten miles from Dulles International Airport. County and state officials came to believe that an asymptomatic traveler on an international flight at Dulles had been bitten by a mosquito, which passed on the infection to the teenagers. It was the tenth such case in a decade (Author's interview, David Gaines [Va. Dept. of Health]; Pastor et al. 2002).

113 Malaria by 1640: Author's interview, Anderson. See also Fischer 1991:14–17. The vice director of the Dutch colony on Delaware Bay suffered a classic malaria attack in 1659 ("confined to my bed between 2 and 3 months, and so severely attacked by tertian ague, that nothing less than death has been expected every other day. . . . All the inhabitants of New Netherland are visited with these plagues" [Letter, Alrichs, J., to Commissioners of the Colonie on the Delaware River, 12 Dec. 1659. In Brodhead ed. 1856–58:vol. 2, 112–14]). See also, Letter, idem, to Burgomaster de Graaf, 16 Aug. 1659. In ibid.: 68–71. Ships came to New England after 1640, but their temporary visits were less likely to spread malaria.

113 Malaria by 1620s: Historians generally maintain that malaria was present in the Chesapeake by the 1680s and possibly by the 1650s (Cowdrey 1996:26–27; Rutman and Rutman 1976:42–43; Duffy 1953:204–07). Kukla (1986:141) suggests that "by 1610 it may have been present to greet Gover-

nor De La Warr, who 'arriv[ed] in Jamestowne [and] . . . was welcomed by a hot and violent ague.' " But this is little more than speculation, as is my own.

114 Quads and dry weather: Author's interviews, Gaines; Chase and Knight 2003.

114 Seasoning: Morgan 2003:180–84 (later improvement); Kukla 1986:136–37; Kupperman 1984:215, 232–36; Gemery 1980:189–96 (improvement); Blanton 1973:37–41; Rutman and Rutman 1976:44–46; Curtin 1968:211–12; Duffy 1953:207–10; Jones 1724:50 ("Climate"); Letter, George Yeardley to Edwin Sandys, 7 Jun. 1620. In KB 3:298 ("seasoned"). See also KB 3:124; 4:103, 191, 4:452; Morgan 2003:158–62, 180–84.

115 Sukey Carter: Carter 1965:vol. 1, 190–94 (all quotes; I omit extraneous material), 221 (death).

117 Costs of servants and slaves: Morgan 2003: 66, 107 (servant pay); Menard 1977:359–60, table 7; U.S. Census Bureau 1975:vol. 2, 1174. Using similar figures, Coelho and McGuire (1997:100–01) estimate that a servant would have to return £2.74 a year to justify the purchase price, but a slave would have to return £3.25. To be sure, the servant would eventually be able to leave his master's employ (Menard looked only at servants with more than four years remaining on their contracts). But the advantages of the slave's permanence wouldn't manifest themselves for years—and Chesapeake Bay with its high mortality rate was not a place where people sought long-term advantages. Such calculations ignore the profits from selling or working slave children. Little evidence exists, though, that slave owners initially understood this potential (Menard 1977:359–60).

117 Adam Smith and slavery: Smith 1979:vol. 1, 99 ("by slaves" [bk. 1, chap. 8, ¶41]); vol. 1, 388 ("domineer" [bk. 3, chap. 2, ¶130). See also vol. 1, 387 (bk. 3, chap. 2, ¶9); vol. 2, 684 (bk. 4, chap. 9, ¶47).

119 English slaves: Guasco 2000: 90–127 (slave censuses, 102, 122). Northwest Africa had a European slave population of about 35,000 in 1580–1680 (Davis 2001:117). Using a mortality estimate of 24–25 percent a year, Davis derived a total European catch of 850,000 in this period. A guess of an average annual total of two thousand seems conservative. Using Davis's mortality ratios, this leads to 48,571 English captives in 1580–1680, hence "tens of thousands." Hebb (1994:139–40) estimates that 8,800 English were enslaved in 1616–42, which would translate into ~25,000 during this period. Many more Italians and Spaniards than English were taken. Plymouth: Laird Clowes et al. 1897–1903:vol. 2, 22–23 ("Between 1609 and 1616, no fewer than four hundred and sixty-six British vessels were captured by [corsairs], and their crews enslaved").

120 Rare but legal English slavery: Guasco 2000:50–63; Friedman 1980. Slaves, mainly prisoners, were sent to England's few galleys.

120 Indentured servants: Galenson 1984 (one-third to one-half, 1); Gemery

1980:esp. table A-7. Most went to Virginia, so the figure there was higher, perhaps "more than 75 percent" (Fischer 1991:227). See also, Tomlins 2001; Menard 1988:105–06.

120 Slaves in 1650: McCusker and Menard 1991:table 6.4; U.S. Census Bureau 1975:vol. 2, 1168.

120 Turn to slavery in 1680s, emergence of England as biggest slaver: Author's interviews, Anderson, Thornton. Numbers: Berlin 2003:table 1; U.S. Census Bureau 1975:vol. 2, 1168. Economics: Menard 1988:108–11, 1977; Galenson 1984:9–13. See also, Eltis and Richardson 2010; Eltis et al. 2009–.

120 Size and profitability of slave trade: Eltis and Engerman 2000 ("tonnage," 129; percent of GDP, 132–34; raw materials, 138). Eltis and Engerman argue that the profits were not oriented toward industrial investment, so the industry had no special role in the Industrial Revolution (136). This contradicts Blackburn's conclusion that "exchanges with the slave plantations helped British capitalism to make a breakthrough to industrialization and global hegemony" (1997:572).

121 Free land and slavery: Smith 1979:vol. 2, 565 ("first master" [bk. 4, chap. 7, §b, ¶2]), Domar 1970. "Wide-open spaces exhibit a bimodal distribution: lots of freedom or coerced labor" (J. R. McNeill, pers. comm.). Morgan (2003:218–22) observes that farmers "solved" the problem by buying vast tracts of land.

121 Price rise in indentured servants as slavery cause: Morgan 2003:chap. 15; Galenson 1984. Morgan locates an effective price rise in increasing trouble with indentured servants in Virginia, Galenson an actual price rise from labor shortages in England.

122 Little Ice Age impact in Scotland: Lamb 1995:199–203; Gibson and Smout 1995:164–71; Flinn ed. 1977:164–86.

122 Scots in Panama: I rely on the fine account in McNeill 2010:106–23 ("of Panama," 123—I have, with McNeill's permission, slightly altered his words). Earlier studies are useful but, in McNeill's phrase, "epidemiologically unaware" (106).

122 "the world": Bannister ed. 1859:vol. 1, 158–59.

124 Founding Carolina: Wood 1996:13–20.

125 Mississippians become confederacies: Snyder 2010:chap. 1; Gallay 2002: 23–24.

125 Slavery in Powhatan, confederacies, and colonists: Smith 2007b:287–88, 298 (examples); Rountree 1990:84, 121 (Powhatan); Snyder 2010:35–40 (Southeast); Woodward 1674:133 (Indians who sell slaves to Virginia). See also, Laubrich 1913:25–47.

126 Flintlocks vs. matchlocks: Snyder 2010:52–55; Chaplin 2001:111–12; Malone 2000:32–35, 64–65.

126 Spanish attack on Carolina: Bushnell 1994:136–38.

127 Carolina slave trade: I am summarizing the argument in Gallay 2002; see also Snyder 2010; Bossy 2009; Laubrich 1913:119–22.

127 Economics of trade: Snyder 2010:54–55 (160 deerskins, "Extreamly" [quoting Thomas Nairne]); Gallay 2002:200–01 (census), 299–308 (export estimate), 311–14 (prices).

128 Massachusetts and New Orleans (footnote): Gallay 2002:308–14 (France); Laubrich 1913:63–102 (France), 122–28 (Massachusetts).

129 Bans on slave imports: Gallay 2002:302–03 (all quotes).

129 Carolina and malaria: McNeill 2010:203–09; Packard 2007:56–61; Coclanis 1991:42–45 (more than three out of four); Wood 1996:63–79 (population, 152); Silver 1990:155–62; Dubisch 1985 (differential mortality, 642); Merrens and Terry 1984 ("ague," 540; "hospital," 549); U.S. Census Bureau 1975:vol. 2, 1168; Childs 1940 (arrival of malaria, chaps. 5–6); Ashe 1917:6 ("Complexions"); Archdale 1822:13. A somewhat similar process occurred in Georgia, which began in 1733 as a free colony (slavery was banned). Scurvy, beriberi, and dysentery, all related to inadequate or contaminated food, were common. Infectious disease was not. The colony became the crown's property in 1752. Slavery was permitted. Malaria and yellow fever quickly followed. Soon it became hard to farm without slaves. The disparity in death rates shrank as Europeans survived and acquired immunities. But it didn't go away. In the 1820s whites in South Carolina were still dying of intermittent, remittent, bilious, and country fevers—the terms then used for malaria—at rates more than four times higher than blacks (Cates 1980).

129 Indian disease deaths: Snyder 2010:65 (Chickasaw), 101–02 (Chakchiuma), 116 ("distressed tribes"); Gallay 2002:111–12 (Quapaw); Laubrich 1913:285–87; Archdale 1822:7 ("answer for").

130 Duffy negativity: E-mail to author, Louis Miller; Webb 2009:21–27; Seixas et al. 2002; Carter and Mendis 2002:572–74; Miller et al. 1976.

131 Sickle-cell: Interviews and e-mail, Spielman; Carter and Mendis 2002:570–71; Livingstone 1971:44–48.

132 Immunities as pivot for slavery: Webb 2009:87–88; Coelho and McGuire 1997; Wood 1996:chap. 3; Dobson 1989; Menard 1977. Some economists have argued that there was little economy of scale in crops climatically suited for New England. But wheat was grown in Piedmont Virginia on big plantations with lots of slaves. Still others have argued that Africans couldn't run away, because their appearance was too distinctive. The obvious retort is that slaves did run away all the time—and that in any case indentured servants could have been branded or tattooed, something already done for criminals. Ultimately, disease counted. "The decimation of a native labor supply in the face of disease, the weakness of the Europeans in their new disease environment and the apparent resistance

of blacks to diseases of hot climates led to the massive importation and exploitation of African slaves" (Dobson 1989:291).

132   Arrival of falciparum: Rutman and Rutman 1980:64–65; idem 1976:42–45.

133   Comparisons of mortality rates: Curtin 1989; 1968:203–08 (48–67 percent, 203; "of the European," 207); Hirsch 1883–86:vol. 1, 220 (malaria in Antilles). Many of the original figures are in Tulloch 1847, 1838.

133   Geography of malaria: My discussion follows McNeill 2010; Webb 2009:chap. 3; Packard 2007:54–78.

134   Falciparum line: Author's interview, National Weather Service (temperatures); Strickman et al. 2000:221.

134   Plantations and South: The debate is summarized in Breeden 1988:5–6. Tara was supposedly in Georgia.

134   Intractable malaria regions: Duffy 1988:35–36; Faust and Hemphill 1948:table 1. Texas had more malaria cases, but also more people.

134   *Quadrimaculatus* habitat and housing: Author's interview, Gaines; Goodwin and Love 1957. The hills don't have to be tall; medical researcher Walter Reed observed that people in the uplands of Washington, D.C., just 200–250 feet above the Potomac, rarely contracted malaria, while "those who live on the low plateau bordering both the Potomac and Anacostia Rivers are affected annually by malarial diseases" (Gilmore 1955:348). See also, Kupperman 1984:233–34.

134   Malaria and culture: Rutman and Rutman 1980:56–58 (all quotes); Dubisch 1985:645–46. Fischer (1991:274–389 passim) makes an extended characterization of Virginia mores.

137   No initial awareness of immunity: Arguing that malaria resistance "must have done a great deal to reinforce the expanding rationale behind the enslavement of Africans," Wood (1996:83–91, quote at 91) and Puckrein (1979:186–93) try to show that Carolina colonists regarded Africans in this way. By contrast, Rutman and Rutman "found no evidence in Virginia to substantiate Wood's thesis" (1976:56). Most historians follow the Rutmans and believe that colonists' views of African immunity came *after* the turn to slavery, not before.

138   Massachusetts disease, slavery: Romer 2009 (8 percent, 118); Dobson 1989:283–84 (health); Massachusetts Body of Liberties (1641):art. 91 (available in many places online).

138   Slavery compared in Argentina and Brazil: Eltis et al. 2009– (2.2 million); Chace 1971 (220,000–330,000 slaves, 107–08; lack of establishment of African culture in Argentina, 121–22; half Argentina African, 126–27); Alden 1963 (half Rio, São Paulo African). Eltis et al. give 75,000 as the number of slaves entering the ports of the Rio Plata; Chace makes clear this is only the registered "pieces," which ignores the much bigger number of illegally imported slaves. By the end of the nineteenth century, Brazil's

great writer, Euclides da Cunha, was celebrating his nation's mixed heritage (Hecht: forthcoming); meanwhile, Argentina's ruling "Generation of Eighty" was boasting that Argentina was the "only great white nation of South America" (Chace 1971:2).

139 Yellow Jack: Much in this section comes from McNeill 2010.

139 Sugar comes to Barbados: McNeill 2010:23–26; Emmer 2006:9–27; Davis 2006:110–16; Blackburn 1997:187–213, 239–31 (slave prices, 230); Sheridan 1994:chap. 7, esp. 128–30; Beckles 1989; Galenson 1982 (slave prices, table 4). I am grateful to the plantation owners in Brazil who let me visit their land to see sugar work.

140 First yellow fever epidemic: McNeill 2010:35, 64 ("populations"); Beckles 1989:118–25; Findlay 1941 (six thousand dead and quarantine, 146); Ligon 1673:21, 25 ("dead," 21).

141 Spread of sugar, ecological ravaging of Caribbean: McNeill 2010:23–33 ("for cultivation," 29); Watts 1999:219–31, 392–402; Sheridan 1994 (production and population figures, 100–02, 122–23); Goodyear 1978:15 (Cuba). Ligon (1673) reported that when the first Europeans landed on Barbados the island was "so overgrown with Wood, as there could be found no Champions [fields], or *Savannas* for men to dwell in" (23).

142 *A. albimanus:* Grieco et al. 2005 (susceptibility to falciparum); Rejmankova et al. 1996 (algal habitat); Frederickson 1993 (habits). Frederickson suggests that it has a preference for cattle "1.6 to 2.1 times greater than that for humans" (14). The gradual replacement of Caribbean cattle by sugar thus increased the risk of malaria.

143 Fourth voyage: During Colón's fourth voyage to the Americas (1502–04) the admiral's nautical career effectively came to an end when he was forced to ground his worm-eaten, sinking ships on Jamaica. To obtain help from Santo Domingo, he asked a trusted lieutenant, Diego Mendez, to canoe 120 miles to Hispaniola. After a brutal journey in the Caribbean summer, Mendez's party made it to shore. Most of the group was too sick to continue to Santo Domingo, Colón's son Hernán wrote later. Mendez nonetheless "left in his canoe to go up the coast of Hispaniola, though suffering from quartan fever" (Colón 2004:322).

143 Environmental changes favor malaria and yellow fever: McNeill 2010:48–50, 55–57; Webb 2009:69–85ff.; Goodyear 1978:12–13 (pots).

143 Caribbean as lethal environment: McNeill 2010:65–68; Webb 2009:83 ("non-immunes"); Curtin 1989:25–30, fig. 1.2, table 1.5. Ligon, who came to Barbados two decades after the first English colonists, found (1673:23) that "few or none of them that first set there, were now living." This may exaggerate. Not many colonists lasted more than a few years, as Ligon said. But that was not only because they died. Many fled to healthier places—Virginia, for one (Sheridan 1994:132–33).

143 Introduction of malaria into Amazon: Cruz et al. 2008 (Madeira survey);

Hemming 2004a:268–70; Requena, F. 1782. Letter to Flóres, M. A. d., 25 Aug. In Quijano Otero 1881:188–97, at 191–95 passim; Orbigny 1835:vol. 3a, 13–36; Edwards 1847:195 ("one case").

144 Guyane: Hecht forthcoming; Ladebat 2008 (coup deportees); Whitehead 1999. I am grateful to Susanna Hecht for drawing this history to my attention.

145 Sugar despotisms: Acemoglu et al. 2001, 2003. "Differences in mortality are *not* the only, or even the main, cause of variation in institutions. For our empirical approach to work, all we need is that they are *a source* of exogenous variation" (Acemoglu et al. 2001:1371). The counterargument is exemplified by Sheldon Watts's claim that the turn to slavery was determined by the "general stagnation of Europe's population growth." In his view, "what *really* mattered were developments in the cosmopolitan core rather than the presence of the frightful country disease, yellow fever, in the Caribbean periphery" (1999:230–34, at 233). But he simply demonstrates that England's population was increasing slowly in the late seventeenth century, not whether the resultant price increase for servants was actually big enough to have any impact. In my view, the contrary has been shown convincingly.

146 Fear of independent institutions: Acemoglu and Robinson forthcoming. After slavery ended in 1834, many sugar planters sold abandoned, marshy land to freed slaves at exorbitant prices. In the next decade, freed Africans created a series of prosperous, self-governing freeholds. Unhappily, they had never learned the techniques, pioneered by Guyana's Indians, to drain the land for long-term cultivation. Because the "Village Movement" deprived British plantations of labor, the colonial government refused to provide the technology and engineering skills for building and maintaining dams and channels that it made available to elites. Unable to keep sugar fields dry, the Village Movement lost its economic base; the freed slaves were forced to return to their plantations (Moore 1999:131–35). Similarly, the wealthy elites feared the small shops opened by many freed slaves. To drive them back into the fields, they imported Portuguese merchants and financed their enterprises with low-interest loans, meanwhile denying all credit to ex-slaves. The ex-slaves soon went under (Wagner 1977:410–11).

146 Stagnation of extractive states: Acemoglu et al. 2002:1266–78 (discouraging settlement, 1271; "entrepreneurs," 1273).

147 British Guyana and Booker Brothers: Rose 2002:157–90 (exports, 186–86); Hollett 1999:chap. 5 (Booker brothers); Moore 1999:136–37 ("their station"); Bacchus 1980:4–30, 217–19 (university); Daly 1975 (fear of education, 162–63, 233–34). On trial in 1823 for fomenting insurrection by teaching slaves the Bible, the missionary John Smith decried plantation owners who believed "that the diffusion of knowledge among the

negroes will render them less valuable as property" (Anon. 1824:78). Indeed, he was charged with informing slaves about "the history of the deliverance of the Israelites" (ibid.:157) and teaching them to read.

148 Disease in U.S. Civil War: Barnes et al. 1990 (35 percent, table 6; 233 percent, table 30; 361,968, table 71; proportion of deaths, xxxviii).

149 "established institutions": The Crittenden-Johnson resolution was adopted in July 1861 by a House vote of 119–2 and in slightly different form by a Senate vote of 30–5.

149 Malaria in American Revolution: Author's interviews, McNeill; McNeill 2010:209–32 ("nearly ruined," 215; "unhealthy swamp," 220; troop levels, 226).

CHAPTER 4 / *Shiploads of Money*

158 Zheng He: Mote 2003:613–17; Levathes 1994 (suppression, 174–81); Finlay 1991 (survey of historians' views, 297–99); Needham et al. 1954–:vol. 4, pt. 3, 486–528ff. (suppressed records, 525). In two books published in 2002 and 2008 a retired submarine commander named Gavin Menzies claimed that Chinese fleets went beyond Africa, reaching the Americas and Europe, dramatically changing world history along the way. Few historians have endorsed this thesis.

159 Chinese "insularity": Author's interviews, Goldstone, Kenneth Pomeranz; Jones 2003:203–05 ("self-engrossment," 205; "from the sea," 203); Goldstone 2000:176–77 ("such voyages," 177); Landes 1999 ("curiosity," 96; "success," 97); Finlay 1991. See also, Braudel 1981–84:vol. 2, 134, vol. 3, 32 ("In the race for world dominion, this [inward turn] was the moment when China lost her position in a contest she had entered without fully realizing it, when she had launched the first maritime expeditions from Nanking in the early fifteenth century"), 485–86, 528–29. Diamond argues that the decision exemplifies a fatal uniformity in China; in fragmented Europe, he says, such a blanket prohibition would have been impossible (Diamond 1999:412–16). As indicated below, China was hardly unified; the blanket prohibition did not stick. Landes charges that Zheng's voyages "reeked of extravagance" (1999:97) as opposed to the more rational, bottom-line-oriented exploration of Europeans. He thus criticizes China's rulers both for shutting down Zheng He's voyages because they were unprofitable and for overspending on those same voyages.

162 Trade bans and tribute trade: Tsai 2002:123–24 (thirty-eight nations), 193–94; So 2000:119–20, 125–27; Deng 1999:118–28; Chang 1983:166–97 (tribute trade), 200–17 (naval decline); Needham et al. 1954–:vol. 4, pt. 3, 527–28 (orders to destroy ships); Kuwabara 1935:97–100 (suppression of foreign families). Confucianism indeed was negative about commercial gain, assigning merchants to the lowest of the "four categories of the

people." But that scorn had relatively little impact in practice, in much the same way that Christian doctrinal scorn for moneylenders and usury did not prevent the emergence of powerful banks. Thus the emperor felt free to begin "tributary" relations with the Ryukyu Islands, an archipelago between Japan and Taiwan that was well known for having good mountain horses, by sending an official to obtain horses in exchange for a "gift" of, among other things, 69,000 pieces of porcelain, one hundred bolts of damask, and almost a thousand iron pots. The tributary gifts from Ryukyu also served as a way to launder Japanese and Southeast Asian goods that were politically inexpedient to acknowledge (Chang 1983:174–78). For the tribute trade with Japan, see Li 2006c:45–47.

163 Composition of *wokou*: Interviews, Li Jinming, Lin Renchuan, Dai Yefeng; Li 2001: 10–13; So 1975:17–36.

164 "were merchants": Lin was referring to a well-known remark by Tang Shu, an official in the Jiajing emperor's court: "Pirates and merchants are both people: when trade is open, pirates become merchants, and when trade is banned, merchants become pirates" (Hu 2006:11.4a–4b; see also, Chang 1983:234). Pirates had periodically plagued the region for two thousand years (Kuwabara 1935:41–42).

164 Fujian geography as factor in maritime trade: author's visits; interviews, Lin, Li; Yang 2002; Clark 1990:51–56 ("be tilled," 52); So 1975:126–27; Deng et al. eds. 1968:vol. 15, "Local Conditions" (The yearly harvest "rarely filled [Fujianese farmers'] bamboo baskets. . . . Therefore calculating individuals saw waves like paths between fields, and relied on masts and sails like ploughs. The wealthy used their riches, and the poor used their bodies, transporting China's goods to countries in foreign lands and trading local products for up to ten times profit. Thus the people were content to place little value on their lives, and one after another rowed across the sea so that it eventually became a habit, and they say there is no better livelihood than this"). To this day many Fujianese are more comfortable in one form or another of Min, an ancient offshoot of Chinese, than standard Mandarin.

166 Yuegang harbor: Author's visits; Li 2001:chap. 1; Lin 1990:170–73; Li, Y. 1563. Request to Establish a County. In Deng et al. eds. 1968:vol. 21, "Writings" ("long time now"); vol. 24, "Collected Stories" (ten-family law). My thanks to Huang Zhongyi and Lin Renchuan for taking me around the remains of Yuegang; to Li Jinming for tolerating two long interviews; and to Kenneth Pomeranz for illuminating discussions.

166 *Wokou* crisis begins: Deng et al. eds. 1968:vol. 18, "Bandit Incursions"; vol. 24, "Collected Stories" (ten-family policy).

167 Zhu Wan: Li 2001:12–13, 24–25; Chang 1983:254–55; So 1975:50–121 (fine, 63); Deng et al. eds. 1968:vol. 18, "Bandit Incursions." I have approximately rendered Zhu's title of "grand provincial coordinator" as "governor."

168 "them to leave": Chang 1983:242.

168 More than twenty thousand died: Chang 1983:246.

169 "hilly ruins": Luo 1983:vol. 2, n.p. ("Records of Eastern Barbarians: Japan").

169 "to the town": Zhuge 1976:n.p. ("Sea Pirates").

170 Twenty-four Generals and end of piracy in Yuegang: Li 2006c (Wu Island, 50); Chang 1983:200–17 (hiring smugglers), 230–34 (officials in smuggling families), 251–58 (battles with Yuegang pirates); So 1975:151–53; Deng et al. eds. 1968:vol. 18, "Bandit Incursions"; vol. 21, "Writings" ("itself moaned"). Li (2001:16) says that Shao actually beheaded Hong.

171 Motives for rescinding ban: One Fujianese official argued that after legalizing international trade the "good people" from Yuegang who were now "scattered abroad" would "return permanently to their homeland and live amongst the rebels. Should any unlawful behavior begin to sprout up, the public will learn of it first and report it to local officials, who could then make a concerted effort to wipe them out" (Li, Y. 1563. Request to Establish a County. In Deng et al. eds. 1968:vol. 21, "Writings").

171 Chinese coins and paper money: Von Glahn 2010 (export coins, 467–68); 2005 ("short-string," 66; *huizi* value, 75); 1996:51–55; Ederer 1964:91–92; Tullock 1957.

172 First European banknotes: Mackenzie 1953:2. They were issued in Sweden, which previously had used heavy copper coins. Very heavy—weighing about forty-three pounds—the Swedish ten-dollar coin is said to be the heaviest ever made. England first tried paper money in 1694.

172 Cowry shells: Johnson 1970. Commodity money like gold is also problematic because if inflation occurs the government has to worry about the ostensible value of the coin becoming less than the actual value of the gold it is made of, which leads to people melting down their change and selling the metal. To forestall this possibility, governments can debase their coins by mixing in less valuable materials. But that creates, in effect, two parallel currencies: a valuable old currency and a less valuable new currency.

174 Cycles of paper and silver: Von Glahn 1996:43–47, 56–82 ("economic realities," 72); Chen et al. 1995; Tullock 1957. Silver use varied widely by location (Pomeranz, pers. comm.).

175 "ring out": Gao Gong, quoted in Quan 1991b.

176 "It never ends": Runan Gazetteer (1608), quoted in Quan 1991b:598; see also, Von Glahn 1996:168.

176 Zhangpu County and Jiajing coins: Von Glahn 1996:86–88, 96–102, 143–57 (Wanli coins not accepted); 220–22 (Gu's economic ideas); Quan 1991b:597 (Gu quote).

177 *Kanyinshi*: Interviews, Li Jinming, Lin Renchuan, Dai Yefeng.

177 Silver in one-tenth: Quan 1991b:573–74. The writer was Jin Xueyan in 1570.

177 Tax reform: Von Glahn details the slow change from a paper-money to a coin to an uncoined silver tax system (1996:75–161 passim). See also, Flynn and Giráldez 2001:262–65; Huang 1981:61–63; Atwell 1982:84–85; Quan 1972b.

178 Decline of silver mining: Von Glahn 1996:114–15; Quan 1991c, 1972b. See also, Atwell 1982:76–79.

178 China seeks silver overseas to finance government: Guo 2002; Qian 1986:69–70; see also, Von Glahn 1996:113–25.

178 Trade-driven diaspora: Guo (2002) says more than 100,000 may have gone out.

178 Chinese in Philippines: Anon. Relation of the Conquest of the Island of Luzon, 20 Apr. 1572. In B&R 3:141–72, at 167–68 (150 in Manila). A lower number—eighty couples, plus, one assumes, children—occurs in Anon. (Martín de Goiti?). Relation of the Voyage to Luzon, 1570. In B&R 3:73–104, at 101. The *Ming Shi* claims that before Legazpi Fujianese "traders of abundant means, several tens of thousand in number," lived in the Philippines. They "took up a long residence there, and did not return home until their sons and grandsons had grown up" (MS 323.211.8370). "Tens of thousand" should be understood figuratively, as "a great number, perhaps as many as ten thousand." My thanks to Devin Fitzgerald for this translation.

178 Discovery of Potosí: Arzáns 1965:vol. 1, 33–39; Gaibrois 1950:11–22; Capoche 1959:77–78; Acosta 1894:308–10; Baquíjano y Carrillo 1793:31–32.

179 Indian metallurgy and smelters: Mann 2005:82–83 (and references therein); Acosta 1894:324–26; Cieza de Léon 1864:388–89 (chap. CIX).

179 Potosí population: Dressing 2007:39–41 (migration restriction); Chandler 1987:529 (list of American cities by population); Arzáns 1965:vol. 1, 43 (more than fourteen thousand in 1546), vol. 3, 286 (160,000 in 1611); López de Velasco 1894:502 (1560s); Acarete du Biscay 1698:2–5 (restrictions); Anon. 1603:377–78; Baquíjano y Carrillo 1793:37–38 (160,000). Potosí's only competitor was Mexico City, which in 1612 had about 145,000 inhabitants (Beltrán 1989:216). At this time, "Seville barely housed 45,000 residents; Paris, 60,000; London and Antwerp, 100,000; and Madrid, 6,000 souls" (Pacheco 1995:274). All these figures are debatable, but they give an idea of relative sizes.

179 Potosí opulence: Author's visit (coat of arms); Arzáns 1965:vol. 1, 250 ("strands of pearl"); Acarete du Biscay 1698:61 (silver paving); Quesada 1890:vol. 1, 178–80 (courtesans), vol. 2, 420 (bidding war). A single peach could cost one hundred pesos (Baquíjano y Carrillo 1793:38).

180 Dependence on imports: Cobb 1949:30–31; López de Velasco 1894:503; Acosta 1894:306; Anon. 1603:373.

180 Mercury amalgamation: Whitaker 1971 ("mountain of Potosí," 105, note 21). My thanks to Bryan Coughlan for helping me with the chemistry.

181 Four thousand a week: Brown 2001:469–70 (Huancavelica); Cole 1985:12 (Potosí). The numbers fluctuated over time; I give typical numbers for the beginning of the program.

181 3–8 million deaths: E.g., Galeano 1997:39. The historian David Stannard (1993:89) wrote that the life span of a mine worker was "about the same as that of someone working at slave labor in the synthetic rubber manufacturing plant at Auschwitz." The actual death rate, though still horrific, seems to have been much smaller (Cole 1985:26).

181 Huancavelica conditions: Brown 2001; Whitaker 1971; Lohmann Villena 1949. A fine summary is Reader 2009: 10–14.

182 Potosí schedule, conditions: Cole 1985:24–25; Acosta 1894:321–23; Acarete du Biscay 1698:50 ("After Six days constant work, the Conductor brings 'em [sic] back the Saturday following." As the laborers staggered to bed, the Potosí governor "causes a review to be made of 'em, to make the owners of the Mines give 'em the Wages that are appointed 'em, and to see how many of 'em are dead, that the [native leaders] may be oblig'd to supply the number that is wanting"); Loaisa 1586:600–03 ("Saturday," 602).

184 Violence in Potosí: Padden 1975:xxviii (bodies in walls); Arzáns 1965:vol. I, 192–93 (first birth); Quesada 1890:vol. I, 387 (city council); Lodena, P. d. Letter to Audiencia de La Plata, 29 Apr. 1604. In Arzáns 1965:vol. I, 258 (tailors). See also, Valenzuela, J. P. d. 1595. Letter to Crown, 8 Apr. In Dressing 2007:38.

185 Montejo and Gudínez: Arzáns 1965:vol. I, 75–92 (all quotes, 75–76).

186 Basque dominance of Potosí: Dressing 2007:65–78, 104–06, 144–45 (Basque gangs as enforcers); Crespo 1956:32–39; Arzáns 1965:vol. I, 186, 249. According to Dressing (2007:128), eleven of the twenty-four members of the city council were Basques.

187 Martínez Pastrana's audit: Dressing 2007:112–31 (council debtors, 128; corruption claims, 130–31); Crespo 1956:39–64 ("not Basque," 39; salary, 48; royal order banning tax cheats, 49–54; council debtors, 50; revolt begins, 63–64).

188 Insurrection in Potosí: Dressing 2007:143–252 (Basque Spanish challenge, 146; saving Verasátegui, 151–53; Manrique's Basque sympathies, fiancée, 164–65, 198–209, 248–49, 285–88; burning down home, 211; "arrogant manner," 230; ransacking homes, 232); Crespo 1956:65 ("cuckolds," 66; saving Verasátegui, 71–73; burning Manrique home, 97–99; riots, 109–12); Arzáns 1965:vol. I, 328–407 (hands and tongue, 330; attempt to kill Manrique, 359–64; viceroy's harsh position, 387–88).

188 Martínez Pastrana, Manrique leave: Dressing 2007:207, 248–49, 277; Crespo 1956:96, 132–33 ("the Basques," 133), 156.

189 Lack of disruption to mines: Dressing 2007:161.

189 1549 shipment: Cobb 1949:30.

189 Silver production: Barrett 1990:236–37 (150,000 tons); Morineau 1985:553–71; Soetbeer 1879:60, 70, 78–79, 82–83 (145,000 tons); Cross 1983:397 (80 percent). Garner says it was "more than 100,000 tons" (1988:898).

190 Inflation and instability: This is the "price revolution" of the late sixteenth century and the "general crisis" of the next, both discussed in Chap. 1.

190 Fall in silver price consequences: Flynn and Giráldez 2008; 2002; 1997. See also below.

191 Yuegang today and past harbor perils: Author's visits; interviews; Li 2008 ("fish scales," 65); Lin 1990:170–73; Deng et al. eds. 1968 (early map of area).

192 Yuegang-Manila trade: Author's interviews, Li Jinming; Li 2001:chap. 7 (number of ships, 86–87); Qian 1986:74 (number of ships); Chaunu 2001:453 (leave in June); Schurz 1939:77 (smuggling); Dampier 1906:vol. 1, 406–07 (description of junk); Morga 1609:vol. 16, 177–83 ("his Majesty," 181); Salazar, D. d. 1588. Relation of the Philipinas Islands. In B&R 7:29–51, at 34. The trade exploded: "In the 64 years from 1580 to 1643, 1,677 Chinese trading vessels went to Manila; an average of 26.2 entered the port each year. Excluding the three years for which there are no records (1590, 1593, 1595) and calculating based on the 61 known years, the actual annual average number of ships entering the port of Manila was 27.5, approximately 13.5 times that prior to the opening of the seas" (Guo 2002:95).

193 Piracy in China Sea: Cevicos, J. 1627. Inadvisability of a Spanish Post on the Island of Formosa. In B&R 22:168–77 (galleys, 168–69); Sotelo, L. 1628. A Synopsis of Juan Cevicos's Discourse Regarding the Dutch Presence in the Seas of Japan and China. In Borao ed. 2001:54–56, at 54–55. See also the letters to the king in idem:57–58.

193 Sangley (footnote): Sande, F. d. 1576. Relation of the Filipinas Islands. In B&R 4:21–97, at 50; Cevicos, J. 1627. Inadvisability of a Spanish Post on the Island of Formosa. In B&R 22:175 ("heathen sangleys").

193 Parián founding, description: Ollé Rodríquez 2006; Schurz 1939:79–82; Bañuelo y Carrillo 1638:69–70; Morga 1609:vol. 16, 194–99; Salazar, D. d. 1583. Affairs in the Philipinas [sic] Islands. In B&R 5:210–55, at 237; idem. 1590. The Chinese, and the Parián at Manila. In B&R 7:212–38, at 220–30. Parián buildings were thrown together from reeds, bamboo, and scraps of wood and tile. Inevitably, the ghetto burned to the ground. It was rebuilt again, in a different place. A few years later it was again consumed by fire and again rebuilt in another location. And again. Each new Parián was bigger than its predecessor.

194 Spanish plans to conquer China, acquiescence to Parián: Ruiz-Stovel

2009 (*gobernadorcillo*, 57); Ollé Rodríquez 2006:40–46 (acquiescence), 2002:39–88 (plans); Guo 2002.

195 King tries to shutter Chinese shops: Felipe II. 1593. Letter to Gómez Pérez Dasmariñas, 17 Jan. In B&R 8:301–11, at 307–08; idem. 1593. Decree on Chinese shops, 11 Feb. In B&R 8:316–18.

196 Chinese drive Spaniards out of business: Bobadilla, D. d. 1640. Relation of the Filipinas Islands. In B&R 29:277–311, at 307–08 ("wooden noses"). Salazar, D. d. 1590. The Chinese and the Parián at Manila. In B&R 7:212–38 ("Spanish tradesman," 226–27).

196 Parián population: Estimates range from ten thousand (1587) to "four to five thousand" (1589) to four thousand (1589) to two thousand (1591) to about one thousand (1588) (in order: Vera, S. d., et al. 1587. Letter to king, 25 Jun. In B&R 6:311–21, at 316; Anon. 1589. Instructions to Gómez Pérez Desmariñas. In B&R 7:141–72, at 164; Vera, S. d. 1589. Letter to king, 13 Jul. In B&R 7:83–94, at 89; Desmariñas, G. P. 1591. Account of the Encomiendas in the Philippine Islands. In B&R 8:96–141, at 96–98; Salazar, D. d. 1588. Relation of the Philipinas [sic] Islands. In B&R 7:29–51, at 34). Discrepancies may be due to not distinguishing between Chinese inside and outside the Parián. One cleric suggested there were three to four thousand in the Parián, four to five thousand on Luzon, and two thousand more during trading times (Salazar, D. d. 1590. The Chinese and the Parián at Manila. In B&R 7:212–38, at 230). See also, Guo 2002:97.

196 Malaria in Manila: DeBevoise 1995:143–45.

196 Chinese silver prices: Ollé Rodríguez 2006 ("for free," 26); Boxer 2001:168–69; Flynn and Giráldez 2001:432–33; Von Glahn 1996:127; Atwell 1982:table 4; Quan 1972d.

197 *"very* rich": Bañuelo y Carrillo 1638:77. (Emphasis mine.)

197 Taxes, levies, freight charges, etc.: Ronquillo de Peñalosa, G. 1582. Letter to Philip II, 16 Jun. In B&R 5:23–33, at 30–31; Salazar, D. d. 1583. Affairs in the Philipinas Islands. In B&R 5:210–55, at 236–40; See also, Letter to M. Enriquez. In B&R 3:291–94.

197 Cartel: Schurz 1939:74–78; Philip II. 1589. Royal Decree Regarding Commerce. In B&R 7:138–40.

197 One-third to one-half silver to China: Flynn and Giráldez 2001:434–37; Quan 1972d. For a dissenting view, see Garner 2006:15–17.

198 Galleons increase in size: Chaunu 2001:198 ("administration itself"); Quan 1972c:470–73; Schurz 1939:194–95.

198 More than fifty tons of silver: Quan 1972d:438–40.

198 Smuggling: Flynn and Giráldez 1997:xxii–xxv (*San Francisco Javier*); Cross 1983:412–13; Schurz 1939:77, 184–87; Álvarez de Abreu, A., ed. 1736. Commerce Between the Philippines and Neuva España. In B&R 30:53–56 (see also 54, note 7); Bañuelo y Carrillo 1638:71 ("been registered"); Garcetas,

M., et al. 1632. Letter from the Ecclesiastical Cabildo to Felipe IV. In B&R 24:245–62, at 254–55.

198 Import quotas: Chaunu 2001:198–200. Their gradual tightening is seen in B&R 6:282, 284; 7:263, 8:313, 12:46, and 30:50–52.

199 Mulberry trees: MS 78.54.1894 ("bolt of silk"); Quan 1972c:453. A somewhat analogous shift took place in Guangdong (Marks 1998:119–21, 181–84).

199 China silk industry: Quan 1972c.

199 Chinese make Spanish fashions: Álvarez de Abreu, A. 1736. Commerce of the Philippines with Nueva España, 1640–1736. In B&R 44:227–313, at 255.

200 Spanish merchants complain of Chinese goods, seek regulatory redress: Álvarez de Abreu, A. 1736. Commerce of the Philippines with Nueva España, 1640–1736. In B&R 44:227–313, at 253–58, 293–95, 303–04.

201 Maluku mutiny: Borao 1998:237–39; MS 323.211.8370–72; Argensola, B. L. d. 1609. *Conqvista de las Islas Malvcas.* In B&R 16:211–318, at 248–61 ("Europa," 258); Morga 1609:vol. 15, 68–72; Dasmariñas, L. P., et al. 1594. Letter to Japanese emperor, 22 Apr. In B&R 9:122–37, at 126–27, 133.

201 Expelling Chinese: Morga, A. d. 1596. Letter to king, 6 Jul. In B&R 9:263–73, at 266; Tello, F. 1597. Letter to Felipe II, 29 Apr. In B&R 10:41–45, at 42; idem:1597. Letter to Felipe II, 12 Aug. In B&R 10:48–50.

201 Benavides: Benavides, M. d. 1603. Letter to Felipe III, July 5. In B&R 12:101–12 ("will remain," 110). Benavides was just a bishop, but he occupied the archbishopric after Salazar's successor died of disease.

202 Gold mountain expedition: Borao 1998:239–42; Morga 1609:vol. 15, 272–76; Salazar y Salcedo, G. d. 1603. Three Chinese Mandarins at Manila. In B&R 12:83–97 (Chinese letter, 87–94); Benavides, M. 1603. Letter from Benavides to Felipe III, 5–6 Jul. In B&R 12:101–26, at 103–06. Beijing officials' suspicions and anger about the incident are detailed in the *Ming Shi-lu* (Wade ed. trans. 2005): Year 30 (Wanli reign), Month 7, Day 27 (12 Sep. 1602); Year 31, Month 11, Day 12 (14 Dec. 1603); Year 32, Month 11, Day 11 (31 Dec. 1604); Year 32, Month 12, Day 13 (31 Jan. 1605).

203 1603 massacre: Chia 2006; Guo 2002; Borao 1998:239–42; Zhang 1968:59–60 (all quotes); Horsley 1950:159; Schurz 1939:86–90; Laufer 1908:267–72; Philips 1891:254; Deng et al. eds. 1968:vol. 18, "Disasters and Achievements" ("In the 31st year of the Wanli reign, 25,000 Chinese were killed in Luzon, eight of every ten from Yuegang"); Wade ed. trans. 2005: Year 32 (Wanli reign), Month 12, Day 13 (31 Jan. 1605); Year 35, Month 11, Day 29 (16 Jan. 1608); Morga 1609:vol. 16, 30–44; B&R 12:138–40, 142–46, 150–52, 153–60, 167–68.

204 Aftermath of massacre: Ollé Rodríguez 2006:44–46 ("Parián," 46); Chang 2000:221–30 (widow suicides); Schurz 1939:91–93; Philips 1891:254; Anon. (Xu Xueju?). 1605. Letter from a Chinese Official to Acuña. In B&R

13:287–91, at 290–91; Laufer 1908:272 ("grow once more"). By 1640 royal officials were again griping that Manila's Spaniards "are always in anxiety about the Chinese, or Sangleys, who number more than 30,000 in Manila" (B&R 30:34).

205 Repeated massacres in Parián: Ruiz-Stovel 2009; Ollé Rodríguez 2006:28–29, 44–45; Chia 2006 (esp. 1686 massacre). In 1709 and 1755 all the Chinese were expelled, but with less bloodshed; the death toll may have been as low as several hundred. The 1820 massacre occurred during a Filipino uprising against foreigners. Firsthand accounts include B&R 29:201–07, 208–58; 32:218–60; 44:146.

206 "nomadic horsemen": Findlay and O'Rourke 2007:xviii.

207 Trade as source of Chinese elite power: Atwell 1982:84–86; Flynn and Giráldez 2002:405; Schell 2001:92.

207 Trade-fueled economic boom: Flynn and Giráldez 2002; Frank 1998:108–11, 160–61; Atwell 1982, 1977; Quan 1972e.

207 Imperial anxieties about Yuegang merchants: Author's interviews, Li; Von Glahn 1996: (merchants as independent power); Qian 1986:75 (merchant cheating); Angeles, J. d. l. 1643. Formosa Lost to Spain. In B&R 35:128–63, at 150 (cheating).

207 China silver prices fall to world level: Flynn and Giráldez 2001:270–72; Pomeranz 2000:272.

208 Mistake of taxing silver weight, not value: Flynn and Giráldez 1997:xxxv–vi. Von Glahn (1996:238) points out that it worked the other way, too—higher silver prices equaled a higher tax burden.

208 Dispute over whether silver helped end Ming: Atwell 2005, 1982; Moloughney and Xia 1989. China also used Spanish silver to buy ginseng and furs from the Manchus, thus funding their enemies (Pomeranz, e-mail to author).

209 Costs to China of silver: Flynn and Giráldez 2001.

CHAPTER 5 / *Lovesick Grass, Foreign Tubers, and Jade Rice*

210 Spread of tobacco in China: Benedict 2011:chap. 1; Brook 2008, 2004 (Wang Pu, 86); Zhang 2006:48.44a–44b ("morning until night"); Jiang and Wang 2006; Lu 1991 (names); Yuan 1995:48–50 (1549 pipes); Goodrich 1938 ("that country," 649); Laufer 1924b. My thanks to Josh D'Aluisio-Guerrieri and Devin Fitzgerald for translations from Chinese sources in this section. Ho (1955:191) says the peanut was the first American introduction, but tobacco caught on faster.

211 Chinese tobacco etiquette: Benedict 2011:chaps. 3, 5; Brook 2004:87–89 ("imagine," 89); Cong ed. 1995:7.1a (poem, attributed to "Mr. Wu"); Lu 1991:1.4a–1.4b ("everywhere," list). My thanks to Prof. Benedict for sending me an early copy of her book.

213 Snuff and Brummell: Laufer 1924b:39–42 ("century," 40); Kelly 2006:110 (Brummell's snuffboxes), 158–61 (one-handed technique), 256 (tea).

214 "hydraulic societies": A clear but harshly critical summary is Blaut 1993:78–90.

215 "staple in Fujian": Crosby 2003:199.

215 China as sweet potato, maize producer: Figures from Food and Agriculture Organization (faostat.fao.org).

215 Introduction of sweet potatoes: Zhang et al. 2007:159 (1590s famine); Song 2007; Shao et al. 2007; Cao 2005:177 (slices); Wang 2004:19–20 (80 percent, 20); Atwell 2001:60–61 (famine); Chen 1980:190–92; Ho 1955:193–94; Goodrich 1938; Xu 1968:vol. 27, 20–21; Chen 1835? ("ground," "a threat"); Anon. 1768? ("length"); Wang 1644:14. Song (2007) and Zhang (2001) discuss the almost simultaneous introduction of maize.

216 Central American origin: D. Zhang et al. 2000.

217 Lin Huailan (footnote): 1888 Dianbai County gazetteer (vol. 20, "Miscellaneous Records"), quoted in Song 2007:34.

217 Rice double-cropping: Ho 1956.

219 Ming-Qing wars, emptying coast: Mote 2003:809–40; Zheng 2001:213–17 (all quotes); Cheng 1990:239–43.

219 Fall in Manila trade: Qian 1986:74; Quan 1972d:445.

219 Zheng Chenggong and Manila: Busquets 2006 ("eight thousand horses," 410); Clements 2004:234–38; Anon. 1663. Events in Manila, 1662–63. In B&R 36:218–60.

220 "from its flow": Mu, T. 1681. Memorial Requesting the Lifting of the Ban on Maritime Trade. Quoted in Quan 1972e:499.

222 Hakka migrate, become shack people: Richards 2005:124–31; Yang 2002:47 ("land left," "to the next"); Leong 1997:43–54, 97–101, 109–25 ("the Hakka cultural group was predominant among the pengmin, especially from the Qing," 125); Osborne 1989:esp. 142–52.

222 Tolerance of sweet potato, maize for bad conditions: Author's interviews, Jiangsu Xuzhou Sweetpotato Research Centre; Song 2007; Mazumdar 2000:67–68; Marks 1998:310–11; Osborne 1989:48–49, 159–60; Ho 1955; Xu 1968:vol. 27, 21 ("them there"—the original has *chi*, which I render as "feet" [1 *chi* = 13.6 in. = 34.5 cm]).

223 Dominance of sweet potatoes and maize: Mazumdar 2000:67; Osborne 1989:188–89; Rawski 1975:67–71; David 1875:vol. 1, 181–95 ("tubers," 188). Shack people also spread tobacco (Benedict 2011:chap. 2).

223 Numbers of shack people: Wang 1997:320–21.

223 Migration wave to west: Rowe 2009:91–95; Richards 2005:112–47ff.; Osborne 1989:240–45; Rawski 1975:64–65.

224 Migrants, American crops help lead to boom: Tuan 2008:138–44; Song 2007; Shao et al. 2007; Lan 2001 (Sichuan); Mazumdar 2000:70; Vermeer 1991 (Shaanxi); Rawski 1975; Ho 1955.

225 China population jump: Lee and Wang 2001:27–40; Wang 1997; Ho 1959:94–95, 101. See also, Frank 1998:167–71.

225 Sweet potato dispersal into Oceania: Montenegro et al. 2008; Ballard et al. eds. 2005; Zhang et al. 2000.

226 Factors increasing population: Rowe 2009 (granaries, 55–57; taxes, 65–69; trade, 55–57, 127–32); Shiue 2005 (disaster relief); Lee and Wang 2001:52–56 (infanticide); Needham et al. 1954–:vol. 6, pt. 6, 128–53 (inoculation). Rowe's well-written, concise book summarizes current understanding about the Qing empire.

227 Hong Liangji bio: Jones 1971 (quotes, 4).

228 Qing occupy Guizhou, push out Miao: Richards 2005:131–37; Elvin 2004:216–44.

229 "flood, drought and plagues": "China's Population Problem" (1793), quoted in DuBary et al. eds. 2000:vol. 2, 174–76.

229 Malthus and reactions: These paragraphs are adapted from Mann 1993:48–49; Malthus 1798:13 ("for man"). See also Standage 2009:126–29.

230 Hong's letter, exile: Jones 1971:156–202. The insurrection was the White Lotus rebellion, set off by a religious movement among China's subalterns, prominent among them Hakka shack people (Hung 2005:164–66).

230 World population and harvest: I am simplifying. World population went up by a factor of 2.16, and wheat, rice, and maize production by, respectively, 2.75, 3.05, and 3.84 (Food and Agricultural Organization data from 2007).

231 Rice price rise: Quan 1972e (Suzhou prices, 485); Marks 1998:232–34 (granaries).

232 Tobacco planting, official concern: Benedict 2011:chap. 2; Tao 2002a ("nearly half," 69), 2002b; Myers and Wang 2002:607–08; Marks 1998:311 (tobacco-planting ban in south China).

234 Rising crop area in 1700–1850: Williams 2006:264; Richards 2005:118. Estimates differ, but the overall trend seems not in dispute.

234 Deforestation through logging (footnote): Yang Chang 2003:44–45; Marks 1998:319–20.

234 Shack people's deforestation leads to erosion: Richards 2005:128–31; Leong 1997:chap. 8; Osborne 1989.

234 Overall ecological risks in lower Yangzi hills: Richards 2005:128–31; Osborne 1989:37–56, 184–86 ("tortoise's back," 49; "future drainage," 87). My thanks to the Chinese farmers who spoke to me about the challenges of rice agriculture.

234 Extra burden of maize: Song 2007:156–58; Osborne 1989:168.

236 "into ravines": Mei 1823:vol. 10, 5a–6a. See also, Osborne 1989:214–15.

236 Rise in floods: Li 1995; Osborne 1989:318–24; Chen 1986; Will 1980:282–85. Marks (1998:328–30) depicts a similar pattern in the south.

237 Flood maps: Central Bureau of Meteorological Sciences 1981.

238 Zhejiang fails to stop clearing, erosion: Osborne 1989:246–57 ("native places," 249); Wang 1850 ("*Why?*").

239 General failures to stop clearance, erosion: Song 2007:158–60; Osborne 1989:23–24, 175, 198, 209–10, 225–26, 257–62. In forty-nine flood-battered counties surveyed by Osborne, twenty-seven blamed shack people for their plight. Of those twenty-seven, twenty-three named the crop responsible for deforestation. Twenty of those twenty-three blamed maize; the other three blamed sweet potatoes (ibid.: 318–24). Some provinces more effectively fought erosion, but were eventually overwhelmed (Will 1980:278–82).

240 Dazhai: Zhao and Woudstra 2007 (slogans, 193); Shapiro 2001:95–114, 137 (calluses, 99; slogans, 96, 107). Zhao and Woudstra credit Dazhai more than Shapiro. Several China scholars—and some Chinese provincial officers—told me that Dazhai had been a fraud from the beginning. None offered proof, though.

241 20 percent: Author's interview, Zhang Liubao (village leader in Zuitou, Shaanxi).

241 Loess Plateau: Mei and Drengne 2001. It covers ~720,000 km²; the nations, ~620,000 km².

241 Soil layers: Author's interview, David Montgomery; Montgomery 2007:21–22.

242 Silt and buildup of Huang He: Mei and Dregne 2001 (one to three inches, forty feet, 12); Will 1980.

242 Huang He management: Pomeranz, e-mail to author; Davids 2006; Elvin 2004:128–40; Dodgen 2001:esp. chaps. 1–3 (Great Wall comparison, 3).

243 1780–1850 floods: Central Bureau of Meteorological Sciences 1981.

243 Erosion boom in Loess Plateau: Wei et al. 2006:13 (one-third, fig. 4—I am approximating). In addition, the level of soil organic matter fell below 1 percent in many areas; typical figures for U.S. farmland are 5 to 8 percent (author's interview, Zhang Zhenzhong, Shaanxi Provincial Institute for Loess Plateau Control).

244 Erosion causes Zuitou migration: Author's interviews, Zuitou.

246 Anti-deforestation programs: Author's interview, Lu Qi, Institute of Desertification Studies, Chinese Academy of Forestry; Yu et al. 2006:236; Levin 2005.

246 "3-3-3" system: Author's interviews, Lu; Gaoxigou officials; Liu Guobing, Research Institute of Water and Soil Conservation in Northwest China, Chinese Academy of Sciences; Xu et al. 2004.

247 Problems with tree planting: Author's visits; Normile 2007; Yu, Yu, and Li, 2006.

CHAPTER 6 / *The Agro-Industrial Complex*

251 Introductory potato facts: Spooner and Hijmans 2001:2101 (species and types [but see later discussion]); Clarkson and Crawford 2001:70–73 (12.5 lbs.); Zuckerman 1998:83 (Marie Antoinette [but see below]); Bourke 1993:90–100 (potato consumption, table 4); Salaman 1985:572–73 (potato war); Kon and Klein 1928 (167-day diet); Gerard 1633: 752 ("knowledge of them"), 925 ("Virginia potato," "common potatoes"). Production rankings from Food and Agriculture Organization (faostat.fao.org). Laufer (1938:15) dismisses the story of Marie Antoinette/Louis XVI as "a good historiette." McNeill (1999:78), Salaman (1985:599), and Langer (1975:55) accept the tale. Cuvier (1861:vol. 2, 15), who knew Antoine-Augustin Parmentier, the supposed provider of the potato flowers, reports that Louis XVI wore the plant in his buttonhole, inspiring the vogue.

253 Giant potato: Anon. 2008. "Lebanese Finds 'Heaviest' Potato," British Broadcasting System, 8 Dec. (http://news.bbc.co.uk/2/hi/middle_east/7771042.stm).

253 Potato as Europe's savior: Standage 2009:120–29; Reader 2009:95–117; McNeill 1999:69 ("and 1950"); Zuckerman 1999:220–28 (his book's subtitle is "How the Humble Spud Rescued the Western World").

254 Drake statue: Reddick 1929.

256 Andean societies: Good overviews include Silverman ed. 2004, D'Altroy, 2002, and esp. Moseley 2001. A popular summary is Mann 2005:chaps. 6–7. See also Gade 1992 ("sophistication," 461), 1975. Eruptions: Siebert and Simkin 2002–.

257 Evolution of *S. tuberosum:* Zimmerer 1998:446–49ff.; Brush et al. 1995:1190; Grun 1990 (four species); Ugent et al. 1987 (thirteen thousand); Ugent et al. 1982 (2000 b.c.).

257 Clay and solanine: Guinea 2006; Browman 2004 (licking); Johns 1986 (adsorption); Weiss 1953. My thanks to Clark Erickson for telling me about these sources.

258 Andean potato treatments: Author's visits; e-mail, Clark Erickson; Yamamoto 1988; Gade 1975:210–14. My thanks to Susanna Hecht for pointing out the parallel between *chuño* and gnocchi.

258 Twenty degrees: Mayer 1994:487.

259 Terraces: Sarmiento de Gamboa 2009:132 ("*andenes*"); Denevan 2001:17–18, 170–211 (extent of terracing, 175); Donkin 1979. My thanks to Clark Erickson and Bill Denevan for helpful discussions.

260 Raised fields: Denevan 2001:24–25, 219, 264–65; Erickson 1994.

260 *Wachos:* Author's interviews and e-mail, Erickson and Denevan; Wilson et al. 2002; Sánchez Farfan 1983:167–69; Bruhns 1981. *Wacho* and *wachu* are the Quechua and Aymara terms; they are known in Spanish as *surcos*.

260 Farming methods: Author's visits; Gade 1975:35–51, 207–10; Rowe 1946:210–16.

261 Potato variety: Brush et al. 1995; Zimmerer 1998 ("United States," 451). The potato center landrace database is at singer.cgiar.org/index.jsp.

262 Potato genetics: Jacobs et al. 2008 ("to accept"); Spooner and Salas 2006:9–23 (overview); Huamán and Spooner 2002 (four species); Spooner and Hijmans 2001 (eight groups); Hawkes 1990. Spooner and Hijmans basically relabeled Hawkes's taxonomy, which described all but one of the cultivars as separate species.

263 Path of potato to Europe: Reader 2009:81–93; Hawkes and Francisco-Ortega 1993 (Canary Islands); Salaman 1985:69–100 (conquistador's revulsion, 69); Laufer 1938:40–62 passim; Roze 1898 (Bauhin, 85–88).

263 Drake: Salaman (1985:144–58), Roze (1898:63–64, 70–74), and, to a lesser extent, McNeill (1999), credit the story. Drake did pick up some potatoes in the Pacific in 1577 (Salaman 1985:147).

264 Potato fears, support: Reader 2009:111–31 (Frederick, 119); Salaman 1985 ("provoke Lust," 106; disease, 108–14; Orthodox, 116; "Popery!," 120); Roze 1898 (establishment, 98; fears, 99, 122–23; "peasants and laborers?," 143). Beeton 1863:585 (potato water). My thanks to Ted Melillo for drawing the last to my attention.

264 Parmentier and France: Standage 2009:121–22; Reader 2009:120–22 (Jefferson, 121); Bouton 1993 (summary of Flour War, xix–xxi); Laufer 1938:63–65; Anon. 1914 (captured five times); Roze 1898:148–82ff. ("Nourish Man," 149; "other countries," 152); Cuvier 1861.

266 European famines, Malthusian trap: Clark 2007:1–8, 19–39; Komlos 1998 ("At least until 1800, but in some places even thereafter, the European demographic system was in a Malthusian homeostatic quasi-equilibrium," 67); Bouton 1993:xix–xxi (food riots); Braudel 1981–84:vol. 1, 74–75 (forty famines, Florence), 143–45 (other quotes); Appleby 1978:102–25ff. (England); Walford 1879:10–12, 266–68 (England).

267 Young's observations: Young 1771:vol. 4, 119–20 ("promoted"), 235–36 (grain), 310 ("in it"). Vandenbroeke (1971:37) cites similar figures for the Netherlands.

267 Four to one (footnote): Atwater 1910:11 (wheat dry matter); Langworthy 1910:10 (potato dry matter). Contemporary breeding has increased the dry matter in both crops a bit.

267 Potato and food supply: Radkau 2008:6 ("interruptus"); Vanhaute et al. 2007:22–23 (10–30 percent); Malanima 2006:111 (calorie supply doubles); Crosby 2003:177 (complementing existing crops), 1995; Clarkson and Crawford 2001:59–79 (40 percent, 59); McNeill 1999 (one-third to one-half land, 79); Komlos 1998; Zuckerman 1999; Masefield 1980:299–301; Langer 1975; Vandenbroeke 1971:38–39 ("food problem"); Connell 1962:60–61.

Potato country: According to the U.N. Food and Agriculture Organization (faostat.fao.org), the top twelve potato consumers, all in the Eastern Hemisphere, stretched in a band from Ireland to the Russian Federation and Ukraine. Thanks to Ted Melillo for drawing my attention to Radkau.

268 Increased reliability: Reader 2009:99 (summer), 118–19 (army); Vandenbroeke 1971:21 (army), 38 (summer crop); McNeill 1999:78 (army); Young 1771:vol. 4, 121–23.

268 Smith quotes: Smith 1979:vol. 1, 176–77 (bk. 1, chap. 11, §n, ¶39). Potatoes and maize were, Smith thought, "the two most important improvements which the agriculture of Europe—perhaps, which Europe itself—has received from the great extension of its commerce and navigation" (vol. 1, 259 [bk. 1, chap. 11, §n, ¶10]).

268 Potato as healthy diet: Zuckerman 1999:6, 31. My sentence about vitamins is a rewritten version of a sentence in Nunn and Qian (2010:169).

269 Potato as cause of population increase: Standage 2009:124–28; Reader 2009:127–29; Clarkson and Crawford 2001:29, 228–33; Zuckerman 1999:220–28; Livi-Bacci 1997:30–34 (doubling); Salaman 1985:541–42; Langer 1975; McKeown et al. 1972; Vandenbroeke 1971:38; Wrigley 1969:162–69; Drake 1969:54–66, 73–75, 157 (Norway). The idea is not new: Alexander von Humboldt said (1822:vol. 2, 440, 449) the potato "has had the greatest influence on the progress of population in Europe. . . . [N]o plant since the discovery of cerealia, that is to say, from time immemorial, has had so decided an influence on the prosperity of mankind as the potato." Livi-Bacci (1997:77–78) argues that this view is "countered by a number of considerations," mainly a decline in grain consumption and real wages. But these declines occurred because farmers were growing more potatoes, which provide better nutrition, and because there were more farmers, which drove down wages. Fogel (2004:3–11) summarizes the debate.

269 Potato examples: Cinnirella 2008:esp. 253–54 (Saxony); Viazzo 2006:182–92, 212–15, 289–92 (Alps); Pfister 1983:esp. 292 (Alps); Brandes 1975:180 (Spain). See also, Reader 2009:94–95.

270 Agricultural revolution: A summary history is Overton 1996.

270 Clover: Kjærgaard 2003. Turnips, too, were important as a fallow crop, because their broad leaves smothered weeds.

271 One-eighth of increase: Nunn and Qian 2010 ("conservative," 37).

271 Chincha Islands and birds: Cushman 2003:56–59; Hutchinson 1950:9–26; Peck 1854a:150–225 (150 feet, 198).

271 Need for nitrogen: Smil 2001:chap. 1. A fine summary in Standage 2009:199–214.

272 Guano on Chinchas: Hutchinson 1950: 14–43 (147 islands, birds, thirty-five pounds), 79–83 (chemical composition).

272 Pre-European use of guano: Julien 1985; Garcilaso de la Vega 1966:vol. 1, 246–47 (pt. 1, bk. 5, chap. 3). Julien and Gade (1975:44) say guano was brought to the highlands; Denevan (2001:35) believes its use was "limited and localized," because of the difficulty in transporting it.

273 Von Humboldt and guano: Fourcroy and Vauquelin 1806 ("they approached," 370).

273 Bone market: Walton 1845:167–68 (lack of interest in guano); Anon. 1822 ("daily bread"), 1829, 1832.

274 Guano mania and Liebig: Cushman 2003:60–62, appendix 1 (export figures); Mathew 1970:112–14; House of Commons 1846:377–78 ("Account of the Number and Tonnage of Vessels . . . engaged in the Guano Trade"); Anon. 1842a (role of Liebig); 1842b:esp. 118, 138–40, 142–44, 146–47 (view of Science); Johnson 1843; Liebig 1840 ("of maize," 81–82). See also, Smil 2001:42. Other sources give somewhat different figures for British guano imports, but there is no dispute about their rapid rise. I have seen four editions of Liebig's book.

274 Beginning of input-intensive agriculture: Melillo 2011; Cushman 2003:37. I have adapted one of Cushman's sentences.

275 Working conditions: Skaggs 1994:chap. 2; Mathew 1977:44–51; Peck 1854a:205–13; Anon. 1853 ("coated with guano," 555).

275 Elías's life: Blanchard 1996.

277 Importing Chinese to Peru: Meagher 2009:94–100 (warehouses), 176–77 (mutinies), 221–24 (more than 100,000, 222); Wu 2009 ("were killed," 47); Skaggs 1994:162–63; Schwendinger 1988:23–26; Mathew 1977:36–43 (eight years, 43); Stewart 1970:82–98. Melillo (2011) sets the context.

277 Mistreatment of Chinese: Meagher 2009:224–29 (cemetery, 226); Wu 2009; Mathew 1977:44–51 (five tons); Stewart 1970 (see, e.g., 21–23, 90–97); Anon. 1856 (torture); Peck 1854a:170, 207–08, 214–16; 1854b ("were digging").

279 Guano monopoly and protest: Skaggs 1994:10–15, 21–30; Mathew 1968:569–74; Markham 1862:308–09 (scorn for Peru); Anon. 1854 ("lower price," 117). Typical U.S. editorials included "The Guano War" (*NYT*, 14 Aug. 1853), "The Guano Question" (*NYT*, 12 Aug. 1852), and "The Guano Question in England" (*NYT*, 29 Sep. 1852).

279 "economic success": Miller 2007:149. I have borrowed Miller's comparison to OPEC, too.

279 Guano Islands Act: Skaggs 1994:172–97 (Navassa), 213, 230–36; Letter, R. S. Bowler to S. Wike, 16 Sep. 1893. In Magoon 1900:15–16 (official list of islands).

280 Industrial monoculture: Pollan (2006:41–48) evocatively describes this transformation.

280 First Green Revolution: Melillo 2011.

280 Comparison of Europe to African nations: Clark 2007:40–50. I am vio-

lently simplifying a complex comparison, but the point is valid. Komlos (1998:68) gives higher figures for European consumption than Clark, but the difference does not alter the comparison.

281 Impact of fertilizers: Smil 2001 (two out of five, xv). Population change: Livi-Bacci 1997:31, World Bank Development Indicators (http://data.worldbank.org/).

281 Guano averts disaster (footnote): Pomeranz 2000:223–25 ("century," 224), 240, appendix B.

281 Two million dead: Zadoks 2008:20–27; Ó Gráda 2000:84–95. Zadoks estimates 750,000 dead in continental Europe, Ó Gráda argues that most estimates of the Irish tally are "one million, or slightly above it" (85). Vanhaute et al. (2007:26) suggest a tally for Europe of "a few hundred thousands" but are not as thorough as Zadoks.

282 Life cycle of blight: Mizubuti and Fry 2006:450–58 (dispersal, 454–55); Judelson and Blanco 2005; Sunseri et al. 2002 (zoospore travel); Jones et al. 1914:11–13, 30–37.

283 Peru initially viewed as blight source: Abad and Abad 2004:682; Andrivon 1996; Bourke 1993:148–49.

283 Mexico as center of diversity, origin: Abad and Abad 2004:682; Grünwald and Flier 2003 (oospores, 174–75); Goodwin et al. 1994 (Mexico to U.S., 11594); Fry et al. 1993:653–55; Hohl and Iselin 1984 (discovery of other type of blight in Europe).

283 Lack of potato in Mexico: Ugent 1968; Humboldt 1822:vol. 2, 76, 399, 439–40, 443–50.

284 Ristaino studies: Gómez-Alpizar et al. 2007 ("the famine," 3310–11); May and Ristaino 2004.

285 Guano ships: Mathew 1977:49; Peck 1854a:159.

285 Blight appears in Europe: Zadoks 2008:9–17 (order more potatoes, 16); Vanhaute et al. 2007:22; Bourke 1993:129–30, 141–49; Decaisne 1846:65–68 (1844 observation); Dieudonné et al. 1845:638 (1845 appearance).

285 Spread of Irish blight: Donnelly 2001:41–47 (one-quarter to one-third loss); Ó Gráda 2000:21–24 (2.1 million acres); Kinealy 1995:31–35 (mid-October), 42–43; Salaman 1985:291–93.

286 Ireland as post-apocalyptic landscape: O'Donnell 2008 (murder rate, 81); Donnelly 2001 (disease, 171–76); Ó Gráda 2007 (rape, 46), 2000 ("jail," 40–41; disease, 91–95; theft, 187–91); Zuckerman 1999:187–219 (lining roads, 193; mantraps, 194–95; diet, 195); Kinealy 1995 (dogs, 173; "creatures," 198).

288 Emigration: Donnelly 2001:178–86 ("in heart," 180); Ó Gráda 1999:104–14, 228–29; Kinealy 1995:chap. 8.

288 English aid, culpability: Reader 2009:176 ("re-established"); Donnelly 2001:233 (Mitchel); Ó Gráda 2000:122–25 (export figures, argument). I am grateful to Charles McAleese for good discussions about this issue.

289 Population impacts: Donnelly 2001:178 (emigrant total); Ó Gráda 2000:5 (worst in history), 229–30.

290 Limits of spores and Irish weather: Aylor 2003:1996 (thirty miles); Sunseri et al. 2002 (5 percent, 444; seventy miles, 449). Not a drop of rain fell in northern Ireland between August 28 and September 13 (Butler et al. 1998). Because the spores could not have survived the sunlit crossing of the Irish Sea, they must have traveled by night.

291 Lumpers and *clachans*: Donnelly 2002:8–10; Zuckerman 1999:141–42; Myers 1998:293, 300–01; Ó Gráda 1994 (about half); Bourke 1993:21 (distribution), 36–42; Salaman 1985:292. Historical descriptions of the Lumper complain of their poor taste; modern descriptions extol their "excellent and rich flavor" (Myers 1998:293). Either tastes have changed or, as one farmer suggested to me, modern plant breeding has produced such tasteless spuds that they are superseded by even the worst past varieties. Using nineteenth-century techniques, Lumpers outproduce modern hybrids (ibid.: 363).

291 Ridged "lazy-bed" fields: Omohundro 2006; Doolittle 2000:chap. 12; Myers 1998 (dimensions, 65; erosion, 88–90); Salaman 1985:232–36, 524 ("lazy root"), 586; Denevan and Turner 1974:27 (temperature difference). I am grateful to Bill Doolittle for many useful talks.

291 Attack on ridged fields: Myers 1998:44, 55–60, 85–86.

292 Disappearance of lazy-beds: Myers 1998:61–66; Murphy 1834:556 ("more clearly").

292 Myers's experiments: Myers 1998 (water in furrows, 153–56; setup, 235–36; blight units, 360; temperature/humidity effects, 365–66, 379–84). Grass grew on the sides of the ridges without interfering with the potatoes; it acted as a kind of fallow, recharging the land even as it was farmed (ibid.: 369–72). Although the furrows are "wasted" space, they allow light to reach the plant understory and provide air circulating; meanwhile, the loss is offset by increased productivity on the ridges.

292 Guano and blight (footnote): Porter 2007 (spore survival); Mizubuti and Fry 2006:451 (survival of sporangia); Aylor 2003:1996 ("through the atmosphere"); Inagaki and Kegasawa 1973 (nematode). The eleven counties were Kerry (Anon. 1842. "Spring Show of the Kerry Farming Society." *British Farmer's Magazine* 6:178–93); Kilkenny (Anon. 1843. Review of *The Irish Sketch Book*. *The Dublin University Magazine* 21:647–56); Meath, Cork (Johnson 1843); Down, Armagh, Louth, Monaghan, Cavan, Kilkenny, Roscommon, Antrim (all from Anon. 1843. On the Celebrated Peruvian Manure Called "Guano." *British Farmer's Magazine* 7:111–24). Another possible explanation for the unusual speed of the blight's spread would be that it was actually introduced a year or two earlier (Bourke 1993:147–48).

294 Myers's conclusions: Myers 1998 (fourteen failures, 63; decline of lazy-bed and failures, 473–75).

295   Failure of scientific explanations: Matta 2009; Zadoks 2008:16–20; Bourke 1993:130–39; Wheeler 1981:321–27 ("our senses," 324); Large 1940:14–19, 27–33, 40–43; Jones et al. 1914:23–33, 58–60. Some historians propose that the potato failures of 1845–47 inflamed already existing discontent, thus contributing to the revolutions of 1848 (Zadoks 2008).

295   Murphy's beetles: Murphy, T. 1862. Letter to *Valley Farmer*, 22 May. Quoted in Tower 1906:26.

295   Spread of beetle in United States: Hsiao 1985:44–45, 71; Tower 1906:25–36; Foster 1876 ("train of cars," 234); Riley 1869:102–03; Walsh 1866. Foster quotes the train story as from the *New York Times* (19 Jul. 1876), but this is incorrect; the source must be another account.

296   Beetle spreads to Europe: J. F. M. Clark 2007:113–16 (trade war, 114); Hsiao 1985:55; Tower 1906:39.

296   Path of beetle to potato: Lu and Lazell 1996; Jacobson and Hsiao 1983; Tower 1906:21–25.

297   "the beetles": Anon. 1875. "The Potato Bug," *NYT*, 2 Jun. Sixteen million: Female fecundity can exceed four thousand, and there are typically two generations in a year (Hare 1990:82–85).

298   Failed efforts to fight beetle: Casagrande 1987:143–44; Riley 1869 ("destroy them," 108); Walsh 1866 (horse-drawn remover, 15).

299   Insect plagues: Essig 1931.

300   Paris Green: J. F. M. Clark 2007:120–24; e-mails to author, Casagrande; Casagrande 1987:144–45; Lodeman 1896:59–69 (London Purple, 65–67); Riley 1869:116.

301   Copper sulfate (discovery, mix with Paris Green): Casagrande 1987:145–46; Large 1940:225–39, 277–79; Lodeman 1896:25–33, 47, 55, 100, 122–23.

302   "clean fields": Pollan 2001:218.

302   Beetle resistance: E-mail to author, Casagrande; Alyokhin et al. 2008 ("management," 400, "production," 407); J. F. M. Clark 2007:124 (first DDT test); Hare 1990:89; Casagrande 1987:146–47; Jacobson and Hsiao 1983 (heterozygosity).

303   Resurgence of blight: Mizubuti and Fry 2006 448–49; Garelik 2002.

CHAPTER 7 / *Black Gold*

307   Dispute over Indians' status: A good summary is Hanke 1994:chap. 1. I discuss this more fully in Chap. 8.

307   Navagero biography: Cicogna 1855 (list of publications, 209–10). The garden may have been inspired by early accounts of the botanical gardens in central Mexico.

307   Team sports in Europe: The sole potential exception was a soccer-like game in Italy, *calcio fiorentino*, recorded as far back as 1530. Europe's second-oldest team sport, polo, was not introduced to Europe until the

nineteenth century. Odds are that the Mesoamerican ball game is the world's oldest continuously played team sport. The court in Paso de la Amada, in the southern tip of Mexico, was constructed in about 1400 b.c. (Hill et al. 1998), whereas polo apparently dates to the time of Christ (Chehabi and Guttmann 2003:385). Lacrosse, indigenous to North America, may also be very old.

308 "Great speed": Navagero 1563:15v–16r; see also, Navagero, A. Letter to G. B. Ramusio, 12 May 1526. In Fabié 1879:378–90, at 389–90.

308 "so elastic": Anghiera 1912:vol. 2, 204–05; Navagero, A. Letter to G. B. Ramusio, 12 Sep. 1525. In Fabié 1879:368–76, at 368–69 (friendship with Martire d'Anghiera).

309 "rather heavy": Oviedo y Valdés 1851–53:165–66 (pt. 1, bk. 6, chap. 2); Covarrubias y Orozco 2006 (lack of word for "bounce," see entries for, e.g., *botar* and *bote*). The first volume of Oviedo's work appeared in 1535; later parts remained unpublished until the nineteenth century.

309 First scientific studies: Condamine 1751a, b.

309 Rubber heats when stretched (footnote): Gough 1805 ("lips," 290).

310 Native rubber uses and methods: Author's interviews and e-mail, John Hemming, Susanna Hecht; Woodroffe 1916:41–46 (tapping, processing); Pearson 1911:59–71 (description of tapping, processing); Johnson 1909:chap. 9 (description of tapping); Spruce 1908:vol. 1, 182–85, 511–15 (tapping); Warren 1851:16 (clothes).

310 Rubber fever: Anon. 1890; Johnson 1893; Coates 1987:29–31; Coslovsky 2005 (import figures, 14, 27).

310 Webster, "effective character": Parton 1865:66.

311 Goodyear and vulcanization: Slack 2003 ("on the spot," 107); Coates 1987:31–33, 36–37. Goodyear's own account (1855) is unreliable.

312 Hancock and vulcanization: Woodruff 1958:chap. 1; Coates 1987:22–28, 33–38; Hancock 1857:91–110 ("little bits," 96). Woodruff quotes two contemporaries who say that Hancock did analyze Goodyear's samples. It is certainly true that Hancock was ungenerous—his otherwise useful autobiography (1857) doesn't even mention Goodyear. Goodyear's patent (No. 3633) and Hancock's patent (No. 10027) are available at the websites of, respectively, the U.S. Patent and Trademark Office and the British Library.

313 Expositions and Goodyear's death: Slack 2003:161–63, 203–10, 230–37; Coates 1987:39–42 ("a postage stamp," 41–42); Bonaparte ed. 1856:542–43.

315 Staudinger: Author's interview, Coughlin; Ringsdorf 2004; Mülhaupt 2004; Morawetz 2002:86–98 ("propaganda," 97).

316 Vulcanization chemistry: A good introduction is Sperling 2006:chaps. 8–9. My thanks to Bryan Coughlin for introducing this book to me.

318 *H. brasiliensis* quality, location: Ule 1905. Four other Amazonian species were also harvested: *H. benthamiana, H. guianensis, Castilla* (or *Castilloa*)

elastica, and *Castilla ulei.* Except for *C. elastica,* they were less significant than *H. brasiliensis.* Some writers have suggested that for this reason one must talk of "rubber booms" in the plural, with *Hevea* and *Castilla* exerting different ecological and economic effects (Santos-Granero and Barclay 2000:chap. 2).

318 Overland vs. river route, rapids: Markham 1871. The waterfalls and rapids could only be negotiated by canoes, and even these frequently capsized, with great loss of life (Anon. 1901).

319 Neville Craig: Fleming 1922:118–19. My thanks to Jamie Owen and Julie Carrington of the Royal Geographic Society for looking up the date of his death and to Robert Charles Anderson for helping me with his Yale alumni records.

320 Craig on the Madeira: Craig 2007 ("theory," 177; "cities," 226; "Parentintins," 237); Hemming 2008:201 (Parentintins). To be fair to Craig, the nasty anti-Italian crack was from a newspaper article; he just quoted it approvingly.

321 Keller's feasts: Keller 1874: 74–77 (turtle), 80–81 (pirarucu and manatee, quote on 81).

321 Agricultural heartland: Mann 2008.

321 Fishing with strychnine: My thanks to Susanna Hecht for a description of this procedure, which is still used today.

322 Overexploitation of rubber trees: Schurz et al. 1925:17–21 (yield); Whitby 1920:5–6 (yield); Labroy 1913:39–47 (average daily production, 47); Pearson 1911:43–44 (overtapping); Smith 1879:108 (killing Belém trees). Average yield figures disguise the high variability among trees. The sources above measured unselected trees; yields are higher today. Belém do Pará means "Bethlehem of the Pará River," the latter being the southern of the two main mouths of the Amazon. Until the twentieth century the city was generally called Pará; Belém is the modern name.

322 1877–79 drought: Davis 2002:79–90, 377–93; Greenfield 2001 (death tally, 45–46). The drought was an El Niño event.

322 Rush into rubber: Santos 1980:66, 83–84 (25,000 estates); Spruce 1908:vol. 1, 507 (rush upstream), 518 ("obtain it").

322 Migrant-driven rise in malaria: Hemming 2004b:268–72 (background); Keller 1874:8 (prevalence in mid-1860s), 40–42 (decreasing on Madeira); Chandless 1866:92 (prevalence in mid-1850s).

322 Malaria and Craig's railroad: Craig 2007:271 (incapacitates half, see also note on 304); 381–83 (refusal of payment); 382–88 (two-thirds, three-quarters sick); 407 (120, more than half sick); 408 ("complete collapse"); 387–403 passim (struggle home).

323 Rubber boom: Overviews include Hemming 2008:175–231, 2004b:261–301; Souza 2001:163–88; Barham and Coomes 1996; Dean 1987; Weinstein 1983; Batista 1976:129–41; Collier 1968.

323 Rubber production, exports, prices: Barham and Coomes 1996:30–32
(New York prices, exports); Santos 1980:52–55, 208–20 (exports, Brazil prices, speculation); Batista 1976:129–40 (Brazil prices); Pearson
1911:214–15 (exports); Anon. 1910 (speculation, "silver"); Fernandes
2008:fig. 2 (London prices); Instituto de Pesquisa Econômica Aplicada
(Brazil) n.d. (exports, available at www.ipeadata.gov.br); U.S. Energy
Information Administration n.d. (U.S. oil prices, www.eia.doe.gov/
pub/international/ielf/BPCrudeOilPrices.xls). Oil prices in 1999 dollars, as per EIA.

324 Colonial Belém: Author's visits; Hemming 2008:48–49, 66, 97–99; Souza
2001:46–47, 61, 91–93; Pearson 1911:20–42 ("smell of it," 22). I am grateful
to Susanna Hecht for touring me through the city.

324 Louche Manaus: Author's visits; Hemming 2008:179–83 ("bubbly,"
182); Jackson 2008:113–15, 252–55; Collier 1968:18–27; Burns 1965; Pearson
1911:93–111.

327 Labor conditions in lowlands: Hemming 2008:198–204; Barham and
Coomes 1996:29–71; Dean 1987:36–41; Woodroffe 1916:49–54; Craig
2007:248–63 ("peons," 251). Some recent scholars argue that past accounts
of universal brutality in the rubber zone are exaggerated and unfair. But
there is little doubt that conditions often were appalling by modern standards.

327 *Caucho* gathering: Santos-Granero and Barclay 2000:23–29; Barham
and Coomes 1996:37–42; Schurz et al. 1925:21; Hardenburg 1913:181–84;
Pearson 1911:156–58; Feldman 2004 (*"sajaduras,"* entry under *quik*, the
rubber tree). My thanks to Lawrence Feldman for drawing my attention to his dictionary and to Scott Sessions for help with translation.
Santos-Granero and Barclay argue that conditions worsened ca. 1900,
when upper-Amazon firms switched from *Castilla* to other *Hevea* species.
Because these could be tapped, rubber barons forced native workers to
stay in one place and walk regular, routinized courses—a violation of
cultural norms. To maintain control, the companies abducted families,
using female hostages as prostitutes.

329 "they left": Da Cunha, E. 1909. "Os Caucheiros," trans. S. B. Hecht. In
Hecht forthcoming. Fitzcarrald was the subject of *Fitzcarraldo*, a 1982 film
by Werner Herzog that is as wondrous artistically as it is unreliable historically. A more reliable biography is Reyna 1941.

329 Rise of Arana: Hecht forthcoming; Goodman 2009:36–41; Hemming
2008:204–07; Jackson 2008:257–61; Lagos 2002 (the most complete biography I have seen); Santos-Granero and Barclay 2000:34–35, 46–55; Stanfield 2001:103–14, 120–23; Collier 1968:27–64; Schurz et al. 1925:364 (22,000
square miles).

330 Abducted Americans in Putumayo: Goodman 2009:17–25; Hardenberg
1913:146–49, 164–81, 195–99 ("syndicate," 178; "the river," 180–81).

330 Hardenburg's crusade, Casement, Arana's fate: E-mail to author, Marie Arana, John Hemming; Hecht forthcoming; Goodman 2009 (Casement); Hemming 2008:207–30; Lagos 2002:68–103ff. (Hardenburg), 301–51 (Casement), 364–65, 377–79 (Arana's death); Stanfield 2001:125–34; Hardenberg 1913:215–64.

332 Scramble for Amazon: Author's visits to Acre, Cobija; interviews, Hecht; Hecht forthcoming; Dozer 1948 ("that valley," 217).

335 Golden Triangle and rubber trees as poppy replacement: Kramer et al. 2009; Shi 2008:23–28; author's interviews, Klaus Goldnick (GTZ, Luang Namtha), Tang Jianwei (Xishuangbanna Tropical Botanical Garden), Nicholas Menzies (Asia Institute), Horst Weyerhaeuser (ICRAF, Vientiane).

336 1,325 acres: Contract between Ban Namma village and Huipeng Rubber, shown to author.

337 Wickham's life: Jackson 2008; Lane 1953–54 (later appearance, pt. 9:7). Jackson, who found Wickham's wife's journal, movingly recounts the family's sufferings.

337 Markham and cinchona: Honigsbaum 2001.

339 Markham, Wickham, and rubber seeds: Jackson 2008:chaps. 8–9 (abandoning relatives, 187; cold reception, 199–202); Hemming 2008:191–95; Dean 1987:7–24 (Wickham's carelessness, 24); Baldwin 1968; Markham 1876 ("permanent supply," 476).

339 Wickham as thief (text and footnote): Jackson 2008:188–93; Dean 1987:22–24 ("century and a half," 23; "human species," 22); Santos 1980:230 note ("international law"), 232.

340 Rubber in Sri Lanka, Malaysia, Indonesia: Jackson 2008:204–05, 265–73; Dean 1987:26–31; Lane 1953–54:pt. 7, 6–7; Large 1940:196–207; Nietner 1880:3 (Sri Lanka coffee industry), 23 ("escaped it," quoting G. H. K. Thwaites); Berkeley 1869 ("three acres," 1157). For simplicity's sake I use modern names—Sri Lanka instead of Ceylon, Malaysia instead of Malaya, and so on.

343 Creation of Fordlândia: Grandin 2009:77–119; Hemming 2008:265–67; Jackson 2008:291–98; Costa 1993:13–17, 21–24, 59–65; Dean 1987:70–76; Davis 1996:338–39. Strictly speaking, Ford did not buy the land but an exclusive concession on it plus a fifty-year tax exemption. Including tires, every car had about eighty pounds of rubber (Costa 1993:133).

344 *M. ulei*.: Money 2006:83–100; Lieberei 2007. I am grateful to Charles Clement for taking me to see some infected trees. The life cycle of the fungus is more complex than my brief summary indicates. For a typical view of the blight at the time, see Schurz et al. 1925:101.

345 Weir, blight, end of Ford's project: Grandin 2009:304–31; Costa 1993:102–06, 127 (Asian percentages); Dean 1987:75–86, 104–07; Agreement for the

Regulation of Production and Export of Rubber, 7 May 1934. In International Labour Office 1943:104–12 ("plant," 110).

345 Natural vs. synthetic rubber: Author's interviews, Rubber Manufacturers Association, Bryan Coughlin; author's visit (condom factory); Davis 1998:134–36.

346 Xishuangbanna: Author's interviews, Tang Jianwei (XTBC), Hu Zhaoyang (Tropical Crops Research Institute); Nicholas Menzies (UCLA); Mann 2009; Sturgeon and Menzies 2008; Stone 2008; Visnawathan 2007; Xu 2006; Shapiro 2001:171–85 ("criticism meetings," 176); Zhang and Cao 1995 (biodiversity). Tang estimates that the planted area in 2008 was about 2,500 square miles. As Spurgeon and Menzies note (2008), the government regarded the Dai and Akha as irrational, backward, and potentially disloyal. The "educated youth" were Han Chinese, who were thought to have the necessary cultural advancement to produce a modern industrial good like latex. By importing Han workers, the government hoped both to produce needed rubber and to "settle" a crucial border with loyal citizens.

348 Expansion into Laos: Author's interviews and e-mails, Jefferson Fox (East-West Center, Hawaii), Yayoi Fujita (University of Chicago), Horst Weyerhaeuser (National Agricultural and Forestry Research Institute, Vientiane), Klaus Goldnick and Weiyi Shi (GTZ, Luang Namtha), Yunxia Li (Macquarie University). Shi 2008 (subsidies); Fujita 2008 (Sing District); Vongkhamor et al. 2007 (2010 total, 6); Fujita et al. 2006; Rutherford et al. 2008:15–16. Sing District covers more than six hundred square miles, but most of it is unusably steep and roadless.

353 Increase in runoff, erosion, water depletion: Ziegler et al. 2009; Guardiola-Claramonte et al. 2008; Stone 2008; Cao et al. 2006 (fog); Wu et al. 2001. See also, Mann 2009.

354 Flights, highway: Fuller 2008; author's visit, interviews with airlines in Bangkok and Vientiane. The road is only two lanes, but that's more than ever before in the region.

354 Potential *M. ulei* disaster, lack of progress on blight: U.S. Department of Defense 2008 (biological weapon); Lieberei 2007 ("dieback of trees," 1); Onokpise 2004 (collection from Brazil); Garcia et al. 2004 (potentially resistant clones); Vinod 2002 (narrowness of genetic base, difficulties of improvement); Weller 1999:table 2 (biological weapon); Davis 1998:123–41 ("synthetic tires," 136).

CHAPTER 8 / *Crazy Soup*

359 Chapel: Alegría 1990:71–77 (first church); Porras Muñoz 1982:130, 399 (Eleven Thousand Martyrs); Gerhard 1978:453–55; Herrera y Tordesillas

1601–15:vol. 1, 344–45 (Dec. 2, bk. 10, chap. 12; Herrera confusedly calls him "Juan Tirado").

359 Garrido's upbringing: E-mails to author, Saunders; Saunders 2010:137–38 (Garrido Sr.); Alegría 1990:15–22; Icaza 1923:vol. 1, 98. My thanks to Matthew Restall, Fritz Schwaller, and especially Alastair Saunders for their generous advice.

359 Restall is skeptical: E-mails to author, Restall; Restall 2000:174, 177. As evidence, Alegría cites Saco's history of American slavery (1879:44), but Saco simply observes that one Portuguese adventurer who made two trips to Africa in the 1480s brought some free Africans to Lisbon. In addition Portugal did not permit free non-Christians to enter the country, so Garrido would have had to convert, probably from Islam. See also, Blackburn 1997:78–79.

360 Tens of thousands of slaves: The classic estimate by Domínguez Ortiz (1952:9) is at most 100,000 for the whole peninsula, "a phenomenon of considerable size, with notable sociological, economic and even ethnic consequences"; later Cortés López estimated about 58,000 in Spain alone (1989:204). For Portugal, see Saunders 2010:19–34.

361 Garrido's arrival: Garrido's *probanza* (testament) of 1538 says "more or less twenty-eight years ago I crossed over to the island of Hispaniola," implying an arrival in 1510, too late to accompany Ponce de León to Puerto Rico in 1508. Alegría believes it most likely that he came between 1503 (the beginning of larger-scale slave importation) and 1508 (the conquest of Puerto Rico). My thanks to Scott Sessions for providing me with a copy of the *probanza* and his translation. A transcription appears in Alegría 1990:127–38.

361 Garrido in Caribbean: Alegría 1990 (Puerto Rico, 29–30; Florida, 37–41; punitive expeditions, 46–47; Cortés, 59–65). Alegría suggests that Garrido sailed with Pánfilo de Narváez, who came to Mexico a year after Cortés; Garrido may be the African called "Guidela" whom Díaz del Castillo (1844:vol. 1, 327) recalls supporting Cortés's takeover of Narváez's force.

362 Attack on Tenochtitlan: The best modern history I know of is Hassig 2006 (failure of first assault, 111–19; 200,000 native allies, 175). Of the four contemporary accounts, the most important is Díaz del Castillo (1844).

362 Smallpox (text and footnote): Hassig 2006:124–30, 186–89; Mann 2005:92–93, 127–29; Crosby 2003:45–51 (a classic account); Restall 2000:178 ("scapegoating"); Durán 1994:563. The evidence for the role of Eguía or Baguía is examined skeptically in Henige 1986.

362 Ambush: Hassig 2006:165–66; Durán 1994:552–54 ("upon it," 553; "then and there," 554); Díaz del Castillo 1844:vol. 2, 82–90 ("distance," 84).

364 Garrido's chapel as graveyard: Díaz del Castillo 1844:vol. 2, 102.

364 Garrido's jobs: Restall 2000:191 ("guard"); Alegría 1990:92–97, 105–07 (expedition); Porras Muñoz 1982:109–10.

364 Garrido's wheat: Alegría 1990:79–85; Gerhard 1978:455–56; López de Gómara 1870:vol. 2, 365 ("much benefit"); Tapia 1539:vol. 2, 592–93 (three kernels); González de León, J. 1538. Statement, 11 Oct. (?). In Alegría 1990:132–33 (experimental farm); Salvatierra, R. 1538. Statement, 4 Oct. (?). In ibid.:134–36 (Garrido's wheat as foundation of Mexican crop). Tapia's figures are different from those of López de Gomara, but the idea is the same. My thanks to Scott Sessions for the translations; I added the exclamation point, which seemed to me to express the chronicler's tone.

365 Folk knowledge: Farmers in central and southern Mexico have told me this several times. I have not seen evidence to prove it.

366 Humankind mixing: I take this point from Crosby (1986:2–3).

366 Migration figures: Eltis et al. 2009–; Horn and Morgan 2005:21–22 (European totals); Eltis 2001; Eltis 1983 ("nineteenth century," 255; European totals, 256). If one includes indentured servants, the ratio of free to unfree becomes even more lopsided (Tomlins 2001:8–9). For a review of Indian numbers, see Denevan 1992a, b; a popular description is Mann 2005.

368 Foundational institution: Here I summarize ideas from a host of scholars, among them Ira Berlin, C. R. Boxer, David Brion Davis, Eugene Genovese, Melville Herskovits, Philip Morgan, Stuart Schwartz, Robert Voeks, Eric Wolf, and Peter Wood (to name only English-language researchers). As Davis (2006:102) puts it, "[B]lack slavery was basic and integral to the entire phenomenon we call 'America.'"

368 Garrido's last years: E-mail to author, Robert C. Schwaller (witchcraft, servant); Alegría 1990:113 (1640s), 127–38 (Garrido's *probanza*); Icaza 1923:vol. 1, 98 (poverty, three children). My thanks to Prof. Schwaller for generously sharing his research on Garrido.

370 Military leaders leave Jerusalem as fighting ends: Albert of Aachen 1120:374–75.

370 Crusaders take Muslim sugar plantations: Ouerfelli 2008:38–41; Ellenbaum 2003; Boas 1999:81–83; Mintz 1986:28–30; Phillips 1985:93–95.

370 Sugarcane genetics: Irvine 1999.

371 Early sugar processing: Galloway 2005:19–21; Daniels 1996:191–92 (500 B.C.), 278–80, 284–96.

371 Sugar in Middle East: Ouerfelli 2008:31–37; Galloway 2005:23–27.

371 "life to it": Pollan 2001:18.

372 Crusaders love sugar, decide to sell it: Ouerfelli 2008:3, 75–76 (sugar in Europe); William of Tyre (*A History of Deeds Done Beyond the Sea* [1182]), quoted in Phillips 1985:93 ("of mankind"); Albert of Aachen 1120:305–06 ("its sweetness").

372 Plantation definition: Craton 1984:190–91.

373 Wage-earning workers: Ouerfelli 2008:287–306 (Sicily, 302–04); Blackburn 1997:76–78. Captive labor was only used in fifteenth-century

Cyprus, where Muslims captured by pirates were put in the fields (Ouer-felli 2008:290).

373 Colón's marriage: Colón 2004:32–33.

373 Rabbits in Porto Santo: Zurara 1896–99:245–47 ("owing to the multitude of rabbits, which are almost without end, no tillage is possible there," 247).

374 Donkey slaughter: Abreu de Galindo 1764:223.

375 Madeira fire: Ca' da Mosto 1895:26 (two days); Frutuoso 1873:61 (seven years), 353, 460–71.

375 Refashioning Madeira for sugar: Vieira 2004:42–48 (factor of more than a thousand, 48; prices, 62–63); Vieira 1998:5–9; Crosby 1986:76–78; Craton 1984:208–09; Greenfield 1977:540–43.

376 Iberian slavery: Blackburn 1997:49–54; Cortés López 1989:esp. 84–88, 140–49, 237–39; Domínguez Ortiz 1952:esp. 17–23 ("sumptuary article," 19).

377 Madeira as springboard for plantation slavery: Vieira 2004:58–74 ("start-ing point," 74); Curtin 1995:24; Crosby 1986:79; Phillips 1985:149; Craton 1984:209–11 (although he argues that São Tomé and Príncipe had "the most potential as pure plantation colonies"); Greenfield 1977:544–48 (Madeira "provided the link" between sugar in the Mediterranean and the American plantations, 537); Frutuoso 1873:655. Fernández-Armesto (1994:198–200) agrees that Madeira's growth was "spectacular" but argues that the Cape Verde Islands, which had more slaves, were where "a new model was introduced: the slave-based plantation economy."

378 São Tomé mosquitoes: Ribiero et al. 1998.

378 São Tomé colonization: Disney 2009:vol. 2, 110–12 (female slaves, 4; Europeans, 111); Magalhães 2008:169–72; Seibert 2006:21–58 (Dutch, 29; bishop, 32; "go out," 52); Thornton 1998:142 (exiled priests); Craton 1984:210–11; Gourou 1963 (two thousand children, 361). The Dutch took the island a third time in 1637 and managed to hang on for a decade.

379 São Tomé sugar: Disney 2009:112–13; Seibert 2006:25–27; Varela 1997:295–98; Vieira 1992:n.p. (31; in 1615 there were just sixty-five plantations); Fru-tuoso 1873:655–56 (I am citing notes from the editor, Alvaro Rodrigues de Azevedo).

379 Lack of yellow fever, malaria, mosquitoes: See, e.g., Capela 1981:64 (absence of vectors); Davidson 1892:vol. 2, 702 ("Malaria is entirely unknown in Madeira. . . . Yellow Fever has never visited this island"); La Roche 1855:141; James 1854:100.

380 Madeira switches from plantation slavery: Disney 2009:90–92; Vieira 1992:n.p.(29–32, 41–42); Rau and de Macedo 1962 (not monoculture, for-eign owners, 23–25); Brown 1903:e21, e32 ("fever, etc.").

380 São Tomé resistance: Seibert 2006:35 (thirty mills); Varela 1997:298–300. See also Harms 2002:283–92.

381 São Tomé's fate: Disney 2009:113–15; Seibert 2006:30–58 passim; Frynas et al. 2003:52–60; Eyzaguirre 1989.

382 Cortés's estate: Barrett 1970:fig. 1. My thanks to Peter Dana, who digitized Barrett's map of the estate to produce the area estimates cited here. See also, Von Wobeser 1988:esp. 38–55.

382 Cortés's activities, return to Spain: Chipman 2005:46 (king's fear of conquistadors); Riley 1972; Barrett 1970:9–13 (mines, ranches, shipyard); Scholes 1958 (tapestries, clothes, 12; investing, 17; shopping mall, 19; gold panning, 20–21; ranches and hog farms, 23; shipbuilding, 26–27); Voltaire 1773:vol. 6, 46 ("cities").

383 *Brief Account:* Las Casas 1992 (quotes, 57, 65)

384 "Catholic faith": *Inter Caetera* (2). Papal Bull of May 4, 1493. In Symcox ed. 2001:34–37, at 36–37. The bull is probably postdated from the summer. The pope issued three similarly worded bulls in response to Colón's voyage, two of which were called *Inter Caetera*.

384 Little interest in evangelization: Simpson 1982:14–21; Konetzke 1958. Spain's long fight against the Moors was a fight against a government, not a religion—"the subjugated Moorish people were guaranteed the free exercise of their Mohammedan belief." Only after Spain's final victory over the Moors in 1492 did "the missionary idea come to the fore" (Konetzke 1958:517–18), because Fernando II and Isabel believed that enforced religious unity would serve the political purpose of unifying their fractious kingdom. Even Colón was not an evangelizer; on his later voyages, at the monarchs' insistence, he brought clergy, but made no effort at conversion.

384 Colón's slaves and Isabel's actions: Simpson 1982:2–5; Las Casas 1951:vol. 1, 419–22; Herrera y Tordesillas 1601–15:vol. 1, 251 (Dec. 1, bk. 7, chap. 14) (Isabel); Cuneo 1495:188 (550); Colón, C. 1494. Memorial to the Sovereigns, Jan. 30. In Varela and Gil eds. 1992:254–73, esp. 260–61 (justifications).

385 *Encomienda* system founded: Guitar 1999, 1998:96–103; Simpson 1982:esp. chap. 1 ("In reality the encomienda, at least in the first fifty years of its existence, was looked upon by its beneficiaries as a subterfuge for slavery," xiii). The system inadvertently reproduced elements from the Triple Alliance and Inka, which also took labor tribute from conquered peoples.

386 Cortés's Indian labor: Von Wobeser 1988:42–44, 55–57, 60 (mill); Riley 1972:273–77; Barrett 1970:86–89, table 11 (two hundred acres); Scholes 1958:18 (three thousand Indian slaves). Scholes's estimate is for the immediate post-conquest period. A 1549 inventory is the only later source for Cortés's Indian slaves (only 186 were left, because most had died in mines).

386 "in the Indies": Colmeiro ed. 1884:202–03.

386 New Laws and reaction: Elliott 2006:132 (Mexico); Hemming 1993:256–59 (Peru); García Icazbalceta ed. 1858–66:vol. 2, 204–19 (text of laws). The

New Laws also abolished the creation of new *encomiendas* and made the old ones not inheritable. The Mexican viceroy proclaimed *"obedezco pero no cumplo"*—I obey the law but do not enforce it.

386 Cortés's deal: Beltrán 1989:22; Riley 1972:278–79; Barrett 1970:78. Three bigger contracts were signed before Cortés's deal, one of them for four thousand slaves. Relatively few slaves were sent across the Atlantic from these efforts (some turned up in Europe). Cortés's deal was the first to *deliver* hundreds of slaves to American plantations (Beltrán 1989:20–24; Rout 1976:37–39).

387 La Isabela teeth: Author's interviews and e-mail, T. Douglas Price; Lyderson 2009. At the time of publication, the find had not been reported in a peer-reviewed journal.

387 Instructions: King and Queen of Spain. 1501. Instructions to Nicolas de Ovando, 16 Sep. In Parry and Keith 1984:vol. 2, 255–58.

387 "cannot be captured": Herrera y Tordesillas 1601–15:vol. 1, 180.

387 Import of Indians, desirability of Africans: Guitar 2006:46, 1998:270–74 (Indians), 278–79 (Africans); Morel 2004:103–04; Las Casas 1992:30 (Bahamas); Anghiera 1912:vol. 2, 254–55, 270–71 (Bahamas).

387 Escalating revolts in Hispaniola: Altman 2007:610–12; Guitar 2006:61–63, 1998:393–403; Boyrie 2005:79–89; Deive 1989:19–75; Scott 1985 ("weapons of the weak"); Ratekin 1954:12 (thirty-four mills); Benzoni 1857:93–95.

388 Cortés and sugar: Von Wobeser 1988:59–64; Barrett 1970:9–17 (estate). His descendants grew sugar there until the nineteenth century, when the newly independent government of Mexico forced them to sell the estate.

388 Sugar rise in Mexico: Von Wobeser 1988:64–69; Acosta 1894:vol. 1, 416 ("Indies"). At about the same time, Portugal was planting sugar in Brazil (Schwartz 1985:15–27)—the industry that went on to wipe out Madeira and São Tomé. Unlike the islands, Mexico had such a big internal market that its producers were not affected by the Brazilian onslaught.

388 Slave imports: Eltis et al. 2009–; Horn and Morgan 2005:21–22 (European totals). Roughly 350,000 went to Brazil; the Spanish Americas received about 300,000, in addition to the approximately 45,000 who had already been shipped over.

388 Africans everywhere in Americas: Peru and Chile: Restall 2000 (Pizarro licenses, 185; Valiente, 187). Chile: Mellafe 1959 (49–50, Valiente); Brazil: Hemming 2004a:140–46; Blackburn 1997:166–74; Schwartz 1988:43–45. Potosí: Assadourian 1966. Lima: Bowser 1974:339. Cartagena: Newson and Minchin 2007:65 (ten to twelve thousand), 136–47 (two thousand, 137). San Miguel de Gualdape: Hoffman 2004:60–83; Anghiera 1912:vol. 2, 258–60; Oviedo y Valdés 1852:vol. 2, 624–32; Herrera y Tordesillas 1615:vol. 2, 307–09 (Dec. 3, bk. 8, chap. 8). Rio Grande: Stern 1991:272 ("drunkenness").

390 Esteban: Goodwin 2008 (death, 335–51); Schneider 2006 (upbringing, 27–28); Ilahiane 2000 (Morocco instability, 7–8); Adorno and Pautz 1999:esp. vol. 2, 18–19, 414–22; Logan 1940; Robert 1929; Niza 1865–68.

394 Cortés's wives, mistresses, children: Hassig 2006:173–74 (capture of Cuauhtemoc); Chipman 2005:passim; Lanyon 2004; López de Gómara 1870:vol. 2, 376.

395 Malinche: Lanyon 1999; Karttunen 1994:1–23; Díaz del Castillo 1844:vol. 1, 84–85.

398 Alessandro de' Medici: Brackett 2005. Brackett suggests that Alessandro was the pope's nephew, rather than a son. But Italian historian Scipione Ammirato heard the story directly from Cosimo de' Medici, Alessandro's successor (1873:12).

398 Cortés's will: Cortés 1548 (other daughter, §33; provision for illegitimate Martín, §23).

398 Cortés vs. Cortés: Lanyon 2004:138–47.

398 Mixing in Hispaniola: Guitar 1999:n.p. (4–5); Schwartz 1997:8–9; 1995:188–89.

399 Pizarro family: Hemming 1993:175–77, 259, 274–77 ("imprisonment," 275–76); Muñoz de San Pedro 1951 (Cortés-Pizarro link).

400 Santiago conquistadors: Mellafe 1959:50–51 (offspring); Schwartz 1995:189 (fifty mestizo children).

400 Mixing in North America (footnote): Colley 2002:233–36 ("empire"); Foster et al. 1998 (Jefferson DNA test); Nash 1999 ("one people," Houston, 11–13).

401 1570, 1640 tallies in Mexico: Bennett 2005:22–23; Beltrán 1989:201–19, esp. tables 6, 10.

402 Morelia and Puebla: Author's visits; Martínez 2008:147 (arbitrage); Verástique 2000:87–130; Hirschberg 1979; Zavala 1947. For a dissenting view on Vasco de Quiroga, see Krippner-Martinez 2000.

403 Church books: Martínez 2008:142.

403 Racial beliefs: Martínez 2008:esp. chaps. 2, 4.

404 Restrictions on mixed-race people: Martínez 2008:147–51; Cope 1994:14–19.

405 *Casta* system and restrictions: Martínez 2008:142–70 ("good mixtures," 162); Katzew 2004:39–61 (categories, 43–44); Cope 1994:24–26, 161–62; Beltrán 1989:153–75. The *casta* system in Argentina is described in Chace 1971:202–08.

408 Moreau de Saint-Méry's racial scheme (footnote): Moreau de Saint-Méry 1797–98:vol. 1, 71–99 ("parts white," 83).

408 Swapping identities: Diego Muñoz: Gibson 1950. Taxes: Schwartz 1995:186. Officials: Love 1967:92–93. Caribbean: Schwartz 1997 ("bad races," 12; "mulattos, and blacks," 15).

408 Mixed marriages and freedom: Mixed children free: Bennett 2005:44–49; Lokken 2001:178–79; Cope 1994:80–82; Love 1967:100–02; Davidson

1966:239–40. Three-quarters of slaves male: Love 1971:84; Carroll 2001:166 (table A.6). Bans on mixed marriages: Love 1971:83–84; Love 1967:99–103. Half of marriages interracial: Lokken 2004:14–16; Valdés 1978:34–44; Love 1971 ("One of the remarkable features of the marriage patterns of persons of African descent in the parish of Santa Veracruz was the fact that [of a total of 1,662 marriages with an African spouse] 847 individuals of non-Negroid ancestry married persons of color," 84); 1967:102–03. Veracruz: Carroll 2001:174 (table A.15).

409   Disappearance of separate groups: Valdés 1978:esp. 57–58, 175–77, 207–09.

409   *Casta* paintings: Martínez 2008:226–38 (museum, 227); Katzew 2004 (more than a hundred sets, 3; quoted captions, figs. 91, 88, 89, 96).

411   Mirra / Catarina's childhood, abduction, unwilling journey to Mexico: Brading 2009 (funeral, 1–2); Bailey 1997:42–48; Castillo Grajeda 1946:29–45 (sexual assault, 42); Ramos 1692:vol. 1, 4a-29b (birth in 1605 "more or less," and noble, Christian childhood, 4b–16a; abduction and journey, 17b-26b). Critic Manuel Toussaint, in his introduction to Castillo Grajeda, says she was born in "1613 or 1614" (10), but gives no source for the claim. Castillo Grajeda does not specify the abuses she was subject to but says the pirate leader "unleashed against Catarina all the furies of hell" during the trip, "ordering her abuse [by his men] in bloody battles."

412   Visions and marriage: Bailey 1997:60 (flowers); Castillo Grajeda 1946:81–83 (feast), 135–36 (staircase of "shimmering clouds," angels); Ramos 1689–92:vol. 2, 36b (nudity).

413   Ramos's condemnation, fate: Brading 2009:10 ("doctrines").

413   Asians jump ship: Slack 2009:39 (60–80 percent, *Espíritu Sancto*); Luengo 1996:99–105 (1565); Beltrán 1989:50 (Legazpi).

414   Lima census (footnote): Cook and Escobar Gamboa eds. 1968:xiii, 524–47.

414   Asian slaves trickle in: Clossey 2006:47 (estimating six hundred a year); Beltrán 1989:49–52; Beltrán 1944:419–21.

414   Ban on Asian servants: Slack 2009:42, 55 (Jesuits).

414   Multicultural militias: Slack 2009:49–52 (samurai); Lokken 2004; Vinson 2000:esp. 91–92. See also, Chace 1971:chap. 8.

415   Catarina wedding night: Bailey 1997:48; Castillo Grajeda 1946:65–69.

416   Puebla ceramics: Author's visits; Slack 2009:44 ("style"); Clossey 2006:45; Mudge 1985.

416   Parián and barbers: Slack 2009:14–16, 43 ("that trade"); Johnson 1998 (Chinese medicine); Anon. 1908:vol. 30, 24 (petition).

418   Confraternity processions: Slack 2009:54; Gemelli Careri 1699–1700:vol. 6, 98–99 ("wounded").

418   Hunger for China: Clossey 2006:42–43 (distance from Mexico—I am almost directly quoting him), 49–51 ("desire," 49).

419 "in the East": Balbuena 2003:89. My thanks to Scott Sessions for helping me with the translation.

419 Mexico City flooding: Candiani 2004; Hoberman 1980.

CHAPTER 9 / *Forest of Fugitives*

421 Calabar and Liberdade: Author's interviews, Salvador (special thanks to Ilê Aiyê). A *quilombo* called Curuzu was the foundation of Liberdade; Calabar, similarly, is now legally part of the larger area called Federação. For other Salvador *quilombos,* see Queiros Mattoso 1986:139–40; Neto 1984. I am grateful to Susanna Hecht for accompanying me to Brazil and acting as translator.

423 More than fifty in United States: Aptheker 1996:151–52.

423 "inventing liberty": Reis 1988. "Escravidão e Invenção da Liberdade" is also the name of his postgraduate program at the Federal University of Bahia.

424 *Simaran:* Arrom 1983.

424 33 to 50 percent mortality in four to five years: Miller 1988:437–41, esp. footnote 221 (see second half on p. 440); Mattoso 1986:43 (6.3 percent/year = 31.5 percent/5 years); Sweet 2003:59–66 (40+ percent/3 years, 60).

424 List of autonomous places: Price ed. 1996:3–4.

424 Suriname war: R. Price 2002; Bilby 1997:664–69 (blood oaths). The first large-scale slave insurrection occurred in 1690 (R. Price 2002:51–52); the treaty was signed in 1762 (ibid.:167–81). But because rebellions dated back to 1674, it seems plausible to call it a hundred-year conflict.

425 "capitulation": Reavis 1878:112–13.

425 Haiti as focus of terror: Reis and Gomes 2009:293; Gomes 2003.

425 Afro-Mexican acknowledgment: Hoffman 2006.

425 U.S. maroon litigation: Koerner 2005.

425 Legend of Aqualtune: Author's visit, Palmares; see, e.g., Schwarz-Bart and Schwarz-Bart 2002:vol. 2, 3–16.

426 Palmares: Classic accounts are Carneiro 1988 and Freitas 1984.

426 Founding in 1605–06: Anderson 1996:551; Kent 1965:165.

426 Palmares location, size: Location: Gomes 2005a:87 (map); Orser 1994:9 (map). Size: Multiple estimates exist, partly because writers differ on what it means to control a territory; I cite an average figure, which readers should understand as merely indicative. See, e.g., Thornton 2008:775 (6,000 km² = 2,300 mi²); Orser and Funari 2001:67 (27,000 km² = ~10,400 mi² [quoting anthropologist Claudi R. Crós]); Orser 1994:9 (65 x 150 km = 9,750 km² = 3,800 mi²); Diggs 1953:63 (1695 estimate of 90 x 50 leagues = 4,500 sq. leagues = 121,680 km² = 47,000 mi²); Ennes 1948: 212 (1694 estimate of 1,060 square leagues = 29,000 km² = 11,000 mi²); Anon. 1678:28

(60 x 60 leagues = 97.000 km² = 38,000 mi²). I assume 1 league = 5.2 km (Chardon 1980 [Spanish and Portuguese units were similar]). Thornton 2008:797 ("outside Europe"). English North America population: U.S. Census Bureau 1975:1168.

426 Macaco, Ganga Zumba: Thornton 2008:776–78; Gomes 2005a:84–87; Anderson 1996:553, 559 (title); Anon. 1678:29–30, 36–38.

427 Slavery and African institutions: Thornton 2010 (attitudes of rulers, 46, 52–53); Klein 2010:57–58; Davis 2006:88–90; Thornton 1998:x (map of sixty states), 74–97 ("African law," 74), 99–100 (slave wars as equivalent to wars of conquest). Wolf 1997:204–31 (pawning, 207–8); Smith 1745:171–90 (Lamb). Thornton (2010:44) points out that African societies had copper, ivory, cloth, and shell currency with which to buy European goods—it was not that they had nothing to sell other than human beings.

428 Purposes of slaves in Africa: Thornton 2008:87–94; Gemery and Hogendorn 1979:439–47 (conditions bad for plantation agriculture).

430 Europeans tap into existing African slave markets: Thornton 2010:42–46 (taking captives without approval, 44–45).

430 Slaves imported to Africa, African demand: Harms 2002:135–37 (imports, all quotes); Lovejoy 2000:57–58.

431 Africans supply, serve on slave ships: Klein 2010:86–87 (crews); Rediker 2008:229–30 (crews), 349; Davis 2006:90 (intermediaries); Thornton 1998:66–71.

431 European inability to raid in Africa (text and footnote): Thornton 1998:chap. 4. "In effect African strength—the capacity to retain territorial integrity—helped foster the slave trade as Europeans established their plantations in the Americas instead of Africa with an elastic supply of coerced African labor" (Eltis 2001:39).

431 Tiny European outposts: Eltis et al. 2009 (estimates); Harms 2002:139–41 (Cape Coast, fewer than ten); 156–60 (Whydah), 203; Postma 1990:62–63 (Elmina).

432 Transformation of slavery: A classic statement of this argument is Lovejoy 2000.

433 Slaves as soldiers: Thornton 1999:138–46 ("prisoners of war," 140).

433 1521 revolt: Guitar 1999:n.p. (14), 1998:361–66; Thornton 1999:141 (military tactics); Deive 1989:33–36; Rout 1976:104–05; Oviedo y Valdés 1851–53:vol. 1, 108–11 (pt. 1, bk. 4, chap. 4).

433 Enriquillo: Altman 2007; Guitar 1999:n.p.; 1998:346–57, 376–86; Thornton 1999:141–42; Deive 1989:36–42; Las Casas 1951:vol. 3, 259–70 (injury to insult, 260); Oviedo y Valdés 1851–53:vol. 1, 140–55 (Africans join, 141). As disease cut Taino numbers, Spain imported slaves from other Caribbean islands. The influx of foreigners threatened Enriquillo's power—they didn't want to be ruled by strangers—a further reason for

his willingness to rebel. Las Casas actually referred to the proverb "tras de cuernos, palos" (after horns, sticks—i.e., adding a beating to infidelity, indicating total victimization). My thanks to Scott Sessions for finding an English near equivalent.

434 Lemba: Guitar 2006:41, 1998:300 (administrators own mills), 396–400 (role of Lemba and other Africans); Landers 2002:234–36 ("able," 234); Deive 1989:49–52.

436 Portuguese fears about Palmares: Lara 2010:8; Gomes 2005; Anderson 1996; Kent 1965:174–75; Blaer 1902; Anon. 1678. The Dutch also feared Palmares (Funari 2003:84).

436 Cultural jumble, including Europeans, in Palmares: Funari 1996:31, 49, note 42. See also, in general, the documents assembled in Freitas ed. 2004.

437 Palmares religion: Vainfas 1996:62–74.

437 Zumbi's life: Gomes 2005a:114–20; Karasch 2002; Freitas 1996; Diggs 1953.

437 1678 treaty: Lara 2010; Anderson 1996:562–63; Anon. 1678.

438 Jorge Velho: Hemming 2004a:362; Freehafer 1970; Board of Missions. 1697. Memorandum, Oct. In Morse ed. trans. 1965:124–26 (translator, "his lusts"); Jorge Velho, D. 1694. Letter to governor, 15 July. In idem 117–18 ("us and ours").

439 Deal cut with Jorge Velho: Ennes 1948:205; Anon. 1693. "Condições adjustadas com o governador dos paulistas Domingos Jorge Velho em 14 de agosto de 1693 para Conquistar e Destruir os Negros de Palmares." In Anon. 1988:65–69.

439 Jorge Velho's march to Palmares: Oliviera 2005; Gomes 2005a:148–61; Hemming 2004a:363; Ennes 1948:208; Anon. (Jorge Velho, D.?) 1693. In Morse ed. 1965:118–26 (quotes, 119).

439 Battle for Macaco: Author's visit, Palmares; Oliviera 2005; Freitas ed. 2004:124–30, 135–37; Anderson 1996:563–64; Freitas 1982:169–88.

442 Zumbi's fate: Freitas ed. 2004:131–34; Anderson 1996:564; Ennes 1948:211.

442 Núñez de Balboa early life, stowing away: Las Casas 1951:vol. 2, 408–15 ("educated man," 408); Altolaguirre y Duvale 1914:xiii–xv ("energetic spirit," xiv); López de Gómara 1922:125; Oviedo y Valdés 1851–53:vol.2, 425–28. Oviedo says that he rolled himself up in a sail, rather than a barrel.

443 Núñez de Balboa seizes power: Araúz Monfante and Pizzurno Gelós 1997:23–27, 100–101 (Indian slavery and gold); Las Casas 1951:vol. 2, 418–31; López de Gómara 1922:vol. 1, 131–37; Altolaguirre y Duvale 1914:xv–lxxxvi; Anghiera 1912:vol. 1, 209–225; Oviedo y Valdés 1851–53:vol.2, 465–78. I have greatly simplified a complex tale of political maneuvering and multiple betrayals.

444 Visit to Comogre: Las Casas 1951:vol. 2, 572–74; López de Gómara

1922:vol. 1, 137–39; Anghiera 1912:vol. 1, 217–23 ("little boats," 221); Oviedo y Valdés 1851–53:vol.3, 9; Núñez de Balboa, V. 1513. Letter to the King, 20 Jan. In Altolaguirre y Duvale 1914:13–25.

444 Expedition to Pacific: Tardieu 2009:43 and note (Nuflo de Olano's reward); Las Casas 1951:vol. 3, 590–97 ("seigneury," 591); López de Gómara 1922:vol. 1, 143–46 (slaves in village, 144); Altolaguirre y Duvale 1914:lxxxviii–xc; Anghiera 1912:vol. 1, 282–87 (an apparently garbled account); Oviedo y Valdés 1851–53:vol.3, 9–12 (partial list of participants).

445 Killing transvestites (footnote): Las Casas 1951:vol. 2, 593–94; Anghiera 1912:vol. 1, 285.

446 First Africans in Panama: Fortune 1967; López de Gómara 1922:vol. 1, 144; Anghiera 1912:vol. 1, 286 (dec. 3, bk. 1, chap. 2); Oviedo y Valdés 1851–53:vol. 3, 45 (bk. 29, chap. 10); Colmenares, R.d. 1516? Memorial against Nuñez de Balboa. In: Altolaguirre y Duvale 1914:150–55, at 155; Ávila, P., et al. 1515. Report to King, 2 May. In: idem:70–72 ("crooked hair," 70).

446 Abandonment of Antigua, foundation of other cities (and footnote): Araúz Monfante and Pizzurno Gelós 1997:45–46; López de Gómara 1922:vol. 1, 159.

447 Núñez de Balboa's fate: López de Gómara 1922:vol. 1, 158; Altolaguirre y Duvale 1914:clxxv–cxc.

447 Nombre de Dios–Panamá road: Tardieu 2009:25–41; Araúz Monfante and Pizzurno Gelós 1997:74–78; López de Gómara 1922:vol. 1, 158–59; Carletti 1701:41–51 ("covers," 43–44; corpses, 49); Requejo Salcedo 1650:78 ("my travels"). Strictly speaking, Carletti was describing the bats in Portobelo on the coast, but they were also plentiful in the forest. Benzoni (1857:142) gives a similar report of the bats.

448 Native population in Panama: Araúz Monfante and Pizzurno Gelós 1997:97; Romoli 1987:22–28; Jaén Suárez 1980 (three thousand people, 77; twenty thousand, 78); Oviedo y Valdés 1851–53:vol. 3, 38 ("uncountable").

448 Importing Indian slaves: Saco 1882:266. See also, Tardieu 2009:46–48.

449 Seven to one: Tardieu 209:48–49; Jaén Suárez 1980:78.

449 Assaults on European colonies: Fortune (1970) collects many accounts, e.g., Benzoni 1857:121.

449 Felipillo: Tardieu 2009:61–63; Pike 2007:245–46; Araúz Monfante and Pizzurno Gelós 1997:134–35; Fortune 1970:pt. 1, 36–38.

450 Bayano's sanctuary: Pike 2007:246–47; Araúz Monfante and Pizzurno Gelós 1997:135–36; Fortune 1970:pt. 2, 33–39; Aguado 1919:vol. 2, 200–13 ("mothers," 201).

450 Ursúa and Bayano: The most important source is Aguado 1919: vol. 2, 200–31 (bk. 9, chaps. 11–13). Modern accounts include Tardieu 2009:chap. 2 ("sudden and sharp," 79); Pike 2007:247–51; Fortune 1970:pt. 2, 40–50. Ursúa was rewarded with the chance to lead an expedition into the Peruvian Amazon, during which he was betrayed and murdered by his subordinates.

451 Unhealthiness of Nombre de Dios: Benzoni 1857:120; Ulloa 1807:93–98; Carletti 1701:42; Gage 1648:369 ("and mariners").

452 Merchants and principal-agent problem: Author's interviews and e-mail, James Boyce, Tyler Cowan, Mark Plummer (economists); Tardieu 2009:108–21; Pike 2007:247.

452 "they meet": Quoted in Tardieu 2009:123–24. See also, Ortega Valencia, P.d. 1573. Letter to the king, 22 Feb. In Wright ed. 1932:46–47. Throwing silver in river: Nichols 1628:281.

452 Drake attacks Nombre de Dios: Fortune 1970:pt. 3, 18–20; Nichols 1628:258–67 ("hight," 264); Nuñez de Prado, J. 1573. Depositions (*probanzas*), Apr. In Wright ed. 1932:54–59; Audiencia of New Granada. 1572. Report to king, 12 Sep. In ibid.:40–41.

453 Failed ambush in Venta de Cruces: Tardieu 2009:126–31; Pike 2007:256–58; Nichols 1628:280–309; Municipal Council of Panamá. 1573. Report to the king, 24 Feb. In Wright ed. 1932:48–51.

454 Attack with French: Nichols 1628:317–25 ("thirty Tun," 318; "of Gold," 323). Testu was wounded and fell behind, too. The pursuing Spaniards killed him on the spot.

455 Inflamed anti-maroon fears, campaign: Tardieu 2009:132–44 (Protestantism, 142); Fortune 1970:pt. 3, 22–34; Royal Officials of Nombre de Dios. 1573. Letter to Crown, 9 May. In Wright ed. 1932:68–70 ("situation promptly," failure to report recovery); Audiencia of Panama. 1573. Report to king, 4 May. In ibid.: 62–67 (failure to report recovery).

456 Deal for freedom: Tardieu 2009:184–246 (details of offer, 185; Portobelo "capitulation," 244–46); Fortune 1970:pt. 3, 34–40 ("the Indies," 39). The leader of the Portobelo maroons was Luis de Mozambique. Pedro Mandinga, who tried to help Drake, was one of his lieutenants.

458 Yanga and Mexico maroons: Rowell 2008 (eleven demands, 6–7); Lokken 2004:12–14 (Guatemala militia); Aguirre Beltrán 1990:128 (origin of Yanga's Bran ethnicity); Carroll 1977 (other examples); Love 1967:97–98; Davidson 1966:245–50 (fashions, 246); Alegre 1842:vol. 2, 10–16 (blood, 11; Bran origin, 12).

460 Miskitu kingdom: Offen 2007 (canes, 274–76), 2002 (clothing and cane, 355); Olien 1987 (racial background and claims, 281–85), 1983 (kingship); Dennis and Olien 1984 (719–20, raids; slavery, 722). My thanks to Prof. Offen for sending his work.

461 U.S. maroons: Sayers et al. 2007 (Great Dismal Swamp); Franklin and Schweninger 2001:86 (thousands in swamp); Aptheker 1996.

463 Black and Red Seminoles: Landers (2002, 1999 [creation of Mosé, 29–60]) has written superbly on the rise of African Florida. Riordan (1996; relations with Creeks, 27–29) and Mulroy (1993; 4 towns, 294) are fine short summaries.

463 Seminole Wars: The wars are far more complex than can be indicated

here, and almost everything about them is subject to argument. For example, the owners of escaped slaves objected to the "capitulation," and Jesup promised not to include recently escaped slaves in the terms, which some have argued rendered the agreement tantamount to a true surrender. Others say, convincingly in my view, that the promise was meaningless because the slaves could not effectively be separated."No matter how mild the system of slavery practiced by the Seminoles, complete freedom was infinitely preferable" (Mulroy 1993:303); "and a half": Giddings 1858:140–41.

465 Haiti: The literature on the revolution is vast. In English the classic studies are by C. L. R. James; Dubois (2005) is a good recent study, available in both French and English. Moreau de Saint-Méry (1797–98) is a fascinating first-hand description of St. Domingue on the eve of revolution. In emphasizing the role of disease, I follow McNeill 2010:236–65 ("graves," 245; "earthworms," 253). See also Davis 2006:chap. 8.

469 Suriname Africans: Price 2011:chap. 1 (25:1, 10), 2002; www.slavevoyages.org (300,000). The U.S. estimate is about 390,000.

470 *A. darlingi* and deforestation: Yasuoka and Levins 2007:453–55; Tadei et al. 1998:333.

470 Suriname maroon wars and treaty: Ngwenyama 2007:59–69; Price 2002:51–52, 167–81; Bilby 1997:667–69 (blood).

471 Stedman: Stedman 2010 (seasoning, 1:102–03; "numerous," candles, 46; impossible to see people, 393; "acquaintances," 100; killed 38, 127; "health," 607). Stedman observed the advantages of acquired immunity: "amongst the Officers & Private men who had *formerly* been in the West indies, none died at all, while amongst the whole number of near 1200 together I Can only Recollect one Single marine who Escaped from Sickness" (607).

472 Logging, mining, and park: tacoba.cimc.com/en/enterprise/tacoba/ tacoba.cimc.com/ en/enterprise/tacoba/ (CIMC website); whc.unesco. org/en/list/1017 (World Heritage site description); Price 2011 (park, 136–40); Alons and Mol eds. 2007:64 (40 percent); Anon. 1998.

472 Maroon population size: Price 2002b.

472 Lack of consultation and IACHR case: Price 2011 (filing petition, 119; Kwinti and park, 136–40). At the time of writing, many legal filings were available at www.forestpeoples.org.

475 Mazagão Velho: Author's visit, interviews; Vidal 2005 (African history of town); Motinha 2005 (inability to repair, 12; abandonment, 25–26); Silva and Tavim 2005:2 (town design); Anderson 1999:28 (1,900).

481 Amazon land rush: This section is adapted from Hecht and Mann 2008.

483 Mojú: Author's visit; author's interviews, Manuel Almeida (Quilombolas Jambuaçu), anonymous informants; Anon. 2006.

CHAPTER 10 / *In Bulalacao*

491 "Bahay Kubo": The plants are *singkamas* (jícama; *Pachyrrhizus erosus*); *talong* (eggplant; *Solanum melongena*); *sigarilyas* (asparagus/winged bean, *Psophocarpus tetragonolobus*); *mani* (peanut, *Archis hypogaea*); *sitaw* (string/yard-long bean, *Vigna spp.*); *patani* (lima/butter bean, *Phaseolus lunatus*); *bataw* (hyacinth bean, *Lablab purpurea*); *kundol* (winter melon, *Benincasa hispida*); *patola* (sponge gourd, *Luffa cylindrica* and *acutangula*); *upo* (wax gourd, *Lagenaria siceraria*); *kalabasa* (kabocha-style squash, *Cucurbita maxima*); *labanus* (radish, *Raphanus sativus*); *mustaza* (mustard, *Brassica juncea*); *sibuyas* (onion, *Allium cepa*); *kamatis* (tomato, *Lycopersicum lycopersicum*); *bawang* (garlic, *Allium sativum*); *luya* (ginger, *Zingiber officinale*); lain *linga* (sesame, *Sesamum orientale*). My thanks to Leonard Co for the botanical identification and translation.

492 Impacts of Philippines exotic species: Department of Environment and Natural Resources (Philippines) and World Fish Center 2006; Lowe et al. 2004 (seven worst invasives).

494 Philippine mahogany: 16 CFR §250.3 (Federal Trade Commission rule, available at: edocket.access.gpo.gov/cfr_2001/janqtr/pdf/16cfr250.2.pdf).

495 Ifugao as landmark: In 1996 the United Nations Educational, Scientific and Cultural Organization saluted the area as a "priceless contribution of Philippine ancestors to humanity" and made it a World Heritage Site (whc.unesco.org/en/list/722). Ifugao is also an international engineering landmark of the American Society of Civil Engineers.

495 Apple snail: Joshi 2005; Caguano and Joshi 2002.

496 500 varieties: Nozawa et al. 2008; Concepcion et al. 2005.

496 Nine new earthworm species: Hong and James 2008; Hendrix et al. 2008:601–02.

497 "their maximum": Quoted in Maher 1973:41.

497 Keesing: Keesing 1962:319 ("innovation"), 322–23 ("mention").

498 First archaeological studies: Acabado 2009; Maher 1972.

499 Five months: Save the Ifugao Terraces Movement 2008:3.

499 Sweet potato in Ifugao: Brosius 1988:97–98; Scott 1958:92–93.

499 Eighth Wonder: Author's interviews, Manila, Ifugao; Harrington 2010; Dumlao 2009. See also the project website at heirloomrice.com, esp. for the annual reports of the Revitalize Indigenous Cordilleran Entrepreneurs, the Filipino side of the project.

503 Mangyan language, culture: Postma ed. 2005.

503 Trade with China: Scott 1984:65–73; Horsley 1950:74–75.

APPENDIXES

514 *Negro* and *preto*: Heywood and Thornton 2007:chap. 6.

517 Zaytun in its heyday: Abu-Lughod 1991:212, 335–36, 350 (population); Clark 1990:46–58; Pearson et al. 2001:187–90, 204–05 (sediment, 190); Polo 2001:211–13 ("profit," 211); Ibn Battuta 1853–58:v. 4, 269–71 ("past counting," 269); Odoric of Pordenone (Hakluyt Goldsmid ed. 1889:vol.9, 133–34) (monks).

518 Pu Shougeng: So 2000:107–22, 301–05; Chen 1983; Kuwabara 1935 (betrayal and siege, 38–40).

519 Islam in Zaytun: Interviews, Ding Yuling, Lin Renchuan; Jin 1982 (translations, intelligibility); Kuwabara 1935:esp. 102–03; Chen 1983 (converts, syncretism, factions). My thanks to Dr. Ding for arranging a tour of Quanzhou's Maritime Museum, of which she is the director.

521 Collapse of Zaytun: Interview, Ding Yuling; Chen 1983; Lin 1990:169 (silting); So 2000:122–29.

WORKS CITED

ABBREVIATIONS

AA	*American Anthropologist*
AAAG	*Annals of the Association of American Geographers*
AHC	*Agricultural History of China* (中国农史)
AHR	*American Historical Review*
AMJTMH	*American Journal of Tropical Medicine and Hygiene*
BAE	Biblioteca de Autores Españoles desde la Formación del Lenguaje hasta Nuestros Días
B&R	Blair and Richardson eds., trans. 1903–09
EB	*Economic Botany*
EHR	*Economic History Review*
HAHR	*Hispanic American Historical Review*
JEH	*Journal of Economic History*
JIH	*Journal of Interdisciplinary History*
JSH	*Journal of Southern History*
JWH	*Journal of World History*
KB	Kingsbury ed. 1999
MMWR	*Morbidity and Mortality Weekly Report*
MS	Zhang et al. eds. 2000
NYT	*New York Times*
P&P	*Past and Present*
PNAS	*Proceedings of the National Academy of Sciences*
QBASVB	*Quarterly Bulletin of the Archaeological Society of Virginia*
VMHB	*Virginia Magazine of History and Biography*
WMQ	*William and Mary Quarterly*
*	*Available gratis on Internet as of* 2011

Abad, Z. G., and J. A. Abad. 2004. "Another Look at the Origin of Late Blight of Potatoes, Tomatoes, and Pear Melon in the Andes of South America." *Plant Disease* 81:682–88.

Abreu de Galindo, J. 1764 (~1600). *The History of the Discovery and Conquest of the Canary Islands*. Trans. G. Glas. London: R. and J. Dodsley and T. Durham.*

Abulafia, D. 2008. *The Discovery of Mankind: Atlantic Encounters in the Age of Columbus*. New Haven, CT: Yale University Press.

Abu-Lughod, J. L. 1991. *Before European Hegemony: The World System A.D. 1250–1350*. New York: Oxford.

Acabado, S. 2009. "A Bayesian Approach to Dating Agricultural Terraces: a Case from the Philippines." *Antiquity* 83:801–14.

Acarete du Biscay. 1698 (1696). *An Account of a Voyage up the River de la Plata and Thence over Land to Peru*. Trans. Anon. London: Samuel Buckley.*

Acemoglu, D., and J. Robinson, Forthcoming. *Why Nations Fail: The Origins of Power, Prosperity and Poverty*. NY: Crown Business.

Acemoglu, D., et al. 2003. "Disease and Development in Historical Perspective." *Journal of the European Economic Association* 1:397–405.

———. 2002. "Reversals of Fortune: Geography and Institutions in the Making of the Modern World Income Distribution." *Quarterly Journal of Economics* 91:1369–1401.

———. 2001. "The Colonial Origins of Comparative Development: An Empirical Investigation." *American Economic Review* 91:1369–1401.

Aceves-Avila, F. J., et al. 1998. "Descriptions of Reiter's Disease in Mexican Medical Texts since 1578." *Journal of Rheumatology* 25:2033–34.

Acosta, J. d. 1894 (1590). *Historia Natural y Moral de las Indias*. 2 vols. Seville: Juan de Leon.*

Adams, H. 1871 (1867). "Captaine John Smith, Sometime Governour in Virginia and Admirall of New England," in C. F. Adams and H. Adams, *Chapters of Erie, and Other Essays*. Boston: James R. Osgood, 192–224.*

Adorno, R., and P. C. Pautz. 1999. *Álvar Núñez Cabeza de Vaca: His Account, His Life and the Expedition of Pánfilo de Narváez*. 3 vols. Lincoln, NE: University of Nebraska Press.

Aguado, P. d. 1919. *Historia de Venezuela*. Madrid: Jaime Ratés, 3 vols.

Aguire Beltrán, G. 1989 (1946). *La Población Negra de México*. Mexico City: Fondo de Cultural Económica.

———. 1944. "The Slave Trade in Mexico." *HAHR* 24:412–31.

Agurto Calvo, S. 1980. *Cuzco. Traza Urbana de la Ciudad Inca*. Proyecto-Per 39, UNESCO. Cuzco: Instituto Nacional de Cultura del Perú.

Albert of Aachen (D'Aix, A.). 1120. "Histoire des Faits et Gestes dans les Régions d'Outre-Mer," in F. M. Guizot, ed., trans., 1824, 21 vols., *Collection des Mémoires Relatifs à l'Histoire de France*. Paris: J.-L.-J. Brière, vol. 20.

Alden, D. 1963. "The Population of Brazil in the Late Eighteenth Century: A Preliminary Study." *HAHR* 43:173–205.

Alegre, F. J. 1842. *Historia de la Compañía de Jesus en Nueva-España*. Mexico City: J. M. Lara, 2 vols.*

Alegría, R. E. 1990. *Juan Garrido, el Conquistador Negro en las Antillas, Florida, México y California, c. 1503–1540*. San Juan: Centro de Estudios Avanzados de Puerto Rico y el Caribe.

Allen, P. G. 2003. *Pocahontas: Medicine Woman, Spy, Entrepreneur, Diplomat*. New York: HarperCollins.

Allison, D. J. 1980. "Christopher Columbus: First Case of Reiter's Disease in the Old World?" *Lancet* 316:1309.

Alonso, L. E., and J. H. Mol., eds. 2007. *A Rapid Biological Assessment of the Lely and Nassau Plateaus, Suriname* (RAP Bulletin of Biological Assessment 43). Arlington, VA: Conservation International.

Altman, I. 2007. "The Revolt of Enriquillo and the Historiography of Early Spanish America." *Americas* 63:587–614.

Altolaguirre y Duvale, A. d. 1914. *Vasco Nuñez de Balboa*. Madrid: Intendencia é Intervención Militares.*

Alyokhin, A., et al. 2008. "Colorado Potato Beetle Resistance to Insecticides." *American Journal of Potato Research* 85:395–413.

Ammirato, S. 1873 (1600–41). *Istorie Fiorentine*. Ed. L. Scarabelli. 3 vols. Turin: Cugini Pomba.*

"Ancient Planters." 1624. "A Brief Declaration of the Plantation of Virginia During the First Twelve Years," in Haile ed. 1998:893–911.

Anderson, R. C. 2006. "Evolution and Origin of the Central Grassland of North America: Climate, Fire and Mammalian Grazers." *Bulletin of the Torrey Botanical Society* 133:626–47, 8–18.

Anderson, R. L. 1999. *Colonization as Exploitation in the Amazon Rain Forest, 1758–1911*. Gainesville: University Press of Florida.

Anderson, R. N. 1996. "The Quilombo of Palmares: A New Overview of a Maroon State in Seventeenth-Century Brazil." *Journal of Latin American Studies* 28:545–66.

Anderson, V. L. 2004. *Creatures of Empire: How Domestic Animals Transformed Early America*. New York: Oxford University Press.

Andrews, G. R. 1980. *The Afro-Argentines in Buenos Aires, Argentina, 1800–1900*. Madison, WI: University of Wisconsin.

Andrivon, D. 1996. "The Origin of *Phytophthora Infestans* Populations Present in Europe in the 1840s: A Critical Review of Historical and Scientific Evidence." *Plant Pathology* 45:1027–35.

Anghiera, P. M. d. (Peter Martyr). 1912 (1530). *De Orbe Novo: The Eight Decades of Peter Martyr D'Anghera*. trans. F. A. MacNutt. 2 vols. New York: G. P. Putnam's Sons.

Anon. 2006. "Quilombo Communities Question the Vale do Rio Doce Actions." *Quilombol@* 16:2. vols.*

———. 1914. "The Potatoes of Parmentier." *The Independent* (New York), 18 May.

———. 1910. "The Secret of London's Rubber Madness." *NYT*, 30 Mar.*

————. 1908 (1635–43). *Actas Antiguas de del Ayuntamiento de la Ciudad de Mexico.* Mexico City: A. Varranza y Comp.

————. 1901. "A Rubber Shipping Port in Brazil." *India Rubber World,* 1 Aug., 327.

————.1890. "How the First Rubber Shoes Found a Market." *India Rubber World and Electrical Trades Review,* 15 Oct., 18.

————.1856. "The Chincha Islands." *Nautical Magazine and Naval Chronicle* 25:181–83.

————.1855. "From the Chincha Islands." *Friends' Intelligencer* 11:110–11.*

————.1854. "The Guano Question." *Farmer's Magazine* 5:117–19.

————.1853. "A Guano Island." *National Magazine* (New York) 3:553–56.*

————.1842a. "Review (Liebig's *Agricultural Chemistry*)." *Farmer's Magazine* 6:1–9.

————.1842b. "Royal Agricultural Society of England. Bristol Meeting." *Farmer's Magazine* 6:115–49.

————.1832. "Importation of Human Bones." *New Monthly Magazine and Literary Journal,* 1 Apr.

————.1829. "Traffic in Human Bones." *Observer* (London), 9 Nov.

————.1824. *An Authentic Copy of the Minutes of Evidence on the Trial of John Smith, a Missionary, in Demerara.* London: Samuel Burton.*

————.1822. "War and Commerce." *Observer* (London), 18 Nov.

————.1678. "Relação das guerras feitas aos Palmares de Pernambuco no tempo do Governador D. Pedro de Almeida de 1675 a 1678." In L. D. Silvas, ed., 1988, *Alguns Documentos para História da Escravidão.* Rio de Janeiro: Ministero da Cultura, 27–44.

————.1635. "A Relation of Maryland." In Hall ed. 1910, 70–112.

————.1603. "Descripción de la Villa y Minas de Potosí." In Espada ed. 1965 (BAE) 183:372–85.

————.1573. "Relacion muy Particular del Cerro y Minas de Potosí y de su Calidad y Labores." In Espada ed. 1965 (BAE) 183:362–71.

Apperson, G. L. 2006 (1914). *The Social History of Smoking.* London: Ballantyne Press.*

Appleby, A. B. 1978. *Famine in Tudor and Stuart England.* Stanford: Stanford University Press.

Aptheker, H. 1939. "Maroons Within the Present Limits of the United States." In Price ed. 2003:151–67.

Araúz Monfante, C. A., and P. Pizzurno Gelós. 1997 (1991). *El Panamá Hispano, 1501–1821.* Panamá: Diario la Prensa, 3rd ed.*

Archdale, J. 1822 (1707). *A New Description of That Fertile and Pleasant Province of Carolina.* Charleston: A. E. Miller.

Archer, G. 1607. "A Relation of the Discovery of Our River from James Fort into the Main." In Haile ed. 1998:101–17.

Arents, G. 1939. "The Seed from Which Virginia Grew." *WMQ* 19:123–29.

Aristotle. 1924. *The Works of Aristotle Translated into English: De Cœlo*. Trans. J. L. Stocks. New York: Oxford University Press.*

Arrom, J. 1983. "Cimarrón: Apuntes sobre sus primeras documentaciones y su probable origen." *Revista Española de Antropología Americana* 13:47–57.

Arzáns de Orsúa y Vela, B. 1965 (1736). *Historia de la Villa Imperial de Potosí*. Ed. L. Hanke and G. Mendoza. 3 vols. Providence: Brown University Press.

Ashe, T. 1917 (1682). *Carolina, or a Description of the Present State of That Country*. Tarrytown, New York: William Abbatt.*

Assadourian, C. S. 1966. *El Tráfico de Esclavos en Córdoba de Angola a Potosí, Siglos XVI–XVII*. Córdoba, Argentina: Universidad Nacional de Córdoba. Cuardernos de Historia 36.

Atwater, H. W. 1910. *Bread and Bread Making*. U.S.D.A. Farmers' Bulletin 389. Washington, DC: Government Printing Office.

Atwell, W. S. 2005. "Another Look at Silver Imports into China, ca. 1635–1644." *JWH* 16:467–90.

———. 2001. "Volcanism and Short-Term Climatic Change in East Asian and World History, c. 1200–1699." *JWH* 12:29–98.

———. 1982. "International Bullion Flows and the Chinese Economy circa 1530–1650." *P&P* 95:68–90.

Aylor, D. E. 2003. "Spread of Plant Disease on a Continental Scale: Role of Aerial Dispersal of Pathogens." *Ecology* 84:1989–97.

Bacchus, M. K. 1980. *Education for Development or Underdevelopment? Guyana's Educational System and Its Implications for the Third World*. Waterloo, Ontario: Wilfrid Laurier University Press.

Bacon, R. 1962 (1267). *The Opus Majus of Roger Bacon*. Trans. R. B. Burke. 2 vols. New York: Russell & Russell.

Baer, K., et al. 2007. "Release of Hepatic *Plasmodium yoelii* Merozoites into the Pulmonary Microvasculature." *PLoS Pathogens* 3:1651–68.*

Bailey, G. A. 1997. "A Mughal Princess in Baroque New Spain." *Anales del Instituto de Investigaciones Estéticas* 71:37–73.

Bailyn, B. 1988 (1986). *The Peopling of British North America: An Introduction*. New York: Vintage.

Balbuena, B. d. 2003 (1604). *La Grandeza Mexicana*. Buenos Aires: Biblioteca Virtual Universal.*

Baldwin, J. T., Jr. 1968. "David B. Riker and *Hevea brasiliensis*." *EB* 22:383–84.

Ballard, C., et al., eds. 2005. *The Sweet Potato in Oceania: A Reappraisal* (Oceania Monograph 56). Sydney: University of Sydney.

Bannister, S., ed. 1859. *The Writings of William Patterson*. London: Judd and Glass.*

Bañuelo y Carrillo, H. 1638. "Bañuelo y Carrillo's Relation." In *B&R* 29:66–85.*

Baquíjano y Carrillo, J. 1793. "Historia del Descubrimiento del Cerro de Potosí." *Mercurio Peruano* 7:25–32 (10 Jan.), 33–40 (13 Jan.), 41–48 (17 Jan.)*

Barbour, P. L. 1963. "Fact and Fiction in Captain John Smith's True Travels." *Bulletin of the New York Public Library* 67:517–28.

Barham, B. L., and O. T. Coomes. 1996. *Prosperity's Promise: The Amazon Rubber Boom and Distorted Economic Development*. Boulder, CO: Westview Press.

Barlow, T. 1681. Brutum Fulmen: *or The Bull of Pope Pius V Concerning the Damnation, Excommunication, and Deposition of Q. Elizabeth*. London: Robert Clavell.*

Barnes, J. K., et al. 1990 (1870). *The Medical and Surgical History of the Civil War*. 15 vols. Wil-mington, NC: Broadfoot Publishing.*

Barrett, W. 1990. "World Bullion Flows, 1450–1800." In J. Tracy, ed., *The Rise of Merchant Empires: Long Distance Trade in the Early Modern World, 1350–1750*. New York: Cambridge University Press, 224–54.

———. 1970. *The Sugar Hacienda of the Marqueses del Valle*. Minneapolis: University of Minnesota Press.

Barth, H. 1857–59. *Travels and Discoveries in North and Central Africa. Being a Journal of an Expedition Undertaken Under the Suspices of H.B.M.'s Government in the Years 1849–1855*. 3 vols. New York: Harper Bros.*

Bartos, P. J. 2000. "The Palacos of Cerro Rico de Potosi, Bolivia: A New Deposit Type." *Economic Geology* 95:645–54.

Batista, D. 1976. *O Complexo da Amazônia*. Rio de Janeiro: Conquista.

Beckles, H. M. 1989. *White Servitude and Black Slavery in Barbados, 1627–1715*. Knoxville: University of Tennessee Press.

Beeton, I. M. 1863 (1861). *The Book of Household Management*. London: Cox and Wyman.

Benedict, C. 2011. *Golden-Silk Smoke: A History of Tobacco in China, 1550–2010*. Berkeley: University of California Press.

Bennett, H. L. 2005 (2003). *Africans in Colonial Mexico: Absolutism, Christianity, and Afro-Creole Consciousness, 1570–1640*. Bloomington: Indiana University Press.

Benzoni, G. 1857 (1572). *History of the New World by Girolamo Benzoni, of Milan*. Trans. W. H. Smyth. London: Hakluyt Society.

Bergh, A. E., ed. 1907. *The Writings of Thomas Jefferson*. 20 vols. Washington, DC: Thomas Jefferson Memorial Association of the United States.

Berkeley, M. J. 1869. "Untitled." *Gardener's Chronicle and Agricultural Gazette*, 6 Nov.

Berlin, I. 2003. *Generations of Captivity: A History of African-American Slaves*. Cambridge, MA: Belknap Press.

Bernáldez, A. 1870 (1513?). *Historia de los Reyes Católicos D. Fernando y Dona Isabel*. 2 vols. Seville: José María Geofrin.

Bernhard, V. 1992. " 'Men, Women and Children' at Jamestown: Population and Gender in Early Virginia, 1607–1610." *JSH* 58:599–618.

Bernstein, W. J. 2008. *A Splendid Exchange: How Trade Shaped the World*. New York: Grove Press.

Bigges, W. 1589. *A Summarie and True Discourse of Sir Frances Drakes West Indian Voyage*. London: Richard Field.*

Bilby, K. 1997. "Swearing by the Past, Swearing to the Future: Sacred Oaths, Alliances, and Treaties among the Guianese and Jamaican Maroons." *Ethnohistory* 44:655–689.

Billings, W. M. 1991. *Jamestown and the Founding of the Nation*. Gettysburg, PA: Thomas Publications.

Billings, W. M., ed. 1975. *The Old Dominion in the Seventeenth Century: A Documentary History of Virginia, 1606–1689*. Chapel Hill: University of North Carolina Press.

Blackburn, R. 1997. *The Making of New World Slavery: From the Baroque to the Modern, 1492–1800*. London: Verso.

Blaer, 1902. "Diario da Viagem do Capitão João Blaer aos Palmares em 1645." *Revista do Instituto Archeologico e Geographico Pernambucano* 10:87–96.

Blair, E. H., and J. A. Robertson, eds., trans. 1903–09. 55 vols. *The Philippine Islands, 1493–1898*. Cleveland: The Arthur H. Clark Co.

Blanchard, P. 1996. "The 'Transitional Man' in Nineteenth-Century Latin America: The Case of Domingo Elias of Peru." *Bulletin of Latin American Research* 15:157–76.

Blanton, W. B. 1973 (1930). *Medicine in Virginia in the Seventeenth Century*. Spartanburg, SC: The Reprint Co.

Blaut, J. M. 1993. *The Colonizer's Model of the World: Geographical Diffusionism and Eurocentric History*. New York: Guilford Press.

Boas, A. J. 1999. *Crusader Archaeology: The Material Culture of the Latin East*. New York: Routledge.

Bohlen, P. J., et al. 2004a. "Ecosystem Consequences of Exotic Earthworm Invasion of North American Temperate Forests." *Ecosystems* 7:1–12.

———. 2004b. "Non-Native Invasive Earthworms as Agents of Change in Northern Temperate Forests." *Frontiers in Ecology and the Environment* 2:427–35.

Bonaparte, N.-J.-C.-P., ed. 1856. *Exposition Universelle de 1855. Rapports du Jury Mixte International*. Paris: Imprimerie Impériale.*

Bond, W. J., et al. 2005. "The Global Distribution of Ecosystems in a World Without Fire." *New Phytologist* 165:525–38.

Borao, J. E. 1998. "Percepciones Chinas sobre los Españoles de Filipinas: La Masacre de 1603." *Revista Española del Pacífico* 8:233–54.*

Borao, J. E., ed. 2001. *Spaniards in Taiwan: 1582–1641*. Taipei: SMC Publishing.*

Bossy, D. I. 2009. "Indian Slavery in Southeastern Indian and British Societies, 1670–1730." In A. Gallay, ed., *Indian Slavery in Colonial America*. Lincoln, NE: University of Nebraska Press.

Bourke, P. M. A. 1993. *"The Visitation of God"? The Potato and the Great Irish Famine*. Dublin: Lilliput Press, 1993.

———. 1964. "Emergence of Potato Blight, 1843–46." *Nature* 203:805–08.

Bouton, C. A. 1993. *The Flour War: Gender, Class and Community in the Late Ancien Régime.* University Park, PA: Pennsylvania State University Press.

Bowser, F. P. 1974. *The African Slave in Colonial Peru, 1570–1650.* Stanford: Stanford University Press.

Boxer, C. R. 2001 (1970). "Plata es Sangre: Sidelights on the Drain of Spanish-American Silver in the Far East, 1550–1700." In Flynn and Giráldez eds. 2001, 165–83.

Boyrie, W. V. 2005. "El Cimarronaje y la Manumisión en el Santo Domingo Colonial. Dos Extremos de una Misma Búsqueda de Libertad." *Clío* (Santo Domingo) 74:65–102.

Brackett, J. K. 2005. "Race and Rulership: Alessandro de' Medici, First Medici Duke of Florence, 1529–1537." In T. F. Earle and K. J. P. Lowe, eds., *Black Africans in Renaissance Europe.* New York: Cambridge University Press, 303–25.

Bradford, W. 1912 (~1650). *History of Plymouth Plantation, 1620–1647.* Boston: Houghton Mifflin.

Brading, D. A. 2009. "Psychomachia Indiana: Catarina de San Juan." *Anuario de la Academia Mexicana de la Historia* 50:1–11.

Brading, D. A., and H. E. Cross. 1972. "Colonial Silver Mining: Mexico and Peru." *HAHR* 52:545–79.

Brain, C. K., and A. Sillen. 1988 "Evidence from the Swartkrans Cave for the Earliest Use of Fire." *Nature* 336:464–66.

Brandes, S. H. 1975. *Migration, Kinship and Community: Tradition and Transition in a Spanish Village.* New York: Academic Press.

Braudel, F. 1981–84 (1979). *Civilization and Capitalism, 15th–18th Century.* Vol. 1: *The Structures of Everyday Life.* Vol. 2: *The Wheels of Commerce.* Vol. 3: *The Perspective of the World.* Trans S. Reynolds. New York: Harper and Row.

Breeden, J. O. 1988. "Disease as a Factor in Southern Distinctiveness." In Savitt and Young eds. 1988:1–28.

Briffa, K. R., et al. 1998. "Influence of Volcanic Eruptions on Northern Hemisphere Summer Temperature over the Past 600 Years." *Nature* 393:450–55.

Bright, C. 1988. *Life out of Bounds: Bioinvasion in a Borderless World.* New York: W. W. Norton.

Brodhead, J. R., ed. 1856–58. *Documents Relative to the Colonial History of the State of New-York.* 2 vols. Albany: Weed, Parsons and Company.*

Brook, T. 2008. *Vermeer's Hat: The Seventeenth Century and the Dawn of the Global World.* New York: Bloomsbury Press.

———. 2004. "Smoking in Imperial China." In S. Gilman and X. Zhou, eds., *Smoke: A Global History of Smoking.* London: Reaktion Books, 84–91.

Brosius, J. P. 1988. "Significance and Social Being in Ifugao Agricultural Production." Ethnology 27:97–110.

Browman, D. 2004. "Tierras comestibles de la cuenca del Titicaca: Geofagia en la prehistoria boliviana." *Estudios Atacameños* 28:133–41.

Brouwer, M. 2005. "Managing Uncertainty Through Profit Sharing Contracts from Medieval Italy to Silicon Valley." *Journal of Management and Governance* 9:237–55.

Brown, A. 1890. *The Genesis of the United States.* 2 vols. Boston: Houghton, Mifflin and Co.

Brown, A. S. 1903 (1889). *Brown's Madeira, Canary Islands and Azores: A Practical and Complete Guide for the Use of Tourists and Invalids.* London: Sampson Low, Marston & Co., 7th ed.*

Brown, K. W. 2001. "Workers' Health and Colonial Mining at Huancavelica, Peru." *Americas* 57:467–96.

Bruhns, K. O. 1981. "Prehispanic Ridged Fields of Central Colombia." *Journal of Field Archaeology* 8:1–8.

Brush, S., et al. 1995. "Potato Diversity in the Andean Center of Crop Domestication." *Conservation Biology* 9:1189–98.

Bullock, W. 1649. *Virginia Impartially Examined, and Left to Publick View, to Be Considered by All Judicious and Honest Men.* London: John Hammond.*

Burns, E. B. 1965. "Manaus, 1910: Portrait of a Boom Town." *Journal of Inter-American Studies* 7:400–21.

Bushnell, A. T. 1994. Situado and Sabana: *Spain's Support System for the Presidio and Mission Provinces of Florida.* Anthropological Papers of the American Museum of Natural History 74. New York: American Museum of Natural History.

Busquets, A. 2006. "Los Frailes de Koxinga." In P. S. G. Aguilar, ed., *La Investigación sobre Asia Pacífico en España.* Colección Española de Investigación sobre Asia Pacífico. Granada: Editorial Universidad de Granada, 393–422.

Butler, C. J., et al. 1998. *Proceedings of the Royal Irish Academy* (Biology and Environment) 96B:123–40 (data at climate.arm.ac.uk/calibrated/rain/).

Byrd, W. 1841 (1728–36). *The Westover Manuscripts: Containing the History of the Dividing Line Betwixt Virginia and North Carolina.* Petersburg, VA: Edmund and Julian C. Ruffin.

"C. T." 1615. *An Aduice How to Plant Tobacco in England and How to Bring It to Colour and Perfection, to Whom It May Be Profitable, and to Whom Harmfull.* London: Nicholas Okes.*

Ca' da Mosto (Cadamosto), A. d. 1895 (~1463). *Relation des Voyages au Côte Occidentale d'Afrique.* Trans. C. Schefer. Paris: Ernest Leroux.*

Cagauan, A. G., and R. C. Joshi. 2002. "Golden Apple Snail *Pomacea spp.* in the Philippines." Paper at 7th ICMAM Special Working Group on Golden Apple Snail, 22 Oct.*

Calloway, C. 2003. *One Vast Winter Count: The Native American West Before Lewis and Clark.* Lincoln: University of Nebraska Press.

Candiani, V. S. 2004. "Draining the Basin of Mexico: Science, Technology and Society, 1608–1808." PhD diss., University of California, Berkely.

Cao, L. (曹玲). 2005. "The Influence of the Introduction of American Cereal

Crops on the Chinese Diet" (美洲粮食作物的传入对我国人民饮食生活的影响). *Agricultural Archaeology* (农业考古) 3:176–81.

Cao, M., et al. 2006. "Tropical Forests of Xishuangbanna, China." *Biotropica* 38:206–09.

Capela, R. A. 1981. "Contribution to the Study of Mosquitoes (Diptera, Culicidae) from the Archipelagos of Madeira and the Salvages." *Arquivos do Museu Bocage* 1:45–66.*

Capoche, L. 1959 (1585). "Relación General de la Villa Imperial de Potosí." In BAE, vol. 122, pp. 69–221.

Carande, R. 1990 (1949). *Carlos V y sus Banqueros*. 3 vols. Barcelona: Crítica, 3rd ed.

Carletti, F. 1701. Ragionamenti de Francesco Carletti sopra le Cose da Lui Vedute ne' suoi Viaggi. Florence: Guiseppe Manni.*

Carroll, P. J. 2001 (1991). *Blacks in Colonial Veracruz: Race, Ethnicity and Regional Development*. Austin: University of Texas Press, 2nd ed.

———. 1977. "Mandinga: The Evolution of a Mexican Runaway Slave Community: 1735–1827." *Comparative Studies in Society and History* 19:488–505.

Carter, L. 1965. *The Diary of Colonel Landon Carter of Sabine Hall, 1752–1778*. Ed. J. P. Greene. 2 vols. Charlottesville: University Press of Virginia.

Carter, R., and K. N. Mendis. 2002. "Evolutionary and Historical Aspects of the Burden of Malaria." *Clinical Microbiology Reviews* 15:564–94.

Casagrande, R. A. 1987. "The Colorado Potato Beetle: 125 Years of Mismanagement." *Bulletin of Entomological Society of America* 33:142–50.

Castellanos, J. d. 1930–32 (1589?). *Obras de Juan de Castellanos*. 2 vols. Caracas: Editorial Sur América.*

Castillo Grajeda, J. d. 1946 (1692). *Compendio de la Vida y Virtudes de la Venerable Catarina de San Juan*. Mexico City: Ediciones Xochitl.

Cates, G. L. 1980. " 'The Seasoning': Disease and Death Among the First Colonists of Georgia." *Georgia Historical Quarterly* 64:146–58.

Centers for Diseases Control and Prevention. 2006. "Locally Acquired Mosquito-Transmitted Malaria: A Guide for Investigations in the United States." *MMWR* 55:1–12.

Central Bureau of Meteorological Sciences (China) (中央气象局气象科学研究院). 1981. *Annual Maps of Precipitation in the Last 500 Years* (中国近五百年旱涝分布图集). Beijing: Cartographic Publishing.

Cervancia, C. R. 2003. "Philippines: Haven for Bees." *Honeybee Science* 24:129–34.

Céspedes del Castillo, G. 1992. *El tabaco en Nueva España: Discurso Leído el Día 10 de Mayo de 1992 en el Acto de su Recepción Pública*. Madrid: Real Academia de la Historia.

Chace, R. E. 1971. "The African Impact on Colonial Argentina." PhD diss., University of California at Santa Barbara.

Champlain, S. d. 1922 (1613). *The Voyages and Explorations of Samuel de Champlain, 1604–1616, Narrated by Himself.* Ed., trans. A. T. Bourne, E. G. Bourne. 2 vols. New York: Allerton Book Co.

Chan, K. S. 2008. "Foreign Trade, Commercial Policies and the Political Economy of the Song and Ming Dynasties of China." *Australian Economic History Review* 48:68–90.

Chanca, D. A. 1494. "Carta a la Ciudad de Sevilla." In C. Jane, ed., trans., 1988 (1930, 1932), *The Four Voyages of Columbus: A History in Eight Documents, Including Five by Christopher Columbus, in the Original Spanish, with English Translations.* New York: Dover, 2nd ed., 20–73.

Chandler, T. 1987. *Four Thousand Years of Urban Growth: An Historical Census.* Lewiston, NY: Edwin Mellen Press, 2nd ed.

Chandless, W. 1866. "Ascent of the River Purûs." *Journal of the Royal Geographical Society of London* 36:86–118.

Chang, P.-T. (張彬村). 2001. "American Silver and Widow Chastity: Cause and Consequence of the Manila Massacre of 1603." In C. Wu, ed. (吳聰敏), *Proceedings of a Symposium in Honor of Prof. Zhang Hanyu* (張漢裕教授紀念研討會論文集). Taipei: National Taiwan University Economics Research Foundation, 205–34.

———. 1990. "Maritime Trade and Local Economy in Late Ming Fukien." In Vermeer ed. 1990, 63–81.

———. 1983. "Chinese Maritime Trade: The Case of Sixteenth-Century Fu-chien (Fukien)." PhD thesis, Princeton University.

Chaplin, J. E. 2001. *Subject Matter: Technology, the Body and Science on the Anglo-American Frontier, 1500–1676.* Cambridge, MA: Harvard University Press.

Chardon, R. 1980. "The Elusive Spanish League: A Problem of Measurement in Sixteenth-Century New Spain." *HAHR* 60:394–402.

Chase, J. M., and T. M. Knight. 2003. "Drought-Induced Mosquito Outbreaks in Wetlands." *Ecology Letters* 6:1017–24.

Chaunu, P. 2001 (1951). "Le Galion de Manille. Grandeur et Décadence d'une Route de la Soie." In Flynn and Giráldez eds. 2001, 187–202.

Chehabi, H. E., and A. Guttmann. 2003. "From Iran to All of Asia: The Origin and Diffusion of Polo." In J. A. Mangan and F. Hong, eds., *Sport in Asian Society: Past and Present.* London: Frank Cass, 384–400.

Chen, C.-N., et al. 1995. "The Sung and Ming Paper Monies: Currency Competition and Currency Bubbles." *Journal of Macroeconomics* 17:273–88.

Chen, D. (陈达生). 1983 (1982). "An Inquiry into the Nature of the Islamic Sects in Quanzhou and the Isbah Disturbance During the Late Yuan Dynasty" (泉州伊斯兰教派与元末亦思巴奚战乱性质试探). In Quanzhou Foreign Maritime Museum and Institute of Quanzhou History, ed., *Symposium on Quanzhou Islam* (泉州伊斯兰教研究论文选). Fuzhou: Fujian People's Publishing House, 53–64.

Chen, G. (陳高傭). 1986 (1939). *Chronological Tables of Natural Disasters in China* (中国历代天灾人祸表). Shanghai: Jinan University Press.

Chen, S. (陈树平). 1980. "Research on the Transmission of Maize and Sweet Potatoes in China" (玉米和番薯在中国传播情况研究). *Social Sciences in China* (中国社会科学) 3:187–204.

Chen, S. (陈世元), ed. 1835? (1768). Record of the Passing-down of the Jin Potato (金薯傳習錄). In Y. Wang (王雲五), and Y. Ji (紀昀) et al., eds., *Continuation of "The Complete Library of the Four Treasuries"* (續修四庫全書). Shanghai: Shanghai Classics Publishing House, vol. 977, pp. 37–79.

———. 1768. "A Factual Account of the Story of Planting Sweet Potatoes in Qinghai, Henan, and Other Provinces" (青豫等省栽種番薯始末寔錄). In Chen ed. 1835, pp. 9b.

Chen, Z. (陳子龍), et al. 1962 (1638) *Collected Writings on Statecraft from the Ming Dynasty* (皇明经世文编). Beijing: Zhonghua Book Company.

Cheng, K.-O. 1990. "Cheng Ch'eng-kung's Maritime Expansion and Early Ch'ing Coastal Prohibition." In Vermeer ed. 1990, 217–44.

Chia, L. 2006. "The Butcher, the Baker, and the Carpenter: Chinese Sojourners in the Spanish Philippines and Their Impact on Southern Fujian (Sixteenth-Eighteenth Centuries)." *Journal of the Economic and Social History of the Orient* 49:509–34.

Childs, S. J. R. 1940. *Malaria and Colonization in the Carolina Low Country, 1526–1696.* Baltimore: Johns Hopkins Press.

Chipman, D. 2005. *Moctezuma's Daughters: Aztec Royalty Under Spanish Rule, 1520–1700.* Austin: University of Texas Press.

Christian, D. 2004. *Maps of Time: An Introduction to Big History.* Berkeley, CA: University of California Press.

Cicogna, E. A. 1855. *Della Vita e Opere di Andrea Navagero: Oratore, Istorico, Poeta Veneziano del Secolo Dicimosesto.* Venice: Andreola.

Cieza de Léon, P. 1864 (1554). *The Travels of Pedro de Cieza de Léon, A.D. 1532–50.* Trans. C. Markham. London: Hakluyt Society.

Cinnirella, F. 2008. "On the Road to Industrialization: Nutritional Status in Saxony, 1690–1850." *Cliometrica* 2:229–57.

Clark, G. 2007. *A Farewell to Alms: A Brief Economic History of the World.* Princeton: Princeton University Press.

Clark, H. R. 1990. "Settlement, Trade and Economy in Fu-chien to the Thirteenth Century." In Vermeer ed. 1990, 35–61.

Clark, J. F. M. 2007. " 'The Eyes of Our Potatoes Are Weeping': The Rise of the Colorado Potato Beetle as an Insect Pest." *Archives of Natural History* 34:109–28.

Clarkson, L. A., and E. M. Crawford. 2001. *Feast and Famine: Food and Nutrition in Ireland, 1500–1920.* Oxford: Oxford University Press.

Clement, C. R. 1999a, b. "1492 and the Loss of Amazonian Crop Genetic Resources." *EB* 53:188–202 (pt. 1), 203–16 (pt. 2).

Clements, J. 2004. *Pirate King: Coxinga and the Fall of the Ming Dynasty.* Thrupp, Stroud (UK): Sutton Publishing Ltd.

Clossey, L. 2006. "Merchants, Migrants, Missionaries and Globalization in the Early-Modern Pacific." *Journal of Global History* 1:41–58.

Clouser, R. A. 1978. *Man's Intervention in the Post-Wisconsin Vegetational Succession of the Great Plains.* Occasional Paper No. 4, Dept. of Geography-Meteorology. Lawrence, KS: University of Kansas.

Coates, A. 1987. *The Commerce in Rubber: The First 250 Years.* Oxford: Oxford University Press.

Cobb, G. B. 1949. "Supply and Transportation for the Potosí Mines, 1545–1640." *HAHR* 29:25–45.

Coclanis, P. A. 1991 (1989). *The Shadow of a Dream: Economic Life and Death in the South Carolina Low Country.* New York: Oxford University Press.

Coelho, P. R. P., and R. A. McGuire. 1997. "African and European Bound Labor in the British New World: The Biological Consequences of Economic Choices." *JEH* 57:83–115.

Cole, J. A. 1985. *The Potosí Mita, 1573–1700: Compulsory Indian Labor in the Andes.* Stanford, CA: Stanford University Press.

Colley, L. 2002. *Captives.* New York: Pantheon.

Colmeiro, M., ed. 1884. *Cortes de los Antiguos Reinos de León y de Castilla.* Madrid: Sucesores de Rivadeneyra.

Colón, C. 1498. Entail of estate, 22 Feb. In Varela and Gil eds. 1992, 353–64.

———. 1493. "Diario del Primer Viaje." In Varela and Gil eds. 1992, 95–217.

Colón, F. 2004 (1571). *The History of the Life and Deeds of Admiral Don Christopher Columbus, Attributed to His Son Ferdinando Colón.* Repertorium Columbianum No. 13. Turnhout, Belgium: Brepols.

Concepcion, R. N., et al. 2005. "Multifunctionality of the Ifugao Rice Terraces in the Philippines." In Indonesian Soil Research Institute. *Multifungsi dan Revitalisasi Pertanian.* Jakarta: ISRI, 51–78.

Condamine, C. M. d. l. 1751a. "Sur la Résine Élastique Nommeé Caoutchouc." *Histoire de l'Académie Royale des Sciences,* 17–22.

———. 1751b. "Mémoire sur une Résine Élastique, Nouvellement Découverte en Cayenne par M. Fresneau." *Mémoires de l'Académie Royale des Sciences,* 319–334.

———. 1745. "Relation Abrégée d'un Voyage Fait dans l'Intérieur de l'Amérique Méridionale." *Mémoires de l'Académie Royale des Sciences,* 391–493.

Cong, C. (陳琮), ed. 1995 (1805). *Tobacco Handbook* (烟草譜). In Y. Wang (王雲五) and Y. Ji (紀昀) et al., eds., *Continuation of "The Complete Library of the Four Treasuries"* (續修四庫全書). Shanghai: Shanghai Classics Publishing House, vol. 1117, pp. 409–81.

Connell, K. H. 1962. "The Potato in Ireland." *P&P* 23:57–71.

Conrad, J. 1999 (1902). *Heart of Darkness.* Calgary, AB: Broadview Press.

Cook, N. D. 2002. "Sickness, Starvation and Death in Early Hispaniola." *JIH* 32:349–86.

———. 1981. *Demographic Collapse, Indian Perú, 1520–1620*. New York: Cambridge University Press.

Cook, N. D., and M. Escobar Gamboa, eds. 1968. *Padrón de los Indios de Lima en 1613*. Lima: Seminario de Historia Rural Andina.

Cope, R. D. 1994. *The Limits of Racial Domination: Plebeian Society in Colonial Mexico City, 1660–1720*. Madison: University of Wisconsin Press.

Cortés, H. 2001 (1971). *Letters from Mexico*. Ed., trans. A. Pagden. New Haven: Yale University Press.

———. 1548. "Testamento de Hernán Cortés." In M. F. Navarrete et al., eds. 1844. *Colección de Documentos Inéditos para la Historia de España*. Madrid: Viuda de Calero, vol. 4, pp. 239–77.

Cortés López, J. L. 1989. *La Esclavitud Negra en la España Peninsular del Siglo XVI*. Salamanca: Ediciones Universidad de Salamanca.

Coslovsky, S. V. 2005. "The Rise and Decline of the Amazonian Rubber Shoe Industry." Unpub. ms. (MIT Working Paper).★

Costa, F. d. A. 1993. *Grande Capital e Agricultura na Amazônia: A Experiência da Ford no Tapajós*. Belém: Universidade Federale do Pará.

Council of the Virginia Company. 1609. "A True and Sincere Declaration of the Purposes and Ends of the Plantation Begun in Virginia." In Haile ed. 1998, 356–71.

Covarrubias y Orozco (Horozco), S. d. 2006 (1611). *Tesoro de la Lengua Castellana o Española*. Ed. I. Arellano and R. Zafra. Madrid: Vervuert.

Cowan, T. W. 1908. *Wax Craft: All About Beeswax: Its History, Production, Adulteration, and Commercial Value*. London: Sampson Low, Marston and Co.★

Cowdrey, A. E. 1996. *This Land, This South: An Environmental History*. Lexington: University Press of Kentucky, 2nd ed.

Craig, A. K., and E. J. Richards. 2003. *Spanish Treasure Bars from New World Shipwrecks*. West Palm Beach, FL: En Rada Publications.

Craig, N. B. 2007 (1907). *Recollections of an Ill-Fated Expedition to the Headwaters of the Madeira River in Brazil*. Whitefish, MT: Kessinger Publishing.★

Crane, E. 1999. *The World History of Beekeeping and Honey Hunting*. New York: Routledge.

Crashaw, W. 1613. "The Epistle Dedicatorie." In A. Whitaker, *Good News from Virginia*. London: William Welpy, 1–23.★

Craton, M. 1984. "The Historical Roots of the Plantation Model." *Slavery and Abolition* 5:190–221.

Craven, A. O. 2006 (1925). *Soil Exhaustion as a Factor in the Agricultural History of Virginia and Maryland, 1606–1860*. Columbia, SC: University of South Carolina Press.

Craven, W. F. 1993 (1957). *The Virginia Company of London*. Baltimore, MD: Genealogical Publishing.

———. 1932. *Dissolution of the Virginia Company: The Failure of a Colonial Experiment*. New York: Oxford University Press.

Crease, R. P. 2003. *The Prism and the Pendulum: The Ten Most Beautiful Experiments in Science*. New York: Random House.

Crespo Rodas, A. 1956. *La Guerra Entre Vicuñas y Vascongados (Potosí, 1622–1625)*. Lima: Tipografía Peruana.

Croft, P. 2003. *King James*. New York: Palgrave Macmillan.

Cronon, W. 1983. *Changes in the Land: Indians, Colonists and Ecology of New England*. New York: Hill and Wang.

Crosby, A. W. 2003 (1973). *The Columbian Exchange: Biological and Cultural Consequences of 1492*. Westport, CT: Praeger.

———. 1995. "The Potato Connection." *Civilization* 2:52–58.

———. 1994. "The Columbian Voyages, the Columbian Exchange, and Their Historians." In A. W. Crosby, *Germs, Seeds & Animals: Studies in Ecological History*. Armonk, NY: M. E. Sharpe.

———. 1986. *Ecological Imperialism: The Biological Expansion of Europe, 900–1900*. New York: Cambridge University Press.

Cross, H. E. 1983. "South American Bullion Production and Export, 1550–1750." In Richards, J. F., ed. *Precious Metals in the Later Medieval and Early Modern Worlds*. Durham NC: Carolina Academic Press, pp. 397–424.

Cruz, R. M. B., et al. 2009. "Mosquito Abundance and Behavior in the Influence Area of the Hydroelectric Complex on the Madeira River, Western Amazon, Brazil." *Transactions of the Royal Society of Tropical Medicine and Hygiene* 103:1174–76.

Cuneo, M. d. 1495. Letter to Gerolamo Annari, 15 Oct. In Symcox ed. 2002, 175–89.

Curtin, P. D. 1995 (1990). *The Rise and Fall of the Plantation Complex: Essays in Atlantic History*. New York: Cambridge University Press.

———. 1989. *Death by Migration: Europe's Encounter with the Tropical World in the Nineteenth Century*. New York: Cambridge University Press.

———. 1968. "Epidemiology and the Slave Trade." *Political Science Quarterly* 83:190–216.

Curtin, P. D., G. S. Brush, and G. W. Fisher, eds. 2001. *Discovering the Chesapeake: The History of an Ecosystem*. Baltimore, MD: Johns Hopkins.

Cushman, G. T. 2003. "The Lords of Guano: Science and the Management of Peru's Marine Environment, 1800–1973." PhD thesis, University of Texas at Austin.

Cuvier. 1861. "Parmentier." In Cuvier, *Recueil des Éloges Historiques Lus dans les Séances Pub-liques de l'Institut de France*. 3 vols. Paris: Firmin Didot Frères, Fils et Cie, 2nd ed., vol. 2, pp. 7–25.

Dale, T. 1615. Letter to "D. M," 18 Jun. In Haile ed. 1998, 841–48.

D'Altroy, T. N. 2002. *The Incas*. Oxford: Blackwell Publishing.

Daly, V. T. 1975. *A Short History of the Guyanese People*. London: Macmillan.

Dampier, W. 1906 (1697–1709). *Dampier's Voyages.* 2 vols. London: E. Grant Richards.*

Daniels, C. 1996. "Agro-Industries: Sugarcane Technology." In Needham et al. 1954–, vol. 6, pt. 3, pp. 1–540.

Darwin, C. R. 1881. *The Formation of Vegetable Mould, Through the Action of Worms, with Observations on Their Habits.* London: John Murray.*

David, J.-P. A. 1875. *Journal de Mon Troisième Voyage d'Exploration dans l'Empire Chinois.* 3 vols. Paris: Librairie Hachette.*

Davids, K. 2006. "River Control and the Evolution of Knowledge: A Comparison between Regions in China and Europe, c. 1400–1850." *Journal of Global History* 1:59–79.

Davidson, A. 1892. *Geographical Pathology: An Inquiry into the Geographical Distribution of Infective and Climatic Diseases.* 2 vols. London: Young J. Pentland.

Davidson, D. M. 1966. "Negro Slave Control and Resistance in Colonial Mexico, 1519–1650." *HAHR* 46:235–53.

Davis, D. B. 2006. *Inhuman Bondage: The Rise and Fall of Slavery in the New World.* Oxford: Oxford University Press.

Davis, M. 2002 (2001). *Late Victorian Holocausts: El Niño Famines and the Making of the Third World.* New York: Verso.

Davis, R. C. 2001. "Counting European Slaves on the Barbary Coast." *P&P* 172:87–124.

Davis, W. 1998. *Shadows in the Sun: Travels to Landscapes of Spirit and Desire.* Washington, DC: Island Press.

———. 1996. *One River: Explorations and Discoveries in the Amazon Rain Forest.* New York: Simon and Schuster.

Deagan, K. A., and J. M. Cruxent. 2002a. *Archaeology at La Isabela: America's First European Town.* New Haven: Yale University Press.

———. 2002b. *Columbus's Outpost Among the Tainos: Spain and America at La Isabela, 1493–1498.* New Haven: Yale University Press.

Dean, W. 1987. *Brazil and the Struggle for Rubber: A Study in Environmental History.* New York: Yale University Press.

DeBary, W. T., et al., eds., trans. 2000. *Sources of Chinese Tradition.* 2 vols. New York: Columbia University Press.

DeBevoise, K. 1995. *Agents of Apocalypse: Epidemic Disease in the Colonial Philippines.* Princeton: Princeton University Press.

De Borja, M. R. 2005. *Basques in the Philippines.* Las Vegas: University of Nevada Press.

Decaisne, M. J. 1846. *Histoire de la Maladie des Pommes de Terre en 1845.* Paris: Librairie Agricole de Dusacq.

De Castro, M. C., and B. H. Singer. 2005. "Was Malaria Present in the Amazon Before the European Conquest? Available Evidence and Future Research Agenda." *Journal of Archaeological Science* 32:337–40.

Decker-Walters, D. 2001. "Diversity in Landraces and Cultivars of Bottle

Gourd (*Lagenaria siceraria;* Cucurbitaceae) as Assessed by Random Amplified Polymorphic DNA." *Genetic Resources and Crop Evolution* 48:369–80.

DeCosta, B. F. 1883. "Ingram's Journey Through North America in 1567–69." *Magazine of American History* 9:168–76.

Defoe, D. 1928 (1724–26). *A Tour through the Whole Island of Great Britain.* 2 vols. New York: Dutton.

Deive, C. E. 1989. *Los Guerrilleros Negros: Esclavos Fugitivos y Cimarrones en Santo Domingo.* Santo Domingo: Fundación Cultural Dominicana.

Del Monte y Tejada, A. 1890. *Historia de Santo Domingo.* 4 vols. Santo Domingo: Garcia Hermanos.

Delaney, C. 2006. "Columbus's Ultimate Goal: Jerusalem." *Comparative Studies in Society and History* 48:260–92.

DeLong, J. B., and Shleifer, A. 1993. "Princes and Merchants: European City Growth before the Industrial Revolution." *Journal of Law and Economics* 36:671–702.

Denevan, W. Forthcoming. "After 1492: The Ecological Rebound." Unpub. ms.
———. 2011. "The 'Pristine Myth' Revisited." *Geographical Review* 101 (10).
———. 2001. *Cultivated Landscapes of Native Amazonia and the Andes.* Oxford: Oxford University Press.
———. 1992a (1976). *The Native Population of the Americas in 1492.* Madison: University of Wisconsin Press, 2nd ed.
———. 1992b. "The Pristine Myth: The Landscape of the Americas in 1492." *AAAG* 82:369–85.

Denevan, W., and B. L. Turner. 1974. "Forms, Functions, and Associations of Raised Fields in the Old World Tropics." *Journal of Tropical Geography* 39:24–33.

Deng, G. 1999. *Maritime Sector, Institutions, and Sea Power of Premodern China.* Westport, CT: Greenwood Press.

Deng, T. (鄧廷祚), et al., eds. 1968 (1762). *Haicheng Gazetteer* (China Gazetteer Collection 92) (海澄縣志, 陳锳等修鄧廷祚等纂, 中國方志叢書). Taipei: Cheng-Wen Publishing Co.

Dennis, P. A., and M. D. Olien. "Kingship Among the Miskito." *American Ethnologist* 11:718–37.

Department of Environment and Natural Resources (Philippines) and World Fish Center. 2006. *Proceedings of the Conference-Workshop on Invasive Alien Species in the Philippines and Their Impacts on Biodiversity.* Quezon City, Philippines, 26–28 July.

De Silva, S. L., and G. A. Zielinski. 1998. "Global Influence of the AD 1600 Eruption of Huaynaputina, Peru." *Nature* 393:455–58.

De Vries, D. P. 1993 (1655). *Voyages from Holland to America, A.D. 1632 to 1644.* Ithaca, NY: Cornell University Library Digital Collections.

De Vries, J. 1984. *European Urbanization, 1500–1800.* Cambridge: Cambridge University Press.

Diamond, J. 1999 (1997). *Guns, Germs, and Steel: The Fates of Human Societies.* New York: W. W. Norton.

Díaz del Castillo, B. 1844 (1568). *The Memoirs of the Conquistador Bernal Diaz del Castillo.* Trans. J. I. Lockhart. 2 vols. London: J. Hatchard and Son.

Dickens, C. 1978 (1861). *Great Expectations.* Oxford: Oxford University Press.

Dieudonné, et al. 1845. "Rapport fait au Conseil Central de Salubrité Publique de Bruxelles sur la Maladie des Pommes de Terre." *Journal de Médicine, de Chirurgie et de Pharmacologie* 3:637–61.*

Diggs, I. 1953. "Zumbi and the Republic of Os Palmares." *Phylon* 14:62–70.

Dillehay, T. D., et al. 2007. "Preceramic Adoption of Peanut, Squash, and Cotton in Northern Peru." *Science* 316:1890–93.

Disney, A. R. 2009. *A History of Portugal and the Portuguese Empire: From Beginnings to 1807.* 2 vols. New York: Cambridge University Press.

Dobson, M. J. 1997. *Contours of Death and Disease in Early Modern England.* Cambridge: Cambridge University Press.

———. 1989. "Mortality Gradients and Disease Exchanges: Comparisons from Old England and Colonial America." *Social History of Medicine* 2:259–97.

———. 1980. " 'Marsh Fever'—The Geography of Malaria in England." *Journal of Historical Geography* 6:357–89.

Dodgen, R. A. 2001. *Controlling the Dragon: Confucian Engineers and the Yellow River in Late Imperial China.* Honolulu: University of Hawai'i Press.

Domar, E. D. 1970. "The Causes of Slavery or Serfdom: A Hypothesis." *JEH* 30:18–32.

Domínguez Ortiz, A. 1952. "La Esclavitud en Castilla durante la Edad Moderna." In A. Domínguez Ortiz, 2003, *La Esclavitud en Castilla durante la Edad Moderna y Otros Estudios de Marginados.* Granada: Editorial Comares, 1–64.

Donegan, K. M. 2002. "Seasons of Misery: Catastrophe and the Writing of Settlement in Colonial America." PhD thesis, Yale University.

Donkin, R. 1979. *Agricultural Terracing in the Aboriginal New World.* Tucson: University of Arizona.

Donnelly, J. S., Jr. 2001. *The Great Irish Potato Famine.* Phoenix Mill, UK: Sutton Publishing.

Doolittle, W. E. 2000. *Cultivated Landscapes of Native North America.* New York: Oxford University Press.

Dowdey, C. 1962. *The Great Plantation: A Profile of Berkeley Hundred and Plantation Virginia from Jamestown to Appomattox.* Charles City, VA: Berkeley Plantation.

Dozer, D. M. 1948. "Matthew Fontaine Maury's Letter of Instruction to William Lewis Herndon." *HAHR* 28:212–28.

Drake, M. 1969. *Population and Society in Norway, 1735–1865.* New York: Cambridge University Press.

Dressing, J. D. 2007. "Social Tensions in Early Seventeenth-Century Potosí." PhD thesis, Tulane University.

Dubisch, J. 1985. "Low Country Fevers: Cultural Adaptations to Malaria in Antebellum South Carolina." *Social Science and Medicine* 21:641–49.

Dubois, L. 2005 (2004). *Avengers of the New World: The Story of the Haitian Revolution*. Cambridge: Harvard University Press.

Duffy, J. 1988. "The Impact of Malaria on the South." In Savitt and Young eds. 1988, 29–54.

———. 1953. *Epidemics in Colonial America*. Baton Rouge: Louisiana State University Press.

Dugard, M. 2006 (2005). *The Last Voyage of Columbus: Being the Epic Tale of Great Captain's Fourth Expedition, Including Accounts of Mutiny, Shipwreck, and Discovery*. New York: Back Bay Books.

Dujardin, J. P., et al. 1987. "Isozyme Evidence of Lack of Speciation Between Wild and Domestic *Triatoma infestans* (Heteroptera: Reduviidae) in Bolivia." *Journal of Medical Entomology* 24:40–45.

Dull, R. A., et al. 2010. "The Columbian Encounter and the Little Ice Age: Abrupt Land Use Change, Fire, and Greenhouse Forcing." *AAAG* 100: 755–71.

Dumlao, A. A. 2009. "Cordillera Heirloom Rice Reaches U.S. Market." *The Philippine Star*, 9 Aug.

Dunn, F. L. 1965. "On the Antiquity of Malaria in the Western Hemisphere." *Human Biology* 37:385–93.

Dúran, D. 1994 (1588?). *The History of the Indies of New Spain*. Trans. D. Heyden. Norman: University of Oklahoma Press.

"E. H." (E. Howes), ed. 1618. *The Abridgement of the English Chronicle, First Collected by M. Iohn Stow*. London: Edward Allde and Nicholas Okes.*

Earle, C. V. 1979. "Environment, Disease, and Mortality in Early Virginia." In Tate and Ammerman 1979, 96–125.

Eastwood, R., et al. 2006. "The Provenance of Old World Swallowtail Butterflies, *Papilio demoleus* (Lepidoptera: Papilionidae), Recently Discovered in the New World." *Annals of the Entomological Society of America* 99:164–68.

Eddy, J. A. 1976. "The Maunder Minimum." *Science* 192:1189–1203.

Ederer, R. J. 1964. *The Evolution of Money*. Washington, DC: Public Affairs Press.

Edwards, C. A. 2004. "The Importance of Earthworms as Key Representatives of Soil Fauna." In C. A. Edwards, ed., *Earthworm Ecology*. Boca Raton, FL: CRC Press, 2nd ed., 3–12.

Edwards, W. H. 1847. *A Voyage Up the River Amazon: Including a Residence at Pará*. New York: D. Appleton.*

Eggimann, G. 1999. *La Population des Villes des Tiers-Mondes, 1500–1950*. Geneva: Libraire Droz.

Ellenbaum, R. 2003 (1995). "Settlement and Society Formation in Crusader Palestine." In T. E. Levy, ed., *The Archaeology of Society in the Holy Land*. New York: Continuum, 502–11.

Elliott, J. H. 2006. *Empires of the Atlantic World: Britain and Spain in America, 1492–1830.* New Haven, CT: Yale University Press.

———. 2002 (1963). *Imperial Spain: 1469–1716.* New York: Penguin Putnam.

Eltis, D. 2002. "Free and Coerced Migrations from the Old World to the New." In: Eltis, D., ed. *Coerced and Free Migration: Global Perspectives.* Stanford: Stanford University Press.

———. 2001. "The Volume and Structure of the Transatlantic Slave Trade: A Reassessment." *WMQ* 58:17–46.

———. 1983. "Free and Coerced Transatlantic Migrations: Some Comparisons." *AHR* 88:251–80.

Eltis, D., and S. L. Engerman. 2000. "The Importance of Slavery and the Slave Trade to Industrializing Britain." *JEH* 60:123–44.

Eltis, D., and D. Richardson. 2010. *Atlas of the Transatlantic Slave Trade.* New Haven: Yale University Press.

Eltis, D., et al. 2009–. *Voyages: The Trans-Atlantic Slave Trade Database,* www.slavevoyages.org.

Elvin, M. 2004. *The Retreat of the Elephants: An Environmental History of China.* New Haven: Yale University Press.

Emmer, P. C. 2006. *The Dutch Slave Trade, 1500–1850.* Trans. C. Emery. New York: Berghahn Books.

Ennes, E. 1948. "The 'Palmares' Republic of Pernambuco, Its Final Destruction, 1697." *The Americas* 5:200–16.

Erickson, C. E. 1994. "Methodological Considerations in the Study of Ancient Andean Field Systems." In N. F. Miller and K. L. Gleason, eds., *The Archaeology of Garden and Field.* Philadelphia: University of Pennsylvania Press, 111–52.

Ernst, A. 1889. "On the Etymology of the Word Tobacco." *AA* 2:133–42.

Espada, M. J. d. l., ed. 1965 (1881–97). *Relaciones Geográficas de Indias: Peru.* BAE, vols. 183–85. Madrid: Atlas.

Esposito, J. J., et al. 2006. "Genome Sequence Diversity and Clues to the Evolution of Variola (Smallpox) Virus." *Science* 313:807–12.

Essig, E. O. 1931. *A History of Entomology.* NY: Macmillan.

Eyzaguirre, P. B. 1989. "The Independence of São Tomé e Principe and Agrarian Reform." *Journal of Modern African Studies* 27:671–78.

Fabié, A. M., ed., trans. 1879. *Viajes for España de Jorge de Einghen, del Baron Leon de Rosmithal de Blatna, de Francesco Guicciardini y de Andrés Navajero.* Libros de Antaño Novamente Dados á Luz por Varios Aficionados, vol. 8. Madrid: Librería de los Bibliófilos.*

Fagan, B. 2002 (2000). *The Little Ice Age: How Climate Made History, 1300–1850.* New York: Basic Books.

Farah, D. 1992. "Light for Columbus Dims: Dominican Project Hits Wall of Resentment." *Washington Post,* 1 Sep., A1.

Faust, E. C., and F. M. Hemphill. 1948. "Malaria Mortality and Morbidity in the United States for the Year 1946." *Journal of the National Malaria Society* 7:285–92.

Faust, F. X., et al. 2006. "Evidence for the Postconquest Demographic Collapse of the Americas in Historical CO_2 Levels." *Earth Interactions* 10:1–11.

Fausz, J. F. 1990. "An 'Abundance of Blood Shed on Both Sides': England's First Indian War, 1609–1614." *VMHB* 98:3–56.

———. 1985. "Patterns of Anglo-Indian Aggression and Accommodation Along the Mid-Atlantic Coast, 1584–1634." In W. W. Fitzhugh, ed., *Cultures in Contact: The European Impact on Native Cultural Institutions in Eastern North America, 1000–1800*. Washington, DC: Smithsonian, 225–68.

———. 1981. "Opechancanough: Indian Resistance Leader." In D. G. Sweet and G. B. Nash, eds. 1981, 21–37.

———. 1977. "The Powhatan Uprising of 1622: A Historical Study of Ethnocentrism and Cultural Conflict." PhD thesis, College of William and Mary.

Feest, C. F. 1973. "Seventeenth Century Virginia Population Estimates." *QBASV* 28:66–79.

Feldman, L. H. 2004. *A Dictionary of Poqom Maya in the Colonial Era*. Thundersley, Essex: Labyrinthos.

Felix, A. F., Jr., ed. 1966. *The Chinese in the Philippines*. Vol. 1: 1570–1770; vol. 2: 1770–1898. Manila: Solidaridad Publishing.

Fernandes, F. T. 2008. "Taxation and Welfare: The Case of Rubber in the Brazilian Amazon." Unpub. ms.*

Fernández-Armesto, F. 2001 (1974). *Columbus: And the Conquest of the Impossible*. London: Phoenix, 2nd ed.

———. 1994 (1987). *Before Columbus: Exploration and Colonisation from the Mediterranean to the Atlantic, 1229–1492*. Philadelphia: University of Pennsylvania Press.

———. 1991. *Columbus*. New York: Oxford University Press.

Findlay, G. M. 1941. "The First Recognized Epidemic of Yellow Fever." *Transactions of the Royal Society of Tropical Medicine and Hygiene* 35:143–54.

Findlay, R., and K. H. O'Rourke. 2007. *Power and Plenty: Trade, War and the World Economy in the Second Millennium*. Princeton: Princeton University Press.

Finlay, R. 1991. "The Treasure Ships of Zheng He: Chinese Maritime Imperialism in the Age of Discovery." *Terrae Incognitae* 23:1–12.

Fischer, D. H. (1991) 1989. *Albion's Seed: Four British Folkways in America*. Oxford: Oxford University Press.

Fisher, J. R. 2003. "Mining and Imperial Trade in Eighteenth-Century Spanish America." In D. O. Flynn et al., eds., *Global Connections and Monetary History, 1470–1800*. Burlington, VT: Ashgate Publishing, 123–32.

Fishwick, M. 1958. "Was John Smith a Liar?" *American Heritage* 9:28–33, 110–11.

Fleming, G. T. 1922. *History of Pittsburgh and Environs, from Prehistoric Days to the Beginning of the American Revolution*. New York: American Historical Society.

Flinn, M. W. 1977. *Scottish Population History: From the 17th Century to the 1930s.* New York: Cambridge University Press.

Flynn, D. O. 1982. "Fiscal Crisis and the Decline of Spain." *JEH* 42:139–47.

Flynn, D. O., and A. Giráldez. 2008. "Born Again: Globalization's Sixteenth-Century Origins." *Pacific Economic Review* 3:359–87.

———. 2002. "Cycles of Silver: Global Economic Unity Through the Mid-Eighteenth Century." *JWH* 13:391–427.

———. 2001 (1995). "Arbitrage, China and World Trade in the Early Modern Period." In Flynn and Giráldez eds. 2001:261–80.

———. 1997. "Introduction." In Flynn and Giráldez, eds. 1997:xv-xl.

———. 1995. "Born with a 'Silver Spoon': The Origin of World Trade in 1571." *JWH* 6:201–21.

Flynn, D. O., and A. Giráldez, eds. 2001. *European Entry into the Pacific: Spain and the Acapulco-Manila Galleons.* Surrey, UK: Ashgate Variorum.

———. 1997. *Metals and Monies in an Emerging Global Economy.* Surrey, UK: Ashgate Variorum.

Fogel, R. W. 2004. *The Escape from Hunger and Premature Death, 1700–2100.* New York: Cambridge University Press.

Food and Agricultural Organization (United Nations). 2003. "Projections of Tobacco Production, Consumption and Trade to the Year 2010." Rome: FAO.*

Forbes, J. D. 2007. *The American Discovery of Europe.* Chicago: University of Illinois Press.

Fortune, A. 1970. "Los Negros Cimarrones en Tierra Firme y su Lucha por la Libertad." *Lotería* (Panamá) 171:17–43 (pt. 1); 172:32–53 (pt. 2); 173:16–40 (pt. 3); 174:46–65 (pt. 4).

———. 1967. "Los Primeros Negros en el Istmo de Panamá." *Lotería* (Panamá) 143:41–64.

Foster, E. A., et al. 1998. "Jefferson Fathered Slave's Last Child." *Nature* 396:27–28.

Foster, G. E. 1876. "The Colorado Potato Beetle." In J. O. Adams et al., *Sixth Annual Report of the Board of Agriculture* (New Hampshire). Concord, NH: Edward A. Jenks, 233–40.*

Fourcroy, A. F., and L. N. Vauquelin. 1806. "Mémoire Sur le Guano, ou Sur l'Engrais Naturel des Îlots de la Mer du Sud, près des Côtes du Pérou." *Mémoires de l'Institut des Sciences, Lettres et Arts: Sciences Mathématiques et Physiques* 6:369–81.*

Frank, A. G. 1998. *ReOrient: Global Economy in the Asian Age.* Berkeley: University of California Press.

Franklin, J. H., and L. Schweninger. 2001. *Runaway Slaves: Rebels on the Plantation.* NY: Oxford University Press.

Frederickson, E. C. 1993. *Bionomics and Control of* Anopheles Albimanus. Washington, DC: Pan American Health Organization.

Freeborn, S. B. 1923. "The Range Overlapping of *Anopheles maculipennis* Meig. and *Anopheles quadrimaculatus* Say." *Bulletin of the Brooklyn Entomological Society* 18:157–58.

Freehafer, V. 1970. "Domingos Jorge Velho, Conqueror of Brazilian Backlands." *The Americas* 27:161–84.

Freitas, D., ed. 2004. *República de Palmares: Pesquisa e Comentários em Documentos Jistóricos do Século XVII*. Maceió: UFAL.

Frelich, L. E., et al. 2006. "Earthworm Invasion into Previously Earthworm-Free Temperate and Boreal Forests." *Biological Invasions* 8:1235–45.

French, H. W. 1992a. "Santo Domingo Journal; For Columbus Lighthouse, a Fete That Fizzled." *NYT*, 25 Sep.

———. 1992b. "Dissent Shadows Pope on His Visit." *NYT*, 14 Oct.

Friedemann, N. S. d. 1993. *La Saga del Negro: Presencia Africana en Colombia*. Bogotá: Pontificia Universidad Javeriana.

Friedman, E. G. 1980. "Christian Captives at 'Hard Labor' in Algiers, 16th–18th Centuries." *International Journal of African Historical Studies* 13:616–32.

Frutuoso, G. 1873 (1591). *As Saudades da Terra Vol. 2: Historia das Ilhas do Porto-Sancto, Madeira, Desertas e Selvaens*. Funchal, Madeira: Typ. Funchalese.*

Fry, W. E., et al. 1993. "Historical and Recent Migrations of Phytophthora Infestans: Chronology, Pathways, and Implications." *Plant Disease* 77:653–61.

Frynas, J. G., et al. 2003. "Business and Politics in São Tomé e Príncipe: From Cocoa Monoculture to Petro-State." *African Affairs* 102:51–80.

Fuente Sanct Angel, R. d. l., and G. Hernández. 1572. "Relación del Cerro de Potosí y su Descubrimiento." In Espada ed. 1965 (BAE) 183:357–61.

Fujita, Y. 2008. "From Swidden to Rubber: Transforming Landscape and Livelihoods in Mountainous Northern Laos." Paper at Social Life of Forests conference, University of Chicago, May 30–31.

Fujita, Y., et al. 2006. "Dynamic Land Use Change in Sing District, Luang Namtha Province, Lao PDR." Vientiane: PRIPODE, Faculty of Forestry, National University of Laos.

Fuller, T. 2008. "A Highway That Binds China and Its Neighbors." *International Herald Tribune*, 30 Mar.*

Fuller, T. 1860 (1662). *The History of the Worthies of England*. 3 vols. London: Thomas Tegg.

Funari, P. P. A. 2003. "Conflict and the Interpretation of Palmares, a Brazilian Runaway Polity." *Historical Archaeology* 37:81–92.

———. 1996. "A Arqueologia de Palmares: Sua Contribuição para o Conhecimento da História da Cultura Afro-Americana." In Reis and Gomes eds. 1996:26–51.

Gade, D. W. 1992. "Landscape, System and Identity in the Post-Conquest Andes." *AAAG* 82:461–77.

———. 1975. *Plants, Man and the Land in the Vilcanota Valley of Peru*. The Hague: Dr. W. Junk.

Gaibrois, M. B. 1950. *Descubrimiento y Fundación del Potosí*. Zaragoza, Spain: Delegación de Distrito de Educación Nacional.

Galeano, E. 1997 (1972). *Open Veins of Latin America: Five Centuries of the Pillage of a Continent*. Boston: Monthly Review Press.

Galenson, D. 1984. "The Rise and Fall of Indentured Servitude in the Americas: An Economic Analysis." *JEH* 44:126.

———. 1982. "The Atlantic Slave Trade and the Barbados Market, 1673–1723." *JEH* 42:491–511.

Gallay, A. 2002. *The Indian Slave Trade: The Rise of the English Empire in the American South, 1670–1717*. New Haven: Yale University Press.

Gallivan, M. D. 2007. "Powhatan's Werowocomoco: Constructing Place, Polity, and Personhood in the Chesapeake, C.E. 1200–C.E. 1609." *AA* 109:85–100.

Gallivan, M. D., et al. 2006. "The Werowocomoco (44GL32) Research Project: Background and 2003 Archaeological Field Season Results." Research Report Series No. 17. Richmond, VA: Department of Historic Resources.*

Galloway, J. H. 2005 (1989). *The Sugar Cane Industry: An Historical Geography from Its Origins to 1914*. New York: Cambridge University Press.

Gallup, J. L., and J. D. Sachs. 2001. "The Economic Burden of Malaria." *AMJTMH* 64 (Supp.):85–96.

Gang, D. 1999. *Maritime Sector, Institutions, and Sea Power of Premodern China*. Contributions in Economics and Economics History 212. Westport, CT: Greenwood Press.

Garcia, D., et al. 2004. "Selection of Rubber Clones for Resistance to South American Leaf Blight and Latex Yield in the Germplasm of the Michelin Plantation of Bahia (Brazil)." *Journal of Rubber Research* 7:188–98.

García-Abásolo, A. 2004. "Relaciones Entre Españoles y Chinos en Filipinas. Siglos XVI y XVII." In Cabrero ed. 2004, vol. 2, 231–48.

García Icazbalceta, J., ed. 1858–66. *Colección de Documentos para la Historia de México*. 2 vols. Mexico City: Antiqua Librería.

Garcilaso de la Vega. 1966 (1609). *Commentaries of the Incas and General History of Peru*. Trans. H. V. Livermore. 2 vols. Austin: University of Texas Press.

Garelik, G. 2002. "Taking the Bite out of Potato Blight." *Science* 298:1702–05.

Garner, R. L. 2007. "Mining Trends in the New World, 1500–1810." Unpub. ms.*

———. 2006. "Where Did All the Silver Go? Bullion Outflows 1570–1650: A Review of the Numbers and the Absence of Numbers." Unpub. ms.*

———. 1988. "Long-Term Silver Mining Trends in Spanish America: A Comparative Analysis of Peru and Mexico." *AHR* 93:898–935.*

Gemelli Careri, G.-F. 1699–1700. *Giro del Mondo*. 6 vols. Milan: Giuseppe Roselli.

Gemery, H. A. 1980. "Emigration from the British Isles to the New World, 1630–1700: Inferences from Colonial Populations." *Research in Economic History* 5:179–231.

————. and Hogendorn, J. S. 1979. "Comparative Disadvantage: The Case of Sugar Cultivation in West Africa." *Journal of Interdisciplinary History* 9:429–49.

Gerard, J. 1633 (1597). *The Herball or Generall Historie of Plantes*. Rev. T. Johnson. London: Adam Islip, Joice Norton and Richard Whitakers.

Gerhard, P. 1978. "A Black Conquistador in Mexico." *HAHR* 48:451–59.

Gibson, A. J. S., and T. C. Smout. 1995. *Prices, Food, and Wages in Scotland, 1550–1780*. New York: Cambridge University Press.

Gibson, C. 1950. "The Identity of Diego Muñoz Camargo." *HAHR* 30:195–208.

Giddings, J. R. 1858. *The Exiles of Florida; or, The Crimes Committed by Our Government Against the Maroons*. Columbus, OH: Follett, Foster and Company.*

Gilmore, H. R. 1955. "Malaria at Washington Barracks and Fort Myer: Survey by Walter Reed." *Bulletin of the History of Medicine* 29:346–51.

Gleave, J. L. 1952. "The Design of the Memorial Lighthouse." In Comite Ejecutivo Permanente del Faro a Colón, *El Faro a Colón*. Ciudad Trujillo (Santo Domingo): Impresora Dominicana, 11–22.

Glover, L. and D. B. Smith. 2008. *The Shipwreck That Saved Jamestown: The Sea Venture Castaways and the Fate of America*. New York: Henry Holt and Company.

Goldstone, J. A. 2000. "The Rise of the West—or Not? A Revision to Socio-Economic History." *Sociological Theory* 18:175–94.

Gomes, F. d. S. 2005. *Palmares*. São Paulo: Editora Contexta.

————. 2003. "Other Black Atlantic Borders: Escape Routes, 'Mocambos,' and Fears of Sedition in Brazil and French Guiana (Eighteenth to Nineteenth Centuries)." *New West Indian Guide* (Leiden) 77:253–87.

Gómez-Alpizar, L., et al. 2007. "An Andean Origin of *Phytophthora infestans* Inferred from Mitochondrial and Nuclear Gene Genealogies." *PNAS* 104:3306–11.

González, R. 2007. "The Columbus Lighthouse Competition: Revisiting Pan-American Architecture's Forgotten Memorial." *ARQ* (Santiago, Chile) 67:80–87.*

González-Cerón, L., et al. 2003. "Bacteria in Midguts of Field-Collected *Anopheles albimanus* Block *Plasmodium vivax* Sporogonic Development." *Journal of Medical Entomology* 40:371–74.

Goodman, D. 2002 (1997). *Spanish Naval Power, 1589–1665: Reconstruction and Defeat*. New York: Cambridge University Press.

Goodman, J. 2009. *The Devil and Mr. Casement: One Man's Battle for Human Rights in South America's Heart of Darkness*. New York: Farrar, Straus and Giroux.

Goodrich, L. C. 1938. "Early Prohibitions of Tobacco in China and Manchuria." *Journal of the American Oriental Society* 58:648–57.

————. 1937. "The Introduction of the Sweet Potato into China." *China Journal* 27:206–08.

Goodwin, M. H., and G. T. Love. 1957. "Factors Influencing Variations in Populations of *Anopheles quadrimaculatus* in Southwestern Georgia." *Ecology* 38:561–70.

Goodwin, R. 2008. *Crossing the Continent, 1527–1540: The Story of the First African-American Explorer of the American South.* New York: HarperCollins.

Goodwin, S. B., et al. 1994. "Panglobal Distribution of a Single Clonal Lineage of the Irish Potato Famine Fungus." *PNAS* 91:11591–95.

Goodyear, C. 1855. *Gum-elastic and Its Varieties: With a Detailed Account of Its Applications and Uses, and of the Discovery of Vulcanization.* 2 vols. New Haven: C. Goodyear.

Goodyear, J. D. 1978. "The Sugar Connection: A New Perspective on the History of Yellow Fever." *Bulletin of the History of Medicine* 52:5–21.

Gough, J. 1805. "A Description of a Property of Caoutchouc, or Indian Rubber." *Memoirs of the Literary and Philosophical Society of Manchester* 1:288–95.*

Gould, A. B. 1984. *Nueva Lista Documentada de los Tripulantes de Colón en 1492.* Madrid: Academia de la Historia.

Gourou, P. 1963. "Une Île Équatoriale: Sâo Tomé de F. Tenreiro." *Annales de Géographie* 72:360–64.

Gradie, C. M. 1993. "The Powhatans in the Context of the Spanish Empire." In Rountree ed. 1993, 154–72.

Grandin, G. 2009. *Fordlandia: The Rise and Fall of Henry Ford's Forgotten Jungle City.* New York: Metropolitan Books.

Grant, V. 1949. "Arthur Dobbs (1750) and the Discovery of the Pollination of Flowers by Insects." *Bulletin of the Torrey Botanical Club* 76:217–19.

Gray, L. C. 1927. "The Market Surplus Problems of Colonial Tobacco." *WMQ* 7:231–45.

Greenfield, G. M. 2001. *The Realities of Images: Imperial Brazil and the Great Drought.* Philadelphia: American Philosophical Society.

Greenfield, S. M. 1977. "Madeira and the Beginnings of New World Sugar Cane Cultivation and Plantation Slavery: A Study in Institution Building." *Annals of the New York Academy of Sciences* 292:536–52.

Gress, D. 1998. *From Plato to NATO: The Idea of the West and Its Opponents.* New York: Free Press.

Grieco, J. P., et al. 2005. "Comparative Susceptibility of Three Species of Anopheles from Belize, Central America to *Plasmodium falciparum* (NF-54)." *Journal American Mosquito Control Association* 21:279–90.

Grun, P. 1990. "The Evolution of Cultivated Potatoes." *EB* 44:39–55.

Grünwald, N. J., and W. G. Flier. 2003. "The Biology of *Phytophthora infestans* at Its Center of Origin." *Annual Review of Phytopathology* 43:171–90.

Guardiola-Claramonte, M., et al. 2008. "Local Hydrologic Effects of Introducing Non-native Vegetation in a Tropical Catchment." *Ecohydrology* 1:13–22.

Guasco, M. J. 2000. "Encounters, Identities and Human Bondage: The Foundations of Racial Slavery in the Anglo-Atlantic World." PhD diss., William and Mary.

Guerra, C. A., et al. 2008. "The Limits and Intensity of *Plasmodium falciparum* Transmission: Implications for Malaria Control and Elimination Worldwide." *PLoS Medicine* 5:e38.*

Guerrero, K. A., et al. 2004. "First New World Documentation of an Old World Citrus Pest, the Lime Swallowtail *Papilio demoleus* (Lepidoptera: Papilionidae), in the Dominican Republic (Hispaniola)." *American Entomologist* 50:227–29.

Guerrero, M. C. 1966. "The Chinese in the Philippines, 1570–1770." In Felix ed. 1966, vol. 1, 15–39.

Guinea, M. 2006. "El Uso de Tierras Comestibles por los Pueblos Costeros del Periodo de Integración en los Andes Septentrionales." *Bulletin de l'Institut Français d'Études Andines* 35:321–34.

Guitar, L. 2006. "Boiling It Down: Slavery on the First Commercial Sugarcane Ingenios in the Americas (Hispaniola, 1530–45)." In Landers and Robinson ed. 2006, 39–82.

———. 1999. "No More Negotiation: Slavery and the Destabilization of Colonial Hispaniola's Encomienda System." *Revista Interamericana* 29:n.p.*

———. 1998. "Cultural Genesis: Relationships Among Indians, Africans and Spaniards in Rural Hispaniola, First Half of the Sixteenth Century." PhD thesis, Vanderbilt University.

Guo, L. (郭立珍). 2002. "The Influences of the Rapid Development of Trade Between China and the Philippines in the Mid-late Ming Dynasty on the Society of Overseas and Ethnic Chinese in Manila" (明朝中后期中菲贸易的迅速发展对马尼拉华侨华人社会的影响). *Journal of Luoyang Normal University* (洛阳师范学院学报) 6:95–97.

Hackett, L. W., and A. Missiroli. 1935. "The Varieties of *Anopheles maculipennis* and Their Relation to the Distribution of Malaria in Europe." *Rivista di Malariologia* 14:3–67.

Haile, E. W., ed. 1998. *Jamestown Narratives: Eyewitness Accounts of the Virginia Colony: The First Decade: 1607–1617*. Champlain, VA: RoundHouse.

Hakluyt, R. 1993 (1584). *A Discourse of Western Planting*, ed. D. B. Quinn and A. M. Quinn. London: Hakluyt Society.

Hall, C. C. 1910. *Narratives of Early Maryland, 1633–1684*. New York: Charles Scribner's Sons.*

Hall, J. A. 1990 (1985). *Powers and Liberties: The Causes and Consequences of the Rise of the West*. Los Angeles: University of California Press.

Hämäläinen, P. 2008. *The Comanche Empire*. New Haven: Yale University Press.

Hamilton, E. J. 1934. *American Treasure and the Price Revolution in Spain, 1501–1650*. Cambridge, MA: Harvard University Press.

Hammett, J. E. 1992. "The Shapes of Adaptation: Historical Ecology of

Anthropogenic Landscapes in the Southeastern United States." *Landscape Ecology* 7:121–35.

Hamor, R. 1615. "A True Discourse of the Present Estate of Virginia." In Haile ed. 1998, 795–841.

Hancock, T. 1857. *Personal Narrative of the Origin and Progress of the Caoutchouc or India-Rubber Manufacture of England.* London: Longman, Brown, Green, Longmans, and Roberts.

Hanke, L. 1994 (1974). *All Mankind Is One: A Study of the Disputation Between Bartolomé de Las Casas and Juan Giné de Sepúlveda on the Religious and Intellectual Capacity of the American Indian.* De Kalb, IL: Northern Illinois University Press.

Hardenburg, W. E. 1913. *The Putumayo: The Devil's Paradise.* London: T. Fisher Unwin.*

Hare, J. D. 1990. "Ecology and Management of the Colorado Potato Beetle." *Annual Review of Entomology* 35:81–100.

Hariot, T. 1588. *A Briefe and True Report of the New Found Land of Virginia.* London: R. Robinson.*

Harms, R. 2002. *The Diligent: A Voyage Through the Worlds of the Slave Trade.* NY: Basic Books.

Harrington, K. 2010. "Rice Riches." *The Spokesman-Review* (Spokane, WA), 17 Mar.

Hashaw, T. 2007. *The Birth of Black America: The First African Americans and the Pursuit of Freedom at Jamestown.* New York: Carroll and Graf.

Hassig, R. 2006 (1994). *Mexico and the Spanish Conquest.* Norman: University of Oklahoma Press, 2nd ed.

Hasteú, E. 1797–1801. *The History and Topographical Survey of the County of Kent.* 12 vols. Canterbury, UK: W. Bristow.*

Hatfield, A. L. 2003. "Spanish Colonization Literature, Powhatan Geographies, and English Perceptions of Tsenacommacah/Virginia." *JSH* 49:245–82.

Hawkes, J. G. 1994. "Origins of Cultivated Potatoes and Species Relationships." In J. E. Bradshaw and G. R. Mackay, eds., *Potato Genetics.* Wallingford, UK: CAB, 3–42.

——. 1990. *The Potato: Evolution, Biodiversity and Genetic Resources.* London: Belhaven Press.

Hawkes, J. G., and J. Francisco-Ortega. 1993. "The Early History of the Potato in Europe." *Euphytica* 70:1–7.

Hays, W. S. T., and S. Conant. 2007. "Biology and Impacts of Pacific Island Invasive Species. 1. A Worldwide Review of Effects of the Small Indian Mongoose, *Herpestes javanicus* (Carnivora: Herpestidae)." *Pacific Science* 61:3–16.

Hazlewood, N. 2005 (2004). *The Queen's Slave Trader: John Hawkyns, Elizabeth I, and the Trafficking in Human Souls.* New York: Harper.

Hebb, D. D. 1994. *Piracy and the English Government, 1616–1642*. Aldershot: Scholar Press.

Hecht, I. W. D. 1969. "The Virginia Colony, 1607–40: A Study in Frontier Growth." PhD thesis, University of Washington.

Hecht, S. B. Forthcoming. *The Scramble for the Amazon: Imperial Contests and the Lost Paradise of Euclides da Cunha*. Chicago: University of Chicago Press.

Hecht, S. B., and C. C. Mann. 2008. "How Brazil Outfarmed the American Farmer." *Fortune*, Jan. 10.

Hegerl, G. C., et al. 2007. "Understanding and Attributing Climate Change." In Solomon 2007, 663–745.

Hemenway, T. 2002. "Learning from the Ecological Engineers: Watershed Wisdom of the Beaver." *Permaculture Activist* 47.★

Hemming, J. 2008. *Tree of Rivers: The Story of the Amazon*. New York: Thames and Hudson.

———. 2004a (1995). *Red Gold: The Conquest of the Brazilian Indians*. London: Pan Books, 2nd ed.

———. 2004b (1995). *Amazon Frontier: The Defeat of the Brazilian Indians*. London: Pan Books, 2nd ed.

———. 1993 (1970). *The Conquest of the Incas*. London: Pan Books, 3rd ed.

Hendrix, P. F., and P. J. Bohlen. 2002. "Exotic Earthworm Invasions in North America: Ecological and Policy Implications." *Bioscience* 52:801–11.

Hendrix, P. F., et al. 2008. "Pandora's Box Contained Bait: The Global Problem of Introduced Earthworms." *Annual Review of Ecology, Evolution and Systematics* 39:593–613.

Heneghan, L., et al. 2007. "Interactions of an Introduced Shrub and Introduced Earthworms in an Illinois Urban Woodland: Impact on Leaf Litter Decomposition." *Pedobiologia* 50:543–51.

Henige, D. 1998. *Numbers from Nowhere: The American Indian Contact Population Debate*. Norman, OK: University of Oklahoma Press.

———. 1986. "When Did Smallpox Reach the New World (and Why Does It Matter)?," in P. E. Lovejoy, ed., *Africans in Bondage: Studies in Slavery and the Slave Trade*. Madison, WI: University of Wisconsin African Studies Program, 11–26.

———. 1978. "On the Contact Population of Hispaniola: History as Higher Mathematics." *HAHR* 58:217–37.

Hernández, J. 2004 (1872–79). *El Guacho Martín Fierro y la Vuelta de Martín Fierro*. Buenos Aires: Stockcero.

Herrera y Tordesillas, A. d. 1601–15. *Historia General de los Hechos de los Castellanos en las Islas i Tierra Firme del Mar Océano*. 4 vols. Madrid: Imprenta Real.★

Heywood, L. M., and Thornton, J. K. 2007. *Central Africans, Atlantic Creoles, and the Foundation of the Americas, 1585–1660*. New York: Cambridge University Press.

Hirsch, A. 1883–86. *Handbook of Geographical and Historical Pathology* (trans. C. Creighton). London: New Sydenham Society, 3 vols.*

Hirschberg, J. 1979. "Social Experiment in New Spain: A Prosopographical Study of the Early Settlement at Pubela de los Angeles, 1531–1534." *HAHR* 59:1–33.

Hitchcock, A. R. 1936. *Manual of the Grasses of the West Indies.* U.S. Dept. of Agriculture Misc. Pub. 243. Washington, DC: Government Printing Office.

Ho, P.-T. (He, B.). 1959. *Studies on the Population of China, 1368–1953.* Cambridge MA: Harvard University Press.

———. 1956. "Early-Ripening Rice in Chinese History." *EHR* 9:200–18.

———. 1955. "The Introduction of American Food Plants into China." *AA* 57:191–201.

Hoberman, L. S. 1980. "Technological Change in a Traditional Society: The Case of the Desagüe in Colonial Mexico." *Technology and Culture* 21:386–407.

Hodge, W. H. 1947. "The Plant Resources of Peru." *EB* 1:119–36.

Hoffman, B. G., and J. Clayton. 1964. "John Clayton's 1687 Account of the Medicinal Practices of the Virginia Indians." *Ethnohistory* 11:1–40.

Hoffman, O. 2006. "Negros y Afromestizos En México: Viejas y Nuevas Lecturas de un Mundo Olvidado." *Revista Mexicana de Sociología* 68:103–35.

Hoffman, P. E. 2004 (1990). *A New Andalucia and a Way to the Orient: The American Southeast During the Sixteenth Century.* Baton Rouge, LA: LSU Press.

Hohl, H. R., and K. Iselin. 1984. "Strains of *Phytophthora infestans* with A2 Mating Type Behavior." *Transactions of the British Mycological Society* 83: 529–30.

Holder, P. 1974 (1970). *The Hoe and the Horse on the Plains: A Study of Cultural Development Among North American Indians.* Lincoln, NE: Bison Press.

Hollett, D. 1999. *Passage from India to El Dorado: Guyana and the Great Migration.* Cranbury, NJ: Associated University Presses.

Homer, S., and R. E. Sylla. 2005. *A History of Interest Rates.* Hoboken, NJ: John Wiley & Sons, 4th ed.

Honigsbaum, M. 2001. *The Fever Trail: In Search of the Cure for Malaria.* New York: Farrar Straus Giroux.

Hong, Y., and S. W. James. 2008. "Nine New Species of Earthworms (Oligochaeta: Megascolecidae) of the Banaue Rice Terraces, Philippines." *Revue Suisse de Zoologie* 115:341–54.

Horn, J. 2010. *A Kingdom Strange: The Brief and Tragic History of the Lost Colony of Roanoke.* New York: Basic Books.

———. 2005. *A Land as God Made It: Jamestown and the Birth of America.* New York: Basic Books.

Horn, J., ed. 2007. *Captain John Smith: Writings with Other Narratives of Roanoke, Jamestown, and the First English Settlement of America.* New York: Library of America.

Horn, J., and P. D. Morgan. 2005. "Settlers and Slaves: European and African

Migrations to Early Modern British America." In E. Mancke and C. Shammas, eds., *The Creation of the British Atlantic World*. Baltimore: Johns Hopkins Press, 19–44.

Horsley, M. W. 1950. "Sangley: The Formation of Anti-Chinese Feeling in the Philippines—A Cultural Study in the Stereotypes of Prejudice." PhD thesis, Columbia University.

Horwitz, T. 2008. *A Voyage Long and Strange: Rediscovering the New World*. New York: Henry Holt.

Hosler, D., et al. 1999. "Prehistoric Polymers: Rubber Processing in Ancient Mesoamerica." *Science* 284:1988–91.

Hourani, G. F. 1995 (1951). *Arab Seafaring in the Indian Ocean in Ancient and Early Medieval Times*. Princeton, NJ: Princeton University Press, 2nd ed.

House of Commons (Great Britain). 1846. "Post-Office./Shipping." In *Accounts and Papers: Twenty-eight Volumes*, vol. 21. In *Parliamentary Papers*, vol. 45. Session 22 Jan.–28 Aug. 1846. London: House of Commons.

Howard, L. O. 1897. "Danger of Importing Insect Pests." In G. M. Hill, ed., *Yearbook of the United States Department of Agriculture*. Washington, DC: Government Printing Office, 529–52.

Hsiao, T. H. 1985. "Ecophysiological and Genetic Aspects of Geographic Variation of the Colorado Potato Beetle." In D. N. Ferro and R. H. Voss, eds., *Proceedings of the Symposium on the Colorado Potato Beetle, 17th International Congress of Entomology*. Amherst, MA: University of Massachusetts, 63–77.

Hu, Z. (胡宗憲). 2006 (1562). *A Maritime Survey: Collected Plans* (籌海圖編). In Y. Ji (紀昀) and X. Lu (陸錫熊) et al., eds., *Wenyuan Publishing House Internet Edition of the Complete Library of The Four Treasuries* (文淵閣四庫全書內網聯版). Hong Kong: Heritage Publishing Ltd.*

Huamán, Z., and D. M. Spooner. 2002. "Reclassification of Landrace Populations of Cultivated Potatoes (Solanum sect. Petota)." *American Journal of Botany* 89:947–65.

Huang, R. 1981. *1587: A Year of No Significance*. New Haven, CT: Yale University Press.

Hudson, C., and C. C. Tesser, eds. 1994. *The Forgotten Centuries: Indians and Europeans in the American South, 1521–1704*. Athens, GA: University of Georgia Press.

Huldén, L., et al. 2008. "Natural Relapses in *vivax* Malaria Induced by Anopheles Mosquitoes." *Malaria Journal* 7:64–75.*

Humboldt, A. v. 1822 (1811). *Political Essay on the Kingdom of New Spain*. Trans. J. Black. 4 vols. London: Longman, Hurst, Rees, Orme, and Brown, 3rd ed.*

Hung, H.-F. 2007. "Changes and Continuities in the Political Ecology of Popular Protest: Mid-Qing China and Contemporary Resistance." *China Information* 21:299–329.

———. 2005. "Contentious Peasants, Paternalist State, and Arrested Capitalism in China's Long Eighteenth Century." In C. Chase-Dunn and E. N.

Anderson, eds., *The Historical Evolution of World-Systems*. New York: Palgrave Macmillan, 155–73.

Hunter, M., and A. Gregory, eds. 1988. *An Astrological Diary of the Seventeenth Century: Samuel Jeake of Rye, 1652–1699*. Oxford: Oxford University Press.

Huntington, E. 1915. *Civilization and Climate*. New Haven: Yale University Press.*

Hunwick, J. O. 1999. *Timbuktu and the Songhay Empire: Al-Sa'dī's Ta'rīkh al-sūdān down to 1613 and Other Contemporary Documents*. Leiden: E. J. Brill.

Hutchinson, G. E. 1950. *The Biogeochemistry of Vertebrate Excretion*. Bulletin of the American Museum of Natural History 96. New York: American Museum of Natural History.

Hutchinson, R. A., and S. W. Lindsay. 2006. "Malaria and Deaths in the English Marshes." *Lancet* 367:1947–51.

Ibn Battuta. 1853–58 (1355). *Voyages d'Ibn Batoutah*. Trans. C. Defrémery et B.R. Sanguinetti. 4 vols. Paris: Imprimerie Impériale.*

Icaza, F. d. A. d. 1923. *Diccionario Autobiografico de Conquistadores y Pobladores de Nueva España*. 2 vols. Madrid: El Adelantado de Segovio.

Ilahiane, H. 2000. "Estevan de Dorantes, the Moor or the Slave? The Other Moroccan Explorer of New Spain." *Journal of North African Studies* 5:1–14.

Inagaki, H., and K. Kegasawa. 1973. "Discovery of the Potato Cyst Nematode, *Heterodera rostochiensis* Wollenweber, 1923, (Tylenchida: Heteroderidae) from Peru Guano." *Applied Entomology and Zoology* 8:97–102.

Ingram, D. 1883 (1582). "Relation of David Ingram." *Magazine of American History* 9:200–08.

International Labour Office. 1943. *Intergovernmental Commodity Control Agreements*. Montreal: International Labour Organization (League of Nations).*

Irvine, J. E. 1999. "*Saccharum* Species as Horticultural Classes." *Journal of Theoretical and Applied Genetics* 98:186–94.

Jackson, J. 2008. *The Thief at the End of the World: Rubber, Power, and the Seeds of Empire*. New York: Viking.

Jacobs, M. M. J., et al. 2008. "AFLP Analysis Reveals a Lack of Phylogenetic Structure Within Solanum Section *Petota*." *BMC Evolutionary Biology* 8:145.*

Jacobson, J. W., and T. H. Hsiao. 1983. "Isozyme Variation Between Geographic Populations of the Colorado Potato Beetle, Leptinotarsa declineata (Coleoptera: Chrysomelidae)." *Annals of the Entomological Society of America* 76:162–66.

Jaén Suárez, O. 1980. "Cinco Siglos de Poblamiento en el Istmo de Panamá." *Lotería* (Panamá) 291:75–94.

James I. 1604. "A Counterblaste to Tobacco." In E. Arber, ed., 1869, *English Reprints*, vol. 8. London:S. I.*

James, J. 1854. *The Treasury of Medicine; or Every One's Medical Guide*. London: Geo. Routledge.*

James, S. W. 1995. "Systematics, Biogeography, and Ecology of Nearctic Earthworms from Eastern, Central, Southern, and Southwestern United States." In P. F. Hendrix, ed., *Earthworm Ecology and Biogeography in North America*. Boca Raton, FL: Lewis, 29–52.

Jansen, E., et al. 2007. "Palaeoclimate." In Solomon 2007, 433–97.

Jefferson, T. 1993 (1781–82). *Notes on the State of Virginia*. Charlottesville, VA: University of Virginia Library Electronic Text Center.*

Jiang, M. and S. Wang. (蒋慕东, 王思明). 2006. "The Spread of Tobacco and Its Influence in China" (烟草在中国的传播及其影响). *AHC* 25:30–41.

Jin, Y. 1982. "The Qur'ān in China." *Contributions to Asian Studies* 17:95–101.

Johns, T. 1986. "Detoxification Function of Geophagy and Domestication of the Potato." *Journal of Chemical Ecology* 12:635–46.

Johnson, C. W. 1843. "On Guano." *Farmer's Magazine* 7:170–74.

Johnson, E. 2005 (1654). *Johnson's Wonder-Working Providence of Sions Saviour in New England*. Boston: Adamant Media.*

Johnson, H. C. S. 1998. "Adjunctive Use of a Chinese Herbal Medicine in the Non-Surgical Mechanical Treatment of Advanced Periodontal Disease on Smokers: A Randomized Clinical Trial." MDS thesis, University of Hong Kong.*

Johnson, H. G. 1893. "The Early American Trade in Pará Rubber." *India Rubber World* 9:41–42.

Johnson, M. 1970. "The Cowrie Currencies of West Africa." *Journal of African History* 9:17–49 (pt. 1), 331–53 (pt. 2).

Johnson, R.(?) 1897 (1609). "*Nova Britannia:* Offering Most Excellent Fruits by Planting in Virginia, Exciting All Such as Be Well Affected to Further the Same." *American Colonial Tracts Monthly* 6.

Johnson, W. H. 1909. *The Cultivation and Preparation of Para Rubber*. London: Crosby Lockwood and Son.

Jones, C. L. 1906. "The Spanish Administration of Philippine Commerce." *Proceedings of the American Political Science Association* 3:180–93.

Jones, E. L. 2003. *The European Miracle: Environments, Economies, and Geopolitics in the History of Europe and Asia*. New York: Cambridge University Press, 3rd ed.

Jones, H. 1724. *The Present State of Virginia*. London: J. Clarke.*

Jones, L. R., et al. 1914. *Investigations of the Potato Fungus Phytophthora Infestans* (Vermont Agricultural Station Bulletin 168). Burlington, VT: Free Press.

Jones, S. M. 1971. "Hung Liang-Chi (1746–1809): The Perception and Articulation of Political Problems in Late Eighteenth Century China." PhD thesis, Stanford University.

Joshi, R. C. 2005. "Managing Invasive Alien Mollusc Species in Rice." *International Rice Research Notes* 30:5–13.

Judelson, H. S., and F. A. Blanco. 2005. "The Spores of Phytophthora: Weapons of the Plant Destroyer." *Nature Reviews Microbiology* 3:47–58.

Julien, C. J. 1985. "Guano and Resource Control in Sixteenth-Century Arequipa." In Masuda, S., et al. eds. *Andean Ecology and Civilization: An Interdisciplinary Perspective on Andean Ecological Complementarity.* Tokyo: University of Tokyo Press, pp. 185–231.

Kalm, P. 1773 (1748). *Travels into North America; Containing Its Natural History, and a Circumstantial Account of Its Plantations and Agriculture in General.* Trans. J. R. Forster. 2 vols. London: T. Lowndes, 2nd ed.*

Kamen, H. 2005. *Spain, 1469–1714: A Society of Conflict.* London: Longman, 3rd ed.

Karttunen, F. 1994. *Between Worlds: Interpreters, Guides and Survivors.* New Brunswick, NJ: Rutgers University Press.

Katzew, I. 2004. *Casta Painting: Images of Race in Eighteenth-Century Mexico.* New Haven: Yale University Press.

Keesing, F. M. 1962. *The Ethnohistory of Northern Luzon.* Stanford: Stanford University Press.

Keller, F. 1874. *The Amazon and Madeira Rivers: Sketches and Descriptions from the Note-Book of an Explorer.* New York: D. Appleton.*

Kelly, I. 2006. *Beau Brummell: The Ultimate Man of Style.* New York: Free Press.

Kelso, W. M. 2006. *Jamestown: The Buried Truth.* Charlottesville, VA: University of Virginia Press.

Kelso, W. M., and B. Straube. 2004. *Jamestown Rediscovery 1994–2004.* Richmond, VA: Association for the Preservation of Virginia Antiquities.

Kent, R. 1965. "Palmares: An African State in Brasil." *Journal of African History* 6:161–75.

Kinealy, C. 1995. *This Great Calamity: The Irish Famine, 1845–52.* Boulder, CO: Roberts Rinehart.

Kingsbury, S. M., ed. 1999 (1906–33). *The Records of the Virginia Company of London.* 4 vols. Westminster, MD: Heritage Books (CD-ROM).

Kirby, J., and R. White. 1996. "The Identification of Red Lake Pigment Dyestuffs and a Discussion of Their Use." *National Gallery Technical Bulletin* 17:56–80.

Kiszewski, A., et al. 2004. "A Global Index Representing the Stability of Malaria Transmission." *AMJTMH* 70:486–98.

Kjærgaard, T. 2003. "A Plant That Changed the World: The Rise and Fall of Clover, 1000–2000." *Landscape Research* 28:41–49.

Klein, H. S. 2010 (1999). *The Atlantic Slave Trade.* New York: Cambridge University Press, 2nd ed.

Koerner, B. 2004. "Blood Feud." *Wired* 13:118–25.*

Kohn, M. Forthcoming. *The Origins of Western Economic Success: Commerce, Finance, and Government in Pre-Industrial Europe.**

Kolb, A. E. 1980. "Early Passengers to Virginia: When Did They Really Arrive?" *VMHB* 88:401–14.

Komlos, J. 1998. "The New World's Contribution to Food Consumption During the Industrial Revolution." *Journal of European Economic History* 27:6–84.

Kon, S. K., and A. Klein. 1928. "The Value of Whole Potato in Human Nutrition." *Biochemical Journal* 22:258–60.

Konetzke, R. 1958. "Points of Departure for the History of Missions in Hispanic America." *Americas* 15:517–23.

Kramer, T., et al. 2009. *Withdrawal Symptoms in the Golden Triangle: A Drugs Market in Disarray.* Amsterdam: Transnational Institute.★

Krech, S. 1999. *The Ecological Indian: Myth and History.* New York: Norton.

Krippner-Martinez, J. 2000. "Invoking 'Tato Vasco': Vasco de Quiroga, Eighteenth–Twentieth Centuries." *Americas* 56:1–28.

Kuchta, D. 2002. *The Three-Piece Suit and Modern Masculinity: England, 1550–1850.* Los Angeles: University of California Press.

Kukla, J. 1986. "Kentish Agues and American Distempers: The Transmission of Malaria from England to Virginia in the Seventeenth Century." *Southern Studies* 25:135–47.

Kupperman, K. O. 2007a. *The Jamestown Project.* Cambridge, MA: Harvard Belknap.

———. 2007b. *Roanoke: The Abandoned Colony.* Savage, MD: Rowman and Littlefield Publishers, 2nd ed.

———. 1984. "Fear of Hot Climates in the Anglo-American Colonial Experience." *WMQ* 41:213–40.

———. 1982. "The Puzzle of the American Climate in the Early Colonial Period." *AHR* 87:1262–89.

———. 1979. "Apathy and Death in Early Jamestown." *Journal of American History* 66:24–40.

Kupperman, K. O., ed. 1988. *Captain John Smith: A Select Edition of His Writings.* Chapel Hill: University of North Carolina Press.

Kuwabara, J. 1935. "P'u Shou-kêng: A Man of the Western Regions, Who Was Superintendent of the Trading Ships' Office in Ch'üan-chou Towards the End of the Sung Dynasty, Together with a General Sketch of the Arabs in China During the T'ang and Sung Eras." *Memoirs of the Research Department of the Toyo Bunko* 7:1–104.

Labroy, O. 1913. *Culture et Exploitation du Caoutchouc au Brésil.* Paris: Société Générale d'Impression.★

Ladebat, P. d. 2008. *Seuls les Morts ne Reviennent Jamais: Les Pionniers de la Guillotine Sèche en Guyane Française sous le Directoire.* Paris: Éditions Amalthée.

Ladurie, E. L. R. 1971 (1967). *Times of Feast, Times of Famine: A History of Climate Since the Year 1000.* Trans. B. Bray. Garden City, NY: Doubleday and Company.

Laird Clowes, W., et al. 1897–1903. *The Royal Navy: A History from the Earliest Times to the Death of Queen Victoria.* 7 vols. London: Sampson Low, Marston and Co.★

Lal, D. 1998. *Unintended Consequences: The Impact of Factor Endowments, Culture, and Politics on Long-Run Economic Performance.* Cambridge, MA: MIT.

Lamb, H. H. 1995 (1982). *Climate, History and the Modern World*. New York: Routledge.

Lampton, D. M., et al. 1986. *A Relationship Restored: Trends in U.S.-China Educational Exchanges, 1978–84*. Washington, DC: National Academy Press.

Lan, Y. (蓝勇). 2001. "The Influences of American Crops Introduced During the Ming and Qing on the Formation of Structural Poverty in Subtropical Mountain Regions" (明清美洲农作物引进对亚热带山地结构性贫困形成的影响). *AHC* 20:3–14.

Landers, J. 2002. "The Central African Presence in Spanish Maroon Communities." In L. M. Heywood, ed., *Central Africans and Cultural Transformations in the American Diaspora*. New York: Cambridge University Press, 227–42.

———. 1999. *Black Society in Spanish Florida*. Urbana: University of Illinois Press.

Landes, D. S. 1999 (1998). *The Wealth and Poverty of Nations: Why Some Are So Rich and Some So Poor*. New York: W. W. Norton.

Lane, E. V. 1953–54. "The Life and World of Henry Wickham." *India Rubber World*, Dec. 5 (pt. 1, 14–17), Dec. 12 (pt. 2, 16–18), Dec 19 (pt. 3, 18–23), Dec. 26 (pt. 4, 5–8), Jan. 2 (pt. 5, 17–19), Jan. 9 (pt. 6, 17–23), Jan. 16 (pt. 7, 7–10), Jan. 23 (pt. 8, 7–10), Jan. 30 (pt. 9, 5–8).

Lane, K. 2002. *Quito 1599: City and Colony in Transition*. Albuquerque: University of New Mexico Press.

Lane, R. 1585–86. "Ralph Lane's Narrative of the Settlement of Roanoke Island." In Quinn and Quinn eds. 1982, 24–45.*

Langer, W. L. 1975. "American Foods and Europe's Population Growth, 1750–1850." *Journal of Social History* 8:51–66.

Langworthy, C. F. 1910. *Potatoes and Other Root Crops as Food*. U.S.D.A. Farmers' Bulletin 295. Washington, DC: Government Printing Office.

Lanyon, A. 2004 (2003). *The New World of Martín Cortés*. Cambridge, MA: Da Capo Press.

———. 1999. *Malinche's Conquest*. St. Leonards, NSW: Allan and Unwin.

Lara, S. H. 2010. "Palmares and Cucaú: Political Dimensions of a Maroon Community in Late Seventeenth-Century Brazil." Paper at 12th Annual Gilder Lehrman Center International Conference at Yale University, 29–30 Oct.*

———. 1996. "Do Singular ao Plural: Palmares, Capitães-do-mato e o Governo dos Escravos." In Reis and Gomes eds. 1996, 81–109.

Large, E. C. 1940. *The Advance of the Fungi*. London: Jonathan Cape.

La Roche, R. 1855. *Yellow Fever, Considered in Its Historical, Pathological, Etiological, and Therapeutical Relations*. 2 vols. Philadelphia: Blanchard and Lea.

Las Casas, B. d. 1992 (1552). *The Devastation of the Indies*. Trans. H. Briffault. Baltimore: Johns Hopkins Press.

———. 1951 (1561). *Historia de las Indias*. 3 vols. Mexico City: Fondo de Cultura Económica.

Laubrich, A. W. 1913. *Indian Slavery in Colonial Times Within the Present Limits of the United States.* Studies in History, Economics and Public Law 134. New York: Columbia University Press.*

Laufer, B. 1938. *The American Plant Migration. Part I: The Potato.* Chicago: Field Museum. Anthropological Series 28.*

———. 1924a. *Tobacco and Its Use in Asia.* Anthropology Leaflet 18. Chicago: Field Museum.*

———. 1924b. *Introduction of Tobacco Into Europe.* Anthropology Leaflet 19. Chicago: Field Museum.*

———. 1908. "The Relations of the Chinese to the Philippine Islands." *Smithsonian Miscellaneous Collections* 50:248–84.

Laufer, B., et al. 1930. *Tobacco and Its Use in Africa.* Anthropology Leaflet 29. Chicago: Field Museum.

Lee, G. R. 1999. "Comparative Perspectives." In M. B. Sussman et al., eds., *Handbook of Marriage and the Family.* New York: Plenum Press, 2nd ed.

Lee, J. Z., and F. Wang. 2001 (1999). *One Quarter of Humanity: Malthusian Mythology and Chinese Realities, 1700–2000.* Cambridge, MA: Harvard University Press.

Lee, K. E. 1985. *Earthworms: Their Ecology and Relationships with Soils and Land Use.* New York: Academic Press.

Legarda, B. J. 1999. *After the Galleons: Foreign Trade, Economic Change and Entrepreneurship in the Nineteenth-Century Philippines.* Manila: Ateneo de Manila University Press.

Léon Guerrero, M. 2000. *El Segundo Viaje Columbino.* PhD thesis, Universidad de Valladolid.*

Leong, S.-T. (Liang, S.-T.). 1997. *Migration and Ethnicity in Chinese History: Hakkas, Pengmin, and Their Neighbors.* Stanford: Stanford University Press.

Leroy, E. M., et al. 2004. "Multiple Ebola Virus Transmission Events and Rapid Decline of Central African Wildlife." *Science* 303:298–99.

Lester, T. 2009. *The Fourth Part of the World: The Race to the Ends of the Earth, and the Epic Story of the Map That Gave America Its Name.* New York: Simon and Schuster.

Levathes, L. 1994. *When China Ruled the Seas: The Treasure Fleet of the Dragon Throne, 1405–1433.* New York: Simon and Schuster.

Levin, S. 2005. "Growing China's Great Green Wall." *Ecos* 127:13.

Lev-Yadun, S., et al. 2000. "The Cradle of Agriculture." *Science* 288:1602–03.

Lewis, C. M., and A. J. Loomie. 1953. *The Spanish Jesuit Mission in Virginia, 1570–72.* Chapel Hill: University of North Carolina Press.

Li, H., et al. 2007. "Demand for Rubber Is Causing the Loss of High Diversity Rain Forest in SW China." *Biodiversity Conservation* 16:1731–45.

Li, J. (李金明). 2008. "The Rise of Yuegang, Zhangzhou and Overseas Chinese from Fujian During the Mid-Ming Dynasty" (明朝中叶漳州月港的兴起与福建的海外移民). In S.-Y. Tang (湯熙勇) et al., eds. *Essays on the*

History of China's Maritime Development (中国海洋发展史论文集). Taipei: Academia Sinica Research Center for Humanities and Social Sciences (中央研究院人文社会科学研究中心), vol. 10, pp. 65–100.

———. 2006a. *Overseas Transportation and Culture Exchange* (三朝平攘錄 交流). Kunming: Yunnan Fine Arts Publishing House.

———. 2006b. "A Theory on the Causes and Nature of the Jiajing Pirate Crisis" (试论嘉靖倭患的起因及性质). In Li 2006a:53–59.

———. 2006c "Smuggling Between Japan and the Ports of Zhangzhou and Quanzhou During the 16th Century" (16世纪漳泉贸易港与日本的走私贸易). In Li 2006a:45–52.

———. 2001. *Zhangzhou Port* (漳州港). Fuzhou: Fujian People's Publishing Co. (福建人民出版社).

Li, X. (李向军). 1995. *Qing Dynasty Disaster Relief Policy* (清代荒政研究). Beijing: China Agricultural Press.

Li, Y., et al. 2007. "On the Origin of Smallpox: Correlating Variola Phylogenics with Historical Smallpox Records." *PNAS* 104:15787–92.

Lieberei, R. 2007. "South American Leaf Blight of the Rubber Tree (Hevea spp.): New Steps in Plant Domestication Using Physiological Features and Molecular Markers." *Annals of Botany* 100:1125–42.

Liebig, J. v. 1840. *Organic Chemistry in Its Applications to Agriculture and Physiology*. trans. L. Playfair. London: Taylor and Walton.*

Ligon, R. 1673. *A True and Exact History of the Island of Barbadoes*. London: Peter Parker.

Lin, R. 1990. "Fukien's Private Sea Trade in the 16th and 17th Centuries." Trans. B. t. Haar. In Vermeer ed. 1990:163–216.

Livi-Bacci, M. 2003. "Return to Hispaniola: Reassessing a Demographic Catastrophe." *HAHR* 83:3–51.

———. 1997. *A Concise History of World Population*. Malden, MA: Blackwell, 2nd ed.

Livingstone, F. B. 1971. "Malaria and Human Polymorphisms." *Annual Review of Genetics* 5:33–64.

———. 1958. "Anthropological Implications of Sickle Cell Gene Distribution in West Africa." *AA* 60:533–62.

Loaisa, R. d. 1586. "Memorial de las Cosas del Pirú Tocantes á los Indios," 5 May. In J. S. Rayon and F. d. Zabálburu, eds., 1889, *Colección de Documentos Inéditos para la Historia de España,* vol. 94.

Lodeman, E. G. 1896. *The Spraying of Plants*. New York: Macmillan and Company.

Logan, R. W. 1940. "Estevanico, Negro Discoverer of the Southwest: A Critical Reexamination." *Phylon* 1:305–14.

Lohmann Villena, G. 1949. *Las Minas de Huancavelica en los Siglos XVI y XVII*. Seville: Escuela de Estudios Hispano-Americanos.

Lokken, P. 2004. "Transforming Mulatto Identity in Colonial Guatemala and El Salvador, 1670–1720." *Transforming Anthropology* 12:9–20.

———. 2004. "Useful Enemies: Seventeenth-Century Piracy and the Rise of Pardo Militias in Spanish Central America." *Journal of Colonialism and Colonial History* 5:2.

———. 2001. "Marriage as Slave Emancipation in Seventeenth-Century Rural Guatemala." *Americas* 58:175–200.

López de Gómara, F. 1870 (1552). *Conquista de México. Cronica General de Las Indias*, pt. 2. 2 vols. Mexico City: I. Escalante.★

López de Velasco, J. 1894 (~1575). *Geografía y Descripción Universal de las Indias*. Madrid: Real Academia de la Historia.★

Lord, L. 2007. "The Birth of America: Struggling from One Peril to the Next, the Jamestown Settlers Planted the Seeds of the Nation's Spirit." *U.S. News & World Report* 142:48–56.

Love, E. F. 1971. "Marriage Patterns of Persons of African Descent in a Colonial Mexico City Parish." *HAHR* 51:79–91.

———. 1967. "Negro Resistance to Spanish Rule in Colonial Mexico." *Journal of Negro History* 52:89–103.

Lovejoy, P. E. 2000 (1983). *Transformations in Slavery: A History of Slavery in Africa*. NY: Cambridge University Press, 2nd ed.

Lowe, S., et al. 2004 (2000). *100 of the World's Worst Invasive Species: A Selection from the Global Invasive Species Database*. Gland, Switzerland: International Union for Conservation of Nature.

Lu, W., and J. Lazell. 1996. "The Voyage of the Beetle." *Natural History* 105:36–39.

Lu, Y. (陸燿). 1991 (~1774). *A Guide to Smoking* (烟譜). In *Complete Collection of Collectanea* (叢書集成續編). Taipei: New Wen Feng Publishing Company, 2nd ed., vol. 86, pp. 675–78.

Luengo, J. M. 1996. *A History of the Manila-Acapulco Slave Trade, 1565–1815*. Tubigon, Bohol, Philippines: Mater Dei Publications.

Luo, Y. (羅日褧). 1983 (1585). *Record of Tribute Guests* (咸賓錄). Beijing: Zhonghua Shuju.

Lyderson, K. 2009. "Who Went with Columbus? Dental Studies Give Clues." *Washington Post*, 18 May.

Lynch, J. 1991. *Spain, 1516–1598: From Nation State to World Empire*. Oxford: Basil Blackwell.

MacKenzie, A. D. 1953. *The Bank of England Note: A History of Its Printing*. London: Cambridge University Press.

Magalhães, J. R. 2008. "O Açúcar nas Ilhas Portuguesas do Atlântico: Séculos XV e XVI." *Varia Historia* (Belo Horizonte) 25:151–75.★

Magoon, C. E. 1900. *Report on the Legal Status of the Territory and Inhabitants of the Islands Acquired by the United States During the War with Spain*. Washington, DC: Government Printing Office.★

Maher, R. F. 1973. "Archaeological Investigations in Central Ifugao." *Asian Perspectives* 16:39–71.

Malanima, P. 2006. "Energy Crisis and Growth, 1650–1850: The European Deviation in a Comparative Perspective." *Journal of Global History* 1:101–21.

Malecki, J. M., et al. 2003. "Local Transmission of *Plasmodium vivax* Malaria— Palm Beach County, Florida, 2003." *MMWR* 52:908–11.

Malone, P. M. 2000 (1991). *The Skulking Way of War: Technology and Tactics Among the New England Indians.* Toronto: Madison Books.

Malthus, T. R. 1798. *An Essay on the Principle of Population.* London: J. Johnson.

Mann, C. C. 2009. "Addicted to Rubber." *Science* 325:564–66.

———. 2008. "Tracing the Ancient Amazonians." *Science* 321:1148–52.

———. 2007. "America: Found and Lost." *National Geographic* 212:32–55.

———. 2005. *1491: New Revelations of the Americas Before Columbus.* New York: Alfred A. Knopf.

———. 1993. "How Many Is Too Many?" *Atlantic Monthly* 271:47–67.

Markham, C. R. 1876. "The Cultivation of Caoutchouc-Yielding Trees in British India." *Journal of the Royal Society of the Arts* 24:475–81.

———. 1871. "On the Eastern Cordillera, and the Navigation of the River Madeira." In British Association for the Advancement of Science, ed., *Report of the Forty-first Meeting.* London: John Murray, 184–85.*

———. 1862. *Travels in Peru and India.* London: John Murray.

Marks, R. B. 2007. *The Origins of the Modern World: A Global and Ecological Narrative from the Fifteenth to the Twenty-first Century.* Lanham, MD: Rowman and Littlefield, 2nd ed.

———. 1998. *Tigers, Rice, Silk, and Silt: Environment and Economy in Late Imperial South China.* New York: Cambridge University Press.

Martin, J. 1622. "How Virginia May Be Made a Royal Plantation." *KB* 3:707–10.

Martin, P. H., et al. 2004. "Forty Years of Tropical Forest Recovery from Agriculture: Structure and Floristics of Secondary and Old-Growth Riparian Forests in the Dominican Republic." *Biotropica* 36:297–317.

Martínez, M. E. 2008. *Genealogical Fictions: Limpieza de Sangre, Religion, and Gender in Colonial Mexico.* Stanford, CA: Stanford University Press.

Masefield, G. B. 1980 (1967). "Crops and Livestock." In E. E. Rich and C. H. Wilson, eds., *The Economy of Expanding Europe in the 16th and 17th Centuries.* Cambridge Economic History of Europe, vol. 4. New York: Cambridge University Press.

Mason, I. L., ed. 1984. *Evolution of Domesticated Animals.* London: Longmans.

Mathew, W. M. 1977. "A Primitive Export Sector: Guano Production in Mid-nineteenth Century Peru." *Journal of Latin American Studies* 9:35–57.

———. 1970. "Peru and the British Guano Market, 1840–1870." *EHR* 23:112–28.

———. 1968. "The Imperialism of Free Trade: Peru, 1820–70." *EHR* 21:562–79.

Matta, C. 2009. "Spontaneous Generation and Disease Causation: Anton de

Bary's Experiments with *Phytophthora infestans* and Late Blight of Potato." *Journal of the History of Biology* 43:459–91.

Mattoso, K. M. d. Q. 1986 (1979). *To Be a Slave in Brazil, 1550–1888*. Trans. A. Goldhammer. New Brunswick, NJ: Rutgers University Press.

Maxwell, H. 1910. "The Use and Abuse of Forests by the Virginia Indians." *WMQ* 19:73–103.

May, K. J., and J. B. Ristaino. 2004. "Identity of the mtDNA Haplotype(s) of *Phytophthora infestans* in Historical Specimens from the Irish Potato Famine." *Mycological Research* 108:1–9.

Mayer, E. 1994. "Recursos Naturales, Medio Ambiente, Tecnología y Desarrollo." In C. Menge, ed., *Perú: El Problema Agrario en Debate, SEPIA V*. Peru: SEPIA-CAPRODA, 479–533.

Mazumdar, S. 2000. "The Impact of New World Food Crops on the Diet and Economy of China and India, 1600–1900." In R. Grew, ed., *Food in Global History*. Boulder, CO: Westview Press.

McCaa, R. 1995. "Spanish and Nahuatl Views on Smallpox and Demographic Catastrophe in Mexico." *JIH* 25:397–431.

McCord, H. A. 2001. "How Crowded Was Virginia in A.D. 1607?" *QBASV* 56:51–59.

McCusker, J. J., and R. R. Menard. 1991 (1985). *The Economy of British America, 1607–1789*. Chapel Hill: University of North Carolina Press.

McDonald, W., ed. 1899. *Select Charters and Other Documents Illustrative of American History, 1606–1775*. New York: Macmillan.★

McKeown, T., et al. 1972. "An Interpretation of the Modern Rise of Population in Europe." *Population Studies* 26:345–82.

McNeill, J. R. 2010. *Mosquito Empires: Ecology and War in the Greater Caribbean, 1620–1914*. New York: Cambridge University Press.

McNeill, W. H. 1999. "How the Potato Changed the World's History." *Social Research* 66:69–83.

Meagher, A. J. 2009. *The Coolie Trade: The Traffic in Chinese Laborers to Latin America, 1847–1874*. Bloomington, IN: Xlibris.

Mei, C., and H. E. Dregne. 2001. "Review Article: Silt and the Future Development of China's Yellow River." *Geographical Journal* 167:7–22.

Mei, Z. (梅曾亮). 1823. "Record of the Shack People" (記棚民事). In Z. Mei, ed., 1855, *Collected Works of the Bojian Studio* (柏梘山房文集). SI:s.n., pp. 10:5a-6a.★

Melillo, E. D. 2011. "The First Green Revolution: Debt Peonage and the Making of the Nitrogen Fertilizer Trade, 1840–1930." Paper at Five-College History Seminar, Amherst College, 11 Feb.

Mellafe, R. 1959. *La Introducción de la Esclavitud Negra en Chile: Trafico y Rutas*. Santiago: Universidad de Chile, Estudios de Historia Economica Americana 2.

Menard, R. R. 1988. "British Migration to the Chesapeake Colonies in the

Seventeenth Century." In L. G. Carr et al., eds., *Colonial Chesapeake Society*. Chapel Hill: University of North Carolina Press, 99–132.

———. 1977. "From Servants to Slaves: The Transformation of the Chesapeake Labor System." *Southern Studies* 16:355–90.

Merrens, H. R., and G. D. Terry. 1984. "Dying in Paradise: Malaria, Mortality and the Perceptual Environment in Colonial South Carolina." *JSH* 50:533–50.

Migge-Kleian, S., et al. 2006. "The Influence of Invasive Earthworms on Indigenous Fauna in Ecosystems Previously Uninhabited by Earthworms." *Biological Invasions* 8:1275–85.

Milhou, A. 1983. Colón y su Mentalidad Mesianica en el Ambiente Franciscanista Español. Cuadernos Colombinos 11. Valladolid: Seminario Americanista de la Universidad de Valladolid.

Miller, H. M. 2001. "Living Along the 'Great Shellfish Bay': The Relationship Between Prehistoric Peoples and the Chesapeake." In Curtin, Brush, and Fisher, eds. 2001, 109–26.

Miller, J. C. 1988. *Way of Death: Merchant Capitalism and the Angolan Slave Trade, 1730–1830*. Madison: University of Wisconsin Press.

Miller, L. H., et al. 1976. "The Resistance Factor to *Plasmodium vivax* in Blacks—The Duffy-Blood-Group Genotype, *FyFy*." *New England Journal of Medicine* 295:302–04.

Miller, S. W. 2007. *An Environmental History of Latin America*. New York: Cambridge University Press.

Mintz, S. 1986 (1985). *Sweetness and Power: The Place of Sugar in Modern History*. New York: Penguin Books.

Mitchell, M. 1964. *Friar Andrés de Urdaneta, O.S.A.* London: Macdonald and Evans.

Mizubuti, E. S. G., and W. E. Fry. 2006. "Potato Late Blight." In B. M. Cooke et al., eds., *The Epidemiology of Plant Diseases*. Dordrecht, The Netherlands, 445–72.

Mokyr, J. 1981. "Irish History with the Potato." *Irish Economic and Social History* 8:8–29.

Moloughney, B., and W. Xia. 1989. "Silver and the Fall of the Ming Dynasty: A Reassessment." *Papers on Far Eastern History* 40:51–78.

Money, N. P. 2007. *The Triumph of the Fungi: A Rotten History*. New York: Oxford University Press.

Montenegro, A., et al. "Modeling the Prehistoric Arrival of the Sweet Potato in Polynesia." *Journal of Archaeological Science* 35:355–67.

Montgomery, D. R. 2007. *Dirt: The Erosion of Civilizations*. Berkeley: University of California Press.

Moore, R. J. 1999. "Colonial Images of Blacks and Indians in Nineteenth Century Guyana." In B. Brereton and K. A. Yelvington, eds., *The Colonial Caribbean in Transition: Essays on Postemancipation Social and Cultural History*. Gainesville: University Press of Florida, 126–58.

Morawetz, H. 2002 (1985). *Polymers: The Origin and Growth of a Science*. Mineola, NY: Dover Publications.

Moreau de Saint-Méry, M. L. E. 1797–98. *Description Topographique, Physique, Civile, Politique et Historique de le Partie Française de l'Isle Saint'Domingue*. 2 vols. Philadelphia: S.I.

Morel, G. R. 2004. "The Sugar Economy of Española in the Sixteenth Century." In S. B. Schwartz, ed., *Tropical Babylons: Sugar and the Making of the Atlantic World, 1450–1680*. Durham: University of North Carolina Press, 85–114.

Morga, A. d. 1609. *Sucesos de las Islas Filipinas*. In B&R 15:25–288, 16:25–210.

Morgan, E. S. 2003 (1975). *American Slavery, American Freedom: The Ordeal of Colonial Virginia*. New York: W. W. Norton, 2nd ed.

Morineau, M. 1985. *Incroyables Gazettes et Fabuleux Métaux: Les Retours de Trésors Américains d'après les Gazettes Hollandaises (XVIe–XVIIIe Siècles)*. New York: Cambridge University Press.

Morison, S. E. 1983 (1970). *Admiral of the Ocean Sea: A Life of Christopher Columbus*. Boston: Northeastern University Press.

Morrow, R. H., and W. J. Moss. 2007. "The Epidemiology and Control of Malaria." In K. E. Nelson and C. M. Williams, *Infectious Disease Epidemiology: Theory and Practice*. Sudbury, MA: Jones and Bartlett, 1087–1138.

Morse, R. M., ed., trans. 1965. *The Bandeirantes: The Historical Role of the Brazilian Pathfinders*. New York: Knopf.

Moseley, M. E. 2001. *The Inca and Their Ancestors: The Archaeology of Peru*. New York: Thames and Hudson, 2nd ed.

Mote, F. W. 2003 (1999). *Imperial China, 900–1800*. Cambridge, MA: Harvard University Press.

Motinha, K. E. F. 2005. "Vila Nova de Mazagão: Espelho de Cultura e de Sociabilidade Portuguesas no Vale Amazônico." Unpub. ms., Congresso Internacional o Espaço Atlântico de Antigo Regime: Poderes e Sociedades, Lisbon 2–5 Nov.*

Mudge, J. M. 1985. "Hispanic Blue-and-White Faience in the Chinese Style." In J. Carswell, ed., *Blue and White: Chinese Porcelain and Its Impact on the Western World*. Chicago: University of Chicago Press.

Mueller, I., et al. 2009. "Key Gaps in the Knowledge of *Plasmodium vivax*, a Neglected Human Malaria Parasite." *Lancet* 9:555–66.

Mülhaupt, R. 2004. "Hermann Staudinger and the Origin of Macromolecular Chemistry." *Angewandte Chemie International Edition* 43:1054–63.

Müller, U. C., and J. Pross. 2007. "Lesson from the Past: Present Insolation Minimum Holds Potential for Glacial Inception." *Quaternary Science Reviews* 26:3025–29.

Mulroy, K. 1993. *Freedom on the Border: The Seminole Maroons in Florida, the Indian Territory, Coahuila, and Texas*. Lubbock, TX: Texas Tech University Press.

Munga, S., et al. 2006. "Association Between Land Cover and Habitat Productivity of Malaria Vectors in Western Kenyan Highlands." *AMJTMH* 74:69–75.

Muñoz de San Pedro, M. 1951. "Doña Isabel de Vargas, Esposa del Padre del Conquistador del Perú." *Revista de Indias* 11:9–28.

Muñoz-Sanza, A. 2006. "La Gripe de Cristóbal Colón. Hipótesis Sobre una Catástrofe Ecológica." *Enfermedades Infecciosas y Microbiología Clínica* 24: 326–34.*

Murphy, E. 1834. "Agricultural Report." *Irish Farmer's and Gardener's Magazine* 1:556–58.*

Myers, M. D. 1998. "Cultivation Ridges in Theory and Practice: Cultural Ecological Insights from Ireland." PhD thesis, University of Texas at Austin.

Myers, R. H., and Y.-C. Wang. 2002. "Economic Developments, 1644–1800." In W. J. Peterson, ed., *The Cambridge History of China, Vol. 9: The Ch'ing Dynasty, Part 1: To 1800*. New York: Cambridge University Press, 563–646.

Nader, H. 1996. "Introduction." In H. Nader, ed., trans., *The Book of Privileges Issued to Christopher Columbus by King Fernando and Queen Isabel*. Repertorium Columbianum No. 2. Los Angeles: University of California Press, 1–58.

Naiman, R. J., et al. 1988. "Alteration of North American Streams by Beaver." *Bioscience* 38:753–62.

Nash, G. B. 1999. "The Hidden History of Mestizo America." In M. Hode, ed., *Sex, Love, Race: Crossing Boundaries in North American History*. New York: New York University Press.

Navagero, A. 1563. *Il Viaggio Fatto in Spagna, et in Francia, dal Magnifico M. Andrea Navagiero, fu Oratore dell'Illustrissimo Senata Veneto, alla Cesarea Maesta di Carlo V*. Venice: Domenico Farri.

Needham, J., et al. 1954–. *Science and Civilisation in China*. 7 vols. New York: Cambridge University Press.

Neill, E. D. 1867. "Ships Arriving at Jamestown, From the Settlement of Virginia Until the Revocation of Charter of London Company." In E. D. Neill, *The History of Education in Virginia During the Seventeenth Century*. Washington, DC: Government Printing Office, 7–11.*

Nelson, L. A. 1994. " 'Then the Poor Planter Hath Greatly the Disadvantage': Tobacco Inspection, Soil Exhaustion, and the Formation of a Planter Elite in York County, Virginia, 1700–1750." *Locus* 6:119–34.

Neto, M. A. d. S. 1984. "Os Quilombos de Salvador." *Princípios* (São Paulo) 8:51–56.

Nevle, R. J., and D. K. Bird. 2008. "Effects of Syn-pandemic Fire Reduction and Reforestation in the Tropical Americas on Atmospheric CO_2 During European Conquest." *Palaeogeography, Palaeoclimatology, Palaeoecology* 264:25–38.

Newson, L. A., and S. Minchin. 2007. *From Capture to Sale: The Portuguese Slave Trade to Spanish South America in the Early Seventeenth Century*. Leiden: Brill.

Ngwenyama, C. N. 2007. Material Beginnings of the Saramaka Maroons: An Archaeological Investigation. Ph.D. thesis, University of Florida.

Nicholls, M., ed. 2005. "George Percy's 'Trewe Relacyon.' " *VMHB* 113:213–75.

Nichols, P. 1628. *Sir Francis Drake Revived*. In Wright ed. 1932:245–326.

Nietner, J. 1880. *The Coffee Tree and Its Enemies, Being Observations on the Natural History of the Enemies of the Coffee Tree in Ceylon*. Colombo: Ceylon Observer Press.*

Niza, M. d. 1865–68 (1539). "Relacion." In J. F. Pacheco, et al., eds. 1865–69, *Colección de Documentos Inéditos Relativos al Descubrimiento, Conquista y Colonizacion de las Posesiones Españolas en América y Occeania*. 42 vols. Madrid: Manuel B. Quirós, vol. 3, 329–50.

Normile, D. 2007. "Getting at the Roots of Killer Dust Storms." *Science* 317:315.

North, D. C., and R. P. Thomas. 1973. *The Rise of the Western World: A New Economic History*. Cambridge: Cambridge University Press.

Nozawa, C., et al. 2008. "Evolving Culture, Evolving Landscapes: The Philippine Rice Terraces." In Amend, T., et al., eds. *Protected Landscapes and Agrobiodiversity Values* (Protected Landscapes and Seascapes, vol. 1: IUCN and GTZ). Heidelberg: Kasparek Verlag, pp. 71–94.*

Nunn, G. E. 1935. "The Imago Mundi and Columbus." *AHR* 40:646–61.

———. 1932. *The Columbus and Magellan Concepts of South American Geography*. Privately printed.

———. 1924. *The Geographical Conceptions of Columbus: A Critical Consideration of Four Problems*. New York: American Geographical Society.

Nunn, N., and N. Qian. Forthcoming. "The Potato's Contribution to Population and Urbanization: Evidence from an Historical Experiment." *Quarterly Journal of Economics*.

Nye, J. 1991. "The Myth of Free-Trade Britain and Fortress France: Tariffs and Trade in the Nineteenth Century." *JEH* 51:23–46.

Oberg, M. L. 2008. *The Head in Edward Nugent's Hand: Roanoke's Forgotten Indians*. Philadelphia: University of Pennsylvania Press.

O'Donnell, I. 2008. "The Rise and Fall of Homicide in Ireland." In S. Body-Gendrot and P. Spierenburg, eds., *Violence in Europe: Historical and Contemporary Perspectives*. New York: Springer.

Odoric of Pordenone. 1846 (1330). "The Eastern Parts of the World Described," trans. H. Yule. In Yule, H., ed., *Cathay and the Way Thither*. London: Hakluyt Society.*

Offen, K. H. 2007. "Creating Mosquitia: Mapping Amerindian Spatial Practices in Eastern Central America, 1629–1779." *Journal of Historical Geography* 33:254–82.

———. 2002. "The Sambo and Tawira Miskitu: The Colonial Origins and Geography of Intra-Miskitu Differentiation in Eastern Nicaragua and Honduras." *Ethnohistory* 49:321–72.

Ó Gráda, C. 2007. "Ireland's Great Famine: An Overview." In C. Ó Gráda, et al., eds., *When the Potato Failed: Causes and Effects of the "Last" European Subsistence Crisis, 1845–1850*. Turnhout, Belgium: Brepols, 43–57.

———. 2000 (1999). *Black '47 and Beyond: The Great Irish Famine in History, Economy and Memory.* Princeton: Princeton University Press.

———. 1994. "The 'Lumper' Potato and the Famine." *History Ireland* 1:22–23.

Olien, M. D. 1987. "Micro/Macro-Level Linkages: Regional Political Structures on the Mosquito Coast, 1845–1864." *Ethnohistory* 34:256–87.

———. 1983. "The Miskito Kings and the Line of Succession." *Journal of Anthropological Research* 39:198–241.

Oliveira, M. L. 2005. "A Primeira Rellação do Último Assalto a Palmares." *Afro-Ásia* 33:251–324.

Ollé Rodríguez, M. 2006. "La Formación del Parián de Manila: La Construcción de un Equilibrio Inestable." In P. S. G. Aguilar, ed., *La Investigación sobre Asia Pacífico en España* (Colección Española de Investigación sobre Asia Pacífico). Granada: Editorial Universidad de Granada, 27–49.

———. 2002. *La Empresa de China: De la Armada Invencible al Galeón de Manila.* Barcelona: Acantilado.

———. 1998. "Estrategias Filipinas Respecto a China: Alonso Sánchez y Domingo Salazar en la Empresa de China (1581–1593)." PhD thesis, Universitat Pompeu Fabra.*

Olofsson, J., and T. Hickler. 2008. "Effects of Human Land-use on the Global Carbon Cycle During the Last 6,000 Years." *Vegetation History and Archaeobotany* 17:605–15.

Omohundro, J. 2006. "An Appreciation of Lazy-Beds." *Newfoundland Quarterly* 99:n.p.*

Onokpise, O. U. 2004. "Natural Rubber, *Hevea brasiliensis* (Willd. Ex A. Juss.) Müll. Arg., Germplasm Collection in the Amazon Basin, Brazil: A Retrospective." *EB* 58:544–55.

Orbigny, A. d. 1835. *Voyage dans l'Amérique Méridionale.* 5 vols. Paris: Pitois-Levrault.*

Orser, C. E. 1994. "Toward a Global Historical Archaeology: An Example from Brazil." *Historical Archaeology* 28:5–22.

———. and Funari, P. P. A. 2001. "Archaeology and Slave Resistance and Rebellion." *World Archaeology* 33:61–72.

Osborne, A. R. 1989. "Barren Mountains, Raging Rivers: The Ecological and Social Effects of Changing Landuse on the Lower Yangzi Periphery in Late Imperial China." PhD thesis, Columbia University.

Ouerfelli, M. 2008. *Le Sucre: Production, Commercialisation et Usages dans la Méditerranée Médiévale.* Boston: Brill.

Overton, M. 1996. *Agricultural Revolution in England: The Transformation of the Agrarian Economy, 1500–1850.* New York: Cambridge University Press.

Oviedo y Valdés, G. F. d. 1851 (1535). *Historia General y Natural de las Indias, Islas y Tierra-Firme del Mar Océano.* 3 vols. Madrid: Academia Real de la Historia.*

Pacheco, W. M. 1995. "El Cerro Rico, una Montaña que Encarna a una Ciudad." In W. M. Pacheco, ed., *El Cerro Rico de Potosí (1545–1995): 450 Años de Explotación*. Potosí: Sociedad Geográfica y de Historia "Potosí," 263–88.

Packard, R. M. 2007. *The Making of a Tropical Disease: A Short History of Malaria*. Baltimore: Johns Hopkins University Press.

Padden, R. C. 1975. "Editor's Introduction." In R. C. Padden, ed., *Tales of Potosí*. Providence, RI: Brown University Press, xi–xxxv.

Parker, G. 2008. "Crisis and Catastrophe: The Global Crisis of the Seventeenth Century Reconsidered." *AHR* 113:1053–79.

———. 1979a. *Spain and the Netherlands, 1559–1659: Ten Studies*. Short Hills, NJ: Enslow Publishers.

———. 1979b (1973). "Mutiny and Discontent in the Spanish Army of Flanders, 1572–1607." In idem. 1979a:106–21.

———. 1979c (1970). "Spain, Her Enemies and the Revolt of the Netherlands 1559–1648." In idem. 1979a:18–44.

Parry, J. H., and R. G. Keith. 1984. *New Iberian World: A Documentary History of the Discovery and Settlement of Latin America to the Early 17th Century*. 5 vols. New York: Times Books.

Parsons, J. T. 1972. "Spread of African Pasture Grasses to the American Tropics." *Journal of Range Management* 25:12–17.

Parton, J. 1865. "Charles Goodyear." *North American Review* 101:65–102.*

Pastor, A., et al. 2002. "Local Transmission of *Plasmodium vivax* Malaria—Virginia, 2002." *MMWR* 51:921–23.

Pearson, H. C. 1911. *The Rubber Country of the Amazon*. New York: The India Rubber World.

Pearson, J. C. 1944. "The Fish and Fisheries of Colonial Virginia." *WMQ* 1:179–83.

Pearson, R., et al. 2001. "Port, City, and Hinterlands: Archaeological Perspectives on Quanzhou and Its Overseas Trade." In A. Schottenhammer, ed., *The Emporium of the World: Maritime Quanzhou, 1000–1400*. Boston: Brill.

Peck, G. W. 1854a. *Melbourne, and the Chincha Islands*. New York: Charles Scribner.*

———. 1854b. "From the Chincha Islands." *NYT*, 7 Jan.

Pederson, D. C., et al. 2005. "Medieval Warming, Little Ice Age, and European Impact on the Environment During the Last Millennium in the Lower Hudson Valley, New York, USA." *Quaternary Research* 63:238–49.

Percy, G. 1625? "A True Relation of the Proceedings and Occurents of Moment Which Have Hap'ned in Virginia from the Time Sir Thomas Gates Was Shipwrack'd upon the Bermudes, Anno 1609, Until my Departure Out of the Country, Which Was in Anno Domini 1612." In Haile ed. 1998, 497–519.

Perdue, T. 2001. *Sifters: Native American Women's Lives*. New York: Oxford University Press.

Perez, B. E. 2000. "The Journey to Freedom: Maroon Forebears in Southern Venezuela." *Ethnohistory* 47:611–34.

Pfister, C. 1983. "Changes in Stability and Carrying Capacity of Lowland and Highland Agro-Systems in Switzerland in the Historical Past." *Mountain Research and Development* 3:291–97.

Philips, G. 1891. "Early Spanish Trade with Chin Cheo (Chang Chow)." *China Review* 19:243–55.

Phillips, W. D. 1985. *Slavery from Roman Times to the Early Transatlantic Trade.* Minneapolis: University of Minnesota Press.

Phillips, W. D., and C. R. Phillips. 1992. *The Worlds of Christopher Columbus.* New York: Cambridge University Press.

Pike, R. 2007. "Black Rebels: The Cimarrons of Sixteenth-Century Panama." *The Americas* 64:243–66.

Pollan, M. 2006. *The Omnivore's Dilemma: A Natural History of Four Meals.* New York: Penguin.

———. 2001. *The Botany of Desire: A Plant's-Eye View of the World.* New York: Random House.

Polo, M. 2001 (1299). *The Travels of Marco Polo.* Trans., ed., W. Marsden and M. Komroff. New York: Modern Library.

Pomeranz, K. 2000. *The Great Divergence: China, Europe, and the Making of the Modern World Economy.* Princeton: Princeton University Press.

Poole, B. T. F. 1974. "Case Reopened: An Enquiry into the 'Defection' of Fray Bernal Boyl and Mosen Pedro Margarit." *Journal of Latin American Studies* 6:193–210.

Porras Muñoz, G. 1982. *El Gobierno de la Ciudad de México en el Siglo XVI.* Mexico City: Universidad Autónoma de México.

Porter, L. D. 2007. "Survival of Sporangia of New Clonal Lineages of *Phytophthora infestans* in Soil Under Semiarid Conditions." *Plant Disease* 91:835–41.

Postma, A., ed. 2005 (1972). *Mangyan Treasures: The Ambahan: A Poetic Expression of the Mangyans of Southern Mindoro, Philippines.* Calapan, Mindoro: Mangyan Heritage Center, 3rd ed.

Postma, J. M. 1990. *The Dutch in the Atlantic Slave Trade, 1600–1815.* NY: Cambridge University Press.

Powars, D. S., and T. S. Bruce. 1999. "The Effects of the Chesapeake Bay Impact Crater on the Geological Framework and Correlation of Hydrogeologic Units of the Lower York-James Peninsula, Virginia." U.S. Geological Survey Professional Paper 1612. Washington, DC: Government Printing Office.*

Price, D. A. 2005 (2003). *Love and Hate in Jamestown: John Smith, Pocahontas, and the Start of a New Nation.* New York: Vintage.

Price, E. O. 2002. *Animal Domestication and Behavior.* Oxford: CABI Publishing.

Price, R. 2011. *Rainforest Warriors: Human Rights on Trial.* Philadelphia: University of Pennsylvania Press.

———. 2002a (1983). *First-Time: The Historical Vision of an African American People.* Chicago: University of Chicago Press, 2nd ed.

———. 2002b. "Maroons in Suriname and Guyane: How Many and Where." *New West Indian Guide* 76:81–88.

Price, R., ed. 1996 (1979). *Maroon Societies: Rebel Slave Communities in the Americas.* Baltimore: Johns Hopkins University Press, 3rd ed.

Proft, J., et al. 1999. "Identification of Six Sibling Species of the *Anopheles maculipennis* complex (Diptera: Culicidae) by a Polymerase Chain Reaction Assay." *Parasitology Research* 85:837–43.

Proulx, N. 2003. *Ecological Risk Assessment of Non-indigenous Earthworm Species.* St. Paul, MN: U.S. Fish and Wildlife Service.

Puckrein, G. 1979. "Climate, Health and Black Labor in the English Americas." *Journal of American Studies* 13:179–93.

Pyne, S. J. 1999. "The Dominion of Fire." *Forum for Applied Research and Public Policy* 15:6–15.

———. 1997a (1995). *World Fire: The Culture of Fire on Earth.* Seattle: University of Washington Press.

———. 1997b (1982). *Fire in America: A Cultural History of Wildland and Rural Fire.* Seattle: University of Washington Press.

———. 1991. "Sky of Ash, Earth of Ash: A Brief History of Fire in the United States." In J. S. Levine, ed., *Global Biomass Burning: Atmospheric, Climatic, and Biospheric Implications.* Cambridge, MA: MIT Press, 504–11.

Pyne, S. J., et al. 1996. *Introduction to Wildland Fire.* New York: John Wiley and Sons, 2nd ed.

Qian, J. (錢江). 1986. "The Development of China-Luzon Trade and Estimated Trade Volume, 1570–1760" (1570–1760 年中國和呂宋貿易的發展及貿易額的估算). *Journal of Chinese Social and Economic History* (中國社會經濟史研究) 3:69–78, 117.

Quan, H. (全漢昇). 1991a. *Research on the Economic History of China* (中國經濟史研究). Taipei: New Asia Institute of Advanced Chinese Studies.

———. 1991b (1967). "Changes in the Purchasing Power of Silver During the Song and Ming Dynasties and the Causes Behind Them" (宋明間白銀購買力的變動及其原因). In Quan 1991a, 571–600.

———. 1991c (1966). "Silver Mining and Taxes in the Ming Dynasty" (明代的銀課與銀產額). In Quan 1991a, 601–23.

———. 1972a. *Collected Essays on the Economic History of China* (中國經濟史論叢). 2 vols. Hong Kong: New Asia Institute of Advanced Chinese Studies.

———. 1972b (1971). "Changes in the Coin-Silver Ratio in Annual Government Revenues and Expenditures from the Song to Ming Dynasty" (自宋至明政府歲入中錢銀比例的變動). In Quan 1972a:vol. 1, 355–68.

———. 1972c (1971). "Chinese Silk Trade with Spanish America from the Late Ming to Mid-Qing" (自明季至清中葉西屬美洲的中國絲貨貿易). In Quan 1972a:vol. 1, 451–73.

———. 1972d. "The Inflow of American Silver to China During the Ming and Qing" (明清間美洲白銀的輸入中國). In Quan 1972a:vol. 1, 435–50.

————. 1972e (1957). "The Relationship Between American Silver and the Price Revolution in 18th-Century China" (美洲白銀與十八世紀中國物價革命的關係). In Quan 1972a:vol. 2, 475–508.

Queiros Mattoso, K. M. d. 1986 (1979). *To Be a Slave in Brazil, 1550–1888* (trans. A. Goldhammer). New Brunswick, NJ: Rutgers University Press.

Quesada, V. G. 1890. *Crónicas Potosinos: Custumbres de la Edad Medieval Hispano-Americana.* 2 vols. Paris: Biblioteca de la Europa y America.

Quijano Otero, J. M. 1881. *Límites de la República de los Estados-unidos de Colombia.* Seville: Francisco Alvarez.

Quinn, D. B. 1985. *Set Fair for Roanoke, 1584–1606.* Chapel Hill: University of North Carolina Press.

————. and A. M. Quinn, eds. 1982. *The First Colonists: Documents on the Planting of the First English Settlements in North America, 1584–90.* Raleigh: North Carolina Department of Cultural Resources.

Rabb, T. K. 1966. "Investment in English Overseas Enterprise, 1575–1630." *EHR* 19:70–81.

Radkau, J. 2008 (2002). *Nature and Power: A Global History of the Environment,* trans. T. Dunlap. NY: Cambridge University Press.

Ramos, A. 1689–92. *Los Prodigios de la Omnipotencia, y Milagros de la Gracia en la Vida de la Venerable Sierva de Dios Catharina de S. Joan.* 3 vols. Puebla (Mexico): Diego Fernández de León.

Ramsdale, C., and K. Snow. 2000. "Distribution of the Genus Anopheles in Europe." *European Mosquito Bulletin* 7:1–26.*

Ratekin, M. 1954. "The Early Sugar Industry in Hispaniola." *HAHR* 34:1–19.

Rau, V., and J. de Macedo. 1962. *O Açucar da Madeira nos Fins do Século XV.* Funchal, Madeira: Junta-Geral do Distrito Autónomo do Funchal.

Rawski, E. S. 1975. "Agricultural Development in the Han River Highlands." *Ch'ing-shih wen-t'i* 3:63–81.

Reader, J. 2009. *Potato: A History of the Propitious Esculent.* New Haven: Yale University Press.

Real Academia Española. 1914. *Diccionario de la Lengua Castellana.* Madrid: Sucesores de Hernando.*

————. 1726–39. 6 vols. *Diccionario de la Lengua Castellana.* Madrid: F. del Hierro.*

Reavis, L. U. 1878. *The Life and Military Services of Gen. William Selby Harney.* St. Louis: Bryan, Brand and Co.*

Reddick, D. 1929. "The Drake Potato Introduction Monument." *Journal of Heredity* 20:173–76.

Rediker, M. 2008 (2007). *The Slave Ship: A Human History.* New York: Penguin.

Reinert, J. F., et al. 1997. "Analysis of the *Anopheles* (*Anopheles*) *quadrimaculatus* Complex of Sibling Species (Diptera: Culicidae) Using Morphological, Cytological, Molecular, Genetic, Biochemical, and Ecological Techniques in an Integrated Approach." *Journal of the American Mosquito Control Association* 13 (Supp.):1-102.

Reis, J. J. 1988. *Escravidão e Invenção da Liberdade: Estudos Sobre o Negro no Brasil.* São Paulo: Editora Brasiliense.

Reis, J. J., and F. d. S. Gomes, 2009. "Repercussions of the Haitian Revolution in Brazil." In D. P. Geggus and N. Fiering, eds., *The World of the Haitian Revolution.* Indianapolis: Indiana University Press, 284–313.

———. 1996. *Liberdade por um Fio: História dos Quilombos no Brasil.* São Paulo: Companhia das Letras.

Reiter, P. 2000. "From Shakespeare to Defoe: Malaria in England in the Little Ice Age." *Emerging Infectious Diseases* 6:1–11.

Rejmankova, E., et al. 1996. "*Anopheles albimanus* (Diptera: Culicidae) and Cyanobacteria: An Example of Larval Habitat Selection." *Environmental Entomology* 25:1058–67.

Requejo Salcedo, J. 1640. *Relación Histórica y Geográfica de la Provincia de Panamá.* In Serrano y Sanz, M., ed. 1908. *Relaciones Históricas y Geográficas de América Central* (Colección Libros y Documentos Referentes a la Historia de América 8). Madrid: Victoriano Suárez, 1–84.*

Restall, M. 2000. "Black Conquistadors: Armed Africans in Early Spanish America." *Americas* 57:171–205.

Reyna, E. 1941. *Fitzcarrald: El Rey de Caucho.* Lima: Taller Grafico de P. Barrantes C.

Reynolds, J. W. 1994. "Earthworms of Virginia (Oligochaeta: Acanthodrilidae, Komarekionidae, Lumbricidae, Megascolecidae, and Sparganophilidae)." *Megadrilogica* 5:77–94.

Ribiero, H., et al. 1998. "Os mosquitos (Diptera: Culicidae) da Ilha de São Tomé." *García de Orta* 22:1–20.*

Rich, B. 1614. *The Honestie of This Age· Prooving by Good Circumstance That the World Was Never Honest Till Now.* London: T. A.

Rich, S. M., and F. J. Ayala. 2006. "Evolutionary Origins of Human Malaria Parasites." In K. R. Dronamraju and P. Arese, eds., *Malaria: Genetic and Evolutionary Aspects.* New York: Springer, 125–46.

Richards, J. F. 2005. *The Unending Frontier: An Environmental History of the Early Modern World.* Berkeley: University of California Press.

Richards, M., and V. Macaulay. 2001. "The Mitochondrial Gene Tree Comes of Age." *American Journal of Human Genetics* 68:1315–20.

Richter, D. K. 2001. *Facing East from Indian Country: A Native History of Early America.* Cambridge, MA: Harvard University Press.

Riley, C. V. 1869. *First Annual Report on the Noxious, Beneficial and Other Insects of the State of Missouri.* Jefferson City, MO: Ellwood Kirby.

Riley, G. M. 1972. "Labor in Cortesian Enterprise: The Cuernavaca Area, 1522–1549." *Americas* 28:271–87.

Ringsdorf, H. 2004. "Hermann Staudinger and the Future of Polymer Research Jubilees—Beloved Occasions for Cultural Piety." *Angewandte Chemie International Edition* 43:1064–76.

Riordan, P. 1996. "Finding Freedom in Florida: Native Peoples, African Americans, and Colonists, 1670–1816." *Florida Historical Quarterly* 75:24–43.

Robert, J. C. 1949. *The Story of Tobacco in America*. Chapel Hill: University of North Carolina Press.

Robert, R. 1929. "Estebanico de Azamor et la Légende des Sept Cités." *Journal de la Société des Américanistes* 21:414.

Roberts, D. R., et al. 2002. "Determinants of Malaria in the Americas." In E. A. Casman and H. Dowlatabadi, eds., *The Contextual Determinants of Malaria*. Washington, DC: Resources for the Future, 35–58.

Rocco, F. 2003. *Quinine: Malaria and the Quest for a Cure That Changed the World*. New York: HarperCollins Perennial.

Rocheleau, D., et al. 2001. "Complex Communities and Emergent Ecologies in the Regional Agroforest of Zambrana-Chacuey, Dominican Republic." *Cultural Geographies* 8:465–92.

Rolfe, J. 1616. "A True Relation of the State of Virginia." In Haile ed. 1998, 865–77.

———. 1614. Letter to Sir Thomas Dale, June (?). In Haile ed. 1998, 850–56.

Romer, R. 2009. *Slavery in the Connecticut Valley of Massachusetts*. Amherst, MA: Levellers Press.

Romoli, K. 1987. *Los de la Lengua Cueva: Los Grupos Indígenas del Istmo Oriental en la Época de la Conquista Española*. Bogotá: Instituto Colombiano de Cultura / Instituto Colombiano de Antropología.

Roorda, E. P. 1998. *The Dictator Next Door: The Good Neighbor Policy and the Trujillo Regime in the Dominican Republic, 1930–45*. Durham, NC: Duke University Press.

Rose, E. A. 2002. *Dependency and Socialism in the Modern Caribbean: Superpower Intervention in Guyana, Jamaica and Grenada, 1970–1985*. Lanham, MD: Lexington Books.

Rountree, H. C. 2005. *Pocahontas, Powhatan, Opechancanough: Three Indian Lives Changed by Jamestown*. Charlottesville: University of Virginia Press.

———. 2001. "Pocahontas: The Hostage Who Became Famous." In Perdue 2001, 14–28.

———. 1996. "A Guide to the Late Woodland Indians' Use of Ecological Zones in the Chesapeake Region." *Chesopiean* 34:1–37.

———. 1993a. "The Powhatans and the English: A Case of Multiple Conflicting Agendas." In Rountree ed. 1993, 173–205.

———. 1993b. "The Powhatans and Other Woodlands Indians as Travelers." In Rountree ed. 1993, 21–52.

———. 1990. *The Powhatan Indians of Virginia: Their Traditional Culture*. Norman: University of Oklahoma Press.

Rountree, H. C., ed. 1993. *Powhatan Foreign Relations, 1500–1722*. Charlottesville: University of Virginia Press.

Rountree, H. C., and E. R. Turner. 1998. "The Evolution of the Powhatan Paramount Chiefdom in Virginia." In E. M. Redmond, *Chiefdoms and Chieftaincy in the Americas*. Gainesville: University Press of Florida.

———. 1994. "On the Fringe of the Southeast: The Powhatan Paramount Chieftaincy in Virginia." In Hudson and Tesser eds. 1994, 355–72.

Rountree, H. C., et al. 2007. *John Smith's Chesapeake Voyages, 1607–1609*. Charlottesville: University of Virginia Press.

Rouse, I. 1992. *The Tainos: Rise and Decline of the People Who Greeted Columbus*. New Haven: Yale University Press.

Rout, L. B. 1976. *The African Experience in Spanish America, 1502 to the Present Day*. New York: Cambridge University Press.

Rowe, J. H. 1946. "Inca Culture at the Time of the Spanish Conquest." In J. H. Steward, ed., *Handbook of South American Indians*. BAE Bulletin 143. 7 vols. Washington, DC: Smithsonian Institution, vol. 2, 183–410.

Rowe, W. T. 2009. *China's Last Empire: The Great Qing*. Cambridge, MA: Belknap Press.

Rowell, C. H. 2008. "The First Liberator of the Americas." *Callaloo* 31:1–11.

Roze, E. 1898. *Histoire de la Pomme de Terre*. Paris: J. Rothschild.*

Rubio Mañé, J. I. 1970. "Más Documentos Relativos a la Expedición de Miguel López de Legazpi a Filipinas." *Boletín del Archivo General de la Nación* 11:82–156, 453–556.

———. 1964. "La Expedición de Miguel López de Legazpi a Filipinas." *Boletín del Archivo General de la Nación* 5:427–98.

Ruddiman, W. F. 2007. "The Early Anthropogenic Hypothesis: Challenges and Responses." *Reviews of Geophysics* 45:RG4001.

———. 2005. *Plows, Plagues and Petroleum: How Humans Took Control of the Climate*. Princeton: Princeton University Press.

———. 2003. "The Anthropogenic Greenhouse Era Began Thousands of Years Ago." *Climatic Change* 61:261–93.

Ruiz-Stovel. L. 2009. "Chinese Merchants, Silver Galleons, and Ethnic Violence in Spanish Manila, 1603–1686." *México y la Cuenca del Pacífico* 12:47–63.

Rule, H. 1962. "Henry Adams' Attack on Two Heroes of the Old South." *American Quarterly* 14:174–84.

Rusconi, R., ed. 1997. *The Book of Prophecies Edited by Christopher Columbus*. Trans. B. Sullivan. Los Angeles: University of California Press.

Rutherford, J., et al. 2008. "Rethinking Investments in Natural Resources: China's Emerging Role in the Mekong Region." Phnom Penh: Heinrich Böll Stiftung, WWF, and International Institute for Sustainable Development.

Rutman, D. B., and A. H. Rutman. 1980. "More True and Perfect Lists: The Reconstruction of Censuses for Middlesex County, Virginia, 1668–1704." *VMHB* 88:37, 74.

———. 1976. "Of Agues and Fevers: Malaria in the Early Chesapeake." *WMQ* 33:31–60.

Ruttner, F. 1988. *Biogeography and Taxonomy of Honeybees*. Berlin: Springer-Verlag.

Rych, B. 1614. *The Honestie of This Age: Proouing by Good Circumstance That the World Was Neuer Honest Till Now*. London: T. A.*

Saco, J. A. 1879. *Historia de la Esclavitud de la Raza Africana en el Nuevo Mundo*. Vol. 1. Barcelona: Jaime Jepús.

Sainsbury, W. N., ed. 1860. *Calendar of State Papers, Colonial Series, 1574–1660* [America and West Indies]. London: Longman, Green, Longman and Roberts.

Saíz, M. C. G. 1989. *Las Castas Mexicanas: Un Género Pictórico Americano*. Milan: Olivetti.

Salaman, R. 1985 (1949). *The History and Social Influence of the Potato*. ed. J. G. Hawkes. New York: Cambridge University Press, rev. ed.

Sale, K. 2006 (1990). *Christopher Columbus and the Conquest of Paradise*. New York: Tauris Parke, 2nd ed.

Samways, M. J. 1999. "Translocating Fauna to Foreign Lands: Here Comes the Homo-genocene." *Journal of Insect Conservation* 3:65–66.

Sánchez Farfan, J. 1983. "Pampallaqta, Centro Productor de Semilla de Papa." In A. M. Fries, ed., *Evolución y Tecnología de la Agricultura Andina*. Cusco: Instituto Indigenísta Interamericano.

Sanders, W. T. 1992. "The Population of the Central Mexican Symbiotic Region, the Basin of Mexico, and the Teotihuacán Valley in the Sixteenth Century." In W. M. Denevan, ed., *The Native Population of the Americas in 1492*. Madison: University of Wisconsin Press, 2nd ed.

Santos, R. 1980. *História Econômica da Amazônia, 1800–1920*. São Paulo: T. A. Queiroz.

Santos-Granero, F., and F. Barclay. 2000. *Tamed Frontiers: Economy, Society, and Civil Rights in Upper Amazonia*. Boulder, CO: Westview Press.

Sanz y Díaz, J. 1967. *López de Legazpi, Alcalde Mayor de México, Conquistador de Filipinas*. Mexico City: Editorial Jus.

Sarmiento de Gamboa, P. 2009 (1572). *History of the Incas*. Trans. B. Bauer and V. Smith. Charleston, SC: Bibliobazaar.* (another translation)

Satow, E. M. 1877. "The Introduction of Tobacco into Japan." *Transactions of the Asiatic Society of Japan* 5:68–84.

Save the Ifugao Terraces Movement. 2008. *The Effects of Tourism on Culture and the Environment in Asia and the Pacific: Sustainable Tourism and the Preservation of the World Heritage Site of the Ifugao Rice Terraces, Philippines*. Bangkok: UNESCO.*

Savitt, T. L., and J. H. Young, eds. 1988. *Disease and Distinctiveness in the American South*. Knoxville: University of Tennessee Press.

Sayers, D. O., et al. 2007. "The Political Economy of Exile in the Great Dismal Swamp." *International Journal of Historical Archaeology* 11:60–97.

Schneider, P. 2006. *Brutal Journey: The Epic Story of the First Crossing of North America*. New York: Holt.

Scholes, F. V. 1958. "The Spanish Conqueror as a Business Man: A Chapter in the History of Fernando Cortés." *New Mexico Quarterly* 28:5–29.

Schoolcraft, H. R. 1821. *Narrative Journal of Travels Through the Northwestern Regions of the United States Extending from Detroit Through the Great Chain of American Lakes, to the Sources of the Mississippi River*. New York: E. & E. Hosford.*

Schurz, W. L. 1939. *The Manila Galleon*. New York: E. P. Dutton.

Schurz, W. L., et al. 1925. *Rubber Production in the Amazon Valley*. Washington, DC: Government Printing Office.

Schwartz, S. B. 1997. "Spaniards, 'Pardos,' and the Missing Mestizos: Identities and Racial Categories in the Early Hispanic Caribbean." *New West Indian Guide* (Leiden) 71:5–19.

———. 1995. "Colonial Identities and the *Sociedad de Castas*." *Colonial Latin American Review* 4:185–201.

———. 1988. *Sugar Plantations in the Formation of Brazilian Society: Bahia, 1550–1835*. New York: Cambridge University Press.

Schwarz-Bart, S., and A. Schwarz-Bart. 2002 (1988). *In Praise of Black Women*. Trans. R.-M. Réjois and V. Vinokurov. 2 vols. Madison: University of Wisconsin Press.

Schwendinger, R. J. 1988. *Ocean of Bitter Dreams: Maritime Relations Between China and the United States, 1850–1915*. Tucson, AZ: Westernlore Publishing.

Scott, J. C. 1985. *Weapons of the Weak: Everyday Forms of Peasant Resistance*. New Haven: Yale University Press.

Scott, W. H. 1984 (1968). *Prehispanic Source Materials for the Study of Philippine History*. Quezon City: New Day Publishers.

Scott, W. R. 1912. *The Constitution and Finance of English, Scottish and Irish Joint-Stock Companies to 1720*. 3 vols. Oxford: Oxford University Press.*

Seibert, G. 2006. *Clients and Cousins: Colonialism, Socialism and Democratization in São Tomé and Príncipe*. Boston: Brill.

Seixas, S., et al. 2002. "Microsatellite Variation and Evolution of the Human Duffy Blood Group Polymorphism." *Molecular Biology and Evolution* 19: 1802–06.

Serier, J.-B. 2000. *Les Barons de Caoutchouc*. Paris: Karthala.

Shao, K., et al. (邵侃;卜风贤). 2007. "Crop Introduction and Spreading in the Ming and Qing Dynasties—A Study on Sweet Potato" (明清时期粮食作物的引入和传播—基于甘薯的考察). *Journal of Anhui Agricultural Sciences* (安徽农业科学) 35:7002–03, 7014.

Shapiro, J. 2001. *Mao's War Against Nature: Politics and the Environment in Revolutionary China.* New York: Cambridge University Press.

Sheridan, R. B. 1994 (1974). *Sugar and Slavery: An Economic History of the British West Indies, 1623–1775.* Kingston: University of the West Indies.

Shi, W. 2008. "Rubber Boom in Luang Namtha: A Transnational Perspective." Vientiane: Deutsche Gesellschaft für Technische Zusammenarbeit.

Shirley, J. W. 1942. "George Percy at Jamestown, 1607–1612." *VMHB* 57:227–43.

Shiue, C. H. 2005. "The Political Economy of Famine Relief in China, 1740–1820." *JIH* 36:33–55.

Siebert, L., and T. Simkin. 2002—. "Volcanoes of the World: An Illustrated Catalog of Holocene Volcanoes and Their Eruptions." Washington, DC: Smithsonian Institution (www.volcano.si.edu/world/).

Silva, M. C. d., and Tavim, J. A. R. S. 2005. "Marrocos no Brasil: Mazagão (Velho) do Amapá em Festa—A Festa de São Tiago." Proceedings of International Conference on "Espaço Atlântico de Antigo Regime: Poderes e Sociedades," Lisbon, 2–5 Nov. 2005.*

Silver, T. 1990. *A New Face on the Countryside: Indians, Colonists and Slaves in South Atlantic Forests, 1500–1800.* New York: Cambridge University Press.

Silverman, H., ed. 2004. *Andean Archaeology.* Malden, MA: Blackwell Publishing.

Simpson, L. B. 1982 (1929). *The Encomienda in New Spain: The Beginning of Spanish Mexico.* Berkeley: University of California Press, 3rd ed.

Skaggs, J. M. 1994. *The Great Guano Rush: Entrepreneurs and American Overseas Expansion.* New York: St. Martin's Press.

Skipton, H. P. K. 1907. *The Life and Times of Nicholas Ferrar.* London: A. R. Mowbray and Co.*

Slack, C. 2003. *Noble Obsession: Charles Goodyear, Thomas Hancock, and the Race to Unlock the Greatest Industrial Secret of the Nineteenth Century.* New York: Hyperion.

Slack, E. R. 2009. "The Chinos in New Spain: A Corrective Lens for a Distorted Image." *JWH* 20:35–67.

Sluiter, E. 1997. "New Light on the '20. and Odd Negroes' Arriving in Virginia, August 1619." *WMQ* 54:395–98.

Smil, V. 2001. *Enriching the Earth: Fritz Haber, Carl Bosch, and the Transformation of World Food Production.* Cambridge, MA: MIT Press.

Smith, A. 1979 (1776). *An Inquiry into the Nature and Causes of the Wealth of Nations.* Oxford: Clarendon Press.

Smith, H. H. 1879. *Brazil: The Amazons and the Coast.* New York: Charles Scribner's Sons.

Smith, J. 2007a (1608). *A True Relation of Such Occurrences and Accidents of Noate as Hath Hapned in Virginia Since the First Planting of that Collony, Which Is Now Resident in the South Part Thereof, Till the Last Returne from Thence.* In Horn ed. 2007, 1–36.*

———. 2007b (1624). *The Generall Historie of Virginia, New-England, and the Sum-*

mer Isles with the Names of the Adventurers, Planters, and Governours from Their First Beginning An. 1584 to This Present 1624. In Horn ed. 2007, 199–670.*

———. 2007c (1630). *The True Travels, Adventures, and Observations of Captaine John Smith, In Europe, Asia, Affrica, and America, from Anno Domini 1593 to 1629.* In Horn ed. 2007, 671–770.*

———. 1998 (1612). *A Map of Virginia.* Charlottesville: Virtual Jamestown, Virginia Center for Digital History, University of Virginia.*

Smith, M. E. 2002. *The Aztecs.* Oxford: Blackwell, 2nd ed.

Smith, W. 1745 (1744). *A New Voyage to Guinea.* London: John Nourse, 2nd ed.*

Snow, K. 1998. "Distribution of Anopheles Mosquitoes in the British Isles." *European Mosquito Bulletin* 1:9–13.*

Snyder, C. 2010. *Slavery in Indian Country: The Changing Face of Captivity in Early America.* Cambridge, MA: Harvard University Press.

So, B. K. L. 2000. *Prosperity, Region, and Institutions in Maritime China: The South Fukien Pattern, 946–1368.* Cambridge, MA: Harvard University Asia Center.

So, K.-W. 1975. *Japanese Piracy in Ming China During the 16th Century.* Lansing: Michigan State University Press.

Soetbeer, A. G. 1879. *Edelmetall-Produktion und Werthverhältniss zwischen Gold und Silber seit der Entdeckung Amerikas bis zur Gegenwart.* Gotha, Germany: Justus Perthes.*

Solomon, S., et al., eds. *Climate Change 2007: The Physical Science Basis.* Working Group I, 4th Assessment Report of the Intergovernmental Panel on Climate Change. New York: Cambridge University Press.*

Somers, G. 1610. Letter to Earl of Salisbury. 15 June. In Haile ed. 1998; 445–46.

Song, J. (宋军令). 2007. "Studies on the Spreading and Growing and Influences of Crops Originated in America During Ming and Qing Dynasties—Focusing on Maize, Sweet Potato and Tobacco" (明情时期美洲农作物在中国的传种及其影响研究—以玉米、番薯、烟草为视角). PhD thesis, Henan University.

Souza, M. 2001. *Breve História da Amazônia.* Rio de Janeiro: AGIR, 2nd. ed.

Spelman, H. 1609. "Relation of Virginia." In Haile ed. 1998, 481–95.

Sperling, L. H. 2006. *Introduction to Physical Polymer Science.* Hoboken, NJ: John Wiley and Sons, 4th ed.

Spooner, D. M., and R. J. Hijmans. 2001. "Potato Systematics and Germplasm Collecting, 1989–2000." *American Journal of Potato Research* 78:237–68; 395.

Spooner, D. M., and A. Salas. 2006. "Structure, Biosystematics, and Genetic Resources." In: J. Gopal and S. M. P. Khurana, *Handbook of Potato Production, Improvement and Post-Harvest Management.* Binghamton, NY: Haworth Press, 1–39.

Spruce, R. 1908. *Notes of a Botanist on the Amazon and Andes.* Ed. A. R. Wallace. 2 vols. London: Macmillan.*

Stahle, D. W., et al. 1998. "The Lost Colony and Jamestown Droughts." *Science* 280:564–67.

Standage, T. 2009. *An Edible History of Humanity.* New York: Walker and Co.

Stanfield, M. E. 2001. *Red Rubber, Bleeding Trees: Violence, Slavery and Empire in Northwest Amazonia, 1850–1933.* Albuquerque: University of New Mexico Press.

Stannard, D. E. 1993 (1992). *American Holocaust: The Conquest of the New World.* New York: Oxford University Press.

Stavans, I. 2001 (1993). *Imagining Columbus: The Literary Voyage.* New York: Palgrave.

Stedman, J. G. 2010 (1796). *Narrative of a Five Years' Expedition Against the Revolted Negroes of Surinam,* ed. R. Price and S. Price. NY: iUniverse. (bowdlerized 1796 ed.)*

Stern, P. 1991. "The White Indians of the Borderlands." *Journal of the Southwest* 33:262–81.

Stevens, R. W. 1894. *On the Stowage of Ships and Their Cargoes, with Information Regarding Freights, Charter-Parties, &c., &c.* New York: Longmans, Green, 7th ed.

Stewart, O. 2002 (1954). *Forgotten Fires: Native Americans and the Transient Wilderness.* Ed. H. T. Lewis and M. K. Anderson. Norman: University of Oklahoma Press.

Stewart, W. 1970 (1951). *Chinese Bondage in Peru: A History of the Chinese Coolie in Peru, 1849–1874.* Westport, CT: Greenwood Press.

Stone, R. 2008. "Showdown Looms Over a Biological Treasure Trove." *Science* 319:1604.

Strachey, W. 1625 (1610). "A True Repertory of the Wrack and Redemption of Sir Thomas Gates, Knight, upon and from the Islands of the Bermudas." In Haile ed. 1998, 381–443.

———. 1612. "The History of Travel into Virginia Britannia: The First Book of the First Decade." In Haile ed. 1998, 567–689.

Strickman, D., et al. 2000. "Mosquito Collections Following Local Transmission of *Plasmodium falciparum* Malaria in Westmoreland County, Virginia." *Journal of the American Mosquito Control Association* 16:219–22.

Striker, L. P. 1958. "The Hungarian Historian, Lewis L. Kropf, on Captain John Smith's *True Travels.*" *VMHB* 66:22–43.

Striker, L. P., and B. Smith. 1962. "The Rehabilitation of Captain John Smith." *JSH* 28:474–81.

Sturgeon, J. C., and N. K. Menzies. 2008. "Ideological Landscapes: Rubber in Xishuangbanna, 1950–2007." *Asian Geographer* 25:21–37.

Sturm, A., et al. 2006. "Manipulation of Host Hepatocytes by the Malaria Parasite for Delivery into Liver Sinusoids." *Science* 313:1287–90.

Sunseri, M. A., et al. 2002. "Survival of Detached Sporangia of Phytophthora infestans Exposed to Ambient, Relatively Dry Atmospheric Conditions." *American Journal of Potato Research* 79:443–50.

Sweet, D. G., and G. B. Nash, eds. 1981. *Struggle and Survival in Colonial America*. Berkeley: University of California Press.

Sweet, J. H. 2003. *Recreating Africa: Culture, Kinship, and Religion in the African-Portuguese World, 1441–1770*. Chapel Hill: University of North Carolina Press.

Symcox, G., ed. 2002. *Italian Reports on America, 1493–1522: Accounts by Contemporary Observers*. Repertorium Columbianum, No. 12. Turnhout, Belgium: Brepols.

———. 2001. *Italian Reports on America, 1493–1522: Letters, Dispatches, and Papal Bulls*. Repertorium Columbianum, No. 10. Turnhout, Belgium: Brepols.

Symonds, W. 1609. *Virginia. A Sermon Preached at White-Chapel, in the Presence of Many, Honourable and Worshipfull, the Adventurers and Planters for Virginia*. London: Eleazar Edgar and William Welby.*

Tadei, W. P., et al. "Ecological Observations on Anopheline Vectors of Malaria in the Brazilian Amazon." *AMJTMH* 1998:325–35.

Tao, W. (陶卫宁) 2003. "Evolution of the Government's Ban on Smoking in the Ming and Qing Dynasties" (明清政府的禁烟及其政策的演变). *Tangdu Journal* (唐都学刊) 19:133–37.

———. 2002a. "Case Studies in Sustainable Development in Agricultural Production Regions—Analysis of the Negative Impacts of Large-Scale Planting of Tobacco in Ruijin and Xincheng During the Qing" (农业生产区域可持续发展个案研究—试析瑞金、新城广植烟草的不良影响). *Journal of Yuncheng College* (运城高等专科学校学报) 20:69–70.

———. 2002b. "The Negative Influence and Inspiration of Tobacco Production in Qing Dynasty" (清代烟草生产的消极影响与启示). *Journal of the Shaanxi Education Institute* (陕西教育学院学报) 18:50–54.

Tapia, A. d. 1539. "Relacion Hecha por el Señor Andrés de Tapia, sobre la Conquista de México." In García Icazbalceta ed. 1858–66, vol. 2, 554–94.

Tardieu, J.-P. 2009. *Cimarrones de Panamá: La Forja de una Identidad Afroamericana en el Siglo XVI*. Madrid: Iberoamericana.

Tate, T. W., and D. L. Ammerman, eds. 1979. *The Chesapeake in the Seventeenth Century: Essays on Anglo-American Society*. Chapel Hill: University of North Carolina Press.

Taviani, P. E. 1996. *Cristoforo Colombo*. 3 vols. Rome: Societá Geografica Italiana.

Taviani, P. E., et al. 1997. *Christopher Columbus: Accounts and Letters of the Second, Third, and Fourth Voyages*. Trans. L. F. Farina and M. A. Beckwith. Nuova Raccolta Colombiana 6. Rome: Istituto Poligrafico e Zecca dello Stato.

Thirsk, J. 2006 (1957). *English Peasant Farming: The Agrarian History of Lincolnshire from Tudor to Recent Times*. Abingdon, UK: Routledge.

Thompson, P. 2004. "William Bullock's 'Strange Adventure': A Plan to Transform Seventeenth-Century Virginia." *WMQ* 61:107–28.

Thornton, J. K. 2010. "African Political Ethics and the Slave Trade." In D. R.

Peterson, ed., *Abolitionism and Imperialism in Britain, Africa and the Atlantic.* Athens: Ohio University Press, 38–62.

———. 2008. "Les États de l'Angola et la Formation de Palmares (Brésil)." *Annales. Histoire, Sciences Sociales* 63:769–97.

———. 1999. *Warfare in Atlantic Africa, 1500–1800.* London: UCL Press.

———. 1998 (1992). *Africa and Africans in the Making of the Atlantic World, 1400–1800.* New York: Cambridge University Press, 2nd ed.

Tiunov, A. V., et al. 2006. "Invasion Patterns of Lumbricidae into the Previously Earthworm-Free Areas of Northeastern Europe and the Western Great Lakes Region of North America." *Biological Invasions* 8:1223–34.

Tomlins, C. 2001. "Reconsidering Indentured Servitude: European Migration and the Early American Labor Force, 1600–1775." *Labor History* 42:5–43.

Tower, W. T. 1906. *An Investigation of Evolution in Chrysomelid Beetles of the Genus Leptinotarsa.* Washington, DC: Carnegie Institution.*

Townsend, C. 2004. *Pocahontas and the Powhatan Dilemma.* New York: Hill and Wang.

Trevelyan, R. 2004 (2002). *Sir Walter Raleigh: Being a True and Vivid Account of the Life and Times of the Explorer, Soldier, Scholar, Poet, and Courtier—The Controversial Hero of the Elizabethan Age.* New York: Holt.

Tsai, S.-S. H. 2002. *Perpetual Happiness: The Ming Emperor Yongle.* Seattle: University of Washington Press, 2nd ed.

Tuan, Y.-F. 2008 (1965). *A Historical Geography of China.* Piscataway, NJ: Aldine Transaction.

Tulloch, A. M. 1847. "On the Mortality Among Her Majesty's Troops Serving in the Colonies During the Years 1844 and 1845." *Journal of the Statistical Society of London* 10:252–59.

———. 1838. "On the Sickness and Mortality Among the Troops in the West Indies," *Journal of the Statistical Society of London* 1:129–42 (pt. 1); 1:216–30 (pt. 2); 1:428–44 (pt. 3).

Tullock, G. 1957. "Paper Money—A Cycle in Cathay." *EHR* 9:393–407.

Turner, E. R. 2004. "Virginia Native Americans During the Contact Period: A Summary of Archaeological Research over the Past Decade." *QBASV* 59:14–24.

———. 1993. "Native American Protohistoric Interactions in the Powhatan Core Area." In Rountree ed. 1993, 76–93.

———. 1982. "A Re-examination of Powhatan Territorial Boundaries and Population, A.D. 1607." *QBASV* 37:45–64.

———. 1973. "A New Population Estimate for the Powhatan Chiefdom of the Coastal Plain of Virginia." *QBASV* 28:57–65.

Ugent, D. 1968. "The Potato in Mexico: Geography and Primitive Culture." *EB* 22:108–23.

Ugent, D., et al. 1987. "Potato Remains from a Late Pleistocene Settlement in Southcentral Chile." *EB* 41:17–27.

————. 1982. "Archaeological Potato Remains from the Casma Valley of Peru." *EB* 36:182–92.

Ule, E. 1905. "Rubber in the Amazon Basin." *Bulletin of the American Geographical Society* 37:143–45.

Ulloa, A. d. 1807 (1743). *A Voyage to South America*, trans. London: J. Stockdale, 2 vols., 5th ed.

U.S. Census Bureau. 1975. *Historical Statistics of the United States, Colonial Times to 1970.* 2 vols. Washington, DC: Government Printing Office.

U.S. Department of Defense. 2008. "Military Critical Technologies List." Washington, DC: Defense Technical Information Center.*

Vainfas, R. 1996. "Deus Contra Palmares. Representações Senhoriais e Ideias Jesuíticas." In Reis and Gomes eds. 1996:60–80.

Valdés, D. N. 1978. "The Decline of the *Sociedad de Castas* in Mexico City." PhD thesis, University of Michigan.

Vallejo, J. 1944. "Una Ficha Para el Diccionario Histórico Español: Cición, Ciciones." *Revista de Filología Española* 28:63–66.

Vandenbroeke, C. 1971. "Cultivation and Consumption of the Potato in the 17th and 18th Century." *Acta Historiae Neerlandica* 5:15–39.

Vanhaute, E., et al. 2007. "The European Subsistence Crisis of 1845–1850: A Comparative Perspective." In C. Ó Gráda, et al., eds., *When the Potato Failed: Causes and Effects of the "Last" European Subsistence Crisis, 1845–1850.* Turnhout, Belgium: Brepols, 15–40.

Varela, C., and J. Gil, eds. 1992 (1982). *Cristóbal Colón: Textos y documentos completos.* Madrid: Alianza Editorial, 2nd rev. ed. (Many texts*)

Varela, H. 1997. "Entre Sueños Efímeros y Despertares: La Historia Colonial de São Tomé y Príncipe (1485–1975)." *Estudios de Asia y África* 32:289–321.

Verástique, B. 2000. *Michoacán and Eden: Vasco de Quiroga and the Evangelization of Western Mexico.* Austin: University of Texas Press.

Vermeer, E. B. 1991. "The Mountain Frontier in Late Imperial China: Economic and Social Developments in the Bashan." *T'oung Pao* 77:300–329.

Vermeer, E. B., ed. 1990. *Development and Decline of Fukien Province in the 17th and 18th Centuries.* New York: E. J. Brill.

Viazzo, P. P. 2006 (1989). *Upland Communities: Environment, Population and Social Structure in the Alps Since the Sixteenth Century.* New York: Cambridge University Press.

Vidal, L. 2005. *Mazagão: La Ville Qui Traverse l'Atlantique du Maro a l'Amazonie.* Paris: Aubier.

Vieira, A. 2004. "Sugar Islands: The Sugar Economy of Madeira and the Canaries, 1450–1650." In S. B. Schwartz, ed., *Tropical Babylons: Sugar and the Making of the Atlantic World, 1450–1680.* Durham: University of North Carolina Press, 42–84.

————. 1998. "As Ilhas do Açúcar: A Economia Açucareira da Madeira e

Canárias nos Séculos XV a XVII." Funchal, Madeira: CEHA-Biblioteca Digital.*

———. 1996. "Escravos com e sem Açúcar na Madeira." In Centro de Estudos de História do Atlântico, ed., *Escravos com e sem Açúcar: Actas do Seminário Internacional*. Funchal, Madeira: CEHA, 93–102.

———. 1992. *Portugal y las Islas del Atlántico*. Madrid: Colleciones Mapfre.

Vinod, K. K. 2002. "Genetic Improvement in Para Rubber (*Hevea brasiliensis* (Willd.) Muell.-Arg.)." In Centre for Advanced Studies in Genetics and Plant Breeding, ed., *Plant Breeding Approaches for Quality Improvement in Crops*. Coimbatore, Tamil Nadu: Tamil Nadu Agricultural University, 378–85.*

Vinson, B. 2000. "Los Milicianos Pardos y la Construcción de la Raza en el México Colonial." *Signos Históricos* 2:87–106.

Visnawathan, P. K. 2007. "Critical Issues Facing China's Rubber Industry in the Era of Market Integration: An Analysis in Retrospect and Prospect." Gota, Ahmedabad: Gujarat Institute of Development Research Working Paper No. 177.*

Voltaire (Arouet, F. M.). 1773 (1756) *Essai sur les Mœurs et l'Esprit des Nations*. 8 vols. Neuchâtel:s.n.*

Von Glahn, R. 2010. "Monies of Account and Monetary Transition in China, Twelfth to Fourteenth Centuries." *Journal of the Economic and Social History of the Orient* 53:463–505.

———. 2005. "Origins of Paper Money in China." In K. G. Rouwenhorst and W. N. Goetzmann, eds., *Origins of Value: The Financial Innovations That Created Modern Capital Markets*. NY: Oxford University Press, 65–89.

———. 1996. *Fountain of Fortune: Money and Monetary Policy in China, 1000–1700*. Berkeley: University of California Press.

Vongkhamor, S., et al. 2007. "Key Issues in Smallholder Rubber Planting in Oudomxay and Luang Prabang Provinces, Lao PDR." Vientiane: National Agriculture and Forestry Research Institute.

Von Wobeser, G. 1988. *La Hacienda Azucarera en la Época Colonial*. Mexico City: Secretaría de Educación Pública.

Wagner, M. J. 1977. "Rum, Policy and the Portuguese: The Maintenance of Elite Superiority in Post-Emancipation British Guiana." *Canadian Review of Sociology and Anthropology* 14:406–16.

Walford, C. 1879. *The Famines of the World: Past and Present*. London: Edward Stanford.*

Walsh, B. D. 1866. "The New Potato Bug." *Practical Entomologist* 2:13–16.

Waltham, T. 2005. "The Rich Hill of Potosi." *Geology Today* 21:187–90.

Walton, W. 1845. "Guano—The New Fertilizer." *Polytechnic Review and Magazine* 2:161–70.

Wang, S. (王思明). 2004. "Introduction of the American-Originated Crops and Its Influence on the Chinese Agricultural Production Structure"

(美洲原产作物的引种栽培及其对=中国农业生 产结构的影响). *AHC* 23: 16–27.

Wang, X. (王象晋). 1644 (1621). *Records of Fragrant Flowers from the Er Ru Pavilion* (二如亭群芳譜).2 vol. S. I.:s.n.

Wang, Y. 1997. "A Study on the Size of the Chinese Population in the Middle and Late Eighteenth Century." *Chinese Journal of Population Science* 9:317–36.

Wang, Y. (汪元方). 1850. "Memorial Requesting a Ban on Shack People Reclaiming Mountains and Blocking Waterways in Order to Prevent Future Calamities" (請禁棚民開山阻水以杜後患疏). In K. Sheng (盛康), ed., 1972, *Collected Writings on Qing Statecraft* (皇朝經世文編續編). Taipei: Wenhai, vol. 39, p. 32.

Warren, J. E. 1851. *Para; Or, Scenes and Adventures on the Banks of the Amazon.* New York: G. P. Putnam.*

Waterhouse, E. 1622. "A Declaration of the State of the Colony and Affaires in Virginia." *KB* 3:541–71.

Watts, P. M. 1985. "Prophecy and Discovery: On the Spiritual Origins of Christopher Columbus's 'Enterprise of the Indies.' " *AHR* 90:73–102.

Watts, S. J. 1999 (1997). *Epidemics and History: Disease, Power and Imperialism.* New Haven: Yale University Press.

Webb, J. L. A. 2009. *Humanity's Burden: A Global History of Malaria.* New York: Cambridge University Press.

Weber, M. 2003 (1904–05). *The Protestant Ethic and the Spirit of Capitalism.* Trans. T. Parsons. New York: Dover.

Wei, J., et al. 2006. "Decoupling Soil Erosion and Human Activities on the Chinese Loess Plateau in the 20th Century." *Catena* 68:10–15.

Weinstein, B. 1983. *The Amazon Rubber Boom: 1850–1920.* Stanford, CA: Stanford University Press.

Weiss, P. 1953. "Los Comedores Peruanos de Tierras: Datos Históricos y Geográficos—Nombres de Tierras Comestibles—Interpretación Fisiológica de la Geofagia y la Pica." *Peru Indigena* 5:12–21.

Weissmann, G. 1998. "They All Laughed at Christopher Columbus." In G. Weissmann, *Darwin's Audubon: Science and the Liberal Imagination.* New York: Basic Books, 149–58.

Weller, R. E., et al. 1999. "Universities and the Biological and Toxin Weapons Convention." *ASM News* 65:403–09.

Wennersten, J. R. 2000. *The Chesapeake: An Environmental Biography.* Baltimore: Maryland Historical Society.

West, T. (Baron de la Warre), et al. 1610. Letter to Virginia Company, 7 Jul. In Haile ed. 1998, 454–64.

Wey Gómez, N. 2008. *The Tropics of Empire: Why Columbus Sailed South to the Indies.* Cambridge, MA: MIT Press.

Wheeler, A. G. 1981. "The Tarnished Plant Bug: Cause of Potato Rot?: An Episode in Mid-Nineteenth-Century Entomology and Plant Pathology." *Journal of the History of Biology* 14:317–38.

Whitaker, A. P. 1971 (1941). *The Huancavelica Mercury Mine: A Contribution to the History of the Bourbon Renaissance in the Spanish Empire.* Westport, CT: Greenwood Press.

Whitby, G. S. 1920. *Plantation Rubber and the Testing of Rubber.* New York: Longmans, Green.*

White, A. 1634. "A Briefe Relation of the Voyage unto Maryland." In Hall 1910, 29–45.

White, G. B. 1978. "Systematic Reappraisal of the *Anopheles maculipennis* Complex." *Mosquito Systematics* 10:13–44.*

Whitehead, N. L. 1999. "Native Peoples Confront Colonial Regimes in Northeastern South America." In F. Soloman and S. B. Schwartz, eds., *The Cambridge History of Native Peoples of the Americas.* Cambridge: Cambridge University Press, 382–442.

Wilentz, A. 1990. "Balaguer Builds a Lighthouse." *Nation* 250:702–05.

Will, P.-E. 1980. "Un Cycle Hydraulique en Chine: La Province du Hubei du XVIe au XIXe siècles." *Bulletin de l'École Française d'Extrême-Orient* 68:261–87.

Williams, D. 1962. "Clements Robert Markham and the Introduction of the Cinchona Tree into British India, 1861." *Geographical Journal* 128:431–42.

Williams, D. J., and D. Matile-Ferraro. 1999. "A New Species of the Mealybug Genus *Cataenococcus* Ferris from Ethiopia on *Ensete Ventricosum,* a Plant Infected by a Virus." *Revue Française d'Entomologie* 21:145–49.

Williams, E. 1650. *Virginia: More Especially the South Part Thereof, Richly and Truly Valued.* London: John Stephenson, 2nd ed.*

Williams, M. 2006. *Deforesting the Earth: From Prehistory to Global Crisis.* Chicago: University of Chicago Press.

———. 1989. *Americans and Their Forests: A Historical Geography.* New York: Cambridge University Press.

Wilson, C., et al. 2002. "Soil Management in Pre-Hispanic Raised Field Systems: Micromorphological Evidence from Hacienda Zuleta, Ecuador." *Geoarchaeology* 17:261–83.

Wilson, E. O. 2006. "Ant Plagues: A Centuries-Old Mystery Solved." In E. O. Wilson, *Nature Revealed: Selected Writings, 1949–2006.* Baltimore: Johns Hopkins University Press, 343–50.

———. 2005. "Early Ant Plagues in the New World." *Nature* 433:32.

Wingfield, E. M. 1608? "A Discourse of Virginia." In Haile ed. 1998, 183–201.

Wither, G. 1880 (1628). *Britain's Remembrancer.* 2 vols. London: Spencer Society.*

Wolf, E. R. 1997 (1982). *Europe and the People Without History.* Berkeley: University of California Press, 2nd ed.

Wood, C. S. 1975. "New Evidence for a Late Introduction of Malaria into the New World." *Current Anthropology* 16:93–104.

Wood, P. H. 1996 (1974). *Black Majority: Negroes in Colonial South Carolina from 1670 Through the Stono Rebellion*. New York: W. W. Norton.

Wood, W. 1977 (1634). *New England's Prospect*. Amherst: University of Massachusetts Press.

Woodroffe, J. F. 1916. *The Rubber Industry of the Amazon, and How Its Supremacy Can Be Maintained*. London: T. Fisher Unwin and Bale, Sons and Danielson.

Woodruff, W. 1958. *The Rise of the British Rubber Industry During the Nineteenth Century*. Liverpool: Liverpool University Press.

Woodward, H. 1674. "A Faithfull Relation of My Westoe Voiage." In A. S. Salley Jr., ed., *Narratives of Early Carolina, 1650–1708*. New York: Charles Scribner's Sons.

World Health Organization. 2010. *World Malaria Report 2010*. Geneva: WHO Press.*

Worster, D. 1994. *Nature's Economy: A History of Ecological Ideas*. New York: Cambridge University Press, 2nd. ed.

Wright, I. A., ed. 1932. *Documents Concerning English Voyages to the Spanish Main, 1569–1580*. London: Hakluyt Society.

Wrigley, E. A. 1969. *Population and History*. New York: McGraw-Hill.

Wu, R. (吴若增). 2009. "Early Stage Chinese Workers in Peru" (早期华工在秘鲁). *Memories and Archives* (档案春秋) 7:47–50.

Wu, S., et al. 2001. "Rubber Cultivation and Sustainable Development in Xishuangbanna, China." *International Journal of Sustainable Development and World Ecology* 8:337–45.

Xu, G. (徐光啟) 1968 (1628). *Complete Treatise on Agricultural Administration* (農政全書). Taipei: The Commercial Press.

Xu, J. 2006. "The Political, Social and Ecological Transformation of a Landscape: The Case of Rubber in Xishuangbanna, China." *Mountain Research and Development* 26:254–62.

Xu, Z., et al. 2004. "China's Sloping Land Conversion Programme Four Years On: Current Situation, Pending Issues." *International Forestry Review* 6: 317–26.

Yamamoto, N. 1988. "Potato Processing: Learning from a Traditional Andean System." In *The Social Sciences at CIP: Report of the Third Social Science Planning Conference*. Lima: International Potato Center, 160–72.

Yang, C. (杨昶). 2002. "The Effect of Ming Dynasty Economic Activities on the Ecological Environment in the South" (明代经济活动对南方生态环境的影响). *Journal of Baoji College of Arts and Science (Social Sciences)* (宝鸡文理学院学报 [社会科学版]) 23:44–49.

Yasuoka, J., and R. Levins. 2007. "Impact of Deforestation and Agricultural Development on Anopheline Ecology and Malaria Epidemiology." *AMJTMH* 76:450–60.

Ye, T., ed. (葉廷芳). 1967 (1825). *Dianbai Gazetteer* (電白縣志). Taipei: Cheng Wen Publishing.

Young, A. 1771. *The Farmer's Tour Through the East of England.* 4 vols. London: W. Strahan.★

Yu, K., D. Li, and D. Li. 2006. "The Evolution of Greenways in China." *Landscape and Urban Planning* 76:223–39.

Yuan, T. (袁庭栋). 1995. *History of Smoking in China* (中国吸烟史话). Beijing: The Commercial Press International.

Zadoks, J. C. 2008. "The Potato Murrain on the European Continent and the Revolutions of 1848." *Potato Research* 51:5–45.

Zamora, M. 1993. *Reading Columbus.* Berkeley: University of California Press.

Zavala, S. 1947. "The American Utopia of the Sixteenth Century." *Huntington Library Quarterly* 10:337–47.

Zhang, D., et al. 2000. "Assessing Genetic Diversity of Sweet Potato (*Ipomoea batatas* [L.] Lam.) Cultivars from Tropical America Using AFLP." *Genetic Resources and Crop Evolution* 47:659–65.

Zhang, D. D., et al. 2007. "Climate Change and War Frequency in Eastern China over the Last Millennium." *Human Ecology* 35:403–14.

Zhang, J. (張景岳). 2006 (1624). *The Complete Works of Jingyue* (景岳全書). In Y. Ji (紀昀) and X. Lu (陸錫熊), et al., eds., *Wenyuan Publishing House Internet Edition of the Complete Library of The Four Treasuries* (文淵閣四庫全書內網聯版). Hong Kong: Heritage Publishing Ltd.★

Zhang, J. (張箭). 2001. "On the Spread of American Cereal Crops" (论美洲粮食作物的传播). *AHC* 20:89–95.

Zhang, J. H., and M. Cao. 1995. "Tropical Forest Vegetation of Xishuangbanna, SW China, and Its Secondary Changes, with Special Reference to Some Problems in Local Nature Conservation." *Biological Conservation* 73:229–38.

Zhang, T. (張廷玉), et al., eds. 2000 (1739). *The Ming History (Ming Shi)* (明史). Academia Sinica Hanji Wenxian Ziliaoku Databases (中央研究院漢籍電子文獻). Taipei: Academia Sinica.★

Zhang, X. (張燮). 1968 (1617). *Studies on the East and West Oceans* (東西洋考). Taipei: The Commercial Press.

Zhao, J., and J. Woudstra. 2007. " 'In Agriculture, Learn from Dazhai': Mao Zedong's Revolutionary Model Village and the Battle Against Nature." *Landscape Research* 32:171–205.

Zheng, Z. 2001 (1992). *Family Lineage Organization and Social Change in Ming and Qing Fujian.* Trans. M. Szonyi. Honolulu: University of Hawai'i Press.

Zhuge, Y. (諸葛元聲). 1976 (1556). *Records of Pingrang Throughout the Three Reigns* (三朝平攘錄). Taipei: Wei-Wen Book & Publishing Co.

Ziegler, A. D., et al. 2009. "The Rubber Juggernaut." *Science* 324:1024–25.

Zimmerer, K. S. 1998. "The Ecogeography of Andean Potatoes." *BioScience* 48:445–54.

Zizumbo-Villarreal, D., and H. J. Quero. 1998. "Re-evaluation of Early Observations on Coconut in the New World." *EB* 52:68–77.

Zuckerman, L. 1999 (1998). *The Potato: How the Humble Spud Rescued the Western World*. New York: North Point.

Zuñiga, M. d. 1814 (1803). *An Historical View of the Philippine Islands. Trans. J. Maver.* 2 vols. London: Black, Parry and Co., 2nd ed.

Zurara (Azurara), G. E. d. 1897–99 (1453). *The Chronicle of the Discovery and Conquest of Guinea*. Trans. C. R. Beazley and E. Prestage. 2 vols. London: Hakluyt Society.

INDEX

MAP CREDITS

Maps by Nick Springer and Tracy Pollock, Springer Cartographics LLC. Sources as follows:

8 Redrawn from Guitar 1998:13.

40 Data collected by *National Geographic* from David G. Anderson, Jeffrey C. Bendremer, Faith Davison, Penelope Drooker, George Hamell, John Hart, Stephen J. Hornsby, Bonnie G. McEwan, Bruce D. Smith, Douglas H. Ubelaker, Marvin T. Smith, Dean R. Snow, Alan Taylor, and John E. Worth. Original map published in *National Geographic,* May 2007. Redrawn with additional data and advice from William Denevan, William Doolittle, Allan Gallay, and William I. Woods. Also consulted: Helen Hornbeck Tanner, *Atlas of Great Lakes Indian History* (1987).

58 Base map by Nick Springer published in *National Geographic,* May 2007; data from sources above, Helen Rountree, Martin D. Gallivan; additional data from Barlow 2003:22. My thanks to William McNulty and the rest of the *NG* cartography staff for allowing Nick and me to adapt these maps.

110 Data from Preservation Virginia, Wetlands Vision (U.K.), Smith 1956. My thanks to Robert C. Anderson and William Thorndale for critiques and suggestions.

136 Data from Kiszewski et al. 2004:488; Webb 2009:87; Gilmore 1955:348; author's interview, Donald Gaines.

238 Redrawn from Central Bureau of Meteorological Sciences (China) 1981.

286 Redrawn from Bourke 1964:806.

328 Data from Hecht forthcoming, pers. comm.; Schurz et al. 1925.

373 Base map redrawn from Galloway 1977:178; additional data from Disney 2009, Ouerfelli 2008, Vieira 1992.

381 Redrawn from Barrett 1970:8.

381 Base map redrawn from Hemming 2004:xx; additional data from Orser 2001:65 (Palmares).

466 Data from Price 2011:6–7 (Suriname, Guyane); Tardieu 2009 (Panama); La Rosa Corso 2003 (Cuba); Lane 2002:chap. 1 (Esmeraldas); Perez 2000:618 (Venezuela); Landers 1999:236 (Florida); Reis and Gomes eds. 1996 (Brazil); Aptheker 1996 (U.S.); Friedemann 1993:70–71 (Colombia); Deive 1989:73 (Hispaniola); Carroll 1977 (Mexico); author's interviews, Fundação Cultural Palmares and Instituto de Terras do Pará.

ILLUSTRATION CREDITS

4, 16, 31, 165, 167, 195, 215, 221 (2), 225, 233, 245 (2), 305, 352 (2), 422, 441, 477 (2), 481, 487, 501, 505 Author's photograph

22 (4), Marconi, P. 1929. *Architettura e Arti Decorative,* 9:100–35.

55 Courtesy Virtual Jamestown (detail, John Smith, *Map of Virginia*)

60 Courtesy of Crawford Lake Conservation Area, Conservation Halton (Ontario)

66 National Portrait Gallery, London (Smith, *True Travels,* 1624)

68, 84 (r), 95, 142, 150 Library of Congress (LC-USZC4-3368, LC-USZC4-3368, G3880 1667 .F3, LC-USZ62-95078, LC-USZC4-9408)

77, 84 (l), 89, 93 (2) Virginia Historical Society (1854.2, Smith, H.L., portrait of George Percy; 1993.192, Anon., Pocahontas, 1616; 1994.65, Merian, *Decima tertia pars Historiae Americanae,* 1634; 1834.1, badge, ca. 1660)

105 Lennart Nilsson / SCANPIX

109 Wellcome Images (V0010519)

135 MGM publicity still

146 Author's collection (E. Riou, *La Guyane Française,* 1867)

158 China Photos / Getty Images

165 Author's collection (1764 gazetteer)

177 Courtesy Bob Reis (anythinganywhere.com)

180 Fundación Cultural Banco Central de Bolivia, Potosí

184 f US 2257.50* Houghton Library, Harvard University (Theodor de Bry, *Collectiones peregrinationum,* 1590)

190 Courtesy Ken and Sue Goodreau, New World Treasures

202 Aizar Raldes / AFP / GettyImages

244 Courtesy Town of Gaoxigou, Shaanxi

255 Author's collection (1916 postcard)

259 Dept. of Rare Books and Manuscripts, Royal Library, Copenhagen (Felipe Guaman Poma de Ayala, El Primer Nueva Corónica y Buen Gobierno [GKS 2234-4[0])

262 International Potato Center (Peru)

276 (2) Courtesy New York Public Library (Alexander Gardner, *Rays of Sunlight from South America*, 1865)

287 (2) Courtesy "Views of the Famine," http://adminstaff.vassar.edu/sttaylor/FAMINE/ (*Illustrated London News*)

293 (t) © National Museums Northern Ireland 2010, Collection Ulster Museum, Belfast (Courtesy of the Trustees of National Museums Northern Ireland)

293 (b) Courtesy Clark Erickson

299 Courtesy Homer Babbidge Library, University of Connecticut

308 Germanisches Nationalmuseum (Christoph Weiditz, *Trachtenbuch*, 1529)

313 (l) Author's collection (Iles, *Leading American Investors*, 1912)

313 (r) Author's collection (Hancock, *Personal Narrative of the Origin and Progress of the Caoutchouc*, 1857)

319 Courtesy Yale University Library

331 Courtesy Biblioteca Luis Ángel Arango del Banco de la República (Colombia)

333 (l), 268 Courtesy Susanna Hecht

333 (r) National Archives (U.K.), FO371/1455

338 Courtesy John Loadman (www.bouncing-balls.com)

342 Author's collection (Falcão, *Album do Acre,* 1906–07)

360 Courtesy Biblioteca Nacional de España (Durán, *Historia de las Indias de Nueva España,* 1587)

363 © Tomás Filsinger 2009

376 Courtesy Huntington Library (Jan van der Straet, *Nova Reperta*, 1584)

402 Courtesy Casa Nacional de Moneda, Potosí

406 (t) Denver Art Museum, Collection of Jan and Frederick Mayer

406 (b), 407 (b) Private collection, Spain

407 (t) Collection of Malú and Alejandra Escandón, Mexico City

415 Basílica de Nuestra Señora de la Merced (Buenos Aires)

CHART CREDITS